HANDBOOK OF APPLICABLE MATHEMATICS

Supplement

HANDBOOK OF APPLICABLE MATHEMATICS

Chief Editor: Walter Ledermann

Editorial Board: Robert F. Churchhouse
Peter Hilton
Emlyn Lloyd
Steven Vajda
Carol Alexander

Volume I: ALGEBRA
Edited by Walter Ledermann, *University of Sussex*
and Steven Vajda, *University of Sussex*

Volume II: PROBABILITY
Emlyn Lloyd, *University of Lancaster*

Volume III: NUMERICAL METHODS
Edited by Robert F. Churchhouse, *University College Cardiff*

Volume IV: ANALYSIS
Edited by Walter Ledermann, *University of Sussex*
and Steven Vajda, *University of Sussex*

Volume V: COMBINATORICS AND GEOMETRY (Parts A and B)
Edited by Walter Ledermann, *University of Sussex*
and Steven Vajda, *University of Sussex*

Volume VI: STATISTICS (Parts A and B)
Edited by Emlyn Lloyd, *University of Lancaster*

SUPPLEMENT
Edited by Walter Ledermann, *University of Sussex*
Emlyn Lloyd, *University of Lancaster*
Steven Vajda, *University of Sussex*
and Carol Alexander, *University of Sussex*

HANDBOOK OF

APPLICABLE MATHEMATICS

Chief Editor: Walter Ledermann

Supplement

Edited by

Walter Ledermann
University of Sussex

Emlyn Lloyd
University of Lancaster

Steven Vajda
University of Sussex

and

Carol Alexander
University of Sussex

A Wiley–Interscience Publication

JOHN WILEY & SONS

Chichester – New York – Brisbane – Toronto – Singapore

Other Wiley Editorial Offices

John Wiley & Sons, Inc., 605 Third Avenue,
New York, NY 10158–0012, USA

Jacaranda Wiley Ltd, G.P.O. Box 859, Brisbane,
Queensland 4001, Australia

John Wiley & Sons (Canada) Ltd, 22 Worcester Road,
Rexdale, Ontario M9W 1L1, Canada

John Wiley & Sons (SEA) Pte Ltd, 37 Jalan Pemimpin   05–04,
Blook B, Union Industrial Building, Singapore 2057

British Library Cataloguing in Publication Data Available:

ISBN 0 471 91825 3

Typeset by Thomson Press (India) Limited, New Delhi
Printed in Great Britain by Courier International, Tiptree, Essex

Contributing Authors

D. J. Bell, University of Manchester Institute of Science and Technology, U.K.

B. Conolly, Queen Mary College, London, U.K.

A. W. Craig, University of Durham, U.K.

A. C. Davison, University of Oxford, U.K.

P. Hilton, State University of New York, Binghamton, U.S.A.

D. Kershaw, University of Lancaster, U.K.

L. Kronsjö, University of Birmingham, U.K.

W. Ledermann, University of Sussex, Brighton, U.K.

C. J. Mitchell, Hewlett-Packard Laboratories, Bristol, U.K.

J. Pedersen, Santa Clara University, California, U.S.A.

F. C. Piper, Royal Holloway & Bedford New College, Egham, U.K.

R. Smith, University of Surrey, Guildford, U.K.

I. Stewart, University of Warwick, Coventry, U.K.

S. Vajda, University of Sussex, Brighton, U.K.

C. Alexander, University of Sussex, Brighton, U.K.

Contents

vii

Introduction to the Handbook of Applicable Mathematics

Today, more than ever before, mathematics enters the lives of every one of us. Whereas, thirty years ago, it was supposed that mathematics was only needed by somebody planning to work in one of the 'hard' sciences (physics, chemistry), or to become an engineer, a professional statistician, an actuary or an accountant, it is recognized today that there are very few professions in which an understanding of mathematics is irrelevant. In the biological sciences, in the social sciences (especially economics, town planning, psychology), in medicine, mathematical methods of some sophistication are increasingly being used and practitioners in these fields are handicapped if their mathematical background does not include the requisite ideas and skills.

Yet it is a fact that there are many working in these professions who do find themselves at a disadvantage in trying to understand technical articles employing mathematical formulations, and who cannot perhaps fulfil their own potential as professionals, and advance in their professions at the rate that their talent would merit, for want of this basic understanding. Such people are rarely in a position to resume their formal education, and the study of some of the available textbooks may, at best, serve to give them some acquaintance with mathematical techniques, of a more or less formal nature, appropriate to current technology. Among such people, academic workers in disciplines which are coming increasingly to depend on mathematics constitute a very significant and important group.

Some years ago, the Editors of the present Handbook, all of them actively concerned with the teaching of mathematics with a view to its usefulness for today's and tomorrow's citizens, got together to discuss the problems faced by mature people already embarked on careers in professions which were taking on an increasingly mathematical aspect. To be sure, the discussion ranged more widely than that—the problem of 'mathematics avoidance' or 'mathematics anxiety', as it is often called today, is one of the most serious problems of modern civilization and affects, in principle, the entire community—but it was decided to concentrate on the problem as it affected professional effectiveness.

There emerged from those discussions a novel format for presenting mathematics to this very specific audience. The intervening years have been spent in putting this novel conception into practice, and the result is the Handbook of Applicable Mathematics.

THE PLAN OF THE HANDBOOK

The 'Handbook' consists of two sets of books. On the one hand, there are (or will be!) a number of *guide books*, written by experts in various fields in which mathematics is used (e.g. medicine, sociology, management, economics). These guide books are by no means comprehensive treatises; each is intended to treat a small number of particular topics within the field, employing, where appropriate, mathematical formulations and mathematical reasoning. In fact, a typical guide book will consist of a discussion of a particular problem, or related set of problems, and will show how the use of mathematical models serves to solve the problem. Wherever any mathematics is used in a guide book, it is cross-referenced to an article (or articles) in the *core volumes*.

There are six core volumes devoted respectively to Algebra, Probability, Numerical Methods, Analysis, Combinatorics and Geometry, and Statistics. These volumes are texts of mathematics—but they are no ordinary mathematical texts. They have been designed specifically for the needs of the professional adult (though we believe they should be suitable for any intelligent adult!) and they stand or fall by their success in explaining the nature and importance of key mathematical ideas to those who need to grasp and to use those ideas. Either through their reading of a guide book or through their own work or outside reading, professional adults will find themselves needing to understand a particular mathematical idea (e.g. linear programming, statistical robustness, vector product, probability density, round-off error); and they will then be able to turn to the appropriate article in the core volume in question and *find out just what they want to know*—this, at any rate, is our hope and our intention.

How then do the content and style of the core volumes differ from a standard mathematical text? First, the articles are designed to be read by somebody who has been referred to a particular mathematical topic and would prefer not to have to do a great deal of preparatory reading; thus each article is, to the greatest extent possible, self-contained (though, of course, there is considerable cross-referencing within the set of core volumes). Second, the articles are designed to be read by somebody who wants to get hold of the mathematical ideas and who does not want to be submerged in difficult details of mathematical proof. Each article is followed by a bibliography indicating where the unusually assiduous reader can acquire that sort of 'study in depth'. Third, the topics in the core volumes have been chosen for their relevance to a number of different fields of application, so that the treatment of those topics is not biased in favour of a particular application. Our thought is that the reader—unlike the typical college student—will already be motivated, through some particular problem or the study of some particular new technique, to acquire the necessary mathematical knowledge. Fourth, this is a handbook, not an encylopedia—if we do not think that a particular aspect of a mathematical topic is likely to be useful or interesting to the kind of reader we have in mind, we have omitted it. We have not set out to include everything known on a particular topic, and we are *not* catering for the professional mathematician! The Handbook has been written as a contribution to the practice of mathematics, not to the theory.

The reader will readily appreciate that such a novel departure from standard textbook writing—this is neither 'pure' mathematics nor 'applied' mathematics as traditionally interpreted—was not easily achieved. Even after the basic concept of the Handbook had been formulated by the Editors, and the complicated system of cross-referencing had been developed, there was a very serious problem of finding authors who would write the sort of material we wanted. This is by no means the way in which mathematicians and experts in mathematical applications are used to writing. Thus we do not apologize for the fact that the Handbook has lain so long in the womb; we were trying to do something new and we had to try, to the best of our ability, to get it right. We are sure we have not been uniformly successful; but we can at least comfort ourselves that the result would have been much worse, and far less suitable for those whose needs we are trying to meet, had we been more hasty and less conscientious.

It is, however, not only our task which has not been easy. Mathematics itself is not easy! The reader is not to suppose that, even with his or her strong motivation and the best endeavours of the editors and authors, the mathematical material contained in the core volumes can be grasped without considerable effort. Were mathematics an elementary affair, it would not provide the key to so many problems of science, technology and human affairs. It is universal, in the sense that significant mathematical ideas and mathematical results are relevant to very different 'concrete' applications—a single algorithm serves to enable the travelling salesman to design his itinerary, and the refrigerator manufacturing company to plan a sequence of modifications of a given model; and could conceivably enable an intelligence unit to improve its techniques for decoding the secret messages of a foreign power. Given this universality, mathematics cannot be trivial! And, if it is not trivial, then some parts of mathematics are bound to be substantially more difficult than others.

This difference in level of difficulty has been faced squarely in the Handbook. The reader should not be surprised that certain articles require a great deal of effort for their comprehension and may well involve much study of related material provided in other referenced articles in the core volumes—while other articles can be digested almost effortlessly. In any case, different readers will approach the Handbook from different levels of mathematical competence and we have been very much concerned to cater for all levels.

THE REFERENCING AND CROSS-REFERENCING SYSTEM

To use the Handbook effectively, the reader will need a clear understanding of our numbering and referencing system, so we will explain it here. Important items in the core volumes or the guidebooks—such as definitions of mathematical terms or statements of key results—are assigned sets of numbers according to the following scheme. There are six categories of such mathematical items, namely:

  (i) Definitions
 (ii) Theorems, Propositions, Lemmas and Corollaries
(iii) Equations and other Displayed Formulae
 (iv) Examples
  (v) Figures
 (vi) Tables

Items in any one of these six categories carry a triple designation a.b.c. of arabic numerals, where 'a' gives the *chapter* number, 'b' the *section* number, and 'c' the number of the individual *item*. Thus items belonging to a given category, for example definitions, are numbered in sequence within a section, but the numbering is independent as between categories. For example, in Section 5 of Chapter 3 (of a given volume), we may find a displayed formula labelled (5.3.7) and also Lemma 5.3.7 followed by Theorem 5.3.8. Even where sections are further divided into *subsections*, our numbering system is as described above, and takes no account of the particular subsection in which the item occurs.

As we have already indicated, a crucial feature of the Handbook is the comprehensive cross-referencing system which enables the reader of any part of any core volume or guide book to find his or her way quickly and easily to the place or places where a particular idea is introduced or discussed in detail. If, for example, reading the core volume on Statistics, the reader finds that the notion of a *matrix* is playing a vital role, and if the reader wishes to refresh his or her understanding of this concept, then it is important that an immediate reference be available to the place in the core volume on Algebra where the notion is first introduced and its basic properties and uses discussed.

Such ready access is achieved by the adoption of the following system. There are six core volumes, enumerated by the Roman numerals as follows:

   I Algebra
  II Probability
 III Numerical Methods
  IV Analysis
   V Combinatorics and Geometry (Parts A and B)
  VI Statistics (Part A and B)

A reference to an item will appear in square brackets and will *typically* consist of a pair of entries [see X, Y] where X is the volume number and Y is the triple designating the item in that volume to which reference is being made. Thus '[see II, (3.4.5)]' refers to equation (3.4.5) of Volume II (Probability). There are, however, two exceptions to this rule. The first is simply a matter of economy!—if the reference is to an item in the same volume, the volume number designation (X, above) is suppressed; thus '[see Theorem 2.4.6]', appearing in Volume III, refers to Theorem 2.4.6 of Volume III. The second exception is more fundamental and, we contend, wholly natural. It may be that we feel the need to refer to a substantial discussion rather than to a single mathematical item (this could well have been the case in the reference to 'matrix', given as an example above). If we judge that such a comprehensive reference is appropriate, then the second entry Y of the reference may carry only two numerals—or even, in an extreme case, only one. Thus the reference '[see I, 2.3]' refers to Section 3 of Chapter 2 of Volume I and recommends the reader to study that entire section to get a complete picture of the idea being presented.

Bibliographies are to be found at the end of each chapter of the core volumes and at the end of each guide book. References to these bibliographies appear in the text as '(Smith (1979))'.

It should perhaps be explained that, while the referencing *within* a chapter of a core volume or *within* a guide book is substantially the responsibility of the author of that part of the text, the cross-referencing has been the responsibility of the editors as a whole. Indeed, it is fair to say that it has been one of their heaviest and most exacting responsibilities. Any defects in putting the referencing principles into practice must be borne by the editors. The successes of the system must be attributed to the excellent and wholehearted work of our invaluable colleague, Carol Jenkins (Mrs Alexander).

Introduction to the Supplement

The Handbook of Applicable Mathematics is not—as we have conceded in the main Introduction—an encyclopedia. Nevertheless, it shares with an encyclopedia the characteristic that *it cannot stand still.* It must try to be as up-to-date as possible, to include recent applications of mathematics in industry, engineering and applied science, and to include areas of mathematics which have recently found application.

Thus, just as the celebrated OED must issue supplements (and even, very occasionally, new editions) so must our much more modest endeavour also seek, in similar ways, to maintain its currency. This objective, then, provides the main motivation for our issuing this supplementary volume to the set of core volumes (I–VI) of the Handbook. All the mathematical topics discussed in the Supplement find application in some area of contemporary interest relevant to the concerns of readers of the Handbook; we adhere, however, to our standard policy of presenting the mathematics independently of its potential and actual applications, relying on our usual comprehensive cross-referencing system to provide the connections with important applications and with other parts of mathematics. We should only add here that Chapter 1, Further Number Theory, is somewhat different in nature from the other chapters, in that its content may well be a necessary prerequisite for some readers to enable them to understand certain other chapters of the Supplement. Indeed, it is precisely this role which the chapter is intended to play and which the editors believe justifies its inclusion.

Having clarified the *main* purpose of the Supplement, we would be less than candid if we did not admit that some of the material to be found in the Supplement should, according to the editors' present opinion, already have been included in the original set of core volumes. (Could there perhaps also be an analogy here with the experience of editors of encyclopedias and dictionaries?) Thus, for example, a substantial part of Chapter 8 falls into this category. The editors have been fortunate to have had at their disposal some very comprehensive reviews of the core volumes, as well as some very valuable private communications relating to the scope and quality of those volumes, and have taken full advantage of all the information made available to them in determining the content of the Supplement.

CHAPTER 1

Further Number Theory

1.1. INTRODUCTION

This chapter is a continuation of the work on number theory presented in I, 4 of this Handbook. The reader will find it helpful to renew his acquaintance with that material before proceeding to a study of the present chapter. But, whenever it is convenient, we shall recall the definitions and results that are relevant for the subsequent discussion.

1.2. BASE OF NOTATION

1.2.1. The Representation Theorem

In this section we are concerned solely with the set $\mathbb{Z}$ of common integers, indeed mostly with non-negative integers.

When we write down the integer

$$n = 108\,467,$$

we are using an abbreviation for the statement that

$$n = 1 \times 10^5 + 0 \times 10^4 + 8 \times 10^3 + 4 \times 10^2 + 6 \times 10 + 7.$$

Generally, every integer that is less than 10^{k+1} can be expressed in the form

$$n = a_k 10^k + a_{k-1} 10^{k-1} + \ldots + a_1 10 + a_0,$$

where each a_i $(0 \le i \le k)$ is a 'digit', that is an integer lying between 0 and 9 inclusive.

In this context there is nothing special about the number 10 and the ensuing decimal notation. Any integer which is greater than unity can be used as what is called the *base of notation*. In fact we have the following result.

THEOREM 1.2.1. *Let s be a fixed integer greater than unity (base of notation). Then every integer which is less than s^{k+1} can be uniquely expressed in the form*

$$n = a_k s^k + a_{k-1} s^{k-1} + \ldots + a_1 s + a_0, \tag{1.2.1}$$

where $0 \le a_i \le s-1$ $(i = 0, 1, \ldots, k)$.

1

Remark 1. By choosing k sufficiently great we can express any positive integer whatsoever in the form (1.2.1).

Remark 2. It is sometimes convenient to abbreviate (1.2.1) by writing

$$n = (a_k, a_{k-1}, \ldots, a_0)_s.$$

The proof of Theorem 1.2.1 rests on the fact that $\mathbb{Z}$ is a Euclidean domain [see I, §4.1.2], that is if n is any integer and if s is an integer greater than unity, then there exists a unique pair of integers q and a satisfying

$$n = qs + a \tag{1.2.2}$$

and

$$0 \le a \le s - 1.$$

Repeated application of (1.2.2) yields a chain of equations

$$n = q_1 s + a_0, \qquad q_1 = q_2 s + a_1, \qquad q_2 = q_3 s + a_2, \ldots. \tag{1.2.3}$$

Evidently

$$n > q_1 > q_2 > \ldots$$

so that ultimately the quotient q must become zero. Thus let k be the least integer such that

$$q_k < s.$$

Then on dividing q_k by s in accordance with (1.2.2) we find that the quotient q_{k+1} is zero. We therefore have to put

$$q_k = a_k. \tag{1.2.4}$$

On eliminating $q_1, q_2, \ldots, q_k$ from the equations (1.2.3) and (1.2.4) we obtain the expansion (1.2.1).

The uniqueness of this expansion is easy to demonstrate. For suppose we had a relation of the form

$$a_k s^k + \ldots + a_1 s + a_0 = b_k s^k + \ldots + b_1 s + b_0, \tag{1.2.5}$$

where $0 \le a_i \le s - 1$ and $0 \le b_i \le s - 1$ $(i = 1, \ldots, k)$. There is no loss of generality in assuming that the two sides of equation (1.2.5) have the same number of terms; for if necessary we may affix a suitable number of initial terms with zero coefficients. In condensed form equation (1.2.5) can be written as

$$Ps + a_0 = Qs + b_0, \tag{1.2.6}$$

where

$$P = a_k s^{k-1} + \ldots + a_1, \qquad Q = b_k s^{k-1} + \ldots + b_1. \tag{1.2.7}$$

We recall that in the quotient and remainder formula (1.2.2) the integers q and a are uniquely determined. Hence we deduce from (1.2.6) that

$$a_0 = b_0$$

and

$$P = Q.$$

On repeating this argument for P and Q and using (1.2.7), we find that $a_1 = b_1$ and so on until we arrive at the conclusion that $a_k = b_k$. This proves the uniqueness of the expansion.

We note that (1.2.3) describes a procedure for obtaining the expansion: the coefficients $a_0, a_1, \ldots, a_k$ are the remainders when first n and then the successive quotients $q, q_1, \ldots$ are divided by s.

EXAMPLE 1.2.1.   Let $n = 108\,467$ (in decimal notation). Let us obtain the expansion of n in the scale of 12 ($s = 12$). Thus

$$
\begin{aligned}
108\,467 &= 9038 \times 12 + \underline{11} \\
9038 &= 753 \times 12 + \underline{2} \\
753 &= 62 \times 12 + \underline{9} \\
62 &= 5 \times 12 + \underline{2} \\
5 &= 0 \times 12 + \underline{5}
\end{aligned}
$$

so that

$$108\,467 = (5, 2, 9, 2, 11)_{12}.$$

1.3. DIVISIBILITY

1.3.1.   The Highest Common Factor (Greatest Common Divisor)

This fundamental concept was discussed in I, §4.1.3. For convenience we recall the customary notation and some of the results established in that earlier chapter.

We say the integer m divides the integer n, and we write

$$m \mid n,$$

if there exists an integer t such that

$$n = tm.$$

When a and b are non-zero integers, we say that t is a common divisor of a and b if

$$t \mid a \quad \text{and} \quad t \mid b.$$

Every pair a, b of integers which are not both zero possesses a *highest common factor* (HCF) or *greatest common divisor* (GCD): this is a positive integer h, written

$$h = (a, b) = (b, a),$$

with the following properties:

(1) h is a common divisor of a and b;

(2) every common divisor of a and b is a common divisor of h.

Attention was drawn to the fact [see I, p. 106] that h can be expressed as a sum of multiples of a and b, that is there exist integers u and v such that

$$h = au + bv. \tag{1.3.1}$$

The integers u and v are not uniquely determined, and they may be negative; indeed, if (1.3.1) holds, then

$$h = a(u + wb) + b(v - wa),$$

where w is an arbitrary integer.

It is interesting to note that the relationship (1.3.1) may be used to characterize the HCF of a and b. In fact we have the following proposition.

PROPOSITION 1.3.1.   *Let a and b be given integers which are not both zero. Then the highest common factor of a and b is the least positive integer that can be expressed in the form*

$$ax + by, \tag{1.3.2}$$

where x and y are suitable integers.

Proof.   Suppose that

$$d = ax_0 + by_0 \tag{1.3.3}$$

is the least positive integer that can be expressed in the form (1.3.2). We assert that d is a factor of a; for, if not, we could write

$$a = qd + r, \quad \text{where } 0 < r < d, \tag{1.3.4}$$

or

$$r = a - qd.$$

On substituting for d from (1.3.3) and collecting terms in a and b we find that

$$r = a(1 - qx_0) + b(- qy_0).$$

Thus r, too, is of the form (1.3.2); but, by (1.3.4), r is positive and smaller than d. This contradicts the minimality of d. Hence we conclude that

$$d \mid a$$

$(r = 0)$. Similarly, it is shown that $d \mid b$, so that d is a common divisor of a and b.

Conversely, it is obvious from (1.3.3) that if $t \mid a$ and $t \mid b$, then $t \mid d$. It follows that d is the greatest common divisor of a and b so that

$$d = h = (a, b) = ax_0 + by_0. \tag{1.3.5}$$

This interpretation of the highest common factor may be useful in determining its value: for we observe that

$$(a, b) = (a + sb, b) = (a, b + ta)$$

whatever the integers s and t (positive or negative); for it is clear that the set of all sums of multiples of a and b is the same as that of sums of multiples of $a + sb$ and b, or of a and $b + ta$. By a convenient choice of s and t and by a repeated application of this procedure, the numbers in the brackets can be reduced to such an extent that the value of the highest common factor becomes evident.

EXAMPLE 1.3.1.

$$(111, 42) = (111 + (-3)42, 42) = (-15, 42) = (-15, 42 + 3(-15))$$
$$= (-15, -3) = (15, 3) = 3.$$

EXAMPLE 1.3.2   Let M be set of positive integers (natural numbers) with the following properties:

(i) if a and b, which need not be distinct, belong to M, so does $a + b$;
(ii) if a and b belong to M and if $a > b$, then $a - b$ belongs to M.

Then each member of M is divisible by the least member of M. (Recall that by the well-ordering principle [see I, p. 33] every collection of natural numbers possesses a least member.) Now let m be the least member and a an arbitrary member of M. Suppose that, on the contrary, a is not divisible by m. Then there exist integers q and r such that

$$a = qm + r, \quad \text{where } q > 0 \text{ and } 0 < r < m.$$

By the property (i) above, we have that $qm \in M$, and it follows from (ii) that

$$r = a - qm \in M.$$

But this contradicts the minimality of m. Hence we conclude that $r = 0$, that is $m \mid a$.

1.3.2.   Coprime Integers

The case in which

$$(a, b) = 1$$

is particularly noteworthy; the integers a and b are then said to be *relatively prime* [see I, p. 106] or *coprime* or simply *prime* to each other. Equation (1.3.5) now implies that there are integers u and v such that

$$1 = au + bv. \tag{1.3.6}$$

Conversely, the equation (1.3.6) ensures that the integers a and b are relatively prime; for it is clear that any common factor of a and b would have to divide 1. Thus we have

PROPOSITION 1.3.2.   *The integers a and b are relatively prime if and only if there exist integers u and v such that* $1 = au + bv$.

EXAMPLE 1.3.3.

$$(21, 46) = (21, 46 + (-2)21) = (21, 4) = (21 + (-5)4, 4)$$

$$= (1, 4) = 1.$$

In order to obtain the relation (1.3.6) we write

$$4 = 46 + (-2)21, \qquad 1 = 21 + (-5)4 = 21 + (-5)\{46 + (-2)21\},$$

whence on collecting terms in 21 and 46 we find that

$$1 = 11 \times 21 + (-5) \times 46.$$

The set of all integers which are coprime to a given integer m enjoys the 'closure' property with regard to multiplicaiton:

PROPOSITION 1.3.3.   *If*

$$(a_1, m) = (a_2, m) = \ldots = (a_h, m) = 1,$$

then

$$(a_1 a_2 \ldots a_h, m) = 1.$$

Proof.   If suffices to prove the result when $h = 2$; the general statement then follows by induction. Thus we assume that there are integers u_1, u_2, v_1, v_2 such that

$$1 = a_1 u_1 + m v_1 \quad \text{and} \quad 1 = a_2 u_2 + m v_2.$$

On multiplying these equations by each other we obtain an equation of the form

$$1 = a_1 a_2 U + mV, \tag{1.3.7}$$

where

$$U = u_1 u_2, \qquad V = u_1 v_2 a_1 + u_2 v_1 a_2 + v_1 v_2 m$$

By Proposition 1.3.2, equation (1.3.7) implies that $(a_1 a_2, m) = 1$, as required.

The following fact about relatively prime numbers has important consequences.

PROPOSITION 1.3.4.   *Let m and n be relatively prime integers and suppose there is an integer t such that n divides tm. Then n divides t.*

Proof.   By (1.3.6) there exist integers u and v such that

$$1 = um + vn. \tag{1.3.8}$$

By hypothesis, there exists an integer q such that

$$tm = qn. \tag{1.3.9}$$

On multiplying (1.3.8) by t and substituting from (1.3.9) we obtain

$$t = n(uq + tv),$$

which shows that indeed n is a factor of t.

1.4. MODULAR ARITHMETIC

1.4.1. Complete Residue Systems

The notion of congruence between integers was introduced in I, 4.3.

We recall that it involves a fixed positive integer m (the *modulus*) and that two integers a and b are said to be *congruent* to each other *modulo m* ('with respect to the modulus m'), if a and b differ by an integral multiple of m, that is if there exists an integer k such that $a - b = mk$. This relationship between a and b is written as

$$a \equiv b \pmod{m}. \tag{1.4.1}$$

If a and b are not congruent to each other modulo m, we write

$$a \not\equiv b \pmod{m},$$

which means that $a - b$ is not a multiple of m.

Congruence is an equivalence relation [see I, 1.3.3]. Accordingly, the set of all integers is partitioned into disjoint classes, two integers belonging to the same class if and only if they are congruent mod m. The structure of the classes may be depicted by the diagram in Figure 1.4.1. On the circle in Figure 1.4.1,

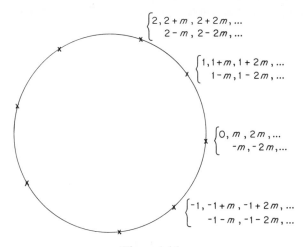

Figure 1.4.1.

we have marked m places, preferably equally spaced. Starting with 0 at one of the places, we enter all the positive integers by going round the circumference of the circle indefinitely often in the counter-clockwise sense, entering consecutive integers at consecutive places; similarly, we enter the negative integers by starting from 0 and going round the circle in the clockwise sense. The integers that occupy the same position on the circle form a *conjugacy class*, and distinct places on the circle correspond to distinct conjugacy classes. Thus there are precisely m conjugacy classes mod m. For this reason the work with conjugacy classes is sometimes referred to as *circular arithmetic*.

It is worth recording that the formula

$$a \equiv 0 \pmod{m}$$

is equivalent to the statement that a is divisible by m.

A different view of conjugacy is based on the quotient-remainder formula (1.2.2): if a is an arbitrary integer, there exist integers q and r (depending on a) such that

$$a = qm + r \quad (0 \le r \le m - 1).$$

Evidently,

$$a = r \pmod{m}.$$

The sitation is summarized in the following.

PROPOSITION 1.4.1.   *Every integer is congruent to one of the numbers*

$$0, 1, \ldots, m - 1 \tag{1.4.2}$$

and two integers are congruent to each other if and only if they are congruent to the same number in the set (1.4.2).

The set (1.4.2) is called a *complete residue system* (the word 'residue' being here synonymous with 'remainder'). More generally, the numbers

$$r_1, r_2, \ldots, r_m$$

form a complete residue system if and only if no two of them are congruent to each other mod m.

The set of residue classes mod m is often denoted by $\mathbb{Z}_m$.

EXAMPLE 1.4.1.   Let a be a fixed residue chosen arbitrarily, and suppose that x runs through the set (1.4.2). Then the map

$$x \rightarrow x + a \pmod{m} \tag{1.4.3}$$

bijectively maps the set (1.4.2) onto itself, that is, it is a permutation of that set. There are m such permutations, one for each residue a.

For instance, when $m = 5$ and $a = 3$, the set

$$0, \ 1, \ 2, \ 3, \ 4,$$

is mapped, by the rule $x \to x + 3 \pmod 5$, into

$$3, \ 4, \ 0, \ 1, \ 2;$$

note that $2 + 3 \equiv 0 \pmod 5$, and so on.

When the value of the modulus is understood, a relation like (1.4.1) is simply referred to as 'a congruence'. It was shown in I, §4.3.1 that congruences (with respect to the same modulus) behave in many ways like equations between numbers: they can be added, subtracted and multiplied but not, in general, divided by one another. Thus if

$$a \equiv b \pmod m \quad \text{and} \quad c \equiv d \pmod m,$$

it follows that

$$a + c \equiv b + d, \qquad a - c \equiv b - d, \qquad ac \equiv bd, \pmod m \qquad (1.4.4)$$

and in particular

$$ka \equiv kb \pmod m, \qquad (1.4.5)$$

where k is an arbitrary integer. But if, conversely, we are given that

$$ka \equiv kb \pmod m$$

it does not follow that $a \equiv b \pmod m$, unless a further condition is satisfied, which are now going to discuss.

PROPOSITION 1.4.2.   (*Cancellation Law*) *If*

$$ka \equiv kb \pmod m \quad and \quad (k, m) = 1,$$

then it follows that $a \equiv b$ (mod m).

Proof.   By hypothesis, $m | k(a - b)$. Applying Proposition 1.3.4 we deduce that $m | a - b$.

EXAMPLE 1.4.2.   We have that

$$34 \equiv 4 \pmod 6, \quad \text{but} \quad 17 \not\equiv 2 \pmod 6,$$

because we tried to cancel the factor 2, which is not coprime to the modulus 6. However,

$$34 \equiv 4 \pmod 5 \quad \text{and} \quad 17 \equiv 2 \pmod 5,$$

because 2 is coprime to 5.

EXAMPLE 1.4.3.   Let p be a prime, and let r be an integer satisfying

$$1 \le r \le p - 1.$$

We recall that the binomial coefficient

$$b_r = p(p-1)\ldots(p-r+1)/r! \tag{1.4.6}$$

is an integer, being equal to the number of ways in which r objects can be selected from a set of p objects [see I, p. 96]. We shall show that b_r is divisible by p: we write (1.4.6) in the form

$$r!b_r = p(p-1)\ldots(p-r+1),$$

which implies that

$$r!b_r \equiv 0 \pmod p. \tag{1.4.7}$$

Each factor of

$$r! = 1 \times 2 \times \ldots \times r$$

is strictly less than p and is therefore coprime to p, because p is a prime number. By Proposition 1.3.3, the product of these factors is also coprime to p, that is

$$(r!, p) = 1 \tag{1.4.8}$$

On applying the cancellation law to (1.4.7), we obtain that

$$b_r \equiv 0 \pmod p.$$

EXAMPLE 1.4.4.   Let p be a prime and let r be an integer satisfying $2 \le r \le p-1$. Then

$$c_r = \binom{p+1}{r} \equiv 0 \pmod p.$$

For we have that

$$c_r r! = (p+1)p(p-1)\ldots(p-r+2),$$

there being r factors on the right. Since $r \ge 2$, one of the factors is p. Hence

$$c_r r! \equiv 0 \pmod p$$

On applying the cancellation rule we obtain that

$$c_r \equiv 0 \pmod p.$$

EXAMPLE 1.4.5.   Let p be a prime and let r be an integer satisfying

$$1 \le r \le p-1.$$

Then

$$d_r = \binom{p-1}{r} \equiv (-1)^r \pmod p.$$

For we have that

$$d_r r! = (p-1)(p-2)\ldots(p-r),$$

there being r factors on the right. Generally,

$$p - k \equiv - k \quad (\mathrm{mod}\ p).$$

Hence

$$d_r r! \equiv (- 1)(- 2) \ldots (- r) = (- 1)^r r! \quad (\mathrm{mod}\ p).$$

On cancelling the factor $r!$ we obtain that

$$d_r \equiv (- 1)^r \quad (\mathrm{mod}\ p).$$

1.4.2. Reduced Residue Systems

We wish to consider in more detail those residues in the set

$$0, 1, \ldots, m - 1$$

which are coprime to m, that is we are interested in the integers s which satisfy

$$0 \le s \le m - 1 \quad \text{and} \quad (s, m) = 1. \tag{1.4.9}$$

The number of integers specified by (1.4.9) is an important function of m.

DEFINITION 1.4.1.   The number of integers which lie between 0 and $m - 1$ (inclusive) and are coprime to m is called the *Euler function* (*totient function*) of m and is denoted by $\phi(m)$.

EXAMPLE 1.4.6.

(i) $\phi(12) = 4$, because among the numbers

$$0, 1, 2, \ldots, 11 \tag{1.4.10}$$

only $1, 5, 7$ and 11 are coprime to 12.
(ii) If p is a prime, then

$$\phi(p) = p - 1, \tag{1.4.11}$$

because in the set

$$0, 1, \ldots, p - 1$$

all numbers except 0 are relatively prime to p.

For a given modulus m put

$$\phi(m) = h$$

and let

$$s_1(= 1), \quad s_2, \ldots, s_h \tag{1.4.12}$$

be all the solutions of (1.4.9). We say that (1.4.12) forms a *reduced residue system* mod m. More generally, the integers

$$z_1, z_2, \ldots, z_h$$

form a reduced residue system mod m if

$$(z_1, m) = 1 \quad (i = 1, \ldots, h) \quad \text{and} \quad z_i \not\equiv z_j \pmod{m} \quad (i \neq j).$$

Whichever reduced residue system is chosen, we can assert that any integer that is prime to m is congruent to one and only one member of that reduced residue system. Unless the contrary is stated, it will be assumed that we are dealing with the reduced residue system defined in (1.4.9).

The Euler function has a very important property: it is *multiplicative* which means that

$$\text{if } (m, m') = 1, \quad \text{then } \phi(mm') = \phi(m)\phi(m').$$

A proof of this fact will be found in any text on number theory, for example Hardy and Wright (1954) p. 53; Dudley (1969) p. 66.

Let

$$m = p_1^{\alpha_1} p_2^{\alpha_2} \ldots p_r^{\alpha_r}$$

be the decomposition of m into powers of distinct primes. Then the multiplicativeness of ϕ implies that

$$\phi(m) = \phi(p_1^{\alpha_1})\phi(p_2^{\alpha_2})\ldots\phi(p_r^{\alpha_r}). \tag{1.4.13}$$

Hence we can find $\phi(m)$ provided that we know the value of m for a prime power. Let p be a prime. Among the integers

$$0, 1, \ldots, p^\alpha - 1 \tag{1.4.14}$$

only the multiples of p fail to be coprime to p; there are $p^{\alpha - 1}$ such multiples in (1.4.14), namely

$$0, p, 2p, \ldots, (p^{\alpha - 1} - 1)p.$$

Hence the number of integers in (1.4.14) which are coprime to p is equal to

$$\phi(p^\alpha) = p^\alpha - p^{\alpha - 1}. \tag{1.4.15}$$

By virtue of the multiplicative property we deduce from (1.4.15) that

$$\phi(m) = \prod_{p \mid m} (p^\alpha - p^{\alpha - 1}) = m \prod_{p \mid m} \left(1 - \frac{1}{P}\right), \tag{1.4.16}$$

where the product is extended over all the distinct primes that divide m.

EXAMPLE 1.4.7.   If p and q are distinct primes, we have that

$$\phi(pq) = (p - 1)(q - 1).$$

Let s_i and s_j be any two (not necessarily distinct) residues coprime to m. By Proposition 1.3.3, $s_1 s_2$ is coprime to m. Hence there exists a unique element s_k in (1.4.12) such that

$$s_i s_j \equiv s_k \pmod{m}. \tag{1.4.17}$$

Thus we can define a binary operation, [see I, §1.5], that is a law of composition for the reduced set of residues: it consists of *ordinary multiplication of integers followed by reduction* mod m. This circumstance provides a fruitful connection between number theory and group theory [see I, §8; in particular, Definition 8.2.5].

THEOREM 1.4.3. *The reduced set of residues* mod m *forms a commutative group of order* $\phi(m)$ *with respect to the binary operation defined in* (1.4.17).

Proof. We shall verify the group axioms in turn:

G1 *closure*: If $(s, m) = 1$ and $(t, m) = 1$, then $(st, m) = 1$ [see (1.4.17)].

G2 *Associative law*: If $(s, m) = (t, m) = (u, m) = 1$, then $(st)u \equiv s(tu) \pmod m$; this follows from the associative law for integers, that is $(st)u = s(tu)$, which in this context may be taken for granted.

G3 *Unit element*: The integer 1 or any integer congruent to 1 serves as a unit for the law of composition defined in (1.4.17).

G4 *Inverse*: Since $(s, m) = 1$, Proposition 1.3.2 tells us that there are integers x and y such that

$$xs + ym = 1.$$

on turning this equation into a congruence mod m we obtain that

$$xs \equiv 1 \pmod m,$$

that is, x is the inverse of s mod m.

Since the equation $st = ts$ holds for all integers, it is obvious that, as a consequence, we have the congruence

$$st \equiv ts \pmod m.$$

Hence the group of reduced residues is commutative (Abelian) [see I, Definition 8.2.2].

We shall now draw attention to some properties which the group of reduced residues has in common with any finite group [see I, §§8.2.2 and 8.2.5].

PROPOSITION 1.4.4. *For a given modulus* m *put* $h = \phi(m)$. *Let* s *run through a reduced system of residues*

$$s_1, s_2, \ldots, s_h \tag{1.4.18}$$

and let a *be any integer that is coprime to* m. *Then the map*

$$s \to as \pmod m \tag{1.4.19}$$

corresponds to a permutation of the set (1.4.18). *There are* $\phi(m)$ *such permutations, one for each of the residue classes which are coprime to* m.

Proof.   It suffices to verify that the map (1.4.19) is bijective [see I, §1.4.2]: indeed

$$as \equiv at \pmod m$$

implies that

$$s \equiv t \pmod m$$

by virtue of the cancellation law [Proposition 1.4.2].

In a finite group the successive powers

$$1, a, a^2, a^3, \ldots$$

of an element a cannot all be distinct. In particular, for each a there exists a least positive integer m, called the *order* or *period* of a such that

$$a^m = 1.$$

The order of a depends on a. But by Lagrange's theorem [see I, Theorem 8.2.3] the order of each element must divide the order of the group to which it belongs.

On applying this result to the groups of residues we obtain the following important theorem:

THEOREM 1.4.5.   (Euler's theorem) *Let m be an integer greater than unity. If s is coprime to m, then*

$$s^{\phi(m)} \equiv 1 \pmod m. \qquad (1.4.20)$$

The case of a prime modulus was already known before Euler's time.

THEOREM 1.4.6.   (Fermat's 'Little theorem') *If p is a prime and if a is not divisible by p, then*

$$a^{p-1} \equiv 1 \pmod p \qquad (1.4.21)$$

and every integer whatsoever satisfies the congruence

$$a^p \equiv a \pmod m. \qquad (1.4.22)$$

Some of the numerical instances of these results can be quite surprising.

EXAMPLE 1.4.8.   When $m = 12$ we have that $\phi(m) = 4$ Hence each of the numbers

$$5^4 - 1, \quad 7^4 - 1 \quad \text{and} \quad 11^4 - 1$$

is divisible by 12; indeed these integers are equal to 624, 2400 and 14 640 respectively.

EXAMPLE 1.4.9.   (Hunter (1964) p. 65) Every integer x satisfies the congruence

$$x^{13} \equiv x \quad (\bmod\ 2730).$$

The reason for this seemingly bizarre statement lies in the fact that

$$2730 = 2 \times 3 \times 5 \times 7 \times 13$$

and that the congruence

$$x^{13} \equiv x$$

holds modulo $2, 3, 5, 7$ and 13, as we shall now verify:

Modulo 2: this means simply that x^{13} and x are either both even or both odd.
Modulo 3: by (1.4.22) we have that $x^3 \equiv x$. Hence

$$x^{12} \equiv x^4 = x^3 \times x \equiv x^2; \quad x^{13} \equiv x^3 \equiv x.$$

Modulo 5: $\qquad\qquad\ x^5 \equiv x; \quad x^{10} \equiv x^2; \quad x^{13} \equiv x^5 \equiv x.$

Modulo 7: $\qquad\qquad\ x^7 \equiv x; \quad x^{13} \equiv x^7 \equiv x.$

Modulo 13: $\qquad\quad\ x^{13} \equiv x.$

We have now proved that $x^{13} - x$ is divisible by 2 and 3 and 5 and 7 and 13 and hence by their product, that is by 2730.

1.4.3. Quadratic Residues

We have seen that addition, subtraction, multiplication and, with some restriction, division can be carried out within the system $\mathbb{Z}_m$ of residues modulo m. It is natural to enquire whether other algebraic operations can also be performed in $\mathbb{Z}_m$.

The simplest case that comes to mind is the following.

PROBLEM.   Given a residue a, does there exist a residue x such that

$$x^2 \equiv a \quad (\bmod\ m); \tag{1.4.23}$$

in other words, has a a square root in $\mathbb{Z}_m$?

At first sight, this appears to be a rather special and unpromising line of investigation. But, in the history of the subject, it was this question which gave rise to some of the most profound and beautiful discoveries in the theory of numbers.

In this short section we shall only be able to mention, mostly without proofs, a few facts which are needed elsewhere in this volume.

For the sake of simplicity we shall restrict ourselves to the case in which the modulus is an odd prime p, and we assume that a is not divisible by p. Thus we are going to study the congruence

$$x^2 \equiv a\,(\bmod\ p), \quad \text{where } (a, p) = 1. \tag{1.4.24}$$

x:	1	2	3	4	5	6
x^2:	1	4	2	2	4	1

Table 1.4.1

It is easy to see that this congruence does not always admit of a solution. For example when $p = 7$, Table 1.4.1 shows all the non-zero residues and their squares (reduced modulo 7). Evidently only the residues 1, 2 and 4 have square roots, namely 1,3 and 2 respectively (recall that $-1 \equiv 6$, etc.). But the congruence (1.4.24) is insoluble when $a = 3$ or 5 or 6.

The difference between square roots and non-square roots is formalized in the following definition.

DEFINITION 1.4.2.   Let p be an odd prime and let a be an integer which is not divisible by p. Then

(i) if there exists an integer x such that $x^2 \equiv a \pmod p$ we say that a is a *quadratic residue* mod p and we write

$$\left(\frac{a}{p}\right) = 1; \tag{1.4.25}$$

(ii) if the congruence $x^2 \equiv a \pmod p$ cannot be solved in $\mathbb{Z}_p$, we say that a is a *quadratic non-residue* mod p and we write

$$\left(\frac{a}{p}\right) = -1. \tag{1.4.26}$$

The bracket symbols which appear in (1.4.25) and (1.4.26) are called the *Legendre symbols* for a and p; its value is referred to as the *quadratic character* of a mod p.

The evaluation of the Legendre symbol is sometimes facilitated by the following result.

THEOREM 1.4.7.   (Euler's criterion) *Let p be an odd prime and let a be an integer which is not divisible by p. Then*

$$a^{\frac{1}{2}(p-1)} \equiv \left(\frac{a}{p}\right) \pmod p \tag{1.4.27}$$

(for a proof see for example Andrews (1971), p. 115.)

We mention a useful corollary.

COROLLARY 1.4.8.   *If a and b are not divisible by p then*

$$\left(\frac{ab}{p}\right) = \left(\frac{a}{p}\right)\left(\frac{b}{p}\right). \tag{1.4.28}$$

Proof. Apply (1.4.27) to the equation

$$(ab)^{\frac{1}{2}(p-1)} = a^{\frac{1}{2}(p-1)}b^{\frac{1}{2}(p-1)}$$

A common type of problem is as follows: given a, determine all odd primes for which a is a quadratic residue.

EXAMPLE 1.4.10. -1 is a quadratic residue modulo the odd prime p if and only if p is of the form $4k+1$, where k is an integer. This follows at once from Theorem 1.4.7, which implies that

$$\left(\frac{-1}{p}\right) = 1$$

if and only if $\frac{1}{2}(p-1)$ is even, that is $p = 4k+1$ for some integer k.

EXAMPLE 1.4.11.   The quadratic character of 5 is more difficult to establish. We refer the reader to the proofs given in the literature (for example Hardy and Wright (1954) p. 76). The result is as follows:

$$\left(\frac{5}{p}\right) = \begin{cases} 1, & \text{if } p = 10k \pm 1 \\ -1, & \text{if } p = 10k \pm 3. \end{cases}$$

We remark that the numbers 1, -1, 3, -3 form a reduced residue system modulo 10; for they consist of 4 $(= \phi(10))$ integers which are incongruent to one another modulo 10.

EXAMPLE 1.4.12.   The results of the two preceding examples in conjunction with Corollary 1.4.8 enable us to find the quadratic character of -5. We have that

$$\left(\frac{-5}{p}\right) = \left(\frac{5}{p}\right)\left(\frac{-1}{p}\right).$$

As we have seen, the character of 5 is determined modulo 10 and the character of -1 is determined modulo 4. It is therefore to be expected that the character of -5 will be determined modulo 20. The eight integers

$$1, \ -1, \ 3, \ -3, \ 7, \ -7, \ 9, \ -9 \tag{1.4.29}$$

may be taken as a reduced system of residues modulo 20, which agrees with the fact that $\phi(20) = \phi(4)\phi(5) = 2 \times 4 = 8$.

Table 1.4.2 gives the values of

$$\left(\frac{5}{p}\right), \ \left(\frac{-1}{p}\right) \ \text{and} \ \left(\frac{-5}{p}\right)$$

for the reduced system of residues mod 20 given in (1.4.29).

| | $p(\bmod 20)$ | | | | | | | |
	1	3	7	9	-9	-7	-3	-1
$\left(\dfrac{5}{p}\right)$	1	-1	-1	1	1	-1	-1	1
$\left(\dfrac{-1}{p}\right)$	1	-1	-1	1	-1	1	1	-1
$\left(\dfrac{-5}{p}\right)$	1	1	1	1	-1	-1	-1	-1

Table 1.4.2

Hence -5 is a quadratic residue modulo the odd prime p if and only if p is of one of the forms

$$20k+1 \quad \text{or} \quad 20k+3 \quad \text{or} \quad 20k+7 \quad \text{or} \quad 20k+9$$

and -5 is a quadratic non-residue in all other cases.

1.5. FINITE FIELDS

1.5.1. The Fields GF(p)

The concept of a field was discussed on several occasions in Volume I, notably in section 2.4.2, where the axioms of a field were stated. For convenience, we recall the definition.

A field F is a collection of elements endowed with two modes of composition: addition, denoted by $a + b$, and multiplication, denoted by ab. These are subject to the following axioms:

F1 F contains at least two distinct elements, one of which is denoted by 0 (the identity element for addition) and another element by 1)the identity element for multiplication).

F2 Addition is commutative and associative, that is

$$a + b = b + a, \quad (a + b) + c = a + (b + c)$$

F3 Multiplication is commutative and associative and every non-zero element possesses a multiplicative inverse. Thus

$$ab = ba, \quad (ab)c = a(bc)$$

and, if $a \neq 0$, there exists a' such that $aa' = 1$.

F4 Addition and multiplication are linked by the distributive law

$$a(b + c) = ab + ac.$$

An important property of fields is expressed in the following.

COROLLARY 1.5.1. *In a field there are no zero divisors; that is if*

$$ab = 0,$$

then either $a = 0$ or $b = 0$ (or both).

The prototype of a field is the set of rational numbers, which is an infinite field. There are fields of various types. In particular, it is an interesting fact that finite fields exist. This was briefly mentioned in I, §4.3.3, and we shall now study the properties of finite fields in more detail.

Let p be a fixed prime (including $p = 2$). Then the complete set of residues

$$\mathbb{Z}_p : 0, 1, 2, \ldots, p - 1 \tag{1.5.1}$$

can be given the structure of a field by defining addition and multiplication as with ordinary integers except that the result is always reduced modulo p so as to cause it to lie between 0 and $p - 1$. It is straightforward to verify that the field axioms are satisfied, including the existence of an inverse; for if a is an element other than zero in the set (1.5.1), then it is coprime to p and according to Theorem 1.4.3 (iv) there exists an element a' such that $aa' \equiv 1 \pmod{p}$ or with our present convention simply $aa' = 1$.

The field displayed in (1.5.1) is denoted by $\mathbb{Z}_p$ (the residues mod p) or else by $GF(p)$, the *Galois field* of order p [see I, §4.3.2].

EXAMPLE 1.5.1.   Let $p = 7$. Then

$$3 + 4 = 0, \qquad 2 - 5 = 4, \qquad 2 \times 6 = 5, \qquad 4^{-1} = 2.$$

Once we have adopted $GF(p)$ as a *ground field* or *prime field*, we can proceed to form algebraical constructions in much the same way as when the ground field consists of the rational or the real numbers. In particular, we may introduce an indeterminate x and consider polynomials

$$f(x) = a_n x^n + a_{n-1} x^{n-1} + \ldots + a_1 x + a_0, \tag{1.5.2}$$

where the coefficcients $a_n, \ldots, a_0$ are elements of $GF(p)$. If a_n is not zero, we say the $f(x)$ is of degree n, and we write

$$\deg f = n.$$

Denoting the zero polynomial simply by 0 we find it convenient to define

$$\deg 0 = -\infty. \tag{1.5.3}$$

Addition and multiplication of polynomials over $GF(p)$ are performed in the same manner as over more familiar fields, having regard to the laws of composition in the ground field. In particular, we mention the following proposition.

PROPOSITION 1.5.2.   (Quotient and remainder formula) *Let $f(x)$ and $g(x)$ be polynomials over $GF(p)$ and suppose that $\deg g \geq 1$. Then there exist unique* *polynomials q(x) and r(x) over* **GF**(p) *such that*

$$f(x) = q(x)g(x) + r(x)$$

and $\deg r < \deg g$.

The concept of divisibility remains unchanged over $GF(p)$; the polynomial $f(x)$ of degree greater than zero is *reducible* over $GF(p)$ if there exist polynomials $f_1(x)$ and $f_2(x)$, each of degree greater than zero, such that the equation

$$f(x) = f_1(x)f_2(x) \tag{1.5.4}$$

holds over $GF(p)$. If $f(x)$ satisfies no such equations, then $f(x)$ is said to be *irreducible* over $GF(p)$.

An interesting question, which is of considerable theoretical importance, concerns the existence of irreducible polynomials over $GF(p)$. We quote the following result without proof [see Dickson (1958) Chapter II].

THEOREM 1.5.3. *For every positive integer n there exists at least one irreducible polynomial of degree n over* **GF**(p).

Obviously a polynomial $f(x)$ of degree two or three is reducible if and only if it has a zero in the ground field; for at least one of the factors would have to be of degree unity, and would be of the form $x - a$, where $a \in GF(p)$, and so $f(a) = 0$.

EXAMPLE 1.5.2.   Let $p = 2$. The field $GF(2)$ consists only of the two elements 0 and 1, with the property that $1 + 1 = 0$.

   (i) The polynomial

$$f(x) = x^3 + x^2 + 1 \tag{1.5.5}$$

   is irreducible over $Gf(2)$ because $f(0) \neq 0$ and $f(1) \neq 0$.
   (ii) The polynomial

$$g(x) = x^3 + x + 1 \tag{1.5.6}$$

   is irreducible over $GF(2)$ because $g(0) \neq 0$ and $g(1) \neq 0$.

1.5.2.   The Fields GF(p^n)

It will soon become apparent that the fields $GF(p)$, which we discussed in the preceding section, are not the only finite fields.

First, we shall mention some of the features that are common to all finite fields. Let F be an arbitrary finite field whose unit element is denoted by 1. The elements

$$1, 1 + 1, 1 + 1 + 1, \ldots$$

all belong to F. We introduce the abbreviation

$$r*1 = 1+1+\ldots+1 \quad (r \text{ terms of } 1), \tag{1.5.7}$$

where r is a positive integer. Note that we do not assert that $r \in F$, although $r*1 \in F$. It is easy to verify that

$$(r \pm s)*1 = (r*1) \pm (s*1) \tag{1.5.8}$$

and

$$(rs)*1 = (r*1)(s*1). \tag{1.5.9}$$

In the familiar fields, for instance the rationals or the reals, the elements (1.5.7) are all distinct; such as field is said to be of *characteristic zero*.

But when F is finite, the elements $r*1\,(r=1,2,\ldots)$ cannot all be distinct. Hence there are positive integers k and l with $k > l$ such that

$$k*1 = l*1$$

and so

$$w*1 = 0, \tag{1.5.10}$$

where $w = k - l$. Let m be the least positive integer such that

$$m*1 = 0. \tag{1.5.11}$$

We claim that m is a prime. For suppose that $m = m_1 m_2$, $m_1 > 1$ and $m_2 > 1$, from (1.5.11) it would then follow that

$$0 = (m_1 m_2 * 1) = (m_1 * 1)(m_2 * 1).$$

Since there are no zero divisors in F, we must have that either $m_1 * 1 = 0$ or else $m_2 * 1 = 0$. But both these equations would contradict the hypothesis that m is the least positive integer satisfying (1.5.11).

Thus with each finite field F there is associated a unique prime p, known as the *characteristic* of F: it has the property that

$$p*1 = 0. \tag{1.5.12}$$

By virtue of (1.5.12) it is evident that the elements $r*1$ introduced in (1.5.7) behave exactly like the elements of $GF(p)$ defined in (1.5.1). We shall accordingly identify $r*1$ with r (the conjugacy class of $r \bmod p$), and we may state the following.

PROPOSITION 1.5.4. *Let F be a field of characteristic p. Then F contains a subfield isomorphic with $GF(p)$* [see I, Definition 8.2.19].

Of course, it may happen that F is identical with $GF(p)$. In any event F can be regarded as a vector space over $GF(p)$ [see I, §5.2]; for if u and v belong to F so does $u + v$, and if a belongs to F so does au. Since F is finite, it is necessarily of finite dimension over $GF(p)$. Hence there exists a basis

$$u_1, u_2, \ldots, u_n$$

of F over $GF(p)$ [see I, §5.4]. An arbitrary element of F is of the form

$$x = a_1 u_1 + a_2 u_2 + \ldots + a_n u_n, \tag{1.5.13}$$

where $a_i \in F (i = 1, 2, \ldots, n)$. We obtain all the elements of F by letting the a_i run independently over the p elements of $GF(p)$. This observation immediately leads to the following proposition.

PROPOSITION 1.5.5. *If F is a finite field of characteristic p, then it consists of p^n elements, where n is a positive integer. We shall call p^n the order of F.*

It still remains to demonstrate that there are indeed fields of order p^n, where n is greater than unity. For this purpose we carry out the following construction.

By Theorem 1.5.3 there exists a polynomial $f(x)$ of degree $n > 1$ which is irreducible over $GF(p)$. Since $f(x)$ remains irreducible if it is multiplied by a non-zero element of $GF(p)$, we may assume that the highest coefficient of $f(x)$ is equal to unity. Thus let

$$f(x) = x^n + c_{n-1} x^{n-1} + \cdots + c_1 x + c_0. \tag{1.5.14}$$

Now we introduce a symbol β which satisfies all the laws of algebra except that

$$\beta^u + c_{n-1} \beta^{n-1} + \ldots + c_1 \beta + c_0 = 0;$$

in other words, we stipulate that

$$f(\beta) = 0. \tag{1.5.15}$$

Since f is irreducible of degree n, this precludes the possibility that β belongs to $GF(p)$; for if f had a zero in the ground field, it would split off a linear factor.

PROPOSITION 1.5.6. *Let $g(x)$ be an arbitrary polynomial over $GF(p)$. Then there exists a polynomial $r(x)$ over $GF(p)$ of degree less than n such that $g(\beta) = r(\beta)$.*

Proof. By Proposition 1.5.2, there exist polynomials $q(x)$ and $r(x)$, where r is of degree less than n, such that

$$g(x) = q(x) f(x) + r(x). \tag{1.5.16}$$

On putting $x = \beta$ and using (1.5.15), we obtain the result required.

Since $f(x)$ is irreducible, its only factors are f itself and unity (or any non-zero constant). It follows that (1.5.15) is the equation of least degree for β over $GF(p)$. For suppose that

$$h(\beta) = 0, \tag{1.5.17}$$

where $h(x)$ is a polynomial of degree less than n. Then

$$HCF(f, h) = 1 \quad \text{or} \quad f.$$

In the first case we should have a relation of the form

$$u(x)f(x) + v(x)h(x) = 1,$$

which leads to a contradiction when we substitute β for x; the second case cannot occur either because it would imply that f divides h, which is impossible because $\deg h < \deg f$. Hence when $\deg h < n$, the equation (1.5.17) can hold only if h is the zero polynomial, that is if all its coefficients are zero.

Next, if $g(x)$ is an polynomial over $GF(p)$ for which

$$g(\beta) = 0,$$

then $g(x)$ is divisible by $f(x)$. This follows by putting $x = \beta$ in (1.5.16) whence $r(\beta) = 0$. Since $\deg r < n$, we deduce that r is the zero polynomial, so that

$$g(x) = q(x)f(x).$$

We now consider the set F of polynomials in β with coefficients in $GF(p)$ and of degree less than n. Thus a typical element of F is of the form

$$u = a_0 + a_1\beta + \ldots + a_{n-1}\beta^{n-1}, \tag{1.5.18}$$

where $a_i \in GF(p)$ $(i = 0, 1, \ldots, n-1)$. We shall show that F forms a field with the usual rules for addition and multiplication, subject only to the further condition (1.5.15); the latter ensures that if u and

$$v = b_0 + b_1\beta + \ldots + b_{n-1}\beta^{n-1}$$

are any elements of F, then

$$uv = c_0 + c_1\beta + \ldots + c_{n-1}\beta^{n-1},$$

where $c_0, c_1, \ldots, c_{n-1} \in GF(p)$. The zero element in F is given by the zero polynomial, that is when $a_0 = a_1 = \ldots = a_{n-1} = 0$ in (1.5.15); for the unit element, the coefficients are $a_0 = 1$, $a_1 = \ldots = a_{n-1}$. It is easy to check that all the field axioms are satisfied by F; only the existence of the inverse requires a little more thought: let u be a non-zero element of F. Then the polynomial $u(x)$, being of degree less than n, is coprime to $f(x)$. Hence there exist polynomials $u_1(x)$ and $f_1(x)$ over $GF(p)$ such that

$$u_1(x)u(x) + f_1(x)f(x) = 1.$$

On putting $x = \beta$ in this equation and using (1.5.15) we find that $u_1(\beta)u(\beta) = 1$, that is

$$\{u(\beta)\}^{-1} = u_1(\beta).$$

Clearly, F is of order p^n because in (1.5.15) the coefficients $a_0, a_1, \ldots, a_{n-1}$ range independently of each other over the p elements of $GF(p)$. It is customary to denote F by

$$GF(p^n) \tag{1.5.19}$$

and to refer to it as the *Galois field* of order p^n. This terminology is justified by the following theorem [see Cohn (1977), Vol. II, §5.7 or Herstein (1964) p. 314].

THEOREM 1.5.7.   *Any two fields of order p^n are isomorphic to each other.*

Surprisingly, this theorem implies that it does not matter which irreducible polynomial of degree n over $GF(p)$ we are using for the construction of $GF(p^n)$.

EXAMPLE 1.5.3.   Let $p = 2$ and $n = 3$. In order to construct a field of order 8 we may use the irreducible polynomial

$$x^3 + x^2 + 1$$

given in (1.5.5). Thus we introduce a symbol β which satisfies the equation

$$\beta^3 = \beta^2 + 1. \tag{1.5.20}$$

Note that since the characteristic is 2, we have that $+1 = -1$. The 8 elements of the field may now be listed as follows

$$0, 1, \beta, \beta + 1, \beta^2, \beta^2 + 1, \beta^2 + \beta, \beta^2 + \beta + 1. \tag{1.5.21}$$

We mention a few manipulations in this field:

$$(\beta^2 + 1) + \beta^2 = 1,$$
$$(\beta^2 + 1) + (\beta^2 + \beta) = \beta + 1,$$
$$(\beta^2 + 1)(\beta^2 + \beta) = \beta^4 + \beta^3 + \beta^2 + \beta = \beta(\beta^3 + \beta^2 + 1) + \beta^2 = \beta^2.$$

EXAMPLE 1.5.4.   As in the previous example we choose $p = 2$ and $n = 3$. But we now select the polynomial $g(x)$ given in (1.5.6) as the basis for constructing a field of order 8. Thus we introduce a symbol α which satisfies the equation

$$\alpha^3 = \alpha + 1, \tag{1.5.22}$$

and the 8 elements are

$$0, 1, \alpha, \alpha + 1, \alpha^2, \alpha^2 + 1, \alpha^2 + \alpha, \alpha^2 + \alpha + 1. \tag{1.5.23}$$

According to Theorem 1.5.7 there should exist a *field-homomorphism T* between the sets (1.5.21) and (1.5.23). (The analogous notion for groups is described in I, Definition 8.2.15.) However, it would not do to define $T(\beta)$ as α, because β and α satisfy different equations, namely (1.5.20) and (1.5.22) respectively. But we may put

$$T(\beta) = \alpha + 1,$$

because $\alpha + 1$ satisfies the same equation as β: in fact

$$(\alpha + 1)^3 + (\alpha + 1)^2 + 1 = \alpha^3 + 3\alpha^2 + 3\alpha + 1 + \alpha^2 + 2\alpha + 1 + 1$$
$$= \alpha^3 + \alpha + 1 = 0.$$

EXAMPLE 1.5.5. The field GF(9) is an extension of GF(3) whose elements we shall denote by 0, 1, -1 ($= 2$). The construction may be based on the polynomial

$$x^2 + 1$$

which is irreducible over GF(3), since it has no zeros in that field. Thus we introduce a symbol γ which satisfies the equation

$$\gamma^2 + 1 = 0.$$

The nine elements of GF(9) can now be listed as follows:

$$0, 1, -1, \gamma, -\gamma, \gamma + 1, \gamma - 1, -\gamma + 1, -\gamma - 1.$$

Most algebraical concepts and results are the same whether the characteristic of the underlying field is zero or a prime. But some formulae assume a rather surprising shape.

PROPOSITION 1.5.8. *Suppose that F is a field of characteristic p and let x and y be indeterminates. Then*

$$(x + y)^p = x^p + y^p \tag{1.5.24}$$

More generally, when n is any positive integer,

$$(x + y)^{p^n} = x^{p^n} + y^{p^n}. \tag{1.5.25}$$

In order to see that the 'wrong' form (1.5.25) of the binomial theorem [I, §3.10] is true in a field of characteristic p, we refer to Example 1.4.3 where it was shown that each of the binomial coefficients

$$\binom{p}{r} \quad (r = 1, 2, \ldots, p - 1) \tag{1.5.26}$$

is congruent to zero (mod p). When regarded as elements of F these coefficients are therefore equal to zero. Hence in the usual binomial expansion

$$(x + y)^p = x^p + \binom{p}{1}x^{p-1}y + \ldots + \binom{p}{p-1}xy^{p-1} + y^p,$$

all except the first and the last terms are zero, which establishes (1.5.24). On raising this equation to the pth power we obtain that

$$(x + y)^{p^2} = (x^p + y^p)^p = x^{p^2} + y^{p^2},$$

and so on.

We can generalize (1.5.24) in a different way by introducing more than two indeterminates. Thus we have that

$$(x_1 + x_2 + \ldots + x_m)^p = x_1^p + x_2^p + \ldots + x_m^p. \tag{1.5.27}$$

When p and n are fixed, it is customary to use the abbreviation

$$q = p^n. \tag{1.5.28}$$

The non-zero elements of every field, whether finite or infinite, form an Abelian group under multiplication, for every non-zero element possesses an inverse, and the other group axioms are easily verified [see I, §8.2.1]. In the case of a finite field this multiplicative group has a very simple structure.

THEOREM 1.5.9. *The non-zero elements of* GF(q) *form a cyclic group of order* $q - 1$, *that is there exists an elements* ρ *called a primitive element of* GF(q), *such that*

$$\rho^{q-1} = 1, \tag{1.5.29}$$

but $\rho^k \neq 1$, *when* $1 \leq k \leq q - 2$. *Hence the complete list of elements of* GF(q) *can be written down as*

$$0, 1, \rho, \rho^2, \ldots, \rho^{q-2}. \tag{1.5.30}$$

EXAMPLE 1.5.6. The multiplicative group of GF(9), which we discussed in Example 1.5.5 is of order 8. So by Lagrange's theorem [see I, Theorem 8.2], the order (period) of an non-zero element can only be 1 and 2 or 4 or 8. The element γ which we used to define GF(9) satisfies

$$\gamma^4 = 1,$$

so γ is not a primitive element. But the element $\gamma + 1$ is primitive, because

$$(\gamma + 1)^2 = \gamma^2 + 2\gamma + 1 = 2\gamma = -\gamma; \qquad (\gamma + 1)^4 = \gamma^2 = -1(\neq 1]).$$

Since an arbitrary non-zero element u of GF(q) can be expressed as

$$u = \rho^k,$$

it follows from (1.5.29) that

$$u^{q-1} = 1.$$

Thus every non-zero element of GF(q) satisfies the equation

$$x^{q-1} = 1,$$

and if we multiply both sides by x we obtain an equation that is satisfied also by zero:

THEOREM 1.5.10. *Every element of* GF(q) *satisfies the equation*

$$x^q = x. \tag{1.5.31}$$

Many authors use this result as the starting point for defining GF(q): regarding (1.5.31) as an equation over GF(p) they introduce GF(q) as the set of all roots of (1.5.31) in a suitable 'extension' field of GF(p). This approach does not involve an irreducible polynomial for the construction of GF(q) and leads to the conclusion that this field is essentially unique. For further details the reader is referred to the text-book literature [for example Cohn (1977) Vol. II, §5.7 or Herstein (1964), p. 315].

1.5.3.   Vector Spaces over GF(q)

We may go one step further and regard GF(q) as the ground field upon which algebraical structures can be built.

Indeed, in I, §5.2 the concept of a vector space V was defined with respect to any ground field F: two vectors $\mathbf{a}$ and $\mathbf{b}$ have a sum $\mathbf{a} + \mathbf{b}$, and a vector $\mathbf{a}$ may be multiplied by a scalar, that is by an element λ of F to yield a unique vector $\lambda\mathbf{a}$.

The notions of linear dependence, basis and dimension apply to any ground field [I, §5.3]. If V is of finite dimension m, then there exist basis vectors $\mathbf{u}_1, \mathbf{u}_2, \ldots, \mathbf{u}_m$ such that an arbitrary vector of V is uniquely given by

$$\mathbf{x} = \lambda_1\mathbf{u}_1 + \lambda_2\mathbf{u}_2 + \ldots + \lambda_m\mathbf{u}_m, \tag{1.5.32}$$

where $\lambda_1, \lambda_2, \ldots, \lambda_m$ are elements of F.

When $F = $ GF(q), it is clear that V consists of $q^m = p^{mn}$ vectors, since each coefficient in (1.5.32) can equal any one of the q elements of GF(q).

A vector space of dimension m over a field of q elements is sometimes denoted by

$$V(m, q).$$

EXAMPLE 1.5.7.   A typical element of $V(m, 2)$ consits of a string of m noughts and ones, like

$$(0, 0, 1, 0, \ldots, 1)_m, \tag{1.5.33}$$

which represent the 2^m vectors of this space.

We recall [see Theorem 1.2.1] that the symbol (1.5.33) was also used to denote an integer s in the binary scale when $0 \leq s \leq 2^m - 1$. In turn, these integers may be regarded as all possible residues when an arbitrary integer is divided by 2^m. Hence we have the result that *the vectors of $V(m, 2)$ are in one-to-one correspondence with the residue classes* mod 2^m.

1.5.4.   Polynomials over GF(q)

The formal properties of polynomials are the same whether the ground field is finite or infinite, and the definitions of divisibility, irreducibility and common factors are unaltered. The quotient-and-remainder formula holds, as does the theorem about unique factorization.

Let

$$P(x) = \alpha_0 + \alpha_1 x + \ldots + \alpha_m x^m \tag{1.5.34}$$

be an irreducible polynomial of degree m over GF(q). We consider the set Φ of residues classes mod $P(x)$. These are in one-to-one correspondence with the polynomials

$$G(x) = \lambda_0 + \lambda_1 x + \ldots + \lambda_{m-1} x^{m-1} \tag{1.5.35}$$

over GF(q) of degree less than m, which are the remainders when an arbitrary polynomial is divided by $P(x)$.

By arguments similar to those used on p. 23, it can be shown that Φ becomes a field if addition and multiplication of polynomials are carried out in the usual way, followed by a reduction mod $P(x)$; this ensures that all operations result in one of the q^m polynomials described in (1.5.35). The zero element of Φ consists of all polynomials divisible by $P(x)$. A non-zero element of Φ is represented by a non-zero polynomial $G(x)$ of degree less than m. Since $P(x)$ is irreducible, the polynomials $G(x)$ and $P(x)$ are necessarily coprime. Hence there exist polynomials $G_1(x)$ and $P_1(x)$ over GF(q) such that

$$G_1(x)G(x) + P_1(x)P(x) = 1.$$

When interpreted as a congruence mod $P(x)$, this equation states that

$$G_1(x)G(x) \equiv 1 \quad (\text{mod } P(x))$$

Hence $G_1(x)$ is the inverse of $g(x)$ in Φ. It is evident from (1.5.36) that Φ consists of $q^m (= p^{mn})$ elements. Hence, by Theorem 1.5.7, Φ is isomorphic with GF(p^{mn}), and we see in retrospect that the polynomial $P(x)$ might have been replaced by any other irreducible polynomial over GF(q) of degree m.

The non-zero elements of Φ form an Abelian group under multiplication of order $q^m - 1$. So each element X of this group satisfies the equation

$$X^{q^m - 1} = 1.$$

[see I, §8.2.5]. Hence $X^{q^m - 1} - 1$ is the zero element of Φ, which means that if $R(x)$ is a non-zero residue mod $P(x)$, then

$$P(x) | (R(x))^{q^m - 1} - 1.$$

Since we have assumed that $m \geq 2$, the indeterminate x is itself one of the possible residues mod $P(x)$. Hence we have the following theorem.

THEOREM 1.5.11.   *Let $P(x)$ be an irreducible polynomial over GF(p^n) of degree $m \geq 2$. Then*

$$P(x) | x^{p^{mn} - 1} - 1. \tag{1.5.36}$$

The exponent of x in this statement need not be the least possible. This suggests the following definition.

DEFINITION 1.5.1. The least positive integer e for which

$$P(x)|x^e - 1$$

is called the *exponent* of $P(x)$.

It is not hard to see that in all cases

$$e|p^{mn} - 1 \qquad (1.5.37)$$

[Dixon (1958) p. 20].

EXAMPLE 1.5.8. Let GF(2) be the ground field, with elements 0 and 1 ($1 + 1 = 0$). The polynomial

$$P(x) = x^4 + x^3 + x^2 + x + 1 \qquad (1.5.38)$$

is irreducible over GF(2). For $P(x)$ has no linear factors, because $P(0) \neq 0$ and $P(1) \neq 0$; neither is a factorization of the type

$$P(x) = (x^2 + ax + 1)(x^2 + bx + 1)$$

possible because it would lead to the equations

$$a + b = 1 \quad \text{and} \quad ab = 1$$

which are incompatible in GF(2). Hence the residues mod $P(x)$ form a field of order 2^4. From Theorem 1.5.11, $P(x)$ dividides $x^{15} + 1$ (recall that $-1 = +1$). But the exponent e of $P(x)$ is not equal to 15. For in GF(2) we have the relation

$$\frac{x^5 + 1}{x + 1} = x^4 + x^3 + x^2 + x + 1,$$

whence

$$P(x)|x^5 + 1$$

Since $e|15$ and evidently $e \neq 3$, it follows that $e = 5$.

W. L.

REFERENCES

Andrews, G. E. (1971). *Number Theory*, W. B. Saunders.
Cohn, P. M. (1977). *Algebra*, vol. 2, John Wiley and Sons.
Dickson, L. E. (1958). *Linear Groups with an Exposition of the Galois Field Theory*, Dover.
Dudley, U. (1969). *Elementary Number Theory*, Freeman.
Hardy, G. H. and Wright, E. M. (1954). *An Introduction to the Theory of Numbers* (3rd edn.), Oxford.
Herstein, I. N. (1964). *Topics in Algebra*, Blaisdell.
Hunter, J. (1964). *Number Theory*, Oliver and Boyd.

CHAPTER 2

Fibonacci and Lucas Numbers

2.1. DEFINITIONS

Leonardo of Pisa, better known as Fibonacci, that is son of Bonacci, a citizen of Pisa, is said to have been the only eminent European mathematician of the middle ages, but he entered the mathematical folklore through another achievement: he asked a question.

Assume the breeding habits of rabbits to be such that each pair produces a new pair every month, starting in the second month of their lives. If we start with one single new-born pair, how many pairs of rabbits will there be after $1, 2, \ldots$ months, if no rabbits ever dies?

The answer is simple enough. At the start of the first month there will be one pair, and they will still be alone at the start of the second month. But at the start of the third month, there will be one more pair.

If there were F_n pairs at the start of the nth month, and F_{n+1} pairs at the start of the next month, then at the beginning of the $(n + 2)$th month there will be those F_{n+1} pairs, but also another F_n pairs, the offspring of the F_n pairs who were there 2 months earlier. Hence

$$F_{n+2} = F_{n+1} + F_n. \tag{2.1.1}$$

To obtain the Fibonacci sequence $F_1, F_2, \ldots$, we start with the term

$$1 \quad 1$$

and add at each stage the last two terms to obtain the next terms, thus

$$1 \quad 1 \quad 2 \quad 3 \quad 5 \quad 8 \quad 13 \quad 21 \quad 34 \quad 55 \ldots$$

The sequence can be extended backwards, using the Fibonacci mechanism (2.1.1) which produces

$$F_0 = 0, \qquad F_{-1} = 1, \qquad F_{-2} = -1$$

and generally

$$F_{-n} = (-1)^{n+1} F_n.$$

Any Fibonacci number can be found by applying (2.1.1) consecutively long enough, but it is convenient to have an explicit formula for F_n, and such a formula can be obtained by observing that (2.1.1) is a recurrence relation. To solve it (see I, 14.13) set $F_n = x^n$, so that (2.1.1) reads

$$x^{n+2} - x^{n+1} - x^n = 0.$$

The solution $x = 0$ is irrelevant in our present context, so that we divide by x^n and solve the 'characteristic equation'

$$x^2 - x - 1 = 0. \tag{2.1.2}$$

The roots of this equation are

$$\frac{1 + \sqrt{5}}{2} = \tau, \quad \text{and} \quad \frac{1 - \sqrt{5}}{2} = \sigma.$$

τ is the golden section number, and

$$\tau = 1.618\ldots, \qquad \tau\sigma = -1, \qquad \tau + \sigma = 1, \qquad \tau - \sigma = \sqrt{5}.$$

τ as well as σ solve (2.1.2), but they are, of course, not integers. However,

$$\alpha\tau^n + \beta\sigma^n$$

also solves (2.1.1), and if α and β are suitably chosen, then we obtain Fibonacci numbers.

α and β must be such that they produce $F_0 = 0$ and $F_1 = 1$. This means

$$\alpha = 1/\sqrt{5} \quad \text{and} \quad \beta = -1/\sqrt{5}.$$

Explicitly, we have

$$F_n = \frac{\tau^n - \sigma^n}{\sqrt{5}} = \frac{\tau^n - \sigma^n}{\tau - \sigma}. \tag{2.1.3}$$

We call $(F_0, F_1) = (0, 1)$ the 'seed' of the Fibonacci sequence. Other seeds produce, by the mechanism (2.1.1), other sequences. For instance, $\alpha = \beta = 1$ produces

$$\tau^n + \sigma^n = L_n, \tag{2.1.4}$$

the Lucas sequence L_i ($i = 1, 2, \ldots$)

$$1 \quad 3 \quad 4 \quad 7\ldots,$$

so-called after the French mathematician of the nineteenth century E. Lucas. This sequence can, of course, also be extended backwards. Note that $L_0 = 2$.

Fibonacci and Lucas numbers are connected by

$$L_n = F_{n+1} + F_{n-1}. \tag{2.1.5}$$

Proof. (2.1.5) holds for $n = 1$, and for $n = 2$ and hence, by (2.1.1), generally.

Another proof of (2.1.5) proceeds by using (2.1.3), (2.1.4) and $\tau\sigma = -1$, as follows:

$$\tau^{n+1} - \sigma^{n+1} + \tau^{n-1} - \sigma^{n-1} = (\tau - \sigma)(\tau^n + \sigma^n).$$

Fibonacci as well as Lucas numbers solve a special case of the more general linear recurrence relation

$$u_{n+k} = a_1 u_{n+k-1} + a_2 u_{n+k-2} + \ldots + a_k u_n.$$

They have been studied extensively, but we restrict ourselves here to the case $k = 2, a_1 = a_2 = 1$.

2.2. EXAMPLES

Fibonacci and Lucas numbers answer a great variety of questions and we now mention some of them.

EXAMPLE 2.2.1. In the species of honey-bees, eggs of females (workers) develop without fertilization into drones (males), so that a drone has only one single parent. The queen's eggs are fertilized by drones and develop into females, that is workers or queens. Thus a female has two parents, one male and one female.

Let us call the parents of a drone its generation 1, consisting of one female, $f_1 = 1$, and no male, $m_1 = 0$. We number the generations backwards in time. The female of generation 1 had a male and a female parent, so that $f_2 = 1, m_2 = 1$. Continuing, we find that generally

$$m_{n+1} = f_n, \qquad m_{n+2} = f_{n+1}, \ldots$$

and

$$f_{n+1} = f_n + m_n, \qquad f_{n+2} = f_{n+1} + m_{n+1}, \ldots$$

This means that

$$f_{n+2} = f_{n+1} + f_n \quad \text{and} \quad m_{n+2} = m_{n+1} + m_n.$$

In either case we observe mechanism (2.1.1). Now

$$m_1 = 0, \quad m_2 = 1; \quad f_1 = 1, \quad f_2 = 1.$$

Therefore

$$m_n = F_{n-1} \quad \text{and} \quad f_n = F_n.$$

Generation n consists of $m_n + f_n = F_n + F_{n-1} = F_{n+1}$ members.

EXAMPLE 2.2.2. A fair coin is tossed until two consecutive heads (HH) appear. Find the probability of the sequence terminating after n tosses.

For $n = 1$, the probability is clearly 0. For $n = 2$, it is $\frac{1}{4}$. For $n \geq 3$, a sequence of length n is either a sequence of length $n - 1$ preceded by a tail (T), or one of length $n - 2$, preceded by HT. The probability of the first case is $\frac{1}{2}$, that of the second case is $\frac{1}{4}$. Therefore the probability of a sequence of length n, P_n say, is

$$P_n = \tfrac{1}{2} P_{n-1} + \tfrac{1}{4} P_{n-2}.$$

The characteristic equation

$$x^2 - \tfrac{1}{2}x - \tfrac{1}{4} = 0$$

has roots $\tfrac{1}{2}\tau$ and $\tfrac{1}{2}\sigma$, so that a general solution is

$$\alpha(\tfrac{1}{2}\tau)^n + \beta(\tfrac{1}{2}\sigma)^n.$$

Because $P_1 = 0$ and $P_2 = \tfrac{1}{4}$, we must have

$$\alpha = 1/(\tau\sqrt{5}) \quad \text{and} \quad \beta = -1/(\sigma\sqrt{5})$$

so that

$$P_n = \frac{\tau^{n-1} - \sigma^{n-1}}{2^n\sqrt{5}} = \frac{F_{n-1}}{2^n}.$$

EXAMPLE 2.2.3.   In how many ways can the integer n be written as a sum of positive integers larger than 1? Call this number T_n.

Write the sums with non-decreasing summands. Some of the sums (possibly none) will have the first term equal to 2, and the others (possibly none) will have the first term larger than 2.

Consider the first type. If we omit the first 2, then there will be T_{n-2} possibilities with total $n - 2$ left. On the other hand, if in the sums of the second type we decrease the first term by 1, then there will be T_{n-1} possibilities with total $n - 1$ left. Therefore

$$T_n = T_{n-1} + T_{n-2}.$$

Since $T_2 = 1$, and $T_3 = 1$, we obtain $T_n = F_{n-1}$.

EXAMPLE 2.2.4.   Consider n people, $P_1, \ldots, P_n (n \geq 3)$ sitting at a circular table. At a given signal, any two neighbours may, but need not, exchange places. Of course, any person can only be in one single exchange at any one time. How many reallocations are possible? Denote their number by H_n.

In order to find H_n, consider first a slightly different problem, where P_i $(i = 1, 2, \ldots)$ sit in a row of n seats, P_1 and P_n in the two seats at the ends of the row, so that there can be no exchange between these two. Denote the number of possible reallocations in this case by G_n.

If P_1 remains seated, then there are $n - 1$ possibilities left, while if P_1 and P_2 exchange places, then the number of possible reallocations is G_{n-2}. Therefore

$$G_n = G_{n-1} + G_{n-2}.$$

Clearly $G_1 = 1$ (no change) and $G_2 = 2$ (no change, or one exchange). Therefore $G_n = F_{n+1}$.

We return to the original problem, to determine H_n. If P_1 and P_n exchange seats, then there are G_{n-2} possibilities left. If they do not, then the number of possibilities is G_n, as above.

We have found that

$$H_n = G_n + G_{n-2} = F_{n+1} + F_{n-1} = L_n$$

by formula (2.1.5).

These two examples were mentioned in V, 15.5.12.

EXAMPLE 2.2.5.   Suppose we want to find the smallest value of a continuous convex function $f(x)$, within a given interval $[A, B]$ of x.

Subdivide the interval into smaller portions, to find a smaller interval within which the argument of the smallest value will lie.

Consider the following procedure. Find two points, C and D, within $[A, B]$, such that $[A, C] = [D, B] < \frac{1}{2}[A, B]$. Compare $f(C)$ and $f(D)$. If $f(C) < f(D)$, then the minimum will be in $[A, D]$, otherwise it will be in $[C, B]$.

Now concentrate on that subinterval which we have found to be of interest, and treat it in the same way, as $[A, B]$ was treated. Thus continue, until n evaluations have been made.

We must now find a criterion for determining the points C and D at the first step, and the analogous points at later steps. Let the length of $[A, B]$ be unity, and denote the ratio $[A, D]/[D, B]$ by r. We might choose r to be τ, and thus carry out the so-called golden section search. Alternatively, we might change r from step to step and make the ratio at the first step equal to F_{n+1}/F_n, at the next step F_n/F_{n-1}, and so on until $F_2/F_1\ (= 1)$. This is the Fibonacci search.

We shall consider that method preferable for which the reduction of the original interval to the final interval is largest.

After n evaluations, the reduction in the golden section search equals τ^n/τ, and in the Fibonacci search it equals F_{n+1}/F_2, which equals, for large n, approximately $\tau^{n+1}/\sqrt{5}$. Now

$$\frac{\tau^{n+1}/\sqrt{5}}{\tau^{n-1}} \approx 1.17.$$

Therefore, by this criterion, the Fibonacci search is preferable.

EXAMPLE 2.2.6.   Fibonacci numbers have appeared in botanical studies. D'Arcy Thompson has remarked that cones have sometimes five rows of scales winding up in one direction and three in another, or perhaps there are eight and five rows respectively, or even thirteen and eight. Similar observations have been made about arrangements of leaves.

Space forbids to enlarge here on the way in which the Fibonacci mechanism is thought to be active in this context.

2.3. RELATIONSHIPS

For many mathematicians and amateurs the fascination of Fibonacci and Lucas numbers is due to the many relationships which these numbers satisfy. The list of such relationships is almost inexhaustible, and we shall mention here a few, which we think are typical.

Relationships can be derived either by using the formula (2.1.1) and induction, or by using the explicit expression (2.1.3) and (2.1.4). Many of them can be derived equally easily by either method, and it is a matter of taste which method is found to be more convenient, or more informative, in a particular case.

We have proved formula (2.1.5) above by both methods. This formula exhibits a connection between Fibonacci numbers and Lucas numbers. We add more connections between these two types.

$$F_{2n} = F_n L_n. \tag{2.3.1}$$

The proof is immediate:

$$\frac{\tau^{2n} - \sigma^{2n}}{\tau - \sigma} = \frac{\tau^n - \sigma^n}{\tau - \sigma}(\tau^n + \sigma^n).$$

$$5F_n^2 - L_n^2 = 4(-1)^{n+1}. \tag{2.3.2}$$

Proof.

$$5\left(\frac{\tau^n - \sigma^n}{\sqrt{5}}\right)^2 - (\tau^n + \sigma^n)^2 = -4\tau^n\sigma^n = 4(-1)^{n+1}.$$

Formula (2.3.2) shows that L_n and F_n cannot have a larger common factor than 2, and that either both are odd, or both are even.

Relationships with binomial coefficients emerge when we expand the powers of $\tau = \frac{1}{2}(1 + \sqrt{5})$ and $\sigma = \frac{1}{2}(1 - \sqrt{5})$ by the binomial theorem [see I, 3.10]. We obtain

$$2^{n-1}F_n = \binom{n}{1} + 5\binom{n}{3} + 5^2\binom{n}{5} + \ldots + 5^r\binom{n}{2r+1} + \ldots \tag{2.3.3}$$

and

$$2^{n-1}L_n = 1 + 5\binom{n}{2} + 5^2\binom{n}{4} + \ldots + 5^r\binom{n}{2r} + \ldots \tag{2.3.4}$$

A further connection between Fibonacci numbers and binomial coefficients is

$$F_n = 1 + \binom{n-2}{1} + \binom{n-3}{2} + \binom{n-4}{3} + \ldots \tag{2.3.5}$$

Proof.   The formula holds for $n = 1$ and for $n = 2$. Assume that it holds for

$$F_m = \sum_0 \binom{m-i-1}{i} \quad \text{and} \quad F_{m+1} = \sum_0 \binom{m-i}{i} = 1 + \sum_0 \binom{m-i-1}{i+1}.$$

Then

$$F_{m+2} = F_{m+1} + F_m$$

$$= 1 + \sum_0 \binom{m-i-1}{i+1} + \sum_0 \binom{m-i-1}{i}$$

$$= 1 + \sum_0 \binom{m-i}{i+1} = \sum_0 \binom{m-i+1}{i}.$$

Thus (2.3.5) holds for F_{m+2} as well. This proves (2.3.5) by induction.

Many relationships of Fibonacci numbers and Lucas numbers derive from a formula which applies to any sequence generated by the mechanism (2.1.1), irrespective of the seed. Denote such a sequence by G_i. Then

$$G_{n+m} = F_{m-1} G_n + F_m G_{n+1} \quad (n = 1, 2, \ldots) \tag{2.3.6}$$

Proof.

$$G_{n+1} = F_0 G_n + F_1 G_{n+1}, \quad \text{because } F_0 = 0 \text{ and } F_1 = 1.$$

Also

$$G_{n+2} = F_1 G_n + F_2 G_{n+1}, \quad \text{because } F_1 = 1 \text{ and } F_2 = 1.$$

Adding, we obtain

$$G_{n+3} = F_2 G_n + F_3 G_{n+1}.$$

Continuing in this manner, we prove (2.3.6) by induction.

To demonstrate the fundamental role of (2.3.6), we derive from it, identifying G_i with F_i and setting $m = n + 1$,

$$F_{2n+1} = F_n^2 + F_{n+1}^2, \tag{2.3.7}$$

and identifying G_i with L_i and setting $m = n + 1$,

$$L_{2n+1} = F_n L_n + F_{n+1} L_{n+1}. \tag{2.3.8}$$

Compare (2.3.7) with

$$F_{2n+1} = F_{n+1} L_{n+1} - F_n L_n \tag{2.3.9}$$

which we derive from $F_{2n+1} = F_{2n+2} - F_{2n}$ and (2.3.1).

A very general formula, referring to a generalized Fibonacci sequence G_i and a (possibly different) generalized sequence H_i, where both G_i and H_i follow (2.1.1) and no assumption is made about seeds, reads

$$G_{n+h} H_{n+k} - G_n H_{n+h+k} = (-1)^n (G_h H_k - G_0 H_{h+k}).$$

We shall not prove this general formula, but only a special case of it, where both G_i and H_i are the Fibonacci sequence F_i, and $h = 1$.

Consider the determinant

$$\begin{vmatrix} F_{n+1} & F_n \\ F_{n+k+1} & F_{n+k} \end{vmatrix} = \begin{vmatrix} F_n & F_{n-1} \\ F_{n+k} & F_{n+k+1} \end{vmatrix} \begin{vmatrix} 1 & 1 \\ 1 & 0 \end{vmatrix}$$

$$\begin{vmatrix} F_{n-1} & F_{n-2} \\ F_{n+k-1} & F_{n+k-2} \end{vmatrix} \begin{vmatrix} 1 & 1 \\ 1 & 0 \end{vmatrix}^2 = \cdots = \begin{vmatrix} F_2 & F_1 \\ F_{k+2} & F_{k+1} \end{vmatrix} \begin{vmatrix} 1 & 1 \\ 1 & 0 \end{vmatrix}^{n-1}$$

that is

$$F_{n+1}F_{n+k} - F_n F_{n+k+1} = (F_{k+1} - F_{k+2})(-1)^{n-1} = F_k(-1)^n.$$

When $k = -1$, we have

$$F_{n+1}F_{n-1} - F_n^2 = (-1)^n, \quad \text{because } F_{-1} = 1 \tag{2.3.10}$$

and when $k = -2$, then

$$F_{n+1}F_{n-2} - F_n F_{n-1} = (-1)^{n+1}, \quad \text{because } F_{-2} = -1. \tag{2.3.11}$$

We add two more formulae, not of particular interest in themselves, but which are used later, in §2.4.

$$3F_n + L_n = 2F_{n+2}, \tag{2.3.12}$$

$$5F_n + 3L_n = 2L_{n+z}. \tag{2.3.13}$$

These formulae are easily proved by induction, since they hold clearly for $n = 1$ and for $n = 2$.

We turn now to formulae involving sums of Fibonacci numbers. We start with a formula for generalized Fibonacci sequences G_i, which follow (2.1.1), independent of the seed.

$$\sum_{i=1}^{n} G_i = G_{n+2} - G_2. \tag{2.3.14}$$

Proof. We have, as a consequence of the mechanism (2.1.1)

$$\sum_{i=1}^{n} G_i = \sum_{i=3}^{n+2} G_i - \sum_{i=2}^{n+1} G_i$$

$$= \left(\sum_{i=1}^{n+2} G_i - G_1 - G_2 \right) - \left(\sum_{i=1}^{n+2} G_i - G_1 - G_{n+2} \right) = G_{n+2} - G_2.$$

$$\sum_{i=1}^{2n} F_i F_{i-1} = F_{2n}^2. \tag{2.3.15}$$

Proof. (2.3.10) can be written

$$(F_n + F_{n-1})F_{n-1} - F_n^2 = (-1)^n.$$

Hence

$$\sum_{i=1}^{2n} F_i F_{i-1} + \sum_{i=1}^{2n} F_{i-1}^2 - \sum_{i=1}^{2n} F_i^2 = \sum_{i=1}^{2n} F_i F_{i-1} - F_{2n}^2 = \sum_{i=1}^{2n} (-1)^i = 0.$$

$$\sum_{i=1}^{n} F_i^2 = F_n F_{n+1}. \tag{2.3.16}$$

Proof.

$$\sum_{i=1}^{n} F_i^2 = \sum_{i=1}^{n} F_i(F_{i+1} - F_{i-1}) = \sum_{i=1}^{n} F_i F_{i+1} - \sum_{i=1}^{n} F_{i-1} F_i = F_n F_{n+1}.$$

Further on the subject of relationships, it may be worth mentioning a connection between Fibonacci and Lucas numbers and hyperbolic functions (see IV, 2.1.3).

By definition,

$$\sinh(nz) = \frac{e^{nz} - e^{-nz}}{2}, \qquad \cosh(nz) = \frac{e^{nz} + e^{-nz}}{2}.$$

Let $z = \frac{1}{2}\ln \tau/\sigma$, then

$$\sinh(nz) = \frac{1}{2}\left[\left(\frac{\tau}{\sigma}\right)^{\frac{1}{2}n} - \left(\frac{\sigma}{\tau}\right)^{\frac{1}{2}n}\right] = \frac{1}{2}\frac{\tau^n - \sigma^n}{(\tau\sigma)^{\frac{1}{2}n}} = \frac{F_n\sqrt{5}}{2i^n} \tag{2.3.17}$$

and

$$\cosh(nz) = \frac{1}{2}\left[\left(\frac{\tau}{\sigma}\right)^{\frac{1}{2}n} + \left(\frac{\sigma}{\tau}\right)^{\frac{1}{2}n}\right] = \frac{1}{2}\frac{\tau^n + \sigma^n}{(\tau\sigma)^{\frac{1}{2}n}} = \frac{L_n}{2i^n}. \tag{2.3.18}$$

(NB In these two formulae i stands, of course, for $\sqrt{-1}$.)

Consequently, any relationship between hyperbolic functions is analogous to one between Fibonacci and/or Lucas numbers. For example,

$$\sinh 2nz = 2 \sinh(nz) \cosh(nz)$$

can be transformed into

$$\frac{F_{2n}\sqrt{5}}{2i^{2n}} = \frac{2F_n\sqrt{5}}{2i^n} \frac{L_n}{2i^n},$$

that is $F_{2n} = F_n L_n$, our formula (2.3.1).

2.4. DIOPHANTINE EQUATIONS

Let us return to formula (2.3.2):

$$\left.\begin{array}{ll} 5F_n^2 - L_n^2 = 4 & \text{when } n \text{ is odd} \\ 5F_n^2 - L_n^2 = -4 & \text{when } n \text{ is even.} \end{array}\right\} \tag{2.3.2}$$

These are two examples of Pell's equation. It can also be shown that these pairs are the only solutions in integers (see Vajda, 1989).

We consider now another pair of Diophantine equations.

$$y^2 - xy - x^2 = \pm 1. \qquad (2.4.1)$$

They are solved by $x = F_n$ and $y = F_{n+1}$, since

$$F_{n+1}^2 - F_n F_{n+1} - F_n^2 = (-1)^n$$

is a simple transformation of (2.3.11), using (2.1.1), and replacing n by $n+1$.

We can write (2.4.1) as

$$(y - \tau x)(y + x/\tau) = \pm 1.$$

The two rectangular hyperbolae in the coordinate plane represented by these two equations have the same asymptotes, and the slope of one of them is τ, the slope of the other is $-1/\tau$.

Again, it can be shown that the only points with integer positive coordinates on the hyperbolae are (F_i, F_{i+1}) (see References).

2.5.   CONVERGENTS

We consider the ratios of consecutive Fibonacci numbers, $F_2/F_1, F_3/F_2$, and so on, using continued fractions (see III, 6.4.5.) as a tool. We shall prove that

$$\lim_{n \to \infty} F_{n+1}/F_n = \tau.$$

We have

$$\frac{F_2}{F_1} = 1, \qquad \frac{F_3}{F_2} = \frac{F_2 + F_1}{F_2} = 1 + \frac{1}{F_2/F_1} = 1 + \frac{1}{1},$$

$$\frac{F_4}{F_3} = \frac{F_3 + F_2}{F_3} = 1 + \frac{1}{F_3/F_2} = 1 + \frac{1}{1 + \frac{1}{1}}$$

and generally

$$\frac{F_{n+1}}{F_n} = 1 + \cfrac{1}{1 + \cfrac{1}{1 + \cfrac{1}{1 + \cfrac{1}{\ddots + \cfrac{1}{1}}}}}$$

We notice that $F_2/F_1, F_3/F_2, \ldots$ are the initial portions of the development of F_{n+1}/F_n. They are called the 'convergents' of the latter ratio.

As n increases, we find

$$\lim_{n \to \infty} F_{n+1}/F_n = 1 + \cfrac{1}{1 + \cfrac{1}{1 + \cfrac{}{\ddots}}},$$

an infinite continued fraction, whose value is f, say. It follows from the structure of this continued fraction that

$$f = 1 + 1/f, \quad \text{that is} \quad f^2 - f - 1 = 0.$$

The roots of this quadratic equation are τ and σ, but since f is obviously positive, we have $f = \tau$.

This follows also from

$$\frac{F_{n+1}}{F_n} = \frac{\tau^{n+1} - \sigma^{n+1}}{\tau^n - \sigma^n},$$

because $|\sigma|$ is less than 1.

We shall now quote results from the theory of continued fractions which apply to our case. In particular, we investigate the manner in which the convergents approach their limit.

$$F_{n+1}/F_n < F_{n-1}/F_{n-2} \quad \text{when } n \text{ is even,} \tag{2.5.1a}$$

$$F_{n+1}/F_n > F_{n-1}/F_{n-2} \quad \text{when } n \text{ is odd.} \tag{2.5.1b}$$

Proof.   Divide (2.3.11) by $F_n F_{n-2}$.

Next, we shall show

$$F_{n+1}/F_n > \tau \quad \text{when } n \text{ is even,} \tag{2.5.2a}$$

$$F_{n+1}/F_n < \tau \quad \text{when } n \text{ is odd.} \tag{2.5.2b}$$

Proof.

$$(-1)^n = F_{n+1}F_{n-1} - F_n^2 = (F_{n-1} + F_n\tau)(F_{n+1} - F_n\tau),$$

because $1 + \tau = \tau^2$. We wrote this as

$$F_{n+1} - F_n\tau = \frac{(-1)^n}{F_{n-1} + F_n\tau} \tag{2.5.3}$$

and (2.5.2a) and (2.5.2b) follow, because $F_{n-1} + F_n\tau$ is positive.

Now divide (2.5.3) on both sides by F_n, and obtain

$$\left| \frac{F_{n+1}}{F_n} - \tau \right| = \left| \frac{1}{F_n F_{n-1} + F_n^2\tau} \right| < \frac{1}{F_n^2}.$$

It can be shown that, of any two consecutive convergents, at least one satisfies

$$\left| \frac{F_{n+1}}{F_n} - \tau \right| < \frac{1}{\sqrt{5F_n^2}},$$

and that 5 is best in the sense that this is not true anymore if 5 is replaced by a larger value.

2.6. FIBONACCI SERIES

The following expansions hold:

$$(m+n)^k = \sum_{i=0}^{\frac{1}{2}(k-1)} \binom{k}{i} (mn)^i (m^{k-2i} + n^{k-2i}) \quad \text{when } k \text{ is odd,}$$

$$(m+n)^k = \sum_{i=0}^{\frac{1}{2}k-1} \binom{k}{i} (mn)^i (m^{k-2i} + n^{k-2i}) + \binom{k}{\frac{1}{2}k} (mn)^{\frac{1}{2}k} \quad \text{when } k \text{ is even, } k \geq 2.$$

Also

$$(m-n)^k = \sum_{i=0}^{\frac{1}{2}(k-1)} \binom{k}{i} (-mn)^i (m^{k-2i} - n^{k-2i}) \quad \text{when } k \text{ is odd,}$$

$$(m-n)^k = \sum_{i=0}^{\frac{1}{2}k-1} \binom{k}{i} (-mn)^i (m^{k-2i} + n^{k-2i}) + \binom{k}{\frac{1}{2}k} (-mn)^{\frac{1}{2}k}$$

$$\text{when } k \text{ is even, } k| \geq 2.$$

If in these equations we set $m = \tau^t$ and $n = \sigma^t$, then we obtain

$$L_t^k = \sum_{i=0}^{\frac{1}{2}(k-1)} \binom{k}{i} (-1)^{it} L_{(k-2i)t} \quad \text{if } k \text{ is odd,} \tag{2.6.1}$$

$$L_t^k = \sum_{i=0}^{\frac{1}{2}k-1} \binom{k}{i} (-1)^{it} L_{(k-2i)t} + \binom{k}{\frac{1}{2}k} (-1)^{\frac{1}{2}tk} \quad \text{if } k \text{ is even,} \tag{2.6.2}$$

$$(5F_t)^k = \sum_{i=0}^{\frac{1}{2}(k-1)} \binom{k}{i} (-1)^{i(t+1)} 5F_{(k-2i)t} \quad \text{if } k \text{ is odd,} \tag{2.6.3}$$

$$(5F_t)^k = \sum_{i=0}^{\frac{1}{2}k-1} \binom{k}{i} (-1)^{i(t+1)} L_{(k-2i)t} + \binom{k}{\frac{1}{2}k} (-1)^{\frac{1}{2}(t+1)k} \quad \text{if } k \text{ is even.} \tag{2.6.4}$$

Now consider

$$\frac{m^k - n^k}{m-n} = \sum_{i=0}^{\frac{1}{2}(k-3)} (mn)^i (m^{k-2i-1} + n^{k-2i-1}) + (mn)^{t(k-1)} \quad \text{if } k \text{ is odd, } k \geq 3,$$

$$\frac{m^k - n^k}{m-n} = \sum_{i=0}^{\frac{1}{2}k-1} (mn)^i (m^{k-2i-1} + n^{k-2i-1}) \quad \text{if } k \text{ is even, } k \geq 2,$$

$$\frac{m^k + n^k}{m + n} = \sum_{i=0}^{\frac{1}{2}(k-3)} (-mn)^i(m^{k-2i-1} + n^{k-2i-1}) + (-mn)^{\frac{1}{2}(k-1)} \quad \text{if } k \text{ is odd}, k \geq 3,$$

$$\frac{m^k - n^k}{m + n} = \sum_{i=0}^{\frac{1}{2}k-1} (-mn)^i(m^{k-2i-1} - n^{k-2i-1}) \quad \text{if } k \text{ is even}, k \geq 2.$$

Setting $m = \tau^t, n = \sigma^t$, we obtain

$$\frac{F_{kt}}{F_t} = \sum_{i=0}^{\frac{1}{2}(k-3)} (-1)^{it} L_{(k-2i-1)t} + (-1)^{\frac{1}{2}(k-1)t} \quad \text{if } k \text{ is odd}, k \geq 3, \qquad (2.6.5)$$

$$\frac{F_{kt}}{F_t} = \sum_{i=0}^{\frac{1}{2}k-1} (-1)^{it} L_{(k-2i-1)t} \quad \text{if } k \text{ is even}, k \geq 2, \qquad (2.6.6)$$

$$\frac{L_{kt}}{L_t} = \sum_{i=0}^{\frac{1}{2}(k-3)} (-1)^{i(t+1)} L_{(k-2i-1)t} + (-1)^{\frac{1}{2}(k-1)(t+1)} \quad \text{if } k \text{ is odd}, k \geq 3, \qquad (2.6.7)$$

$$\frac{F_{kt}}{L_t} = \sum_{i=0}^{\frac{1}{2}k-1} (-1)^{i(t+1)} F_{(k-2i-1)t} \quad \text{if } k \text{ is even}, k \geq 2. \qquad (2.6.8)$$

Consider also

$$m^k + n^k = (m+n)^k + \sum_{i=1}^{[\frac{1}{2}k]} (-1)^i \frac{k}{i} (mn)^i (m+n)^{k-2i} \binom{k-i-1}{i-1}.$$

If we set $m = \tau^t, n = \sigma^t$, we obtain

$$L_{kt} = L_t^k + \sum_{i=1}^{[\frac{1}{2}k]} \frac{k}{i} (-1)^{i(t+1)} L_t^{k-2i} \binom{k-i-1}{i-1}. \qquad (2.6.9)$$

When k is even, then the last term will be $2(-1)^{\frac{1}{2}t(t+1)}$, and when k is odd, then the right-hand side will be divisible by L_t. On the other hand, if $m = \tau^t$, $n = -\sigma^t$, then

$$F_{tk} = (\sqrt{5})^{k-1} F_t^k + \sum_{i=1}^{\frac{1}{2}(k-1)} \frac{k}{i} (-1)^{it} (\sqrt{5})^{k-2i-1} F_t^{k-2i} \binom{k-i-1}{i-1} \quad \text{when } k \text{ is odd,}$$

$$\qquad (2.6.10)$$

$$L_{tk} = (\sqrt{5})^k F_t^k + \sum_{i=1}^{\frac{1}{2}k} (-1)^{it} (\sqrt{5})^{k-2i} F_t^{k-2i} \binom{k-i-1}{i-1} \quad \text{when } k \text{ is even.} \quad (2.6.11)$$

Observe that some of these equations give us some insight into questions of divisibility. In particular, (2.6.5) and (2.6.6) show that F_{kt} is divisible by F_t. For odd k this follows also from (2.6.10). From (2.6.7) we see that L_{kt} is divisible by L_t when k is odd. This follows also from (2.6.9). Moreover, (2.6.9) shows also that L_{kt} is not divisible by L_t if k is even.

We shall deal with problems of such divisibility in §2.7.

2.7. DIVISIBILITY

We have just seen that the equations of §2.6 give us some insight into divisibility questions of Fibonacci and Lucas numbers. In the present section we approach these problems more directly.

First, we point out that F_i and F_{i+1} are relatively prime for any i. If this were not so, then because of (2.1.1) we would conclude that F_{i-1} and F_i have also a common factor different from 1, and working backwards this would also apply to F_1 and F_2, which is clearly not the case.

We shall now present our main conclusions in a number of theorems.

THEOREM 2.7.1.   *If s is divisible by t, then F_s is divisible by F_t.*

Proof. F_{2t} is divisible by F_t, see formula (2.3.1). In (2.3.6), let G_i be F_i and set $m = t, n = rt$, thus

$$F_{(r+1)t} = F_{t-1}F_{rt} + F_t F_{rt+1}.$$

Hence, if F_{rt} is divisible by F_t, then so is $F_{(r+1)t}$. This proves Theorem 2.7.1 by induction.

THEOREM 2.7.2.   *If $(s; t) = d$, then $(F_s; F_t) = F_d$.*

(Here $(a; b)$ denotes the largest (greatest) common divisor of a and b.)

Proof.   The largest (greatest) common divisor of s and t can be found by the Euclidean algorithm (see I, Example 14.3.3).

Let $s = p_0 t + r_1$ $(0 \le r_1 < t)$. Then $(F_s; F_t) = (F_{p_0 t + r_1}; F_t)$ and by, (2.3.6),

$$(F_{p_0 t + r_1}; F_t) = (F_{r_1}F_{p_0 t - 1} + F_{r_1 + 1}F_{p_0 t}; F_t). \qquad (2.7.1)$$

Because, by Theorem 2.7.1, F_t divides $F_{p_0 t}$, (2.7.1) is equal to

$$(F_{r_1}F_{p_0 t - 1}; F_t).$$

Since $F_{p_0 t - 1}$ and $F_{p_0 t}$ are relatively prime, so are $F_{p_0 t - 1}$ and F_t. Thus we have shown that

$$(F_s; F_t) = (F_{r_1}; F_t).$$

Continuing in parallel with the Euclidean algorithm, we obtain eventually

$$(F_s; F_t) = F_d.$$

THEOREM 2.7.3.   *(Converse of Theorem 2.7.1) If F_s is divisible by F_t, then s is divisible by t $(t \ne 2)$.*

(We exclude $t = 2$, because every integer is divisible by $F_2 = 1$.)

Proof. By assumption,

$$(F_s; F_t) = F_t.$$

Denote $(s; t)$ by d. We must show that t equals d. By Theorem 2.7.2, it follows from $(s; t) = d$ that $(F_s; F_t) = F_d$. But if, by assumption, $(F_s; F_t) = F_t$, then t equals d, hence $(s; t) = t$, which means that s is divisible by t.

THEOREM 2.7.4. (*Converse of Theorem 7.2*) If $(F_s; F_t) = F_d$, then $(s; t) = d$.

Proof. If $(F_s; F_t) = F_d$, then both F_s and F_t must be divisible by F_d. Therefore, by Theorem 2.7.3, both s and t are divisible by d. It remains to be shown that d is, in fact, the largest, not just some common divisor of s and t.

Now if the largest (greatest) common divisor were not d, but some multiple rd $(r > 1)$, then by Theorem 2.7.2 we would have $(F_s; F_t) = F_{rd}$, and not F_d, as we have assumed.

2.8. PRIME FACTORS

We introduce this section by quoting facts about factors of binomial coefficients, in order to avoid interruptions of the main argument when we study Fibonacci and Lucas numbers.

Let p be a prime. Then

$$\binom{p}{n} \equiv 0 \pmod{p} \quad \text{for } 1 \le n \le p - 1 \tag{2.8.1}$$

(see Example 1.4.3),

$$\binom{p-1}{n} \equiv (-1)^n \pmod{p} \quad \text{for } 1 \le n \le p - 1 \tag{2.8.2}$$

(see Example 1.4.5),

$$\binom{p+1}{n} \equiv 0 \pmod{p} \quad \text{for } 2 \le n \le p - 1 \tag{2.8.3}$$

(see Example 1.4.4).

We shall also have occasion to refer to the result in the theory of quadratic residues, which states that

$$5^{\frac{1}{2}(p-1)} \equiv 1 \pmod{p}$$

if and only if p is a prime of the form $5t \pm 1$, and

$$5^{\frac{1}{2}(p-1)} \equiv -1 \pmod{p}$$

if and only if p is a prime of the form $5t \pm 2$, where t is a positive integer (Theorem 1.4.7 and Example 1.4.11).

We proceed to prove formulae concerning congruences of Fibonacci numbers and of Lucas numbers modulo an odd prime p.

From (2.3.3), (2.8.1) and Fermat's theorem (1.4.6),

$$F_p \equiv 5^{\frac{1}{2}(p-1)} \pmod{p}$$

and hence

$$F_p \equiv 1 \pmod{p} \quad \text{if the prime } p \text{ is of the form } 5t \pm 1, \qquad (2.8.4a)$$

$$F_p \equiv -1 \pmod{p} \quad \text{if the prime } p \text{ is of the form } 5t \pm 2. \qquad (2.8.4b)$$

In (2.3.3), set $n = p - 1$. Then

$$2^{p-2}F_{p-1} \equiv -[1 + 5 + 5^2 + \ldots + 5^{\frac{1}{2}(p-3)}] = -\tfrac{1}{4}[5^{\frac{1}{2}(p-1)} - 1] \pmod{p}.$$

Therefore, if p is of the form $5t \pm 1$, and hence $5^{\frac{1}{2}(p-1)} \equiv 1 \pmod{p}$,

$$F_{p-1} \equiv 0 \pmod{p} \quad \textit{if } p \text{ is a prime of the form } 5t \pm 1. \qquad (2.8.5)$$

Alternatively, set $n = p + 1$ in (2.3.3). Then

$$2^p F_{p+1} = \binom{p+1}{1} + 5\binom{p+1}{3} + \ldots + 5^{\frac{1}{2}(p-1)}\binom{p+1}{p}.$$

Thus, by (2.8.3),

$$2^p F_{p+1} \equiv 1 + 5^{\frac{1}{2}(p-1)} \pmod{p}.$$

Consequently, if p is of the form $5t \pm 2$, then using Fermat's theorem

$$F_{p+1} \equiv 0 \pmod{p} \quad \text{if the prime } p \text{ is of the form } 5t \pm 2. \qquad (2.8.6)$$

From (2.3.4) and (2.8.1), we obtain

$$L_p \equiv 1 \pmod{p}. \qquad (2.8.7)$$

In (2.3.4) set $n = p - 1$, then

$$2^{p-2}L_{p-1} \equiv 1 + 5 + 5^2 + \ldots + 5^{\frac{1}{2}(p-1)} = \tfrac{1}{4}[5^{\frac{1}{2}(p+1)} - 1] \pmod{p}$$

and hence, if p is a prime of the form $5t \pm 1$, $5^{\frac{1}{2}(p+1)} \equiv 5 \pmod{p}$ and

$$2^{p-2}L_{p-1} \equiv \tfrac{1}{4}(5 - 1) = 1 \pmod{p}, \quad 2^{p-1}L_{p-1} \equiv 2 \pmod{p}.$$

Consequently

$$L_{p-1} \equiv 2 \pmod{p} \quad \text{if the prime } p \text{ is of the form } 5t \pm 1. \qquad (2.8.8a)$$

On the other hand, if in (2.3.4) we set $n = p + 1$, then $2^p L_{p+1} \equiv 1 + 5^{\frac{1}{2}(p+1)}$ (mod p) and it follows that

$$L_{p+1} \equiv -2 \pmod{p} \qquad (2.8.8b)$$

when the prime p is of the form $5t \pm 2$.

Resuming, we can state the following theorem.

THEOREM 2.8.1.

2 is a factor of all F_{3i} (since $F_3 = 2$, and by Theorem 2.7.1)
5 is a factor of all F_{5i} (since $F_5 = 5$, and by Theorem 2.7.1)
A prime p of the form $5t \pm 1$ is a factor of F_{p-1} (see (2.8.5))
A prime p of the form $5t \pm 2$ is a factor of F_{p+1} (see (2.8.6))

Observe the phrase 'a factor of'. We use it in the last two instances, because F_{p-1} and F_{p+1} are, respectively, not necessarily the smallest Fibonacci numbers which contain the prime p as a factor. This is seen in the following two examples.

EXAMPLE 2.8.1. $29 = 5 \times 6 - 1$, and $F_{28} = 317811$ does indeed contain 29 as a factor; but so does the smaller $F_{14} = 377 = 29 \times 13$.

EXAMPLE 2.8.2. $17 = 5 \times 3 + 2$, and F_{18} does contain 17 as a factor, but so does $F_9 = 2 \times 17$.

The question of whether any integer, prime or composite, is the factor of some Fibonacci number will be answered, affirmatively, in Theorem 2.9.1, and negatively for Lucas numbers in the remark following it.

In the first part of this section we have dealt with the problem of finding a Fibonacci number which contains a given prime factor. Now we turn to the question, in some sense converse to the previous one, which prime factors can a given Fibonacci number contain?

To answer this question, we shall once more have to refer to the theory of quadratic residues.

Consider again the formula (2.3.2)

$$5F_n^2 - L_n^2 = 4(-1)^{n+1}.$$

Thus for odd n we have

$$4 \equiv -L_n^2 \pmod{5F_n^2}.$$

This implies that -1 is a quadratic residue modulo those primes which divide F_n.

Now -1 is a quadratic residue of 2, and of primes of the form $4t + 1$ (see Example 1.4.10). Therefore Fibonacci numbers of odd order can only contain such primes as factors.

EXAMPLE 2.8.3. $F_{15} = 2 \times 5 \times 61$.

For odd n we have also

$$4 \equiv 5F_n^2 \pmod{L_n^2}.$$

5 is a quadratic residue of 2 and of primes of the form $10t \pm 1$ (see Example 1.4.11), so that Lucas numbers of odd order can only have such prime factors.

EXAMPLE 2.8.4. $L_{15} = 4 \times 11 \times 31$.

When n is even, then

$$4 \equiv -5F_n^2 \quad (\text{mod } L_n^2).$$

Because -5 is a quadratc residue modulo 2 and modulo those primes of the form $20t + 1$, $20t + 3$, $20t + 7$, or $20t + 9$, (see Example 1.4.11) these are the primes which can appear as prime factors of a Lucas number of even order.

EXAMPLE 2.8.5. $L_{24} = 2 \times 47 \times 1103$.

2.9.   RESIDUE PERIODS

Consider the smallest residues modulo m of a Fibonacci sequence. For instance, modulo 8 we have

$$0 \ 1 \ 1 \ 2 \ 3 \ 5 \ 0 \ 5 \ 5 \ 2 \ 7 \ 1|0 \ 1 \ 1...$$

After 12 terms, the sequence is repeated. Such repetition is bound to occur, because apart from the irrelevant pair 0 0 there are only $m^2 - 1$ possible pairs of successive terms, and once a pair is repeated, the complete sequence is also repeated. It follows that the residue 0, with which we started, will also reappear, hence we get the following theorem.

THEOREM 2.9.1. *Every integer is the factor of some and hence of infinitely many, Fibonacci numbers.*

This argument does not apply to Lucas numbers, because L_0 is not zero. For example, the cycle modulo 8 of Lucas residues repeats also after 12 terms, but does not contain 0. No Lucas number is divisible by 8.

We call the length of the cycle of residues its period, and we denote it by $r(m)$. The following is a table of periods for selected moduli.

Modulus m	2	3	4	5	6	7	8	9	10	15	20	25	30
$r(m)$ for F_i	3	8	6	20	24	16	12	24	60	40	60	100	120
$r(m)$ for L_i	3	8	6	4	24	16	12	24	12	8	12	20	24

All periods, except those for $m = 2$ are divisible by 2. Also, periods of Fibonacci numbers and of Lucas numbers, for the same modulus, are equal for powers of a prime $p \neq 5$, while those for powers of 5 are five times as large for Fibonacci numbers as they are for Lucas numbers. It can be shown that this is true for any modulus.

Any value in the cycle is repeated after $r(m)$ terms, but might be repeated earlier, as we have seen in the example for modulus 8 above. If a term is repeated for the first time after k terms, then k is a factor of m. This can be seen as follows.

In (2.3.6), let $G_i = F_i$, and write s for n and t for m, thus

$$F_{s+t} = F_{t-1}F_s + F_tF_{s+1}.$$

Therefore, if $F_s = 0 \pmod m$ and $F_t = 0 \pmod m$, then $F_{s+t} \equiv 0 \pmod m$.

Similarly, $F_{s-t} = F_{-t-1}F_s + F_{-t}F_{s+1}$, and this is again divisible by m if F_s and $F_{-t}(=(-1)^{t+1}F_t)$ are divisible by m for $s > t$. This proves the following theorem (see Example 1.3.2).

THEOREM 2.9.2   *The subscripts for which Fibonacci numbers are divisible by m are integer multiples of some number k.*

We shall now study how the period modulo m depends on m. The period modulo 2 is 3: 0 1 1|0 1 1|..., and the period modulo 5 is

20: 0 1 1 2 3 0 3 3 1 4 0 4 4 3 2 0 2 2 4 1|0 1 1...

Concerning other primes, we have the following.

THEOREM 2.9.3.   *If the prime p is of the form $5t \pm 1$, then the period modulo p divides (or is equal to) $p - 1$.*

The follows from (2.8.5) and (2.8.4).

The smallest prime p of the form $5t \pm 1$ which has a period smaller than $p - 1$ is $p = 29$, with period 14.

THEOREM 2.9.4.   *If the prime p is of the form $5t \pm 2$, then the period modulo p divides (or is equal to) $2p + 2$.*

Proof.   (2.8.6) states that $F_{p+1} \equiv 0 \pmod p$, but by (2.8.4b) $F_p \equiv -1 \pmod p$ and is not congruent to $1 \pmod p$, therefore the period is not $p + 1$, nor any factor of it. But if

$$F_p \equiv -1 \pmod p \quad \text{and} \quad F_{p+1} \equiv 0 \pmod p,$$

then

$$F_{p+2} \equiv -1 \pmod p \quad \text{and} \quad F_pF_{p+2} \equiv 1 \pmod p.$$

It follows from (2.3.6) that

$$F_{2p+3} = F_pF_{p+2} + F_{p+1}F_{p+3}$$

and hence $F_{2p+3} \equiv 1 \pmod p$.

Together with Theorem 2.7.1 and (2.8.6), $F_{2p+2} \equiv F_{p+1} \equiv 0 \pmod p$. This proves Theorem 2.9.4.

The smallest prime p of the form $5t \pm 2$ which has a priod smaller than $2p + 2$ is $p = 47$, with period 32.

If we know the period modulo a prime p, then the period modulo a power of p follows from Theorem 2.9.5.

THEOREM 2.9.5.   *Let p be a prime. If t is the largest integer for which $r(p^t) = r(p)$, then for $s > t$*

$$r(p^s) = p^{s-t} r(p).$$

For a modulus which is the product of prime powers, the following theorem holds.

THEOREM 2.9.6.   *The period modulo $m = \prod_i(p_i^{e_i})$, where the p_i are distinct primes, and the e_i are positive integers, equals the least common multiple of $r(p_i^e)$.*

Proof.   Let n be a factor of m. Then, by Theorem 2.9.2, the cycle of smallest residues modulo m repeats, after $cr(n)$ terms, c being a positive integer.

It might be interesting to know whether there are moduli such that the cycle of Fibonacci residues, or of Lucas residues, contains all possible smallest residues, and if so, whether it contains all of them equally often. This is of some interest if the sequence of residues is used as a source for pseudo-random numbers.

THEOREM 2.9.7.   *The cycle of smallest residues is uniform, that is all possible smallest residues appear in it equally often, for Fibonacci numbers if m is a power of 5, but never for Lucas numbers.*

In a similar context one would wish to know which seed for a generalized Fibonacci sequence leads to the longest period. We prove Theorem 2.9.8.

THEOREM 2.9.8.   *The seed of the ordinary Fibonacci sequence, that is $(0, 1)$, leads to the longest possible period.*

Proof.   The cycle of Fibonacci residues repeats, say, after k terms. This means that

$$F_{m+k-1} \equiv F_{m-1} \pmod{m} \quad \text{and} \quad F_{m+k} \equiv F_m \pmod{m}.$$

Then, by (2.3.6)

$$G_{n+m+k-1} \equiv G_{n+m-1} \quad \text{and} \quad G_{n+m+k} \equiv G_{n+m} \pmod{m}.$$

This means that the sequence of any generalized Fibonacci residues repeats, at the latest, after k terms, and possibly earlier, for the same modulus.

S. V.

REFERENCES

An extensive study of the topic of the present chapter, with historical references and a detailed list of sources is contained in:

Vajda, S. (1989). *Fibonacci and Lucas numbers and the Golden Section: Theory and Applications*, Ellis Horwood.
The Fibonacci Quarterly published by the Fibonacci Association is a journal devoted to the study of integers with special properties.

CHAPTER 3

Cryptography and Cryptanalysis

3.1. INTRODUCTION

The need to keep certain messages secret has been appreciated for many years. If the communicants are able to use non-interceptable means of transmission then, obviously, their messages are automatically secure. However, if this is not the case, then one common way of securing the contents of a message is by transforming it prior to transmission. This is the objective of a cipher system, see Figure 3.1.1. The art (and science) of designing such systems is called cryptography.

Before we can begin our discussion of cryptography we must introduce a few basic definitions and some notation.

The idea of a cipher system is to disguise confidential information in such a way that its meaning is unintelligible to an unauthorized person. The information to be concealed is called the *plaintext* (or just the *message*) and the operation of disguising it is known as *encryption* or *enciphering*. The enciphered message is called the *ciphertext* or *cryptogram*. Thus, referring to Figure 3.1.1, the enciphering key $k(E)$ determines an enciphering function $f_{k(E)}$ such that, if m is the message, the cryptogram c is given by $c = f_{k(E)}(m)$. The person who enciphers the message is known as the *encipherer*, while the person to whom he sends the cryptogram is called the *recipient* or *receiver*. The process of applying a key to translate back from the ciphertext to the plaintext is known as *deciphering*. So, once again referring to Figure 3.1.1, the deciphering key $k(D)$ determines the deciphering function $f_{k(D)}$ such the $m = f_{k(D)}(c)$. *Cryptanalysis*

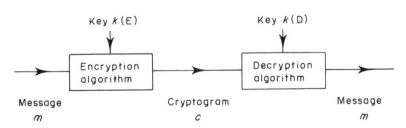

Figure 3.1.1. Cipher system

involves either trying to determine the plaintext from the ciphertext without knowledge of the key or trying to deduce the key being used.

There are two fundamentally different types of cipher systems.

DEFINITION 3.1.1.   A cipher system is called *conventional* (or *symmetric*) if $k(D)$ can be easily determined from knowledge of $k(E)$. Thus, in this case, there is essentially only one key and we often speak of a single key k.

DEFINITION 3.1.2.   A cipher system is called *public key* (or *asymmetric*) if it is computationally infeasible to determine $k(D)$ from $k(E)$.

There are a number of aspects to the design of a good cipher system. They include

designing the cipher algorithm,
deciding how to use it,
incorporating it into the existing communications system,
devising a key management scheme.

Clearly most of these aspects are non-mathematical, and thus it is important not to give the impression that cryptography is a branch of mathematics. Cryptography is an interdisciplinary subject incorporating computer science, engineering, mathematics and statistics.

Although we have introduced cryptography as a means of providing privacy, modern cryptography has many other applications. These include the provision of message integrity, user identification, digital signatures and access control.

Examples of Enciphering Algorithms

EXAMPLE 3.1.1.   Caesar ciphers [see §1.4]: Assign the elements of Z_{26} to the alphabet as follows: $A = 0, B = 1, \ldots, Z = 25$. For each $\lambda \in Z_{26}$, let f_λ be the permutation $\alpha \to \alpha + \lambda \pmod{26}$. Clearly there are 26 such permutations. If we regard the choice of λ as determining an enciphering key then the deciphering key is $26 - \lambda \pmod{26}$. Thus if the enciphering function is addition of $\lambda \pmod{26}$ then the deciphering function is addition of $26 - \lambda \pmod{26}$ which is, of course, the inverse of the enciphering function.

EXAMPLE 3.1.2.   Simple substitution ciphers: If θ is any permutation of $\{A, B, \ldots, Z\}$ then the enciphering algorithm replaces each letter α by $\theta(\alpha)$. It is known that there are 26! permutations.

Since the aim of cryptography is to provide security we need some criteria under which to assess a cipher system. It has become customary to assume that the attacker has

(a) complete knowledge of the algorithm,
(b) a large volume of ciphertext,
(c) plaintext corresponding to some of the ciphertext.

There are, of course, systems with secret proprietary enciphering and deciphering algorithms, and keeping the algorithm secret obviously enhances the security level. (In this context we may regard an algorithm as a class of functions depending on a parameter, called a key.) However, it is widely felt that algorithms should be strong enough to offer adequate protection even when they are known. The three assumptions listed above are often referred to as the *worst case conditions*.

One obvious implication of the acceptance of the worst case conditions is that the protection of the message relies on the secrecy of the deciphering key. Anyone who knows $k(D)$ and the algorithm has the same knowledge as the genuine receiver and can decrypt the message. So one potential attack, if the algorithm is known, is to try all possible keys and to eliminate all incorrect ones.

To withstand this type of attack a large number of keys are required. However, the precise size depends on the particular application and the sensitivity of the secret information.

DEFINITION 3.1.3.   The time for which the messages must remain secret is called the *cover time* of the system.

EXAMPLE 3.1.3.   Example of an exhaustive key search: Suppose the interceptor knows that a Caesar cipher is being used and intercepts OJYVT. Since there are only 26 keys he can try all possible values for $k(D)$. This gives the following set of 26 possible messages:

OJYVT	XSHEC	GBQNL
PKZWU	YTIFD	HCROM
QLAXV	ZUJGE	IDSPN
RMBYW	AVKHF	JETQO
SHCZX	BWLIG	KFURP
TODAY	CXMJH	LGVSQ
UPEBZ	DYNKI	MHWTR
VQFCA	EZOLJ	NIXUS
WRGDB	FAPMK	

On the assumption that the message was in English, he would then deduce that the correct decipherment was TODAY. From this he would know that the enciphering key was 21 and be able to decipher the rest of the message.

Example 3.1.3 shows quite clearly why systems with a small number of keys are vulnerable. However, it is important to realize that, although essential, a

Type of attack	Key size
A human effort at 1 key/s	25 bits
A processor at 10^6 keys/s	45 bits
1000 such processors	55 bits
10^6 such processors	65 bits

Table 3.1.1

large key space does not guarantee security. The 'classical' counterexample is the monalphabetic simple substitution cipher system of Example 3.1.2. Here there are $26! \approx 4 \times 10^{26}$ keys (a very large number!) clearly no one will attack such a system by attempting an exhaustive key search. However, the widely known expected frequency distributions of single letters, bigrams, etc. in the English language provide a very simple, and highly successful, statistical attack which can (and often is) applied by young school children.

Since any cipher system with a known algorithm is prone to an exhaustive key search attack, the size of the key space provides an upper bound on the level of security attainable. If a key is an h-bit-tuple then there are 2^h keys. Thus, provided we have some idea of the time needed to try one key, the required cover time gives us a lower bound for the key length. Table 3.1.1 gives approximate values for the key length to guarantee that a complete exhaustive key search will take a year.

Suppose that an attacker knows the enciphering algorithm and a known plaintext/ciphertext pair m and c. Then $c = f_k(m)$ and the attacker wants to know k. We have already seen that he might try all possibilities for k. However, if we look back at Example 3.1.1, then we see that, in that instance, a key search is unnecessary as he need only solve a simple equation.

EXAMPLE 3.1.4.   Suppose a Caesar cipher was used and that the message letter C becomes a ciphertext letter H. If the key is λ then we know $8 \equiv 3 + \lambda \pmod{26}$, i.e. $\lambda = 5$.

We have already seen that ensuring there are sufficiently many keys to prevent an exhaustive search is by no means enough to provide any assurances about the security.

The designer actually needs to set the attacker mathematical problems for which no solution technique exists that can be performed in less than the cover time. An exhaustive key search is merely the most naive of the solution techniques available. Thus one obvious consideration for the designer is the time complexity function of the problem involved. Complexity Theory [see Chapter 9] is another fascinating area on the mathematics/computer science interface. Optimal algorithms for solving a number of mathematical problems can be found in Knuth's monumental work (Knuth (1981)) while Garey and

Johnson is a standard reference for complexity theory (Garey and Johnson (1979)).

3.2.   THE SHANNON APPROACH

In this section we discuss the information theoretic approach of Shannon (1948, 1949). This work has had a fundamental influence on the subject and its importance cannot be overemphasized. Another important paper on this topic is that of Hellman (1977) while Khinchin's book (Khinchin (1957)) is one of the standard texts in the area.

3.2.1.   Definition of a Cipher System

Let M denote the (finite) set of all possible messages, C be the (finite) set of all possible cryptograms and K the (finite) set of all possible keys. Each key k of K, together with the enciphering algorithm, determines a mapping $t_k: M \to C$.

Although it is not always the case in practice, we will simplify our discussion by assuming that no pair of distinct keys determine the same transformation. Under this assumption if $K = \{k_1, k_2, \ldots, k_n\}$ then we will let t_i denote the mapping determined by k_i and even identify t_i with k_i. If, for a given message m and transformation $t_i, c = t_i(m)$ then, clearly, knowledge of c and t_i must determine m uniquely. Thus each transformation must have a left inverse, i.e. each transformation must be injective [see I, §1.4.2].

We can now give Shannon's definition of a cipher system.

DEFINITION 3.2.1.   A *cipher system* is a (finite) family T of injective mappings from a (finite) set of messages M into a (finite) set of cryptograms C. For each $t_i \in T$, the probability of t_i being chosen is denoted by p_i. Similarly, for each message m_i, there is an a priori probability of m_i being sent. We note this by $p(m_i)$.

On the assumption that the interceptor has complete knowledge of the cipher system being used, the difference between the recipient's and interceptor's knowledge (prior to use of the system) is that, whereas the recipient knows which transformation is to be used, the interceptor knows only the a priori probability of each transformation.

DEFINITION 3.2.2.   A cipher system is said to be *closed* if each $t_i \in T$ maps M onto C, i.e. if each t_i is surjective.

Since each t_i is injective, Definition 3.2.2 and I, §1.4.2 give the following.

PROPOSITION 3.2.1.   *A cipher system is closed if and only if $|M| = |C|$.*

Now suppose that T_1 and T_2 are two cipher systems such that M_i and C_i are the message and cryptogram space of $T_i(i = 1, 2)$. If $C_1 \subseteq M_2$ then $T_2 T_1 = \{t_i^{(2)} t_j^{(1)} | t_i^{(2)} \in T_2, t_j^{(1)} \in T_1\}$ is a set of mappings $M_1 \rightarrow C_2$. The process of enciphering by using the mappings in $T_2 T_1$ is called *superenciphering*. If $t = t_i^{(2)} t_j^{(1)}$, then the probability of choosing t is the sum of the products $p_a^{(2)} p_b^{(1)}$ taken over those pairs a, b such that $t = t_a^{(2)} t_b^{(1)}$.

Let M and C both be the English alphabet and identify each letter with an integer in Z_{26}. So, for example, $A = 0, B = 1, C = 2, \ldots, Z = 25$.

EXAMPLE 3.2.1.   Additive ciphers (also known as Caesar ciphers) [see Example 3.3.1]. Let $t_i : x \rightarrow x + i \pmod{26}$. Then $T = \{t_i | i \in Z_{26}\}$. For an additive cipher system there are 26 mappings. Assume that each t_i is equally likely, i.e. $p_i = 1/26$.

EXAMPLE 3.2.2.   Multiplicative ciphers: Let $s_i : x \rightarrow xi \pmod{26}$. Then $S = \{s_i | i \in Z_{26}$ and $(i, 26) = 1\}$. Since $|S| = 12$, for a multiplicative cipher system there are 12 mappings [see Proposition 1.4.4] Assume that each s_i is equally likely, i.e. $p_i = 1/12$.

EXAMPLE 3.2.3.   Affine ciphers: Affine ciphers are obtained by super-enciphering additive and multiplicative ciphers.

If $t_i : x \rightarrow x + i \pmod{26}$ and $s_j : x \rightarrow xj \pmod{26}$ then $t_i s_j : x \rightarrow xj + i \pmod{26}$.

Each transformation has associated probability $\frac{1}{12} \times \frac{1}{26}$.

In each of these three examples $M = C$. Systems with this property are both common and interesting.

DEFINITION 3.2.3.   A cipher system with $M = C$ is called *endomorphic*.

For any endomorphic cipher system we can superencipher T with itself to obtain the product TT which we write as T^2.

DEFINITION 3.2.4.   An endomorphic cipher system T is called *idempotent* if $T^2 = T$.

Note that our earlier examples of additive, multiplicative and affine ciphers are all idempotent. In fact they are groups with the product of mappings as the binary operation [see I, §8.2].

3.2.2.   Secrecy

If an interceptor attacks a system by trying all possible mappings then, clearly, he will try those with the highest associated probabilities first. However, this attack can only succed if he has some way of recognizing the correct mapping. If he tries to decipher a cryptogram with an arbitrarily chosen mapping, then there are three possible outcomes:

(a) correct decipherment,
(b) incorrect decipherment,
(c) no decipherment possible.

If (c) occurs, then he will know that that particular mapping was not used and eliminate it. However, if the system is closed, then (c) can never occur. So if the system is closed, then one of (a) and (b) must occur. If the cryptanalyst were unable to distinguish between (a) or (b) then he would be unable to identify the correct mapping. This is, clearly, a necessary property for a system to be secure and is the basic idea behind Shannon's definition of perfect secrecy.

3.2.3. Pure Ciphers

DEFINITION 3.2.5. If T is a cipher system with message space M and cryptogram space C then T is *pure* if, for any $t_i, t_j, t_l \in T, t_i t_j^{-1} t_l \in T$. If T is not pure then it is *mixed*.

The concept of a pure cipher is crucial. Before discussing its importance we note that the following are easily proven (Beker and Piper (1982)).

PROPOSITION 3.2.2. *An idempotent cipher system is pure.*

PROPOSITION 3.2.3. *A pure cipher system is closed.*

If the system T is idempotent, then T is a permutation group on M and the orbits under T form a partition of M. Clearly if the system is not idempotent, then T cannot be a group. However, if T is pure, then we can define equivalence relations on both M and C and thereby partition their elements into equivalence classes.

If T is a pure cipher system with message space M and cryptogram space C, we define a relation $\sim$ on M by $m_1 \sim m_2$ if and only if $\exists t_i, t_j \in T$ with $m_1 = t_i^{-1} t_j(m_2)$. It is easy to show:

PROPOSITION 3.2.4. $\sim$ *is an equivalence relation on M* [see I, §1.3.3].

DEFINITION 3.2.6. The equivalence class $M(m)$ of m under $\sim$ is called the *message residue class* of m.

DEFINITION 3.2.7. For any $m \in M$ the set $C(m) = \{t(m) | t \in T\}$ is called the *cryptogram residue class* of m.

Note the significance of these two definitions. If m is encrypted to give c, and c is intercepted, then $M(m)$ is the set of all messages which would be obtained by deciphering c using all transformations. Similarly $C(m)$ is the set of all cryptograms which might represent m. Thus $M(m)$ is the set of all messages which can be encrypted to give any particular $c \in C(m)$. So if we intercept c we will be unable to distinguish, with certainty, between the correct message and any other message in $M(m)$.

The following is easily proved.

PROPOSITION 3.2.5.   *Let T be a pure cipher system with message space M and cryptogram space C. Then*

(i) $|M(m)| = |C(m)|$ *for all* $m \in M$.

(ii) $|C(m)| \big| |T|$ *for all* m.

(iii) *If* $|T|/|C(m)| = g_m$, *then for any* $m_i \in M(m)$ *and* $c_j \in C(m)$ *there are exactly* g_m *mappings* $t \in T$ *with* $t(m_i) = c_j$.

EXAMPLE 3.2.4.   Suppose that we use the additive cipher system to encipher the message $m = $ CAT. Then $M(m) = C(m) = \{$CAT, DBU, ECV, FDW, GEX, HFY, IGZ, JHA, KIB, LJC, MKD, NLE, OMF, PNG, QOH, RPI, SQJ, TRK, USL, VTM, WUN, XVO, YWP, ZXQ, AYR, BZS$\}$.

3.2.4.   Perfect Secrecy

Suppose that we have a cipher system T with a finite message space $M = \{m_1, m_2, \ldots, m_n\}$, a finite cryptogram space $C = \{c_1, c_2, \ldots, c_u\}$ and transformations $t_1, t_2, \ldots, t_h$. Suppose that, for any m_i, the a priori probability of m_i being transmitted is $p(m_i)$. If the cryptanalyst intercepts a particular cryptogram c_j, then, for each message m_i, he can, at least in principle, use Bayes' rule [see II, §16.4] to calculate the a posteriori probability $p_j(m_i)$ that m_i was transmitted. (Thus $p_j(m_i)$ is the probability that m_i was transmitted given that c_j was received.)

DEFINITION 3.2.8.   The system T is said to have *perfect secrecy* if, for every message m_i and every cryptogram c_j, $p_j(m_i) = p(m_i)$.

Thus, if T has perfect secrecy, the cryptanalyst who intercepts c_j has obtained no further information to enable him to decide which message was transmitted. Clearly perfect secrecy is highly desirable.

For any cryptogram c_j we will let $p(c_j)$ denote the probability of obtaining c_j (from any message), and the probability of obtaining c_j if message m_i is transmitted by $p_i(c_j)$. Thus if, as usual, we let p_f be the probability of choosing transformation t_f or, equivalently, key k_f, $p_i(c_j) = \sum p_f$, where the summation is over all those f for which $c_j = t_f(m_i)$.

The following theorem is a simple application of Bayes' theorem [see II, §16.4].

PROPOSITION 3.2.6. *A necessary and sufficient condition for perfect secrecy is that*

$$p_i(c_j) = p(c_j) \quad \text{for all } m_i \text{ and } c_j.$$

If each key is equiprobable then it is now easy to see that if T has perfect secrecy, then there is a constant, v say, such that there are exactly v keys which send a given message m onto a given cryptogram c. Even if all keys are not equally likely, perfect secrecy must mean that if a cryptogram c is intercepted, then, for any message m_i, $p_i(c) = p(c) \neq 0$. Thus for each m_i there is at least one transformation sending m_i onto c. This leads to the following important result.

PROPOSITION 3.2.7 *In a system with perfect secrecy, $|T| \geq |M|$.*

EXAMPLE 3.2.5. It is easy to construct perfect secrecy for systems with $|T| = |M|$. We exhibit an example with $|M| = |C| = |T| = 5$, for which all messages and keys are equiprobable. We represent the system as a bipartite graph (Figure 3.2.1).

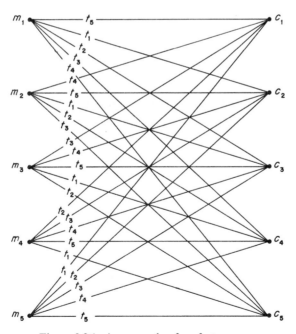

Figure 3.2.1. An example of perfect secrecy

Random sequence k_1 k_2 k_3 ... k_n

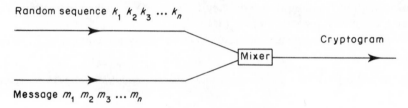

Message m_1 m_2 m_3 ... m_n

Figure 3.2.2. The one-time-pad

We can now give necessary and sufficient conditions for certain systems to offer perfect secrecy.

PROPOSITION 3.2.8. *Let T be a cipher system in which the number of messages, the number of keys and the number of cryptograms are all equal. Then T has perfect secrecy if and only if*

(i) *there is exactly one mapping sending each message to each cryptogram; and*
(ii) *all keys are equally likely.*

EXAMPLE 3.2.6.   The one-time-pad: There is one particular system which offers perfect secrecy and which deserves special mention; the *one-time-pad*. In this system there is an upper bound, n say, on the number of 'characters' in a message. Furthermore the length of the keys, which are all equally likely, is at least as large as n. If the message $m = m_1 m_2 ... m_n$ is to be enciphered, then a random sequence $k_1 k_2 ... k_n$ of the characters is selected. The enciphering is then fully described by Figure 3.2.2.

If the message and random sequence are both binary, then the mixer is likely to be an exclusive-or gate. (An exclusive-or gate will output a 0 if the two entries are identical and a 1 if they differ, i.e. it performs addition modulo 2.)

3.2.5.   Cryptanalysis

EXAMPLE 3.2.7.   An example of how to break a cipher: Knowledge of the cipher system being used implies knowledge of the a priori probability of each possible mapping being used. If the system is breakable then, as soon as he obtains part of a cryptogram, the interceptor tries to calculate the a posteriori probability for each mapping. In this small example we assume that an additive cipher was used and that the interceptor obtains the cryptogram KWVAQ one letter at a time. Table 3.2.1 shows the a posteriori of which key was used after each letter was received. (These ,probabilities were computed using known statistical properties of the English language.) After the fifth letter, he has eliminated every key except one.

Residue class	$n = 1$	$n = 2$	$n = 3$	$n = 4$	$n = 5$
kwvaq	0.008				
lxwbr	0.040				
myccs	0.024	0.028			
nzydt	0.067	0.001			
oazeu	0.075	0.014			
pbafv	0.019				
qcbgw	0.001				
rdchx	0.060	0.070			
sediy	0.063	0.257	0.427	0.182	
tfejz	0.091	0.003			
ugfka	0.028	0.052			
vhglb	0.010				
wihmc	0.024	0.128			
xjind	0.002				
ykjoe	0.020				
zlkpf	0.001	0.001			
amlqg	0.082	0.072	0.004		
bnmrh	0.015				
consi	0.028	0.202	0.515	0.818	1
dpotj	0.043				
eqpuk	0.127	0.044			
frqvl	0.022	0.058			
gsrwm	0.020	0.015			
htsxn	0.061	0.052	0.046		
iutyo	0.070	0.001			
jvuzp	0.002				

Table 3.2.1

Note: In this example the interceptor reached the situation where one key had probability 1. Clearly for systems which large key spaces the simplistic approach outlined in the example is not practical. Nevertheless each extra part of the cryptogram is likely to eliminate further keys. One of our main concerns in the rest of this section will be to determine the amount of cryptogram needed before a unique solution can be expected. This is a very important concept, called the *unicity distance*.

3.2.6. Entropy and Equivocation

In our standard scenario the cryptanalyst re-evaluates the a posteriori probability of all of the mappings each time that he intercepts some more of the cryptogram. In Example 3.2.7 the situation was reached where one particular mapping had probability 1 (and he was absolutely confident that he knew the key). In general, of course, he will not reach this situation and will be forced to make decisions on much less conclusive evidence. The cryptanalyst's confidence can be measured by the *conditional entropy* or, as Shannon called it, *equivocation*.

For a cipher system we can define both the message entropy $H(M)$ and the key entropy $H(K)$. We will concentrate on $H(M)$. The message entropy reflects the confidence with which we can predict that a given message will be transmitted. (The crucial property of the entropy function is that its value is 0 precisely when one event has associated probability 1 and it takes its maximum value whan all events are equally likely.) This is not quite the right measure if we wish to cryptanalyse a system. Instead we want to know how confidently we can deduce that a given message was sent, given that we have intercepted some ciphertext. To do this we must define equivocation.

Notation:

    C' is any set of cryptograms
    C'_n is the set of cryptograms of length n in C'

For any c and m

    $p(c, m)$ is the probability that m was sent and c was received
    $p_c(m)$ is the probability that m was sent, given that c was received

DEFINITION 3.2.9.   The *message equivocation* $H_{C'}(M, n)$ is defined by:

$$H_{C'}(M, n) = - \sum_{\substack{m \in M \\ c \in C'_n}} p(c, m) \log p_c(m).$$

    Key equivocation is defined similarly.

Clearly the equivocations are 'conditional entropies' which are monotonically decreasing functions of n.

Of particular interest is the situation where $C' = \{c\}$, i.e. where one cryptogram is intercepted. Here we abuse notation slightly and let C'_n be the 'cryptogram' consisting of the first n symbols of c.

EXAMPLE 3.2.8.   In Example 3.2.7, if logarithms are taken to base 10, then $H_c(M, 1) = 1.26$, $H_c(M, 2) = 0.948$, $H_c(M, 3) = 0.394$, $H_c(M, 4) = 0.206$ and $H_c(M, 5) = 0$.

3.2.7. Random Ciphers

We will now introduce Shannon's concept of a random cipher. Suppose that we have an alphabet A with x symbols.

DEFINITION 3.2.10.   A *random cipher* system S has the following properties:

(1) All keys are equiprobable.
(2) Every message is an n-tuple of elements of A, for some fixed n.
(3) S has x^n messages, i.e. every possible n-tuple is a message.
(4) The messages of S can be divided into two groups:
    (i) the meaningful messages,
    (ii) the meaningless messages.
(5) All meaningful messages have approximately equal probabilities which are significantly higher than the negligible probabilities of the meaningless messages.

Notation: Let $r_0 = \log_2 x$ and let $r_n = H_c(M, n)/n$, where $H_c(M, n)$ is calculated using logorithms to base 2. It is then straightforward to show that there are $2^{r_n n}$ meaningful messages.
    The following result can be proved.

PROPOSITION 3.2.9.   *If a random cipher system S is used and a cryptogram c is deciphered using a randomly chosen key, then the probability of obtaining a meaningful message is $2^{-n(r_0 - r_n)}$.*

Proposition 3.2.9 shows that the discrepancy between r_0 and r_n is important. If, for instance, $r_0 = r_n$ then every message is meaningful and, obviously, the probability of obtaining a meaningful message is 1.

DEFINITION 3.2.11.   The *redundancy* of the system is $d_n = r_0 - r_n$.

Clearly the redundancy is a measure of the likelihood that a message chosen at random will be meaningful. Since there are $2^{H(K)}$ keys, the expected number of meaningful decipherments when we try all keys is $2^{H(K) - nd_n}$. If $H(K) \gg nd_n$, then there will be a large number of meaningful decipherments and, consequently, a low probability of recognizing or guessing the correct message. However, if there were only one meaningful decipherment then, as soon as a meaningful decipherment was obtained, we would know we had the correct message.
    In most practical situations the constraints of a random cipher system are not satisfied, e.g. messages do not all have the same length. However, it may happen that, for any n, the system obtained by considering only the messages of length n resembles a random system. For many systems, for example those using the English language, it can be shown that there exists an r such that $r_n \to r$ as $n \to \infty$. The corresponding value for d_n is denoted by d, i.e. $d = r_0 - r$.

DEFINITION 3.2.12. r is the *true rate* of the language.

DEFINITION 3.2.13.   If a cipher system S has $2^{H(K)}$ keys and the true rate of the language is r then the the *unicity distance* of S is the smallest integer n_0 greater than or equal to the value $H(\mathrm{K})/d$.

The unicity distance is a very important concept. It is the smallest integer such that the expected number of meaningful messages obtained by deciphering a cryptogram using all keys is at most one.

EXAMPLE 3.2.9.   For a simple substitution cipher there are 26! keys [see Example 3.1.2]. Thus $H(\mathrm{K}) = \log_2 (26!) \approx 88.4$. The number of letters in the alphabet is 26 so $r_0 = \log_2 26 \approx 4.7$. In order to evaluate the unicity distance we need a value for the true rate r of the language. There is no universally accepted value for this. One estimate (Hellman (1974)) puts $r = 1$ while another (Deavours (1977)) has $r = 1.5$. These give $n_0 = 23.9$ and $n_0 = 27.6$ respectively. Thus we would expect a unique meaningful decipherment for any cryptogram with 28 or more letters.

We end our discussion of Shannon's approach by noting that if a system S is used to protect messages which are shorter than the unicity distance then an exhaustive key search will not uniquely identify the correct message.

3.2.8.   Comment

Whilst it is of great mathematical and theoretical interest, Shannon's notion of perfect secrecy is of limited practical significance. Virtually all cipher systems in use are not perfect in the Shannon sense, although many of them are practically secure. This is because, although the interceptor often has enough information to deduce the plaintext from the ciphertext, no technique is known to solve the mathematical problem posed within the cover time. This informal notion of computational difficulty has been the basis of virtually all recent practical cipher systems.

However, it would clearly be desirable to provide a more formal, mathematical definition of this notion of difficulty. This could then be used to provide a definition of practical security. In the last few years a small number of (theoretical) ciphers have been devised which are 'provably' difficult to break in the following sense: If a method exists to break the cipher in a feasible time, then a polynomial-time algorithm exists to solve some other problem which is generally believed to be difficult. For example, the family of algorithms of Blum and Micali, which we mention in the next section, have the property that a practical algorithm to break them could be transformed into an efficient algorithm to solve the discrete logarithm problem.

3.3.   STREAM CIPHERS

As we saw in the last section, the one-time-pad offers perfect secrecy. However, the key for such a system is a randomly generated seqeuence of the same length as the message. Clearly no two parties can generate identical random sequences simultaneously and, in most situations, the key management problem makes the one-time-pad virtually unusable. Stream ciphers may, in some sense, be regarded as practical modifications of the one-time-pad.

3.3.1.   Stream Ciphers and the One-time-pad

In order to implement a stream cipher, it is first necessary to design a pseudo-random binary sequence generator, the output from which is used to encrypt the plaintext data on a bit-by-bit basis. More precisely, if the message consists of a sequence of n bits, $m_1 m_2 \ldots m_n$ say, then it is combined with n consecutive output bits from the sequence generator, $s_1 s_2 \ldots s_n$ say, so that the ciphertext $c_1 c_2 \ldots c_n$ is defined by: $c_i = m_i + s_i \pmod 2$, $1 \le i \le n$ [see §1.4].

The important fact to note is that, once the key is chosen, each bit of ciphertext c_i is a function only of the values of m_i and i. Similarly, when decrypting, each bit of plaintext m_i is a function solely of c_i and i.

We can represent the principles diagrammatically, as in Figure 3.3.1.

Stream ciphers are widely used in military and paramilitary applications for encrypting both data and digitized speech. In these areas of application, which were until recently the most commonly recognized uses for encryption techniques, they probably form the dominant type of cipher technique.

One reason for their dominance is the relative ease with which good sequence generators may be designed and implemented. However, the chief reason for their dominance in this area is the fact that they do not propagate errors. The sort of channels used for tactical military and paramilitary data and digitized speech traffic have a strong tendency to be of poor quality. So any cipher system which would increase an already relatively high error rate, would almost certainly render channels, which are usable for clear data, unusable for enciphered data.

From Figure 3.3.1 it is clear that the effectiveness of a stream cipher depends upon the properties of the keystream sequence. For the rest of this section we will discuss the type of properties which a 'good' keystream sequence should have and consider ways in which such sequences can be generated.

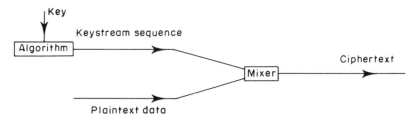

Figure 3.3.1. A stream cipher

3.3.2. Shift Registers

Traditionally, linear feedback shift registers (LFSRs) have been major components in most practical sequence generators. This is due to their cheapness, ease of implementation and the fact that their underlying mathematics is well understood. The main modern reference for stream ciphers is Rueppel's excellent book (Rueppel (1986)), while shift register sequences are discussed in Golomb (1982) and Selmer (1966). There is also a brief discussion in Beker and Piper (1982).

Figure 3.3.2 shows a shift register with feedback. Each of the squares labelled $S_0 S_1, \ldots, S_{n-1}$ is a binary storage element, which might be a flip-flop (bistable) position on a delay line or some other memory device. These n binary storage elements are called the *stages* of the shift register and, at any given time, their contents are called its *state*. The state of a shift register is a binary n-tuple which we may, at various times, regard as the binary expression of an integer. Clearly a shift register with n stages has 2^n possible states. If, for each i, s_i is the entry in S_i we will either write the state as $s_0 s_1 \ldots s_{n-1}$ or as the n-tuple $(s_0, s_1, \ldots, s_{n-1})$.

At time intervals, which are determined by a master clock, the contents of S_i are transferred into S_{i-1} for all i with $1 \le i \le n-1$. However, to obtain the new value for location S_{n-1} we compute the value of a given function $f(s_0, s_1, \ldots, s_{n-1})$ of all the present terms in the register, and transfer this into S_{n-1}.

DEFINITION 3.3.1.   The state of a shift register when $t = 0$ is called its *initial state*.

DEFINITION 3.3.2.   The function $f(s_0 s_1, \ldots, s_{n-1})$ is called the *feedback function* of the shift register.

DEFINITION 3.3.3.   The feedback function of a shift register is called *linear* if $f(s_0 s_1, \ldots, s_{n-1}) = f_0 s_0 + f_1 s_1 + \ldots + f_{n-1} s_{n-1}$ with each $f_i = 0$ or 1 and arithmetic is modulo 2. The constants $f_0, \ldots, f_{n-1}$ are called *feedback coefficients*.

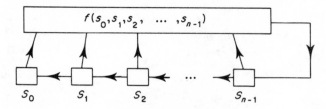

Figure 3.3.2. A shift register

DEFINITION 3.3.4. If S is a shift register with feedback function f, then the sequence (s_t) defined by $s_t = S_0(t)$ is said to be *generated* by S. It is also called the *output sequence*.

Note: (s_t) is completely determined by f and the initial state.
    Before giving a simple example we need some general properties of sequences.

DEFINITION 3.3.5. (s_t) is *periodic* if $\exists m > 0$ such that $s_{t+m} = s_t$ for all $t \geq 0$.

DEFINITION 3.3.6. If (s_t) is periodic then the smallest positive integer p such that $s_{t+p} = s_t$ for all t is called the *period* of (s_t).

DEFINITION 3.3.7. If (s_t) has period p then $s_0, s_1, \ldots, s_{p-1}$ is called a *generating cycle* for (s_t).

Note: Any sequence generated by a finite state machine is either periodic or ultimately periodic (i.e. periodic if the first 'few' terms are ignored).
    In this section we shall concentrate on binary sequences and will not always give the most general definitions.

EXAMPLE 3.3.1. Suppose we have a five-stage shift register with feedback function $s_0 + s_3$. If the initial state is 01010 then the sequence of states is as shown in Table 3.3.1.

Since the state with $t = 31$ is the same as that with $t = 0$, the sequence of states has period 31. Clearly this implies that the output sequence also has period 31. Its generating cycle is

    0 1 0 1 0 1 1 1 0 1 1 0 0 0 1 1 1 1 1 0 0 1 1 0 1 0 0 1 0 0 0.

In this example a five-stage register was used to generate a sequence of period 31. It is easy to show that this is the maximum possible period for a five-stage register. In fact we have

$t = 0$	01010	$t = 8$	01100	$t = 16$	11100	$t = 24$	10010
1	10101	9	11000	17	11001	25	00100
2	01011	10	10001	18	10011	26	01000
3	10111	11	00011	19	00110	27	10000
4	01110	12	00111	20	01101	28	00001
5	11101	13	01111	21	11010	29	00010
6	11011	14	11111	22	10100	30	00101
7	10110	15	11110	23	01001	31	01010

Table 3.3.1

PROPOSITION 3.3.1.   *If* (s_t) *is generated on an n-stage shift register with linear feedback then the period of* (s_t) *is at most* $2^n - 1$.

DEFINITION 3.3.8.   A sequence of period $2^n - 1$ generated on an *n*-stage LFSR is called an *m-sequence*.

It should be noted here that if (s_t) is generated by an *n*-stage LFSR with feedback coefficients $f_0, f_1, \ldots, f_{n-1}$ then (s_t) is a recursive sequence [see I, §14.12] satisfying the recurrence relation

$$s_{t+n} = \sum_{i=0}^{n-1} f_i s_{t+i}.$$

The theory of linear recursive sequences enables us to deduce many properties of these sequences.

PROPOSITION 3.3.2.   *If* (s_t) *is generated on LFSR of length n then knowledge of any 2n consecutive bits is sufficient to determine the feedback coefficients.*

If we adopt the shift register approach, then it is helpful to introduce the concept of a characteristic polynomial.

DEFINITION 3.3.9.   If S is an *n*-stage LFSR with feedback coefficients $f_0, f_1, \ldots, f_{n-1}$ then the polynomial

$$f(x) = f_0 + f_1 x + \ldots + f_{n-1} x^{n-1} + x^n$$

is called the *characteristic polynomial* of S. It is also referred to as a *characteristic polynomial* for any sequence (s_t) generated by S.

Notation: The set of all binary sequences (s_t) with $f(x)$ as a characteristic polynomial is denoted by $\Omega(f)$. It is easy to show, (see Selmer (1966)) that

PROPOSITION 3.3.3.   *If $f(x)$ has degree n then $\Omega(f)$ is an n-dimensional vector space over* GF(2) [see Example 1.5.7].

There is a well-developed theory relating the algebraic properties of $f(x)$ and various properties of the sequences in $\Omega(f)$. These are discussed in the books of Golomb and Selmer. Before we can state one of the most significant of these results, we must define the exponent of a polynomial [see Definition 1.5.1].

DEFINITION 3.3.10.   The *exponent* of the binary polynomial $f(x)$ is the value e such that $f(x)|x^e + 1$ but $f(x)|x^r + 1$ for any positive integer $r < e$.

PROPOSITION 3.3.4.   *Let $f(x)$ be a binary polynomial of degree n and let*

$(s_t) \in \Omega(f), (s_t) \neq (0)$. *Then (s_t) is an m-sequence if and only if the exponent of $f(x)$ is $2^n - 1$.*

One of the main reasons why the theory of LFSRs is so important to the study of periodic binary sequences is the following simple lemma.

PROPOSITION 3.3.5.   *If (s_t) is a binary sequence of period p then $(s_t) \in \Omega(x^p + 1)$.*

This implies that every periodic binary sequence can be generated on an LFSR.

It is perhaps worthwhile to pause for a moment and see the relevance of the last few results to the theory of stream ciphers. If we assume that the system is subject to a known plaintext attack, then we note that knowledge of a corresponding message/ciphertext pair of bits implies knowledge of the corresponding bit of the keystream sequence. Thus, in a known plaintext attack, part of the keystream sequence is known. If a finite state machine is used as the sequence generator, then the keystream sequence will be periodic and therefore, by Proposition 3.3.5, could have been generated on an LFSR. But if that LFSR had length n then, by Proposition 3.3.2, knowledge of $2n$ consecutive bits would determine the feedback coefficients and enable the attacker to determine the remainder of the keystream sequence. This, in turn, would reveal the rest of the message. Thus the size of the 'equivalent' LFSR is very important. This leads us to the following important definition.

DEFINITION 3.3.11.   Let (s_t) be a binary sequence of period p. The length n of the smallest LFSR which can generate (s_t) is called the *linear complexity* (or *linear equivalence*) of (s_t).

From Propositions 3.3.1 and 3.3.5 we have

PROPOSITION 3.3.6.   *If (s_t) has period p and linear complexity n then $\log_2(p + 1) \leq n \leq p$.*

From the cryptographic viewpoint we do not want knowledge of 'part' of any keystream sequence to enable an attacker to be able to determine the rest of it. This, for the reasons given earlier, leads us to the requirement that a keystream sequence should have

  (i) long period,
  (ii) large linear complexity.

3.3.3.   Pseudo-randomness

If we consider the stream cipher as a modification of the one-time-pad then the keystream sequence is the 'substitute' for the random sequence. As such it needs to 'destroy' the statistics of the plaintext. Clearly, as the following simple example shows, having long period and large linear complexity is not sufficient for this.

EXAMPLE 3.3.2. If (s_t) has period p with generating cycle $s_0 = s_1 = \ldots = s_{p-2} = 0, s_{p-1} = 1$, then (s_t) has linear complexity p.

Golomb suggested the following definition for a pseudo-random binary sequence.

DEFINITION 3.3.12. A binary sequence of period p is called a *PN-sequence* if it satisfies the following postulates.

R1 If p is even then the cycle of length p shall contain an equal number of zeros and ones. If p is odd then the number of zeros shall be one more or one less than the number of ones.

R2 In the cycle of length p, half the runs have length 1, a quarter have length 2, an eighth have length 3 and, in general, for each i for which there are at least 2^{i+1} runs, $1/2^i$ of the runs have length i. Moreover, for each of these lengths, there are equally many runs of zeros as there are runs of ones.

R3 The our-of-phase autocorrelation is a constant. (The out-of-phase autocorrelation is a measure of the 'agreement' between the sequence and non-zero phase shifts of it. For a definition of the out-of-phase autocorrelation see Beker and Piper (1982).)

The following are easily proved.

PROPOSITION 3.3.7. *Any PN-sequence has period $2^x - 1$ for some positive integer x.*

PROPOSITION 3.3.8. *Any m-sequence is a PN-sequence.*

However, although it was widely conjectured to be true, the converse is untrue. There is a PN-sequence of period 127 which is not an *m*-sequence (see Cheng (1981)).

It would be very interesting mathematically (although probably irrelevant cryptographically), to know all PN-sequences. With this in mind, it is perhaps worth digressing to point out that this problem is easily translated into a problem about symmetric two-designs or, equivalently, difference sets. The connection between these (seemingly) unrelated areas was established by Golomb (1982) and exploited further by Piper and Walker (1984).

For many years *m*-sequences were considered to be 'good' keystream sequences. However, Propositions 3.2.9 and 3.3.5 show quite clearly that an

m-sequence of period $2^n - 1$ has the minimum possible linear complexity and is, therefore, totally unsuitable.

So far our discussion has concentrated on properties of infinite periodic binary sequences. Although this is valid, in the sense that this is (almost always) what keystream sequences are, it overlooks one very important point; messages have finite length. Thus only finite subsequences of the keystream sequences will be used and, in practice, these subsequences are likely to be very short in comparison with the period. Thus, although obviously relevant, the 'global' properties of the infinite periodic keystream sequence are probably less important than the 'local' properties of the finite subsequences used to encipher messages. Thus the mathematical analysis of entire sequences is replaced by a number of statistical tests which are applied to numerous subsequences of the infinite sequence. A detailed discussion of a number of appropriate tests can be found in Kimberley (1986).

3.3.4. Linear Complexity Profiles

The general idea that local properties are more important than global ones was taken one step further by Rueppel (1984) who made the following definitions.

DEFINITION 3.3.13. If $s_0 s_1 \ldots s_{n-1}$ is a finite sequence of length n then its *local linear complexity* is the length of the shortest LFSR which can generate it.

DEFINITION 3.3.14. Let $s_0 s_1, \ldots, s_{n-1}$ be any finite sequence of length n. For any $0 \le i \le n - 1$ let $n(i)$ be the local linear complexity of $s_0 s_1, \ldots, s_i$. The pairs of integers $(i, n(i)), i = 0, \ldots, n - 1$, are called the *linear complexity profile* of $s_0 s_1, \ldots, s_{n-1}$.

It is easy to show that $n(i)$ is a non-decreasing function of i. Furthermore if a subsequence (of the keystream sequence) of length i is used to encipher a message, then the value of its local linear complexity is more relevant than the global linear complexity of the complete keystream sequence. If an attacker is trying to use knowledge of some corresponding consecutive plaintext/ciphertext bits to determine the complete message then, for a message of length i, he needs only $2n(i)$ bits. Thus it is the value of the local linear complexity, and not the global linear complexity, which determines the volume of known plaintext/ciphertext needed for this type of attack to succeed.

Rueppel showed that the expected value of the local linear complexity of a randomly selected binary sequence of length n is close to $\frac{1}{2}n$ and Rueppel's work has been exploited independently by Carter and Niederreiter to introduce further statistical tests based on a sequence's linear complexity profile. These tests involve comparing the properties of a given sequence with those which might be expected from a random sequence (Carter (1989), Niederreiter (1988)).

The implementation of Carter's test is made possible by the existence of a well-known elegant algorithm for determining the local linear complexity of a finite sequence. (This is the celebrated Berlekamp–Massey algorithm, see Massey (1969).)

Thus we can extend our list of the main requirements of a good keystream sequence to

(1) long period,

(2) good linear complexity profile (which implies large linear complexity),

(3) good local statistical properties.

One popular, practical, easily implementable way of achieving this is to combine a number of LFSRs using non-linear logic. Some suggested methods employ feed-forward logic, as in the systems of Bruer (1984), Herlestam (1983), Jennings (1983) and Siegenthaler (1984), whereas others incorporate elements of non-linear feedback logic, as in the systems of Chambers and Jennings (1984), Gollman (1986) and Smeets (1986).

A number of the techniques used involve interesting applications of polynomials over GF(2). There are a number of elegant mathematical results which imply guaranteed minimum linear complexities if LFSRs are combined in specific ways. However, although mathematics is a powerful tool when discussing period and linear complexity, it cannot, in general, be used to predict the statistical properties of the sequences.

The emphasis on the use of LFSRs in most practical systems is undoubtedly due to their cheapness and to the fact that they are both fast and easy to implement. The assessment of the security level of the resultant sequences relies on statistical testing rather than any precise mathematical quantification of the security level.

This typifies the practical approach we have taken so far in assessing the merits of stream cipher sequence generators. We have considered particular properties which sequence generators must have in order to resist certain types of cryptanalytic attack.

3.3.5. Cryptographically Strong Pseudo-random Bit Generators

A more desirable approach might be to construct a formal definition of secure sequence generator, which would replace the listing of an arbitrary number of individual properties. Following this direction of thought, Shamir (1981) introduced the concept of a cryptographically strong pseudo-random sequence. This was modified by Blum and Micali (1984) to produce the following informal description of a *Cryptographically Strong Pseudo-Random Bit (CSPRB) Generator*.

A *CSPRB generator* is a program G that, upon receiving as input a random number i (called the *seed*), outputs a sequence of pseudo-random bits $s_0, s_1, s_2, \ldots$. G possesses the following properties:

(1) The bits s_i are easy to generate, i.e. each s_i is output in time polynomial in the length of the seed.
(2) The bits s_i are unpredictable, i.e. given G and $s_0, s_1, s_2, \ldots, s_{n-1}$, but not the seed, it is computationally infeasible to predict s_n with better than probability $\frac{1}{2}$. (Here n is polynomial in the length of the seed.)

Blum and Micali then describe a possible implementation of a CSPRB generator based on the assumption that the discrete logarithm problem is intractible. (The discrete logarithm problem is discussed in §3.5.1.) We note that condition (2) implies the three requirements listed on p. 74.

As yet there do not appear to be any examples of CSPRB generators used in practical systems. There is no doubting the desirability of the concept, but the difficulty of finding examples which are practical to implement has been a problem. However, it appears that this situation is changing and CSPRB generators will eventually be implemented in real systems. Indications of progress in this direction are given in a paper presented at Eurocrypt 1988 (Schnorr (1988)).

3.4. CONVENTIONAL BLOCK CIPHERS

Whereas in a stream cipher the plaintext is enciphered bit by bit, in a block cipher system the plaintext is grouped into 'blocks' of constant size, s bits say, and enciphered block by block. In most situations the resultant block of ciphertext also has size s and we will usually assume this to be the case. Thus each key for a block cipher determines a permutation of the set of binary s-tuples. Mathematicians view binary s-tuples in many different ways: as elements of $V(s, 2)$ (the s-dimensional vector space over GF(2) [see Example 1.5.7]), as the integers modulo 2^s, as binary polynomials of degree at most $s - 1$, etc. For symmetric key systems it is usual to work in $V(s, 2)$, but modular arithmetic plays a leading role in asymmetric (or public) key cryptography [see §1.4]. In this section we will use the vector space notation and terminology, as introduced in I, §5. We assume $M = C = V(s, 2)$ and $K = V(t, 2)$, i.e. keys are binary t-tuples.

DEFINITION 3.4.1. A *block cipher* is a mapping from $V(s, 2) \times V(t, 2)$ onto $V(s, 2)$ such that, for each $\mathbf{k} \in V(t, 2)$, the mapping $V(s, 2) \times (\mathbf{k}) \to V(s, 2)$ is bijective [see I, §§1.4 and 5.11].

EXAMPLE 3.4.1. Suppose each $\mathbf{k} \in K$ determines a non-singular $s \times s$ binary matrix $\mathbf{A}_k$ [see I, 6.4] and let the mapping from $V(s, 2) \times V(t, 2)$ to $V(s, 2)$ be given by $\mathbf{m} \times \mathbf{k} \to \mathbf{m}\mathbf{A}_k$, where $\mathbf{m}$ is a row vector [see I, §5.11].

The cipher system in Example 3.4.1 is called a *linear system* and has the obvious weakness that knowledge of s corresponding plaintext/ciphertext blocks may be sufficient to easily determine the matrix A_k, i.e. the key, by simple matrix inversion techniques. (If the s plaintext blocks are not linearly independent, then it will still almost certainly be possible to drastically reduce the number of possible keys using simple linear algebra.) One of the principal objectives in designing a block cipher system is to try to ensure that the system is not vulnerable to attacks which are faster than an exhaustive key search. Although there are a few obvious ground rules, e.g. that each ciphertext bit should depend on each bit of the key and corresponding plaintext block, it is usually difficult to prove that this objective has been achieved.

3.4.1. Feistel Ciphers

One of the most widely discussed types of block cipher is the Feistel cipher (Feistel, Notz and Smith (1975)).

EXAMPLE 3.4.2.   The Feistel cipher: In a Feistel cipher the block size has to be even. If we have $s = 2n$ then each s-bit message block $\mathbf{m}$ is divided into two n-bit blocks and written

$$\mathbf{m} = (\mathbf{m}_0, \mathbf{m}_1).$$

Each key $\mathbf{k}$ defines a set of 'subkeys' $\mathbf{k}_1, \mathbf{k}_2, \ldots, \mathbf{k}_h$ for some fixed integer h, and each subkey $\mathbf{k}_i$ determines a function f_i of the vectors of $V(n, 2)$ into themselves. Any message $\mathbf{m}$ is enciphered in h 'rounds' using the following rule:

$$\text{At round 1:}\quad \mathbf{x}_0 = (\mathbf{m}_0, \mathbf{m}_1) \rightarrow \mathbf{x}_1 = (\mathbf{m}_1, \mathbf{m}_2),$$
$$\vdots \qquad\qquad\qquad \vdots$$
$$\text{At round } i:\quad \mathbf{x}_{i-1} = (\mathbf{m}_{i-1}, \mathbf{m}_i) \rightarrow \mathbf{x}_i = (\mathbf{m}_i, \mathbf{m}_{i+1}),$$
$$\vdots \qquad\qquad\qquad \vdots$$
$$\text{At round } h:\quad \mathbf{x}_{h-1} = (\mathbf{m}_{h-1}, \mathbf{m}_h) \rightarrow \mathbf{x}_h = (\mathbf{m}_h, \mathbf{m}_{h+1})$$

where $\mathbf{m}_{i+1} = \mathbf{m}_{i-1} + f_i(\mathbf{m}_i)$ for each i. The ciphertext is then the $2n$ bit block $\mathbf{m}_h, \mathbf{m}_{h+1}$.

To decipher, we note that, since all 'addition' is bit-wise modulo 2, the equation $\mathbf{m}_{i+1} = \mathbf{m}_{i-1} + f_i(\mathbf{m}_i)$ can also be written as $\mathbf{m}_{i-1} = \mathbf{m}_{i+1} + f_i(\mathbf{m}_i)$. Thus if we reverse the two halves of the ciphertext block and apply the encipherment procedure, but using the subkeys and hence the f_i, in reverse order, we have

$$\mathbf{x}_h = (\mathbf{m}_{h+1}, \mathbf{m}_h) \rightarrow \mathbf{x}_{h-1} = (\mathbf{m}_h, \mathbf{m}_{h-1}) \quad \text{at round 1}$$
$$\vdots$$
$$\mathbf{x}_i = (\mathbf{m}_{i+1}, \mathbf{m}_i) \rightarrow \mathbf{x}_{i-1} = (\mathbf{m}_i, \mathbf{m}_{i-1}) \quad \text{at round } h+1-i$$
$$\vdots$$
$$\mathbf{x}_1 = (\mathbf{m}_{2+1}, \mathbf{m}_1) \rightarrow \mathbf{x}_0 = (\mathbf{m}_1, \mathbf{m}_0) \quad \text{at round } h.$$

Thus we can decipher provided that we can reproduce each f_i at the appropriate moment. It is important to note that we do not require the functions to have any special properties. In particular they do not need to be reversible. Typical values for s and h are 64 and 16 respectively (these are the values used in the DES algorithm described below).

From its description it should be clear that, except for certain very special functions f_i (which can easily be avoided!), there will be no obvious 'simple' mathematical expression relating corresponding plaintext/ciphertext pairs. It should also be clear that it is easy to choose the f_i so that each ciphertext bit depends on every message and key bit.

In 1977 the National Bureau of Standards published the Data Encryption Standard (DES) (NBS (1977)). The DES is a 16-round Feistel cipher with 64-bit blocks and a 56-bit key. Every detail of the algorithm is contained in the original publication and so it is literally a public algorithm. (The details of the algorithm can be found in either the original publication or in a number of the standard cryptography textbooks. We will not give them here.) The use of DES was made mandatory for Federal departments and agencies if they wished to encipher non-classified material and its use by commercial and private organizations was encouraged. It has since (effectively) become an international standard for financial transactions.

3.4.2. Cryptanalysis

The publication of DES was seen by many as a 'challenge' to break it. Academic and professional cryptographers assumed that there must be a faster way of attacking it than trying all 2^{56} possible keys. Literally hundreds of papers have been written about DES, but all have failed to produce an attack which reduces the size of the effective exhaustive key search by a factor which is greater than two (Hellman *et al.* (1976)).

One line of attack adopted by a number of mathematicians has been to assume that DES contains some type of 'trapdoor', knowledge of which would considerably shorten an attack. They then invent possible trapdoors, some very ingenious, and proceed to show that DES does not possess them.

As an illustration we introduce the idea of linear structures (Evertse (1988)). It provides an illustration of how the application of some standard results and arguments from linear algebra can result in potential attacks on block cipher systems.

DEFINITION 3.4.2.   A *linear structure* of block cipher $T: V(s, 2) \times V(t, 2) \rightarrow V(s, 2)$ is a pair (V, W) where V is a subspace of $V(s, 2) \times V(t, 2)$ and W is a subspace of $V(s, 2)$ [see I, §5.5] such that, for each pair $(\mathbf{m}, \mathbf{k}) \in V$, each $\mathbf{m}' \in V(s, 2)$ and each $\mathbf{k}' \in V(t, 2)$, we have

$$T(\mathbf{m} + \mathbf{m}', \mathbf{k} + \mathbf{k}') + T(\mathbf{m}, \mathbf{k}) + T(\mathbf{m}', \mathbf{k}') + T(\mathbf{0}_s, \mathbf{0}_t) \in W.$$

The linear structure is *trivial* if either $V = \{(\mathbf{0}_s, \mathbf{0}_t)\}$ or $W = V(s, 2)$.

Note: Every block cipher has the trivial linear structures. Furthermore a linear block cipher is a linear structure.

The following is easily proved (see Evertse (1988)).

PROPOSITION 3.4.1.   *Let* $T:V(s,2) \times V(t,2) \to V(s,2)$ *be a block cipher and let* (V,W) *be a linear structure of* T. *Further let* A, B *be any linear mappings with domains* $V(s,2) \times V(t,2)$ *and* $V(s,2)$ *respectively, such that* $\ker(A) = V$ *and* $\ker(B) = W$. [*see* I, §5.11]. *Then there exists a linear mapping* $F_1 : V(s,2) \times V(t,2) \to V(s,2)$ *and a (not necessarily linear) mapping* $F_2 : \mathrm{Im}(A) \to \mathrm{Im}(B)$, *both easily computable from* T, A *and* B, *such that* $BT(\mathbf{m},\mathbf{k}) = F_2 A(\mathbf{m},\mathbf{k}) + F_1(\mathbf{m},\mathbf{k})$ *for all* $\mathbf{m} \in V(s,2)$ *and* $\mathbf{k} \in V(t,2)$.

If a block cipher has a linear structure then it is vulnerable to the following type of known plaintext attack which is usually faster than an exhaustive key search.

(This attack is a special example of the following type of general attack. Suppose that the system has 2^k keys but that an attacker has some way of determining k_1 bits of the key used before trying to find the remaining $k - k_1$ bits. Then the total number of trials needed for an exhaustive key search is $2^{k_1} + 2^{k-k_1}$ which is, usually, considerably smaller than 2^k. Sometimes the attacker may only be able to limit the k_1 bits to x choices, in which case the key search will be $2^{k_1} + x2^{k-k_1}$, which is still likely to be less than 2^k.)

Notation: A_1, A_2, H_1, H_2 are the linear mappings defined as follows:

$$A(\mathbf{m},\mathbf{k}) = A_1(\mathbf{m}) + A_2(\mathbf{k})$$
$$F_1(\mathbf{m},\mathbf{k}) = H_1(\mathbf{m}) + H_2(\mathbf{k})$$

Let $n = \dim(\ker(A_2))$. Clearly $n \leq t$.

The Known Plaintext Attack

Suppose the cryptanalyst has a plaintext/ciphertext pair $\mathbf{m}, \mathbf{c}$ and wants to find the key $\mathbf{k}$ with $T(\mathbf{m},\mathbf{k}) = \mathbf{c}$. He does the following:

(1) For all $\mathbf{k}' \in \mathrm{Im}(A_2)$ he tries to solve

$$\left. \begin{aligned} A_2(\mathbf{k}) &= \mathbf{k}' \\ H_2(\mathbf{k}) &= B(\mathbf{c}) + F_2(A_1\mathbf{m} + \mathbf{k}') + H_1(\mathbf{m}) \end{aligned} \right\} \qquad (3.4.1)$$

   with $\mathbf{k} \in V(t,2)$.
(2) For all $\mathbf{k}' \in \mathrm{Im}(A_2)$ for which equations (3.4.1) have a solution $\mathbf{k} \in V(t,2)$, he checks whether $T(\mathbf{m},\mathbf{k}) = \mathbf{c}$.

If there are x values for $\mathbf{k}'$ for which (3.4.1) has a solution and if the kernel of the mapping $\mathbf{k} \rightarrow (A_2(\mathbf{k}), H_2(\mathbf{k}))$ has dimensions $n_1 \leq n$, he will find the key in about $2^{t-n} + x2^{n_1}$ encryptions which, in general, is less than the 2^t encryptions needed for an exhaustive key search.

EXAMPLE 3.4.3.   For any integer n let $\mathbf{1}_n$ and $\mathbf{0}_n$ denote the vectors of length n which have each entry equal to 1 and 0 respectively.

If we denote the DES mapping by $\mathrm{DES}(\mathbf{m}, \mathbf{k})$ then it has the property $\mathrm{DES}(\mathbf{m} + \mathbf{1}_{64}, \mathbf{k} + \mathbf{1}_{56}) = \mathrm{DES}(\mathbf{m}, \mathbf{k}) + \mathbf{1}_{64}$ for all $\mathbf{m}, \mathbf{k}$. Hence $\langle (\mathbf{1}_{64}, \mathbf{1}_{56}), \{\mathbf{0}_{64}\} \rangle$ is a linear structure of DES. (This is the complementation property discovered by Hellman *et al.*) In the above known plaintext attack, $x = 2$ and the 2 values for $\mathbf{k}'$ are of the form $\mathbf{k}, \mathbf{k} + \mathbf{1}_{56}$. Thus there is an attack on DES involving 2^{55} encryptions.

This, it must be stressed, is just one way in which linear algebra has been used to either look for weaknesses in DES or determine desirable properties for a 'good' block cipher.

Just as with stream ciphers, it is probably true that statistical testing is a crucial element in the assessment of the security of conventional block ciphers. Such tests are designed to check that there are no obvious, exploitable relations between plaintext, key and ciphertext. One common test, for instance, uses a fixed key to first encrypt a plaintext block and then encrypt the same plaintext with one bit changed. It then compares the two ciphertext blocks to (hopefully) ascertain that there is no obvious correlation.

3.5. PUBLIC KEY SYSTEMS

For a public key (or asymmetric) system it must be computationally infeasible to deduce the deciphering key $k(D)$ from knowledge of the enciphering key $k(E)$ and the enciphering algorithm. The design of such a system is a challenge to which many mathematicians and computer scientists have responded.

Before looking at the mathematics behind some of the proposed asymmetric key systems it is worth noting that such systems were first proposed in the exciting, and aptly named, paper 'New directions in cryptography' (Diffie and Hellman (1976)). In that paper the authors considered two other seemingly impossible (at least before their paper!) problems:

(1) The key distribution problem: How can two people who have never met and do not trust each other, but want to communicate privately using a conventional (or symmetric) key system, agree in advance on a key that will be known to them but to no one else?

(2) The signature problem: Can digital electronic messages be signed by the author so that (a) no one can forge the signature and (b) he cannot later deny sending it?

3.5.1. One-way Functions

Diffie and Hellman recognized the potential importance of the use of one-way functions, i.e. functions which were 'easy' to implement but 'hard' to invert. One of the areas where it was natural to look was that of NP-complete problems (see, for example, Garey and Johnson (1979) or Aho, Hopcroft and Ullman (1974)). However, in the context of public key cryptography, most NP-complete problems were not appropriate since someone, i.e. the genuine recipient, has to be able to compute both $k(E)$ and $k(D)$.

Thus the concept of a trapdoor one-way function was needed, i.e. a function which allowed someone in possession of specific secret information (the trapdoor) to compute the inverse function.

The following examples of one-way functions have assumed central roles in modern cryptography.

EXAMPLE 3.5.1.   The Knapsack function: Let $A = \{a_1, a_2, \ldots, a_n\}$ be a set of n distinct integers.

Notation: If $\mathbf{a} = (a_1, a_2, \ldots, a_n)$ is an integer n-tuple and $\mathbf{x} = (x, x_2, \ldots, x_n)$ is a binary n-tuple then we write

$$\mathbf{a}\cdot\mathbf{x} \quad \text{for} \quad \sum_{i=1}^{n} a_i x_i.$$

If $\mathbf{a} = (a_1, a_2, \ldots, a_n)$ then we can define a mapping $\mathbf{x} \to \mathbf{a}\cdot\mathbf{x}$ from the set of binary n-tuples $\{0, 1\}^n$ to the set of integers $\mathbb{Z}$. Clearly, given $\mathbf{a}$ and $\mathbf{x}$, it is easy to compute $\mathbf{a}\cdot\mathbf{x}$. However, for general $\mathbf{a}$, the problem of determining $\mathbf{x}$ from $\mathbf{a}$ and $\mathbf{a}\cdot\mathbf{x}$ is known to be NP-complete. Thus, for general $\mathbf{a}$, the function $\mathbf{x} \to \mathbf{a}\cdot\mathbf{x}$ is a one-way function from $\{0, 1\}^n$ to $\mathbb{Z}$.

EXAMPLE 3.5.2.   Multiplication of two large primes: If p and q are any two large primes, then computing $n = pq$ is easy. However, it is, generally, extremely difficult to determine p or q from n. The time complexity function of the best-known algorithms is about $\exp[\sqrt{\log n \log(\log n)}]$ for large n.

EXAMPLE 3.5.3.   Exponentiation in a finite field: For any prime power q the multiplication group of $GF(q)$ is cyclic [see Theorem 1.5.9]. If q is a large prime power and a is a primitive element of $GF(q)$ then, for any non-zero element y in $GF(q)$, there exists an $x < q$ with $y = a^x$ in $GF(q)$. If q is a large prime this last equation can be replaced by $y \equiv a^x \pmod{q}$.

DEFINITION 3.5.1.   If $y = a^x$ in $GF(q)$ then x is called the *discrete logarithm* of y to the base a in $GF(q)$.

The mapping $x \rightarrow a^x \pmod{p}$, for a large prime p, is a one-way function. (Computing $a^x \pmod{p}$ requires at most $2\log_2 x$ multiplications whereas the best-known algorithms for extracting logarithms modulo p require a precomputation of the order of $\exp[\sqrt{\log p \log(\log p)}]$ operations.)

It is worth noting that arithmetic in $GF(2^n)$ for large n can be performed faster than arithmetic over large primes. (This is because arithmetic in these fields can be performed with LFSRs.) Thus a number of mathematicians prefer to use the one-way function $x \rightarrow a^x$ in $GF(2^n)$. However, the speed with which the arithmetic can be performed in finite fields of characteristic two [see §1.5] means that care must be taken to ensure that the exponent n is large enough (see Coppersmith (1984)).

3.5.2. Some Public Key Systems

In this section we will introduce the two well-known asymmetric key systems; the Merkle–Hellman system and the RSA system. The Merkle–Hellman system has been broken and is no longer a serious candidate for public acceptance. However, we include it to illustrate how trapdoor one-way functions may be used.

EXAMPLE 3.5.4. The Merkle–Hellman system: As we saw in Example 3.5.1, for arbitrary $A = \{a_1, a_2, \ldots, a_n\}$ the mapping $\mathbf{x} \rightarrow \mathbf{x} \cdot \mathbf{a}$ is a one-way function from $\{0, 1\}^n \rightarrow \mathbb{Z}$. However, if A has the property that, for all k, $a_k > \sum_{i=1}^{k-1} a_i$ then it is easy to compute $\mathbf{x}$ from $\mathbf{x} \cdot \mathbf{a}$. (Note that for the special case $a_1 = 1$, $a_2 = 2, \ldots, a_i = 2^{i-1}, \ldots, a_n = 2^{n-1}$, computing $\mathbf{x}$ from $\mathbf{x} \cdot \mathbf{a}$ is nothing more than finding the binary expansion of $\mathbf{x} \cdot \mathbf{a}$)

DEFINITION 3.5.2. If $A = \{a_1, a_2, \ldots, a_n\}$ has the property $a_k > \sum_{i=1}^{k-1} a_i$ for all k, then A is called a *superincreasing set*.

In order to obtain a key for the Merkle–Hellman system we choose a super-increasing sequence $B = \{b_1, b_2, \ldots, b_n\}$. We then choose two suitable large numbers w, m with $(w, m) = 1$ and form $A = \{a_1, a_2, \ldots, a_n\}$ where $a_i \equiv wb_i \pmod{m}$ for $i = 1, 2, \ldots, n$. The set A will no longer be superincreasing and will appear to be randomly chosen.

Our cipher system now enciphers a n-bit message block $\mathbf{x}$ as the integer $\mathbf{x} \cdot \mathbf{a}$, where $\mathbf{a} = (a_1, \ldots, a_n)$. An interceptor will know $\mathbf{a}$, because A is the public enciphering key, and will know $t = \mathbf{x} \cdot \mathbf{a}$. However, he is now faced with the problem of determining $\mathbf{x}$ which, since $\mathbf{x} \rightarrow \mathbf{x} \cdot \mathbf{a}$ is one way, is hard.

As the designers of the system, we have the secret knowledge that A was obtained from B using w and m. This trapdoor information allows us to decipher t. To do so we first compute w^{-1} such that $ww^{-1} \equiv 1 \pmod{m}$. (Note that w^{-1} exists because $(w, m) = 1$, see Theorem 14.3.) We then compute $w^{-1}t \pmod{m}$.

Since $\mathbf{a} = w\mathbf{b} \pmod{m}$ we have

$$w^{-1}t \pmod{m} \equiv w^{-1}(\mathbf{x} \cdot w\mathbf{b}) \pmod{m} \equiv \mathbf{x} \cdot \mathbf{b} \pmod{m}.$$

But since B was superincreasing, computing $\mathbf{x}$ from knowledge of $w^{-1}t$ and B is easy. Thus we can decipher the cryptogram.

EXAMPLE 3.5.5.   A numerical example using the Merkle–Hellman system: Choose

$$B = (2, 6, 9, 19, 41, 79, 159, 320, 643, 1310), \qquad w = 1053, \qquad m = 2719.$$

Then

$$A = \{2106, 880, 1320, 974, 2388, 1617, 1568, 2523, 48, 897\}.$$

If $\mathbf{x} = 0111001100$, then $t = 880 + 1320 + 974 + 1568 + 2523$, i.e. $t = 7265$. To decipher, we first compute $w^{-1} = 1996$. Thus $w^{-1}t \equiv 1996.7265 \equiv 513 \pmod{2719}$.

Thus, whereas the interceptor has to determine $\mathbf{x}$ from 7265 and an (apparently) randomly chosen set, we can find $\mathbf{x}$ from 513 and the superincreasing set $\{2, 6, 9, 19, 41, 79, 159, 320, 643, 1310\}$.

To illustrate how easy it is to find $\mathbf{x}$ for a superincreasing set, we will decipher 513. If $\mathbf{x} = (x_1 x_2 \ldots x_{10})$ then clearly, since $513 < 643$ and $513 < 1310$, $x_9 = x_{10} = 0$. Also, since B is superincreasing, $\sum_{i+1}^{7} x_i < 320$. Thus $x_8 = 1$. Now $513 - 320 = 193$ must be $\sum_{i+1}^{7} x_i a_i$. Thus again since B is superincreasing, since $a_7 = 159 < 193$, we have $x_7 = 1$. Since $193 - 159 = 34$, we have $x_5 = x_6 = 0$. Similarly $x_4 = 1$ and, since $34 - 19 = 15$, $x_3 = x_2 = 1$, $x_1 = 0$. Thus $\mathbf{x} = (011100110)$.

This example is, of course, too small for the full extent of the interceptor's difficulties to be apparent. However, the case of decipherment using the secret key (i.e. knowledge of w and m) should be clear.

Note: This system (and natural extensions of it) has been systematically attacked. First, Shamir showed how to take the public set A and evaluate a pair of w', m' which would convert A back to a superincreasing set. Even though there was no guarantee that this would be the original B, this was sufficient to enable the interceptor to decipher all cryptograms. (Shamir (1982), (1984)).

The system as described in this section could be 'improved' by choosing another pair, w_1, m_1 say, and letting the key be the set obtained by multiplying the elements of A by $w_1 \pmod{m_1}$. In fact this process can be iterated many times. However, all such systems have now been broken (Brickell (1984)). In the reference quoted, Brickell announced the breaking of a system with $n = 100$, which used 40 iterations, in about one hour of Cray-1 time.

EXAMPLE 3.5.6.   The RSA system. For an RSA system we generate two large primes p and q and compute their product n. The RSA system is then a block cipher in which the plaintext and ciphertext blocks are integers between 0 and $n - 1$.

The value of n is public knowledge and the enciphering key is an integer e such that $(e, (p - 1)(q - 1)) = 1$. (Note that $(p - 1)(q - 1)$ is the value of the Euler totient function $\phi(n)$ [see Definition 1.4.1 and Example 1.4.6].) A message block m is then enciphered as $c \equiv m^e \pmod{n}$.

For large n, it is generally accepted that, unless the factors of n are known, $m \to m^e \pmod{n}$ is a one-way function and thus it is very difficult to determine m from knowledge of m^e, n and e.

However, since we know the factors of n we can compute $p - 1$ and $q - 1$ and then, since e was chosen so that $(e, (p - 1)(q - 1)) = 1$, compute the integer d such that $de \equiv 1 \pmod{(p - 1)(q - 1)}$. It is important to note here that Euclid's algorithm for computing d is well known and easy to implement [see I, §4.1.3]. The reason that the interceptor cannot find d is that the modulus of the equation, i.e. $(p - 1)(q - 1)$, is unknown.

Euler's theorem [see Theorem 1.4.5] asserts that, for any integer n and any integer a with $(a, n) = 1$, $a^{\phi(n)} \equiv 1 \pmod{n}$. This can be used to show that $a^{k\phi(n) + 1} \equiv a \pmod{n}$ for all a. But since $\phi(n) = (p - 1)(q - 1)$ and $ed \equiv 1 \pmod{(p - 1)(q - 1)}$, we have $ed = k\phi(n) + 1$, for some k. Thus $c^d \equiv m^{ed} = m^{k\phi(n) + 1} \equiv m \pmod{n}$.

In other words, d is the (secret) deciphering key and the cryptogram c is deciphered by forming $c^d \pmod{n}$.

EXAMPLE 3.5.7. A numerical example using the RSA system: as a very small example, suppose $p = 17$ and $q = 31$ so that $n = pq = 527$ and $\phi(n) = (p - 1)(q - 1) = 480$. If $e = 7$ is chosen then $d = 343$ ($7 \times 343 = 2401 = 5 \times 480 + 1$). And if $m = 2$ then

$$c = m^e \pmod{n} = 2^7 \pmod{527} = 128.$$

Note again that only the public information $\{e, n\}$ is required for enciphering m. To decipher, the secret key d is needed to compute

$$m = c^d \pmod{n} = 128^{343} \pmod{527} \equiv 2 \pmod{527}$$

Note: The RSA system exploits the fact that finding large primes (typically the minimum size of primes suggested for current applications is about 256 bits) is computationally easy whereas factoring the product of two such primes is (apparently) computationally infeasible. However, we must point out that no one has succeeded in showing that, in general, breaking the RSA system is equivalent to factoring n. It is not inconceivable that there might be some method for calculating eth roots modulo n without being able to factor n. It is, however, worth noting that the ability to discover d from e and n implies the ability to factor n. This is true because knowledge of a multiple of $(p - 1)(q - 1)$ enables us to find p and q.

Before discussing some of the current mathematical problems being considered by the cryptographers who are concerned with RSA we will digress to discuss the Diffie–Hellman key exchange scheme. This makes sense since the same type of fundamental problems arise.

3.5.3. The Diffie–Hellman Key Exchange

Suppose that A and B wish to establish a common key so that they can use a symmetric key cipher. They choose a large integer q, either a prime or a power of 2, and a primitive element a in $GF(q)$ [see §1.5.2].

A and B randomly select integers x_A and x_B respectively in the range 1 to $q - 1$. A computes $y_A = a^{x_A}$ in $GF(q)$ and sends it to B while B computes $y_B = a^{x_B}$ in $GF(q)$ and sends it to A. Both A and B can then compute $K_{AB} = y_A^{x_B} = y_B^{x_A} = a^{x_A x_B}$ in $GF(q)$.

We have already discussed the difficulty of the discrete logarithm problem and, provided q is large enough, any interceptor who obtains y_A and y_B will still not be able to compute K_{AB}. However, as we noted earlier, the work of Coppersmith indicates that extreme caution is required if $q = 2^n$, and in this case the exponent n needs to be very large if the scheme is to be practically secure.

Of course it should not be overlooked that A needs some assurance that his received value of y_B actually originated from B, and vice-versa.

3.5.4. Related Mathematical Problems

Anyone interested in implementing either the RSA public key system or the Diffie–Hellman key exchange protocol has to be concerned with both the security issues and the problems associated with the speed and cost of implementation. For the Diffie–Hellman protocol, the main problem is (probably) the choice between using a large prime modulus or a power of 2. These issues are discussed in Odlyzko (1985).

Implementors of RSA need to be conversant with the latest progress on the factorization of large numbers. It is currently possible to factorize products of two large primes if these primes have certain special properties. This means that prime numbers for use in RSA need to be chosen with care; the current list of known properties to avoid, and a method to choose 'strong' primes avoiding these properties, is given in Gordon (1984). It is conceivable that new factoring algorithms could be discovered which add to the list of 'weak' primes. Such a discovery, even if the likelihood of choosing such a prime were very small, could have a very serious impact on existing RSA users! They also need faster algorihms for modular multiplication, together with algorithms which reduce the number of multiplications when raising a number to a given power. Some interesting predictions concerning factorization can be found in Pomerance, Smith and Tuler (1988).

Given that a 512-bit modulus is to be used (and this appears to be generally accepted as necessary for an acceptable security level), then the fastest chips claim throughput speeds of about 64 kbit/s. This is significantly slower than the throughout speeds for DES chips. Thus there are still many situations where RSA is too slow to be used for data encryption and its main proposed uses tend to be for digital signatures and key encryption. Before we move on to discussing digital signatures and personal identification protocols, we mention an excellent recent article: The First Ten Years of Public Key Cryptography (Diffie (1988)).

3.6. PERSONAL IDENTIFICATION AND SIGNATURES

3.6.1. Signature Schemes

As we have already noted, the concept of a digital signature was first discussed by Diffie and Hellman (1976). Since then there have been many proposed implementations.

According to ISO 'the term digital signature is used to indicate a particular authentication technique used to establish the origin of a message in order to settle potential disputes over what message (if any) was sent'. A signature is also used to prove the identity of the sender. During the signing process the signer uses information which is known only by him. (It is the use of this information which is accepted as proof of the identify of the signer.) During the verifying process, public information is used to check that the signer's private information was used. It is, therefore, crucial that knowledge of the public information should not enable anyone to deduce the signer's private information or, in any other way, allow false signatures to be generated.

One popular solution of the digital signature problem is the use of those special public key systems for which the deciphering process can be applied to a coded version of the plaintext. In the only practical examples of such systems, deciphering and enciphering involve (essentially) the same process. Furthermore the most popular example is the use of RSA, where enciphering and deciphering both involve modular exponentiation. Here the signer's secret information is the secret RSA key, i.e. the number which, in normal applications, he uses to decipher cryptograms sent to him.

EXAMPLE 3.6.1. Digital signatures using RSA: If we let (n, e) denote A's public enciphering key and let d be the secret deciphering exponent then, in order to see how the RSA scheme can be used for digital signatures, we merely note that if x is any known number than only A can compute $y \equiv x^d \pmod{n}$. However, anyone who knows the public key e can compute $y^e \pmod{n}$ and, if the answer is x, will know that y originated from A.

In the above example of a signature scheme, we have ignored two problems. The first is the problem of messages which are longer than a single RSA block. The approach normally adopted for using RSA to encipher messages, namely dividing the message into appropriate length blocks, is not immediately applicable to digital signatures. If each block of a message is individually signed, then a malicious interceptor could rearrange, duplicate or omit blocks from a message in an undetectable way. The second is the need for messages x to contain sufficient 'redundancy'. If any message x is acceptable, then a forger can simply choose y at random, calculate $x \equiv y^e \pmod{n}$ and then claim that y is A's signature on x.

There are a number of ways of overcoming these problems. One of the most popular is the application of a 'hashing function' to the message to reduce it to the length of a single block prior to signature. The signed hashed version is then appended to the message. This hashing function needs to be public (so that signatures can be checked) and 'one-way'. By 'one-way' here we mean that, given a message and a hashed version of that message, it should not be feasible to generate a new message having the same hashed value.

Because of the way in which digital signatures may be used, hash functions are normally required to satisfy a slightly stronger property, called collision-freedom. This property requires that it should be computationally infeasible to generate a pair of messages with the same hash value.

A number of candidates for hash functions exist: one such function is based on the use of modular squaring and is described in Girault (1988).

3.6.2. Identification

There are many situations where identification is necessary but where the two parties concerned are (potential) adversaries. In these situations it is necessary to make it impossible for B to impersonate A even after B has observed and verified arbitrarily many proofs of identity provided by A. In Fiat and Shamir (1987) a number of typical instances are listed. Their list includes:

(i) 'passports (which are often inspected and photocopied by hostile governments)',
(ii) 'military command and control systems (whose terminals may fall into enemy hands)',
(iii) 'computer passwords (which are vulnerable to hackers and wire tappers)'.

The identification procedure will be interactive. In this situation the person identifying himself, A say, has some secret knowledge K known only to him. A then proves his identity to B by performing certain functions which establish that he knows K. However, in order to prevent B, or anyone intercepting the communications, from impersonating A at a later date, the method of using K must be such that it reveals nothing about K itself. A typical conversation between A and B might be set of questions from B to A followed by answers from A to B 'proving' to B that A possesses K. In most instances B will make an immediate decision about whether or not he accepts A's identity and, if he does not believe that it is A, will deny the forger the service, access or reply that he requires. If, however, decisions are not to be made in real time, then it is important to note that, in most identification protocols, B can 'invent' a sequence of questions and answers which will prove to a third party that A

identified himself. Thus most identification schemes are not true signature schemes. However, the design of such schemes presents some very interesting mathematical problems. One solution has been presented by Fiat and Shamir.

EXAMPLE 3.6.2.   The Fiat–Shamir identification scheme: the security of this scheme relies on the fact that finding square roots modulo a large composite number is difficult unless its factorization is known. However, the situation is slightly different to that for RSA. Whereas no one has shown that breaking RSA is equivalent to factoring n the following result proves that this scheme is secure provided factoring is hard.

PROPOSITION 3.6.1.   *Any algorithm to find square roots modulo n can be used to factorize n.*

Note: There are efficient algorithms for finding square roots, modulo primes. [see §1.4.3]. If n is a large composite number of the form $n = pq$, with p and q distinct primes, then all the known practical algorithms for finding square roots modulo n require knowledge of p and q (see Knuth (1981)).

We will not give a detailed desciption of the complete Fiat–Shamir scheme. Instead we will merely outline a simpler earlier version which illustrates the basic underlying mathematical idea. In the description we will assume that the scheme is being used to issue ID cards.

A Simplified Version

The first part of the procedure involves the card issuer generating two large strong primes p and q. The product $n = pq$ is public, in the sense that its value is known to all participants in the scheme. (Note that this implies that the card issuer must be trusted by all participants.)

When issuing a card, the card issuer puts the identifying information of the holder into a format suitable for the card. If we call these data I then it is likely that I will need to be transformed (using a one-way function f) for use in the identification calculations. However, before f is applied to I, I is concatenated with a small number c chosen so that $f(I, c)$ is a square modulo n. (This is done by trying various values for c until one 'works'. This should not involve too many trials as roughly 25 per cent of all numbers are squares modulo n.) If we write $v = f(I, c)$ then I, c, f (and hence v) are all made public to anyone who needs them. However, since the factors of n are unknown to everyone except the card issuer, only the card issuer will be able to compute u such that $u^2 \equiv v \pmod{n}$. The card issuer computes a value for u and this is then stored on the card. It is this value u which is going to be used to prove that the card was issued by the card issuer.

When the card is presented to a terminal, the terminal can certainly compute *v*. It could, of course, merely challenge the card to produce *u* and then readily check that $u^2 \equiv v \pmod{n}$. Unfortunately this naive protocol could only be used once because it involves disclosing the value of *u* and enables any interceptor, or the terminal staff, to impersonate the card. Thus we have to find a dialogue between the card and terminal which proves that the card holds *u* but does not reveal any information about is value.

The following simple but ingenious sequence illustrates how this can be achieved. (Recall that *f* is public, that the terminal knows *n* and is given *I* and *c* so that it can compute *v*.)

Step 1 The card generates a random number *r* and computes $x \equiv r^2 \pmod{n}$.

Step 2 The card sends *x* and $y \equiv vx^{-1} \pmod{n}$ to the terminal.

Step 3 The terminal checks that $v \equiv xy \pmod{n}$.

Step 4 The terminal then decides whether to ask the card for the square root of *x* (mod *n*) or for the square root of *y* (mod *n*). It then requests one of the square roots from the card.

Step 5 The card responds by sending *r* if asked for the square root of *x* and $s \equiv ur^{-1} \pmod{n}$ if asked for the square root of *y*.

Step 6 The terminal checks either that $r^2 \equiv x \pmod{n}$ or $s^2 \equiv y \pmod{n}$, depending upon the request in Step 4.
If the check in Step 6 fails then the terminal rejects the card.

Some Comments

1. If both *r* and *s* were revealed by the card then $u \equiv rs \pmod{n}$ would also be revealed. Thus the random numbers *r* must be large enough so that there is no (realistic) chance of using the same value twice.
2. The terminal has no way of knowing whether the card constructed *x* or *y* independently of *u*.
3. Even if the card does not know *u*, it has a 50 per cent chance of being able to answer any one of the terminal's challenges correctly. If the procedure is repeated *k* times, then the chance of the card being correct each time if *u* is unknown is 1 in 2^k. So, for instance, if $k = 20$ the chances are about one in a million.
4. As pointed out earlier, a record of the total dialogue between the card and terminal cannot be used to 'prove' the authenticity of the card. The terminal could always 'manufacture' a dialogue in which the card was always asked for the square root of the square which was generated independently of *v*.

Fiat and Shamir give a faster version of this protocol in which (essentially) each card stores a number of modular square roots. They also suggest how their scheme might be used for digital signatures.

3.7. OTHER MATHEMATICAL THEMES

There are a number of areas of mathematical applications to cryptography which we have not covered here. We now mention a few of the more mathematically interesting of these topics.

3.7.1. Extensions to the Shannon Approach

The Shannon approach to cryptography has been extended in a number of different ways. One of the most significant extensions relates to message authentication. Shannon's theory only deals with secrecy of messages, although, in practice, protection against fraudulent messages, or illicit modifications to genuine messages, is often at least as important. In a number of recent papers (notably Simmons (1985)) Simmons and other authors have shown how the Shannon theory can be extended to give 'perfect' protection against such threats.

3.7.2. Threshold Schemes and Key Distribution Patterns

A threshold scheme is an extension of an idea used by manufacturers of large vaults with complex locks. It is often necessary for n authorized persons to hold physical keys to a lock with the property that at least k of them must use their keys simultaneously for the locks to be opened. A threshold scheme provides a method of encrypting a message and distributing 'part keys' to n authorized recipients, so that at least k of them must pool their part keys in order to read the message. Various examples of threshold schemes have been proposed (see, for example, Kranakis (1986)).

A key distribution pattern uses a somewhat similar idea to simplify key distribution in very large networks, where it is required to give (in advance) every one of n users the capability to exchange secure messages with any other user. One possible solution is to generate $\frac{1}{2}n(n-1)$ keys, one for each pair, and distribute these appropriately. A proposed altenative solution is to give each user a set of 'subkeys' with the property that for any pair of users the set of subkeys they hold in common is known to no other single user. The subkeys held in common between each pair of users can then be used to construct a secret key for communication, which will then be known to no other users. For further details see, for example, Mitchell and Piper (1987).

It is interesting to note that important examples of both threshold schemes and key distribution patterns can be constructed using block designs. Moreover, finite geometries are also of application in constructing 'perfect' authentication schemes.

3.7.3. Zero-knowledge Proofs

The Fiat–Shamir identification scheme described in Example 3.6.2 is one example of a cryptographic protocol, i.e. an exchange of messages to perform a security function. In the case of the Fiat–Shamir protocol, the purpose of the protocol is to prove the identity of one party to another by proving possession of a secret. The intriguing notion of proving one's identity by showing the possession of a secret (without revealing it) has led to the definition and development of zero-knowledge protocols.

Informally, a zero-knowledge protocol is one in which one party (the prover) can prove to another party (the verifier) the possession of a secret in such a way that no useful knowledge about the secret is revealed to observers of the protocol; the formal definition is rather more complex. There has been an explosion of interest in such protocols over the last two or three years; the degree of interest is shown in the high percentage of papers devoted to this topic in conferences such as Crypto 88.

We do not discuss this topic further here both because of space limitations and because of the degree of mathematical sophistication involved. Zero-knowledge protocols, along with provably secure ciphers, may well become very significant over the next few years.

F. C. P. and C. J. M.

October 1988

REFERENCES

This list includes a number of books on cryptography together with those papers referred to in this chapter. It should be noted that most of the books were not written for mathematicians; those with the most mathematical content are Kranakis (1986), Koblitz (1987) and Welsh (1988).

Aho, A. V., Hopcroft, J. E. and Ullman, J. D. (1974). *The Design and Analysis of Computer Algorithms*, Addison-Wesley.
Beker, H. J. and Piper, F. C. (1982). *Cipher Systems*, Van Nostrand Reinhold.
Blum, M. and Micali, S. (1984). How to Generate Cryptographically Strong Sequences of Pseudo-random Bits, *SIAM J. Comput.* **13**, 850–864.
Brickell, E. F. (1984). Solving Low Density Knapsacks, *Advances in Cryptology: Proceedings of Crypto 83*, Plenum Press, 25–37.
Bruer, J-O. (1984). On Pseudo Random Sequences as Crypto Generators, *Proceedings of the 1984 International Zurich Seminar on Digital Communications*, IEEE, 157–161.
Carter, G. (1989). Aspects of local linear complexity, Ph.D. Thesis, University of London.
Chambers, W. G. and Jennings, S. M. (1984). Linear Equivalence of Certain BRM Shift-register Sequences, *Electronics Letters* **20**, 1018–1019.
Cheng, U. (1981). Properties of Sequences, Ph.D. Thesis, University of Southern California.
Coppersmith, D. (1984). Fast Evaluation of Logarithms in Fields of Characteristic two, *IEEE Trans. Inf. Theory* IT-**30**, 587–594.
Davies, D. W. and Price, W. L. (1984). *Security for Computer Networks*, John Wiley and Sons.
Deavours, C. A. (1977). Unicity Points in Cryptanalysis, *Cryptologia* **1**, 46–68.
Denning, D. E. R. (1983). Cryptography and Data Security, Addison-Wesley.
Diffie, W. (1988). The First Ten Years of Public-key Cryptography, *Proc. IEEE* **76**, 560–577.

Diffie, W. and Hellman, M. E. (1976). New Directions in Cryptography, *IEEE Trans. Inf. Theory* IT-**22**, 644–654.

Evertse, J-H. (1988). Linear Structures in Blockciphers, *Advances in Cryptology: Proceedings of Eurocrypt 87*, Springer-Verlag, 249–266.

Feistel, H., Notz, W. A. and Smith, J. L. (1975). Some Cryptographic Techniques for Machine-to-machine Data Communications, *Proc. IEEE* **63**, 1545–1554.

Fiat, A. and Shamir, A. (1987). How to Prove Yourself: Practical Solutions to Identification and Signature Problems, *Advances in Cryptology: Proceedings of Crypto 86*, Springer-Verlag, 186–194.

Garey, M. R. and Johnson, D. S. (1979). *Computers and Intractability*, W. H. Freeman and Co.

Girault, M. (1988). Hash Functions using Modulo-N Operations, *Advances in Cryptology: Proceedings of Eurocrypt 87*, Springer-Verlag, 217–226.

Gollman, D. (1986). Linear Complexity of Sequences with Period p^n, paper presented at Eurocrypt 86.

Golomb, S. W. (1982). *Shift Register Sequences* (revised edition), Aegean Park Press.

Gordon, J. (1984). Strong RSA Keys, *Electronics Letters* **20**, 514–516.

Hellman, M. E. (1974). The Information Theoretic Approach to Cryptograpy, Stanford University Report MEH76-1.

Hellman, M. E. (1977). An Extension of the Shannon Theory Approach to Cryptography, *IEEE Trans. Inf. Theory* IT-**23**, 289–294.

Hellman, M. E. *et al.* (1976). Results of an Initial Attempt to Cryptanalyze the NBS Data Encryption Standard, Stanford University Report SEL76-042.

Herlestam, T. (1983). On the Complexity of Certain Crypto Generators, *Proceedings of IFIP/Sec 83*, North Holland, 305–308.

Jennings, S. M. (1983). Multiplexed Sequences: Some Properties of the Minimum Polynomial, *Cryptography: Proceedings, Burg Feuerstein 1982*, Springer-Verlag, 189–206.

Khinchin, A. (1957). *Mathematical Foundations of Information Theory*, Dover Publications.

Kimberley, M. E. (1986). Statistics in Cryptology, M.Sc. Dissertation, Brunel University.

Knuth, D. E. (1981). *The Art of Computer Programming. Volume 2: Seminumerical algorithms* (2nd edn), Addison-Wesley.

Koblitz, N. (1987). *A Course in Number Theory and Cryptography*, Springer–Verlag.

Konheim, A. G. (1981). *Cryptography: A Primer*, John Wiley and Sons.

Kranakis, E. (1986). *Primality and cryptography*, B. G. Teubner/John Wiley and Sons.

Massey, J. L. (1969). Shift-register Synthesis and BCH Decoding, *IEEE Trans. Inf. Theory* IT-**15**, 122–127.

Meyer, C. H. and Matyas, S. M. (1982). *Cryptography; A New Dimension in Computer Data Security*, John Wiley and Sons.

Mitchell, C. J. and Piper, F. C. (1987). The Cost of Reducing Key-storage Requirements in Secure Networks, *Computers and Security* **6**, 339–341.

NBS (1977). Data Encryption Standard, *Federal Information Processing Standards Publication 46*, National Bureau of Standards.

Niederreiter, H. (1988). The Probabilistic Theory of Linear Complexity, *Advances in Cryptology: Proceedings of Eurocrypt 88*, Springer-Verlag, 191–209.

Odlyzko, A. M. (1985). Discrete Logarithms in Finite Fields and their Cryptographic Significance, *Advances in Cryptology: Proceedings of Eurocrypt 84*, Springer-Verlag, 224–314.

Piper, F. C. and Walker, M. (1984). Binary Sequences and Hadamard Designs, *Pitman Research Notes in Mathematics* **114**, 67–76.

Pomerance, C., Smith, J. W. and Tuler, R. (1988). A Pipe-line Architecture for Factoring Large Integers with the Quadratic Sieve Algorithm, *SIAM J. Comput.* **17**, 387–403.

Rueppel, R. A. (1984). New Approaches to Stream Ciphers D.Sc. Dissertation, Swiss Federal Institute of Technology, Zurich.

Rueppel, R. A. (1986). *Analysis and Design of Stream Ciphers*, Springer-Verlag.

Schnorr, C. P. (1988). On the Construction of Random Number Generators and Random Function Generators, *Advances in Cryptology: Proceedings of Eurocrypt 88*, Springer-Verlag, 225–232.

Selmer, E. S. (1966). *Linear Recurrence Relations over Finite Fields*, Department of Mathematics, University of Bergen.

Shamir, A. (1981). On the Generation of Cryptographically Strong pseudo-random Sequences, *Proceedings of the 8th Colloquium on Automata, Languages and Programming, Acre, Israel*, Springer-Verlag, 544–550.

Shamir, A. (1982). A Polynomial Time Algorithm for Breaking the Basic Merkle–Hellman Cryptosystem, *Proceedings of the 23rd Annual Symposium on Foundations of Computer Science*, IEEE, 145–152.

Shamir, A. (1984). A Polynomial-time Algorithm for Breaking the Basic Merkle–Hellman Cryptosystem, *IEEE Trans. Inf. Theory* IT-**30**, 699–704.

Shannon, C. E. (1948). A Mathematical Theory of communication, *Bell System Technical Journal* **27**, 379–423 and 623–656.

Shannon, C. E. (1949). Communication Theory of Secrecy Systems, *Bell System Technical Journal* **28**, 656–715.

Siegenthaler, T. (1984). Correlation Immunity of Nonlinear Combining Functions for Cryptographic Applications, *IEEE Trans. Inf. Theory* IT-**30**, 776–780.

Simmons, G. (1985). Authentication Theory/Coding Theory, *Advances in Cryptology: Proceedings of Crypto 84*, Springer-Verlag, 411–431.

Smeets, B. (1986). A Note on Sequences Generated by Clock Controlled Shift Registers, *Advances in Cryptology: Proceedings of Eurocrypt 85*, Springer-Verlag, 142–148.

Welsh, D. (1988). *Codes and Cryptography*. Oxford University Press.

CHAPTER 4

Catalan Numbers and Their Various Uses

4.0. INTRODUCTION

Probably the most prominent among the special integers which arise in combinatorial contexts [see I, §3.10] are the binomial coefficients

$$\binom{n}{r}.$$

These have many uses and many different interpretations. Here we would like to stress one particular interpretation in terms of paths on the integral lattice in the coordinate plane. Thus a path is a sequence of points $P_0 P_1 \ldots P_m, m \geq 0$, where each P_i is a lattice point (that is, a point with *integer* coordinates) and $P_{i+1}, i \geq 0$, is obtained by stepping one unit east or one unit north of P_i. We say that this is a *path from P to Q* if $P_0 = P, P_m = Q$. It is now easy to see

THEOREM 4.0.1. *The number of paths from $(0,0)$ to (a,b) is the binomial coefficient*

$$\binom{a+b}{a}.$$

Proof. We may denote a path from $(0,0)$ to (a,b) as a sequence consisting of a *E*s (*E* for East) and b *N*s (*N* for North) in some order. Thus the number of such paths is the number of such sequences. A sequence consists of $(a+b)$ symbols and is determined when we have decided which a of the $(a+b)$ symbols should be *E*s. Thus we have

$$\binom{a+b}{a}$$

choices of symbol.

93

It should now be clear, by an evident translation of the origin, that if $c \leq a, d \leq b$, then the number of paths from $P(c,d)$ to $Q(a,b)$ is the binomial coefficient

$$\binom{a+b-c-d}{a-c}.$$

We will present another family of special integers, called the *Catalan numbers* (after the nineteenth-century Belgian mathematician Eugène Charles Catalan) which are closely related, both algebraically and conceptually, to the binomial coefficients. These Catalan numbers have many combinatorial interpretations, of which we will emphasize three, in addition to their interpretation in terms of paths on the integral lattice. Since all these interpretations remain valid if one makes a conceptually obvious generalization of the original definition of Catalan numbers, we will present these interpretations in this generalized form. Thus the definitions we are about to give depend on a parameter p, which is an integer ≥ 2. The original Catalan numbers—or, rather, characterizations of these numbers—are obtained by taking $p = 2$.

Let us then present three very natural combinatorial concepts, which arise in rather different contexts, but which turn out to be equivalent.

Let p be a fixed integer ≥ 2. Then we define three sequences of positive integers, depending on p, as follows:

a_k: $_pa_0 = 1$, $_pa_k = $ number of p-ary trees [see V, §6.2] with k source-nodes, $k \geq 1$.

b_k: $_pb_0 = 1$, $_pb_k = $ number of ways of associating k applications of a given p-ary operation, $k \geq 1$.

c_k: $_pc_0 = 1$, $_pc_k = $ number of ways of subdividing a convex polygon [see V, §1.1.4] into k disjoint $(p + 1)$-gons by means of non-intersecting diagonals, $k \geq 1$.

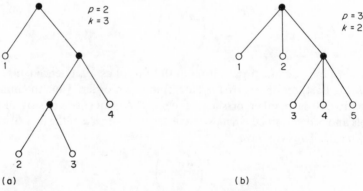

Figure 4.0.1. (a) A binary tree with 3 source-nodes ($\bullet$) and 4 end-nodes ($\bigcirc$) (b) A ternary tree with 2 source-nodes ($\bullet$) and 5 end-nodes ($\bigcirc$)

$$p = 2 \qquad\qquad\qquad\qquad\qquad\qquad\qquad p = 3$$
$$k = 3 \qquad\qquad\qquad\qquad\qquad\qquad\qquad k = 2$$

$(s_1((s_2 s_3)s_4))$ $\qquad\qquad\qquad\qquad\qquad\qquad (s_1 s_2(s_3 s_4 s_5))$

(a) $\qquad\qquad\qquad\qquad\qquad\qquad\qquad\qquad$ (b)

Figure 4.0.2. (a) An expression involving 3 applications of a binary operation applied to 4 symbols (b) An expression involving 2 applications of a ternary operation applied to 5 symbols

(a) $\qquad\qquad\qquad\qquad\qquad\qquad\qquad\qquad$ (b)

Figure 4.0.3. (a) A 5-gon subdivided into three 3-gons by 2 diagonals (b) A 6-gon subdivided into two 4-gons by 1 diagonal

(As indicated, we suppress the 'p' from the symbol if no ambiguity need be feared.) Note that, if $k \geq 1$,

  (i) a p-ary tree with k source-nodes has $(p-1)k + 1$ end-nodes and $pk + 1$ nodes in all (Figure 4.0.1);
  (ii) the k applications of a given p-ary operation are applied to a sequence of $(p-1)k + 1$ symbols (Figure 4.0.2);
  (iii) the polygon has $(p-1)k + 2$ sides and is subdivided into k disjoint $(p+1)$-gons by $(k-1)$ diagonals (Figure 4.0.3).

A well-known and easily proved result (for the case $p = 2$ see Sloane (1973)) is the following:

THEOREM 4.0.2. $_p a_k = {}_p b_k = {}_p c_k$.

The reader will probably have no trouble in seeing how the trees of Figure 4.0.1 are converted into the corresponding expressions of Figure 4.0.2 and vice versa. However, relating the trees of Figure 4.0.1 (or the parenthetical expressions of Figure 4.0.2) to the corresponding dissected polygons of Figure 4.0.3 is more subtle (and the hint given in Sloare (1973), in the case $p = 2$,

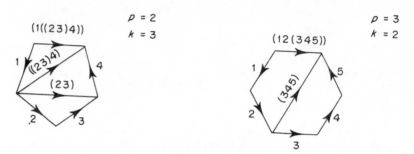

Figure 4.0.4. The dissected polygons of Figure 4.0.3 with diagonals and last side labelled

is somewhat cryptic), so we will indicate the proof that $b_k = c_k$. Thus, suppose that we are given a rule for associating k applications of a given p-ary operation to a string of $(p-1)k+1$ symbols $s_1, s_2, \ldots, s_{(p-1)k+1}$. Label the successive sides (in the anticlockwise direction) of the convex $[(p-1)k+2]$-gon with the integers from 1 to $(p-1)k+1$, leaving the top horizontal side unlabelled. Pick the first place along the expression (counting from the left) where a succession of p symbols is enclosed in parentheses. If the symbols enclosed run from s_{j+1} to s_{j+p}, draw a diagonal from the initial vertex of the $(j+1)$st side to the final vertex of the $(j+p)$th side, and label it $(j+1,\ldots,j+p)$. Also imagine the part $(s_{j+1}\ldots s_{j+p})$ replaced by a single symbol. We now have effectively reduced our work to a set of $(k-1)$ applications and our polygon to a set of two polygons, one a $(p+1)$-gon, the other a $[(p-1)(k-1)+2]$-gon, so we proceed inductively to complete the rule for introducing diagonals. Moreover, the eventual label for the last side will correspond precisely to the string of $(p-1)k+1$ symbols with the given rule for associating them, that is, to the original expression. Figure 4.0.4 illustrates the labelling of the dissections of the pentagon and hexagon that are determined by the corresponding expressions of Figure 4.0.2. The converse procedure, involving the same initial labelling of the original sides of a dissected polygon, and the successive labelling of the diagonals introduced, leads to an expression which acts as label for the last (horizontal) side.

We now illustrate a slightly more complicated situation.

EXAMPLE 4.0.1.   Let $p = 3, k = 4$ and consider the expression

$$(s_1(s_2 s_3(s_4(s_5 s_6 s_7)s_8))s_9).$$

The dissection of the convex 10-gon associated with this expression, with the appropriate labelling, is shown in Figure 4.0.5. The corresponding ternary tree is shown in Figure 4.0.6, which also demonstrates how the tree may be associated directly with the dissected polygon. The rule is 'Having entered a $(p+1)$-gon through one of its sides, we may exit by any of the other p sides'.

In the case $p = 2$, any of a_k, b_k, c_k may be taken as the definition of the kth

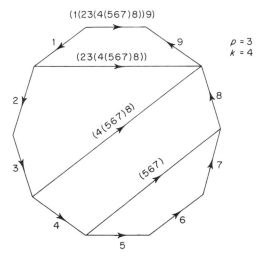

Figure 4.0.5. The polygonal dissection associated
with $(s_1(s_2s_3(s_4(s_5s_6s_7)s_8))s_9)$

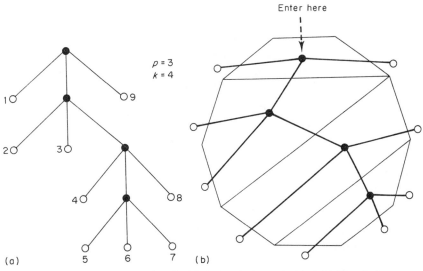

Figure 4.0.6. (a) The tree associated with $(s_1(s_2s_3(s_4(s_5s_6s_7)s_8))s_9)$ (b) Associating a tree
with a dissected polygon (in this case the tree shown in (a) and the polygon of
Figure 4.0.5) Note that ● = interior node = source node, ○ = exterior node = end-
node

Catalan number. Thus we may regard any of $_pa_k, _pb_k, _pc_k$ as defining the *generalized kth Catalan number.* Moreover, it is known (see Klarner (1970)) that

THEOREM 4.0.3.

$$_pa_k = \frac{1}{k}\binom{pk}{k-1} = \frac{1}{(p-1)k+1}\binom{pk}{k} \quad (k \geq 1).$$

The 'classical' proof of this result is rather sophisticated. One of our principal aims here is to give an elementary proof of Theorem 4.0.3, based on yet a fourth interpretation of the generalized Catalan numbers in terms of paths on the integral lattice. We lay particular stress on the flexibility of this fourth interpretation, which we now describe.

We say that a path from P to Q is *p-good* if it lies entirely below the line $y = (p-1)x$. Let $d_k = _pd_k$ be the number of p-good paths from $(0, -1)$ to $(k, (p-1)k-1)$. (Thus, by our convention, $d_0 = 1$.) We extend Theorem 4.0.2 to assert

THEOREM 4.0.4. $_pa_k = _pb_k = _pc_k = _pd_k.$

Proof.   First we restate (with some precision) the definitions of b_k and d_k, which we will prove equal.

b_k = number of expressions involving k applications of a p-ary operation (on a word of $k(p-1)+1$ symbols). Let E be the set of all such expressions.

d_k = number of good paths from $(0, -1)$ to $(k, (p-1)k-1)$; by following such a path by a single vertical segment, we may identify such paths with certain paths (which we will also call 'good') from $(0, -1)$ to $(k, (p-1)k)$. Let P be the set of all such paths.

We set up functions $\Phi: E \to P$, $\Psi: P \to E$, which are obviously mutual inverses. Thus Φ is obtained by removing all final parentheses* and (reading the resulting expression from left to right) interpreting a parenthesis as a horizontal move and a symbol as a vertical move. The inverse rule is Ψ. It remains to show (i) that Im Φ consists of elements of P (and not merely of arbitrary paths from $(0, -1)$ to $(k, (p-1)k)$); and (ii) that Im Ψ consists of elements of E (that is, of well-formed, meaningful expressions).

(i) Let E be an expression. We argue by induction on k that ΦE is a good path. Plainly ΦE is a path from $(0, -1)$ to $(k, (p-1)k)$, so we have only to

---

*This converts our expression into a meaningful reverse Polish expression (MRPE) (see Hillman *et al.* (1987)). Note that the closing parentheses are, strictly speaking, superfluous, provided that we know that we are dealing with a p-ary operation. It is not even necessary to know this if we are sure that the expression is well-formed, since the 'arity' is $((\sigma - 1)/\kappa) + 1$, where σ is the number of symbols and κ is the number of opening parentheses.

prove it good. This obviously holds if $k = 1$. Let $k \geq 2$ and let E involve k applications. There must occur somewhere in E the section

$$(s_{j+1}s_{j+2}\ldots s_{j+p})$$

Call this the *key* section. We wish to show that if (u, v) is a point on the path ΦE which is *not* the endpoint, then $v < (p - 1)u$. Let E' be the expression obtained from E by replacing the key section by s. Then E' involves $(k - 1)$ applications. Now if A is the point on ΦE corresponding to the parenthesis of the key section, then the inductive hypothesis tells us that $v < (p - 1)u$ if (u, v) occurs prior to A. Assume now that (u, v) is not prior to A. There are then two possibilities:

If E ends with s_{j+p}, so that E' ends with s, then $\exists(u', v')$, the last point of $\Phi E'$, such that $u' = u - 1$, $v' > v - p + 1$, $v' = (p - 1)u'$. Hence $v - p + 1 < (p - 1)(u - 1)$, so $v < (p - 1)u$.

If E does not end with s_{j+p}, so that E' does not end with s, then $\exists(u', v')$, not the last point of $\Phi E'$, such that $u' = u - 1$, $v' \geq v - p + 1$, $v' < (p - 1)u'$ (by the inductive hypothesis). Hence $v - p + 1 < (p - 1)(u - 1)$, so $v < (p - 1)u$.

This completes the induction in the Φ-direction. Figure 4.0.7, which gives a special, but not particular, case, may be helpful in following the argument.

(ii) Let $\mathscr{P}$ be a good path to $(k, (p - 1)k)$. We argue by induction on k that $\Psi\mathscr{P}$ is a well-formed expression. This holds if $k = 1$, since then there is only one good path $\mathscr{P}$ and $\Psi\mathscr{P}$ is $(s_1 s_2 \ldots s_p$. Let $k \geq 2$. Then since the path climbs altogether $k(p - 1) + 1$ places in k jumps, there must be a jump somewhere (not at the initial point) of $\geq p$ places. Let A be a point on $\mathscr{P}$ where the path takes one horizontal step followed by p vertical steps, bringing it to C. Let B be the point one step above A (so that B is not on $\mathscr{P}$). Then B is on $y = (p - 1)x$ if and only if C is on $y = (p - 1)x$, that is, if and only if C is the endpoint of $\mathscr{P}$; otherwise B is below $y = (p - 1)x$. Let $\mathscr{P}_1, \mathscr{P}_2$ be the parts of $\mathscr{P}$ ending in A and beginning in C respectively. Let $\mathscr{P}'_2$ be the translate of $\mathscr{P}_2$, given by

$$u' = u - 1, \qquad v' = v - (p - 1),$$

and let $\mathscr{P}'$ be the path consisting of $\mathscr{P}_1$, followed by AB, followed by $\mathscr{P}'_2$. It is trivial that $\mathscr{P}'$ is a good path to $(k - 1, (p - 1)(k - 1))$. Thus, by the inductive hypothesis, $\Psi\mathscr{P}'$ is a well-formed expression involving $(k - 1)$ applications. Let s be the symbol in $\Psi\mathscr{P}'$ corresponding to AB. Replace s by $(s_{j+1}s_{j+2}\ldots s_{j+p})$, where these symbols do not occur in $\Psi\mathscr{P}'$. Then the new expression is well-formed and is obviously $\Psi\mathscr{P}$. This completes the induction in the Ψ-direction so that Theorem 4.0.3 is established.

We then proceed to calculate d_k. We will base our proof that

$$_p d_k = \frac{1}{k}\binom{pk}{k - 1}, \quad k \geq 1,$$

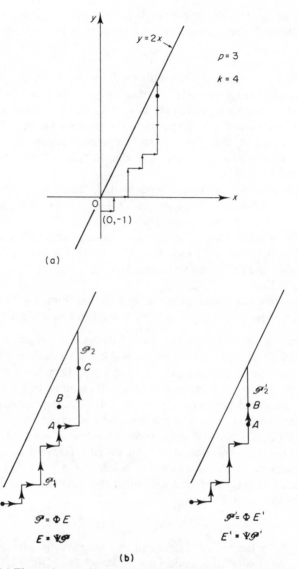

Figure 4.0.7. (a) The path associated with the expression $E = (s_1(s_2s_3(s_4(s_5s_6s_7)s_8))s_9)$
(b) The inductive step in either direction

on a *generalized Jonah formula* (Theorem 4.2.1), relating (generalized) binomial coefficients to certain generalizations of the quantities $_pd_k$.

Thus $_pd_{qk}, q \le p - 1, k \ge 1$, is defined as the *number of p-good paths from* $(1, q - 1)$ *to* $(k, (p - 1)k - 1)$, so that it follows immediately that $_pd_{0k} = {}_pd_k$. The original Jonah formula concerned the case $p = 2$ and was restricted to $q = 0$ (and the parameter n appearing in the formula was restricted to being an integer

$\geq 2k$ instead of an arbitrary real number). The quantities $_p d_{qk}$ play a crucial role in counting p-good paths and it turns out to be no more difficult to calculate d_{qk} than to calculate d_k. The counting of p-good paths is achieved in Section 4.3.

By considering the last application of the given p-ary operation, it is clear that $_p b_k$ satisfies the recurrence relation (generalizing the familiar formula in the case $p = 2$)

$$_p b_k = \sum_{i_1 + i_2 + \ldots + i_p = k - 1} {_p b_{i_1}} {_p b_{i_2}} \cdots {_p b_{i_p}}, \quad k \geq 1. \tag{4.0.1}$$

In Section 4.2 we also enunciate a similar recurrence relation for $_p d_k$, thus providing an alternative proof that $_p b_k = {_p d_k}$. We show in outline how (4.0.1) leads to a proof of Theorem 4.0.3 via generating functions and the theory of Bürmann–Lagrange series (see Hurwitz and Courant (1922)).

The case $p = 2$, dealing with the Catalan numbers as originally defined, is discussed in Section 4.1. In that special case we have available an elegant proof that

$$_2 d_k = \frac{1}{k} \binom{2k}{k-1}, \quad k \geq 1,$$

using André's Reflection Method (André (1887)); however, we have not discovered in the literature a means of generalizing this method effectively.

The authors wish to thank Richard Guy, Richard Johnsonbaugh, Hudson Kronk and Tom Zaslavsky for very helpful conversations and communications during the preparation of this article.

4.1.   CATALAN NUMBERS AND THE BALLOT PROBLEM

In the late nineteenth century the so-called *ballot problem* was exercising the minds of mathematicians and probabilists. We suppose an election is held in which there are two candidates, A and B, and we further suppose that A receives a votes, B receives b votes with $a > b$, so that A is elected. The question raised is this—what is the probability that, throughout the counting of the votes, A stays ahead of B? This translates easily to a question about paths on the integral lattice in the coordinate plane. Each path from $(0, 0)$ to (a, b) represents a possible sequence of counting of the votes. Now there are

$$\binom{a + b}{a}$$

such paths, so we need to know how many of these paths remain below the line $y = x$ except at the initial point $(0, 0)$. This number is the number of 2-good paths from $(1, 0)$ to (a, b).

The problem was first solved by Bertrand, but a very elegant solution was given in 1887 by the French mathematician Désiré André (see André (1887), Hilton and Pedersen (1989)). In fact, his method allows one easily to count the number of 2-good paths from (c, d) to (a, b), where (c, d), (a, b) are any two lattice points below the line $y = x$. Since $p = 2$ throughout this discussion, we will speak simply of *good* paths; any path from (c, d) to (a, b) in the complementary set will be called a *bad* path. We will assume from the outset that both good paths and bad paths exist; it is easy to see that this is equivalent to assuming

$$d < c \leq b < a.$$

(Note that this condition is certainly satisfied in our ballot problem, provided the losing candidate receives at least one vote!)

Now the number of paths from (c, d) to (a, b) is the binomial coefficient

$$\binom{(a + b) - (c + d)}{a - c};$$

thus it suffices to count the number of *bad* paths from (c, d) to (a, b); let us call those points P and Q. If $\mathscr{P}$ is a bad path, let it first make contact with the line $y = x$ at F; let $\mathscr{P}_1, \mathscr{P}_2$ denote the subpaths PF, FQ, so that, using juxtaposition for path composition, $\mathscr{P} = \mathscr{P}_1 \mathscr{P}_2$; and let $\bar{\mathscr{P}}_1$ be the path obtained from $\mathscr{P}_1$ by *reflecting in the line* $y = x$. Then, if $\bar{\mathscr{P}} = \bar{\mathscr{P}}_1 \mathscr{P}_2$, $\bar{\mathscr{P}}$ is a path from $\bar{P}(d, c)$ to $Q(a, b)$; and the rule $\mathscr{P} \mapsto \bar{\mathscr{P}}$ sets up a one-to-one correspondence between the set of bad paths from P to Q and the set of paths from $\bar{P}$ to Q. It follows that there are

$$\binom{(a + b) - (c + d)}{a - d}$$

bad paths from P to Q, and hence

$$\binom{(a + b) - (c + d)}{a - c} - \binom{(a + b) - (c + d)}{a - d} \tag{4.1.1}$$

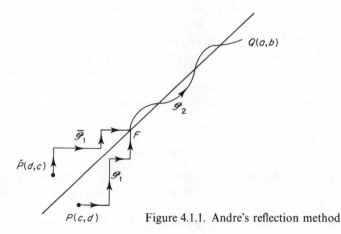

Figure 4.1.1. Andre's reflection method

good paths from P to Q. The argument, illustrated in Figure 4.1.1, is called *André's reflection method.*

Notice, in particular, that if $c = 1$, $d = 0$, this gives the number of good paths from $(1, 0)$ to (a, b) as

$$\binom{a+b-1}{a-1} - \binom{a+b-1}{a} = \frac{a-b}{a+b}\binom{a+b}{a}.$$

Thus the probability that a path from $(0, 0)$ to (a, b) proceeds first to $(1, 0)$ and then continues as a good path to (a, b) is $(a - b)/(a + b)$; this then is the solution to the ballot problem.

We return now to Catalan numbers. The kth Catalan number c_k is expressible as $_2d_k$, the number of good paths from $(0, -1)$ to $(k, k - 1)$. According to (4.1.1) this is

$$\binom{2k}{k} - \binom{2k}{k+1} = \frac{1}{k}\binom{2k}{k-1}.$$

Thus the kth Catalan number c_k is given by the formula

$$c_0 = 1, \quad c_k = \frac{1}{k}\binom{2k}{k-1} \quad (k \geq 1). \tag{4.1.2}$$

Notice that an alternative expression is

$$c_k = \frac{1}{k+1}\binom{2k}{k} \quad (k \geq 0). \tag{4.1.3}$$

Notice, too, that by translation c_k may also be interpreted as the number of good paths from $(1, 0)$ to $(k + 1, k)$.

André's reflection method does not seem to be readily applicable to obtaining a formula corresponding to (4.1.2) in the case of a general $p \geq 2$. Indeed, in the next section, where we calculate $_pc_k, p \geq 2$, we do *not* obtain a general closed formula for the number of p-good paths from (c, d) to (a, b). We calculate explicitly the number $_pd_{qk}$ of p-good paths from $(1, q - 1)$ to $P_k(k, (p - 1)k - 1)$ for $q \leq p - 1$. This enables us, by translation, to calculate the number of p-good paths from any lattice point C below the line $y = (p - 1)x$ to P_k. For if $C = (c, d)$, then the number of p-good paths from C to P_k is the number of p-good paths from $(1, q - 1)$ to P_{k-c+1}, i.e. $_pd_{q,k-c+1}$, where $q = d + 1 - (p - 1)(c - 1)$. On the other hand, although a convenient explicit formula when the terminal of the path is an arbitrary lattice point below the line $y = (p - 1)x$ does not seem available, we will nevertheless derive in Section 4.3 an *algorithmic formula* in the form of a finite sum, depending on the quantities d_{qk}.

The standard procedure for calculating c_k (we here revert to the original case $p = 2$) is based on the evident recurrence relation

$$c_k = \sum_{i+j=k-1} c_i c_j \quad (k \geq 1, c_0 = 1). \tag{4.1.4}$$

This relation is most easily seen by considering b_k and concentrating on the *last* application of a given binary operation within a certain expression. If we form the power series

$$S(x) = \sum_{k=0}^{\infty} c_k x^k, \tag{4.1.5}$$

then (4.1.4) shows that S satisfies

$$xS^2 - S + 1 = 0, \qquad S(0) = 1,$$

so that

$$S = \frac{1 - \sqrt{1 - 4x}}{2x}. \tag{4.1.6}$$

It is now straightforward to expand (4.1.6) and thus to obtain confirmation of the value of c_k already obtained in (4.1.2). As we will demonstrate, this analytical method for calculating c_k does generalize, but the generalization is highly sophisticated and we prefer to emphasize a much more elementary, combinatorial argument to calculate $_p c_k$.

4.2.   GENERALIZED CATALAN NUMBERS

Let p be a fixed integer ≥ 2, and let P_k be the point $(k, (p-1)k - 1)$, $k \geq 0$. We define the numbers $_p d_{qk} = d_{qk}$, as follows.

DEFINITION 4.2.1.   Let $q \leq p - 1$. Then $d_{q0} = 1$ and d_{qk} is the number of p-good paths from $(1, q - 1)$ to P_k, if $k \geq 1$.

We note that, as mentioned in the Introduction, we have

$$_p c_k = {}_p d_{0k} \quad (k \geq 0), \tag{4.2.1}$$

where $_p c_k$ is the kth (generalized) Catalan number. For, if $k \geq 1$, every p-good path from $(0, -1)$ to P_k must first proceed to $(1, -1)$. We further note that we have the relation

$$d_{p-1,k} = d_{0,k-1} \quad (k \geq 1); \tag{4.2.2}$$

for an easy translation argument shows that the number of p-good paths from $(1, p - 2)$ to P_k is the same as the number of p-good paths from $(0, -1)$ to P_{k-1}. Thus $d_{p-1,1} = 1 = d_{00}$ if $k = 1$, and, if $k \geq 2$, the number of p-good paths from $(0, -1)$ to P_{k-1} is, as stated above, the number of p-good paths from $(1, -1)$ to P_{k-1}, that is, $d_{0,k-1}$.

We may write $_p d_k$ for $_p d_{0k}$, so that, if $k \geq 0$,

$$_2 d_k = c_k, \tag{4.2.3}$$

the kth Catalan number, and

$$_pC_k = {_pd_k} = {_pd_{0,k}} = {_pd_{p-1,k+1}}, \tag{4.2.4}$$

the generalized kth Catalan number. We will also suppress the 'p' from these symbols and from the term 'p-good' if no ambiguity need be feared.

We now enunciate two fundamental properties of the numbers d_{qk}, for a fixed $p \geq 2$.

Recall that, for any *real* number n, the binomial coefficient

$$\binom{n}{0}$$

may be interpreted as 1, and the binomial coefficient

$$\binom{n}{r}$$

may be interpreted as the expression

$$\frac{n(n-1)\ldots(n-r+1)}{r!},$$

provided r is a positive integer (see, for example, Hilton and Pedersen (1989), also [I, 3.10]). Using this interpretation, we prove

THEOREM 4.2.1.   (Generalized Jonah Formula) *Let n be any real number. Then, if $k \geq 1$, and $q \leq p-1$,*

$$\binom{n-q}{k-1} = \sum_{i=1}^{k} d_{qi} \binom{n-pi}{k-i}. \tag{4.2.5}$$

Proof.   We first assume that n is an integer $\geq pk$, so that $(k, n-k)$ is a lattice point not below the line $y = (p-1)x$, and partition the paths from $(1, q-1)$ to $(k, n-k)$ according to where they first meet the line $y = (p-1)x$. The number of these paths which first meet this line where $x = i$ is $\gamma_i \delta_i$, where

γ_i = number of paths from $(1, q-1)$ to $(i, (p-1)i)$ which stay below

$y = (p-1)x$ except at the endpoint

= number of good paths from $(1, q-1)$ to P_i

= d_{qi}

and

δ_i = number of paths from $(i, (p-1)i)$ to $(k, n-k)$

$= \binom{n-pi}{k-i}.$

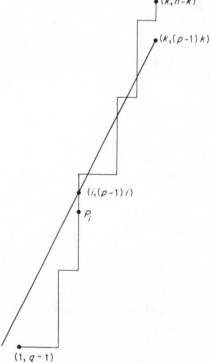

Figure 4.2.1 Partitioning   paths   from $(1, q-1)$ to $(k, n-k)$ according to where they first meet the line $y = (p-1)x$

Since i ranges from 1 to k and there are

$$\binom{n-q}{k-1}$$

paths in all from $(1, q-1)$ to $(k, n-k)$, formula (4.2.5) is proved in this case (see Figure 4.2.1).

We now observe that each side of (4.2.5) is a polynomial in n over the rationals $\mathbb{Q}$ of degree $(k-1)$, and we have proved that these two polynomials agree for infinitely many values of n. They therefore agree for all *real* values of n.

We may use Theorem 4.2.1 to calculate the value of d_{qk}; recall that $q \leq p-1$.

THEOREM 4.2.2.   *If* $k \geq 1$,

$$d_{qk} = \frac{p-q}{pk-q}\binom{pk-q}{k-1}.$$ (4.2.6)

Proof. Since, obviously, $d_{q1} = 1$, formula (4.2.6) holds if $k = 1$. We therefore assume $k \geq 2$. We first substitute* $n = pk - 1$ into (4.2.5), obtaining

$$\binom{pk-q-1}{k-1} = \sum_{i=1}^{k} d_{qi}\binom{pk-pi-1}{k-i}$$

We next substitute $n = pk - 1$ in (4.2.5), but now replace k by $(k-1)$, obtaining

$$\binom{pk-q-1}{k-2} = \sum_{i=1}^{k-1} d_{qi}\binom{pk-pi-1}{k-i-1}.$$

(Recall that $k \geq 2$.) Now we have the universal identity

$$\binom{n}{r+1} = \frac{n-r}{r+1}\binom{n}{r}.$$

Hence

$$\binom{pk-pi-1}{k-i} = \frac{pk-pi-k+i}{k-i}\binom{pk-pi-1}{k-i-1} = (p-1)\binom{pk-pi-1}{k-i-1},$$

so that

$$\binom{pk-q-1}{k-1} - (p-1)\binom{pk-q-1}{k-2} = d_{qk}\binom{-1}{0} = d_{qk}.$$

Finally,

$$\binom{pk-q-1}{k-1} - (p-1)\binom{pk-q-1}{k-2} = \binom{pk-q}{k-1}\left[\frac{pk-q-k+1}{pk-q} - \frac{(p-1)(k-1)}{pk-q}\right]$$

$$= \frac{p-q}{pk-q}\binom{pk-q}{k-1},$$

and (4.2.6) is proved.

Note that nothing prevents us from taking q negative in Theorems 4.2.1 and 4.2.2. Taking $q = 0$ we obtain the values of the generalized Catalan numbers, namely,

COROLLARY 4.2.3.

$$_p d_0 = 1 \quad \text{and} \quad _p d_k = \frac{1}{k}\binom{pk}{k-1} \quad (k \geq 1).$$

We come now to the second fundamental property of the numbers d_{qk}, for a fixed $p \geq 2$.

---

*It is amusing to note that the *combinatorial* argument in the proof of Theorem 4.2.1 required $n \geq pk$, and we are here substituting a value of n less than pk.

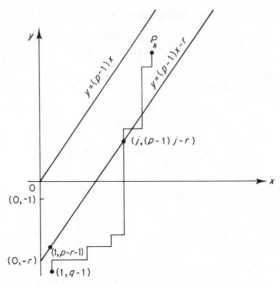

Figure 4.2.2. Proof of Theorem 4.2.4

THEOREM 4.2.4.   (Recurrence relation for d_{qk}) *Choose a fixed* $r \geq 1$. *Then, if* $k \geq 1$, *and* $q < p - r$,

$$d_{qk} = \sum_{i,j} d_{p-r,i} d_{q+r,j}, \quad \text{where } i \geq 1, j \geq 1, i+j = k+1. \quad (4.2.7)$$

Proof.   The proof proceeds by partitioning the good paths from $(1, q - 1)$ to P_k according to where they first meet the line $y = (p - 1)x - r$ (see Figure 4.2.2). We suppress the details of the argument, which resemble those for Theorem 4.2.1. However, notice that formula (4.2.7) generalizes the familiar recurrence relation (4.1.4),

$$c_k = \sum_{i+j=k-1} c_i c_j \quad (k \geq 1),$$

for the Catalan numbers, in the light of (4.2.3) and (4.2.4)—we have merely to substitute $p = 2$, $q = 0$, $r = 1$. In fact, we may use Theorem 4.2.4 to express d_{qk} in terms of the generalized Catalan numbers $_p d_i$, at the same time generalizing (4.2.4). Specifically,

THEOREM 4.2.5.   *If* $q \leq p - 1$, *and if* $k \geq 1$, *then*

$$d_{qk} = \sum_{i_1 + i_2 + \ldots + i_{p-q} = k-1} {}_p d_{i_1} {}_p d_{i_2} \ldots {}_p d_{i_{p-q}}. \quad (4.2.8)$$

Proof.   The case $q = p - 1$ is just (4.2.4), so we assume $q < p - 1$. Then, by

Theorem 4.2.4, with $r = 1$,

$$d_{qk} = \sum_{i,j} d_{p-1,i} d_{q+1,j}, \quad \text{where } i \geq 1, j \geq 1, i + j = k + 1$$

$$= \sum_{i_1,j_2} {}_p d_{i_1} d_{q+1,j_2}, \quad \text{where } j_2 \geq 1, i_1 + j_2 = k,$$

using (4.2.4). Iterating this formula, we obtain

$$d_{qk} = \sum_{i_1 + i_2 + \ldots + i_{p-q-1} + j_{p-q} = k} {}_p d_{i_1}\, {}_p d_{i_2} \cdots {}_p d_{i_{p-q-1}}\, {}_p d_{p-1,j_{p-q}}, \quad \text{where } j_{p-q} \geq 1.$$

A second application of (4.2.4) yields the formula (4.2.8).

Setting $q = 0$ in (4.2.8), and again using (4.2.4), yields

COROLLARY 4.2.6. *If $k \geq 1$,*

$$_p d_k = \sum_{i_1 + i_2 + \ldots + i_p = k - 1} {}_p d_{i_1}\, {}_p d_{i_2} \cdots {}_p d_{i_p}.$$

Since the numbers $_p b_k$ are easily seen to satisfy the same relation, Corollary 4.2.6 provides another proof of the equality of $_p d_k$ with the kth generalized Catalan number. It also shows us that, if $S(x)$ is the power series

$$S(x) = \sum_{k=0}^{\infty} {}_p d_k x^k,$$

then

$$x S^p = S - 1. \tag{4.2.9}$$

Klarner (1970) attributes to Pólya and Szegö (1954) the observation that we may invert (4.2.9) to obtain

$$S(x) = 1 + \sum_{k=1}^{\infty} \frac{1}{k} \binom{pk}{k-1} x^k;$$

indeed, this is the solution given to Problem 211 in Polya and Szegö (1954), p. 125, based on the theory of Bürmann–Lagrange series. Thus Corollary 4.2.6 leads to a different, but far more sophisticated, proof of the values of the generalized Catalan numbers.

Note that these values could, of course, have been obtained without introducing the quantities $_p d_{qk}$ for $q \neq 0$ (see Corollary 4.2.3). However, these quantities have an obvious combinatorial significance and satisfy a recurrence relation (Theorem 4.2.4) much simpler than that of Corollary 4.2.6. In the light of (4.2.1) the quantities $_p d_{qk}$ themselves deserve to be regarded as generalizations of the Catalan numbers—though certainly not the ultimate generalization! We will see in the next section that they are useful in certain important counting algorithms.

4.3. COUNTING *p*-GOOD PATHS

We will give, in this section, an algorithm for counting the number of *p*-good paths from (c, d) to (a, b), each of those lattice points being assumed below the line $y = (p - 1)x$. As explained in Section 4.1, it suffices (via a translation) to replace (c, d) by the point $(1, q - 1)$, with $q \leq p - 1$. Further, in order to blend our notation with that of Section 4.2, we write $(k, n - k)$ for (a, b), so that $n < pk$. We assume there are paths from $(1, q - 1)$ to $(k, n - k)$, i.e. that $k \geq 1$, $n - k \geq q - 1$.

Of course, there are

$$\binom{n - q}{k - 1}$$

paths from $(1, q - 1)$ to $(k, n - k)$. It is, moreover, easy to see that there exist *p*-bad paths from $(1, q - 1)$ to $(k, n - k)$ if and only if $n - k \geq p - 1$, so we assume this—otherwise, there are only *p*-good paths and we have our formula, namely,

$$\binom{n - q}{k - 1}.$$

To sum up, we are counting the *p*-good paths from $(1, q - 1)$ to $(k, n - k)$ under the (non-trivializing) assumption that $1 \leq k \leq n - p + 1 \leq p(k - 1)$; notice that this composite inequality implies that, in fact, $k \geq 2$, so that we assume

$$2 \leq k \leq n - p + 1 \leq p(k - 1). \tag{4.3.1}$$

There are

$$\binom{n - q}{k - 1}$$

paths in all, so we count the *p*-bad paths. Let $l = [(n - k)/(p - 1)]$, where $[x]$ is the integral part of x. Then (4.3.1) ensures that $l \geq 1$, and a *p*-bad path from $(1, q - 1)$ to $(k, n - k)$ will meet the line $y = (p - 1)x$ first at some point $Q_j(j, (p - 1)j)$, where $j = 1, 2, \ldots, l$.

Since there are d_{qj} path from $(1, q - 1)$ to Q_j which stay below the line $y = (p - 1)x$ except at Q_j, and since there are

$$\binom{n - pj}{k - j}$$

paths from Q_j to $(k, n - k)$, it follows that there are

$$d_{qj}\binom{n - pj}{k - j}$$

bad paths from $(1, q - 1)$ to $(k, n - k)$ which first meet the line $y = (p - 1)x$ at

Q_j. Thus the number of p-bad paths from $(1, q-1)$ to $(k, n-k)$ is

$$\sum_{j=1}^{l} d_{qj}\binom{n-pj}{k-j},$$

(see Figure 4.3.1).

We have proved

THEOREM 4.3.1.   *The number of p-good paths from $(1, q-1)$ to $(k, n-k)$, under the conditions (4.3.1), is*

$$\binom{n-q}{k-1} - \sum_{j=1}^{l} d_{qj}\binom{n-pj}{k-j}, \quad \text{where } l = \left[\frac{n-k}{p-1}\right].$$

We may express this differently, however, and in a way which will be more convenient if l is relatively large. For if we invoke Theorem 4.2.1 we see that

COROLLARY 4.3.2.   *The number of p-good paths from $(1, q-1)$ to $(k, n-k)$ under the conditions (4.3.1), is*

$$\sum_{j=l+1}^{k} d_{qj}\binom{n-pj}{k-j}, \quad \text{where } l = \left[\frac{n-k}{p-1}\right].$$

It is interesting to observe that the binomial coefficients entering into the formula for Corollary 4.3.2 are certainly 'generalized', since $k-j > n-pj$ if $j > l$. Thus if $l+1 \le j \le k$, then

$$\binom{n-pj}{k-j} = 0,$$

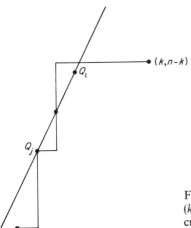

Figure 4.3.1. A bad path from $(1, q-1)$ to $(k, n-k)$, $n < pk$. Notice that Q_l is the last possible crossing point of the line $y = (p-1)x$ for such a path

so long as $n - pj \geq 0$, i.e. $j \leq [n/p]$. Since $n/p > (n-k)/(p-1)$, it follows that $[n/p] \geq [(n-k)/(p-1)]$, so that we may improve Corollary 4.3.2, at least formally, by replacing l by $m = [n/p]$. Thus

COROLLARY 4.3.3.   *The number of p-good paths from* $(1, q-1)$ *to* $(k, n-k)$, *under the conditions* (4.3.1), *is*

$$\sum_{j=m+1}^{k} d_{qj} \binom{n-pj}{k-j}, \quad \text{where } m = \left[\frac{n}{p}\right].$$

We give an example.

EXAMPLE 4.3.1.   Let $q = 0$, $k = 5$, $p = 3$, $n = 9$. The inequalities (4.3.1) are certainly satisfied and $m = 3$. Thus the number of 3-good paths from $(1, -1)$ to $(5, 4)$ is

$$d_{04} \binom{-3}{1} + d_{05} \binom{-6}{0} = -3d_{04} + d_{05}.$$

Now d_{0j} is the jth (generalized) Catalan number for $p = 3$; thus, by Corollary 4.2.3,

$$d_{04} = \frac{1}{4} \binom{12}{3} = 55, \qquad d_{05} = \frac{1}{5} \binom{15}{4} = 273.$$

Thus the number of 3-good paths is 108. Notice that the total number of paths is

$$\binom{9}{4} = 126.$$

Notice also that, in this case, $l = [(n-k)/(p-1)] = 2$, so that there is an advantage in replacing l by m.

   Circumstances will determine whether it is more convenient to use Theorem 4.3.1 or Corollary 4.3.3. In the example above there is little to choose. However, whereas in Theorem 4.3.1 each term has an obvious combinatorial meaning, it is difficult to assign a combinatorial meaning to the terms in Corollary 4.3.3; the binomial coefficient

$$\binom{n-pj}{k-j} \quad (j \geq m+1),$$

does not count the paths from Q_j to $(k, n-k)$, since there are none—might it be said to count 'phantom paths'?

Historical note

   The versatile Belgian mathematician Eugène Charles Catalan (1814–1894) defined the numbers named after him in connection with his solution of the problem of dissecting a polygon, by means of non-intersecting diagonals, into triangles; thus his definition of the kth Catalan number is our c_k. In fact, the problem of enumerating such dissections had already been solved, but not in convenient or perspicuous form, by Segner in the eighteenth century, and Euler (see Euler (1758/9) in the Bibliography) had immediately obtained a simpler solution involving generating functions—we have sketched this approach in Section 4.1. Catalan's own solution, achieved almost simultaneously with that of Binet around 1838, was even simpler, not requiring the theory of generating functions, which was regarded at the time as rather 'delicate' in view of its (apparent) dependence on the notion of convergence of series.

A host of other interpretations of the Catalan numbers have been given, largely inspired by work in combinatorics and graph theory, and have been collected by Gould (see Gould (1979) in the Bibliography). These interpretations have led to generalizations, of which the most obvious and important, treated in this article, have been studied by probabilists (e.g. Feller) and others. The ballot problem itself leads to an evident generalization, considered by Feller, in which the line $y = x$ is replaced by the line $y = (p - 1)x$.

An interesting generalization of a somewhat different kind is treated by Wenchang Chu (see Wenchang Chu (1987) in the Bibliography).

P. H. and J. P.

REFERENCES

André, D. (1887). Solution directe du problème résolu par M. Bertrand, *Comptes Rendus* **105**, 436–437.

Hillman, A. P., Alexanderson, G. L. and Grassl, R. M. (1987). *Discrete and Combinatorial Mathematics*, Dellen/Macmillan.

Hilton, Peter, and Pedersen, Jean (1989). Extending the Binomial Coefficients to Preserve Symmetry and Pattern, in *Symmetry 2: Unifying Human Understanding*, Pergamon Press.

Hilton, Peter, and Pedersen Jean (1990). The Ballot Problem and Catalan Numbers, *Nieuw Archief voor Wiskunde* (to appear).

Hurwitz, A. and Courant, R. (1922). *Allgemeine Funktionentheorie und elliptische Funktionen. Geometrische Funktionentheorie*, Springer, 128.

Klarner, David A. (1970). Correspondences between Plane Trees and Binary Sequences, *J. of Comb. Theory* **9**, 401–411.

Pólya, G. and Szegö, G. (1954). *Aufgaben und Lehrsätze aus der Analysis*, Vol. I, Springer, 125.

Sloane, N. J. A. (1973). *A Handbook of Integer Sequences*, Academic Press.

BIBLIOGRAPHY

[Hillman, Alexanderson and Grassel (1987), Klarner (1970) and Sloane (1973) in the References may also be regarded as general sources on Catalan numbers.]

Alter, R. (1971). Some Remarks and Results on Catalan Numbers, *Proc. 2nd Louisiana Conf. on Comb., Graph Theory and Comp.* 109–132.

André, D. (1887). Solution directe du problème résolu par M. Bertrand, *Comptes Rendus* **105**, 436–437.

Brown, W. G. (1965). Historical Note on a Recurrent Combinatorial Problem, *Amer. Math. Monthly* **72**, 973–977.

Chorneyko, I. Z. and Mohanty, S. G. (1975). On the Enumeration of Certain Sets of Planted Trees, *J. Combin. Theory Ser.* B **18**, 209–221.

Wenchang Chu, (1987). A New Combinatorial Interpretation for Generalized Catalan Numbers, *Discrete Math.* **65**, 91–94.

Chung, K. L. and Feller, W. (1949). On Fluctuations in Coin-tossing, *Proc. Nat. Acad. Sci., U.S.A.* **35**, 605–608.

Cong, T. T. and Sato, M. (1982). One-dimensional Random Walk with Unequal Step Lengths Restricted by an Absorbing Barrier, *Discrete Math.* **40**, 153–162.

Conway, John H. and Guy, Richard K. (1989). *The Book of Numbers*, Sci. Amer. Library, W. H. Freeman.

Dershowitz, N. and Zaks, S. (1980). Enumeration of Ordered Trees, *Discrete Math.* **31**, 9–28.

Donoghey, R. and Shapiro, L. W. (1977). Motzkin Numbers, *J. Combin. Theory Ser.* **A23**, 291–301.

Eggleton, Roger B. and Guy, Richard K. (1988). Catalan Strikes Again! How Likely is a Function to be Convex?, *Math. Mag.* **61**, 211–219.

Erdélyi, A. and Etherington, I. M. H. (1941). Some Problems of Non-associative Combinatorics, II MR4 p. 68 (D43) (1941) *Edinburgh Math. Notes* **32**, 7–12.

Euler, L. (1758/9). Novi Commentarii Academiae Scientarium Imperialis Petropolitanque **7**, 13–14.

Good, I. J. (1960). Generalizations to Several Variables of Lagrange's Expansion, *Proc. Camb. Phil. Soc.* **56**, 367–380.

Gould, Henry W. (1979). *Bell and Catalan Numbers, Research Bibliography of Two Special Number Sequences*, available from the author (Department of Mathematics, West Virginia University, Morgantown, WV 26506). The 1979 edition sells for $3.00 and contains over 500 references pertaining to Catalan numbers.

Gould, Henry W. *Combinatorial Identities*, available from the author (Department of Mathematics, West Virginia University, Morgantown, WV 26506).

Guy, Richard K. (1958). Dissecting a Polygon into Triangles, *Bulletin of the Malayan Mathematical Society* **5**, 57–60.

Guy, Richard K. (1984). A Medley of Malayan Mathematical Memories and Mantissae, *Math. Medley* (Singapore), **12**, Mp/1, 9–17.

Hoggatt, V. E. Jr. and Bicknell, Marjorie (1976). Pascal, Catalan, and General Sequence Convolution Arrays in a Matrix, *Fibonacci Quarterly*, April, 135–143.

Hoggatt, V. E. Jr. and Bicknell, Marjorie (1976). Sequences of Matrix Inverses from Pascal, Catalan, and Related Convolution Arrays, *Fibonacci Quarterly*, October, 224–232.

Hoggatt, V. E. Jr. and Bicknell, Marjorie (1976). Catalan and Related Sequences Arising from Inverses of Pascal's Triangle Matrices, *Fibonacci Quarterly*, December, 395–405.

Klamkin, M. S. (1962). Problem 4983, *Amer. Math. Month.* **69**, 930–931.

Kuchinski, M. Catalan Structures and Correspondences, M.Sc. thesis, Department of Mathematics, West Virginia University, Morgantown, WV 26506.

Motzkin, Th. (1948). Relations between Hypersurface Cross Ratios, and a Combinatorial Formula for Partitions of a Polygon, and for Non-associative Products, *Bull. Amer. Math. Soc.* **54**, 352–360.

Pólya, G. (1956). On Picture-writing, *Amer. Math. Month.* **63**, 689–697.

Raney, G. N. (1960). Functional Composition Patterns and Power Series Inversion, *Trans. Amer. Math. Soc.* **94**, 441–451.

Rogers, D. G. (1978). Pascal Triangles, Catalan Numbers and Renewal Arrays, *Discrete Math.* **22**, 301–310.

Sands, A. D. (1978). On Generalized Catalan Numbers, *Discrete Math.* **21**, 218–221.

Memoirs by Eugène-Charles Catalan relevant to the theme of Catalan numbers may be found in *Journal de Mathématiques pures et appliquées de Liouville*, (1), **III**, 508–816; **IV**, 91–94, 95–99; **VI**, 74 (1838–41).

For details of Catalan's life and work see Les travaux mathématiques de Eugène-Charles Catalan, *Annuaire de L'Académie Royale des Sciences, des Lettres et des Beaux-Arts de Belgique*, Brussels (1896), 115–172.

CHAPTER 5

Integral Equations

5.1. INTRODUCTION

The researches of Fredholm and Volterra on integral equations may be said to mark the conception of functional analysis, since, in retrospect at least, it seems a short step to abstract from their results the notion of an operator and then to formulate a theory of linear operators. Indeed the preface to Banach's '*Theorie des opérations linéaires*' opens with '*La théorie des opérations, créée par V. Volterra,...*'.

The term *integral equation* was first used by Du Bois-Raymond in 1883. Although Volterra had, according to Tricomi (1957), anticipated Fredholm's method of solution, the general theory of integral equations began with the publication of Fredholm's seminal paper of 1903.

An integral equation is an operator equation in which the unknown function occurs under the sign of integration. Linear equations constiute the most important and best-understood class, and the greater part of this chapter will be devoted to outlining theories for different types of them. A description of numerical methods for the solution of integral equations as well as a summary of the theory is contained in III, §10. Results from linear algebra will provide a useful guide, since theorems which had been proved in a finite dimensional situation often remain true, with suitable modifications, in the more general context.

A list of standard references is given at the end of the chapter and more general conditions for these theories to hold will be found there. In particular the books by Cochrane (1972), Mikhlin (1957), Smithies (1958) and Tricomi (1957) will be found to expand on much of the material presented here.

The Lebesgue integral [see IV, §4.9] will be used throughout this chapter.

DEFINTION 5.1.1.   The function f will be said to be in $L^2[a, b]$ if the Lebesgue integral

$$\int_a^b |f(t)|^2 \, dt$$

exists. This will be abbreviated to $f \in L^2[a, b]$.

117

DEFINITION 5.1.2.

(a) A *non-linear Fredholm integral equation of the second kind* has the form

$$f(x) = g(x) + \int_a^b G[x, t; f(t)] \, dt \quad (a \le x \le b), \tag{5.1.1}$$

(b) A *non-linear Volterra integral equation of the second kind* has the form

$$f(x) = g(x) + \int_a^x G[x, t; f(t)] \, dt \quad (a \le x \le b). \tag{5.1.2}$$

In each case the functions g and G are known.

The equations are said to be *linear* if G depends linearly on f, that is to say if it can be written as

$$G[x, t; f] = k(x, t) f. \tag{5.1.3}$$

A Volterra equation is a special case of a Fredholm equation, since if we assume in (5.1.1) that

$$G[x, t; f] = 0 \quad \text{for } t > x, \tag{5.1.4}$$

then it becomes the Volterra equation.

It seems natural to distinguish the two types of equation in this way; however, it is probably historically accurate to call the general non-linear Fredholm equation *Urysohn's equation* since Fredholm's investigations were confined to the linear case.

The emphasis in this chapter is on equations of the second kind. Those of the *first kind* are of the form

$$g(x) = \int_a^b G[x, t; f(t)] \, dt \quad (a \le x \le b), \tag{5.1.5}$$

where f has to be found when all the other functions are known. In §5.5 we discuss those of Fredholm type, and in §5.6, briefly, the ones which correspond to the Volterra equation. Particular equations, which sometimes arise in practice, can often be solved explicitly. Some examples will be given in §5.8. [See also IV, §13 for a general treatment of transform methods.]

A Fredholm equation has a great deal in common with a boundary value problem for an ordinary differetial equation [see IV, 2.5.1], and a Volterra equation with an initial value problem. We illustrate these observations by the following examples.

EXAMPLE 5.1.1.   The solution of the differential equation

$$f''(x) = F[x, f(x)] \quad (0 \le x \le 1), \quad \text{subject to } f(0) = f(1) = 0,$$

satisfies the Fredholm equation

$$f(x) = -\int_0^x (1-x)t\,F[t,f(t)]\,dt - \int_x^1 x(1-t)F[t,f(t)]\,dt \quad (0 \le x \le 1).$$

This can be written concisely as

$$f(x) = -\int_0^1 \gamma(x,t)F[t,\,f(t)]\,dt,$$

where

$$\gamma(x,t) = \begin{cases} (1-x)t & (0 \le t \le x), \\ (1-t)x & (x \le t \le 1). \end{cases}$$

The function γ will occur frequently in this chapter.

EXAMPLE 5.1.2.   The solution of

$$f''(x) = F[x,f(x)] \quad (0 \le x \le 1), \quad \text{subject to } f(0) = f_0, \quad f'(0) = f'_0,$$

satisfies the Volterra equation

$$f(x) = f_0 + xf'_0 + \int_0^x (x-t)F[t,f(t)]\,dt \quad (0 \le x \le 1).$$

The reformulation of a problem as an integral equation is sometimes an advantage when a numerical solution is required. Also it forms part of the standard procedure in the proof of the existence of a solution of an ordinary differential equation by Picard iteration.

DEFINITION 5.1.3.   (Liouville–Neumann series) Define the sequence of functions $\{f_n\}$ by the iteration

$$f_{r+1}(x) = g(x) + \int_a^b G[x,t;f_r(t)]\,dt \quad (a \le x \le b, r = 0, 1, \ldots), \qquad (5.1.6)$$

where

$$f_0(x) = 0 \quad (a \le x \le b).$$

The *Liouville–Neumann* series is

$$\sum_{r=1}^\infty [f_r(x) - f_{r-1}(x)] \quad (a \le x \le b), \qquad (5.1.7)$$

in other words

$$\lim_{n \to \infty} f_n(x) \quad (a \le x \le b).$$

If the series converges (that is to say if the sequence of functions $\{f_n\}$ has a limit), then the sum (limit) will be a function f which, if it belongs to a class for which the equation is defined, will be a solution.

Since Volterra's equation is a special case of Fredholm's this definition covers both equations.

DEFINITION 5.4.5.   *G* satisfies a *uniform Lipschitz condition* in a region *D* of $\mathbb{R}^2$ if there exists a positive constant *L*, the *Lipschitz constant*, such that

$$|G[x,t;f_1] - G[x,t;f_2]| \le L|f_1 - f_2| \quad \text{for } (x,t) \in D \quad \text{and} \quad f_1, f_2 \in (-\infty, \infty)$$
(5.1.8)

[IV, 7.11.1].

THEOREM 5.1.1.   *Let G satisfy a uniform Lipschitz condition in* $D = \{(x,t): a \le x, t \le b\}$, *and let g and G (regarded as a function of x and t) be continuous in* $[a,b]$. *Then, for* $a \le x \le b$,

   (a) *the Liouville–Neumann series for* (5.1.1) *converges if* $L < 1$, *and its sum is a function which is continuous in* $[a,b]$ *and which satisfies* (5.1.1);
   (b) *the Liouville–Neumann series for* (5.12) *converges for any value for L, its sum is continuous in* $[a,b]$ *and satisfies* (5.1.2).

It is usually difficult or even impossible to determine the Lipschitz constant. The value of this theorem is not so much that it gives an easily verifiable test for the existence of a solution but rather an assurance that such equations can have solutions.

5.2.   FREDHOLM INTEGRAL EQUATIONS

DEFINITION 5.2.1.   An *inhomogeneous Fredholm linear integral equation of the second kind* is of the form

$$f(x) = g(x) + \lambda \int_a^b k(x,t) f(t)\, dt \quad (a \le x \le b);$$
(5.2.1)

the function k is called the *kernel*, and g the *free term*.

The problem is to find f when k and g are given. The scalar λ may be known or, when g is the zero function, may have to be found and will give rise to an eigenvalue problem.

Fredholm (1903) presented his theory first for equations with kernels which are continuous in the square $a \le x, t \le b$. His explicit solution will be given in Theorem 2.1. Shortly afterwards, in 1907, Schmidt gave other proofs of what are now known as the Fredholm theorems which do not depend on determinants [see I, §§6.8–6.14] and which are valid for a wider class of kernels.

It will be convenient to have a notation for certain determinants which arise both here and in §5.4.

DEFINITION 5.2.2.

$$k\begin{pmatrix} y_1, y_2, \ldots, y_r \\ z_1, z_2, \ldots, z_r \end{pmatrix} = \begin{vmatrix} k(y_1, z_1) & \ldots & k(y_1, z_r) \\ \ldots & \ldots & \ldots \\ k(y_r, z_1) & \ldots & k(y_r, z_r) \end{vmatrix} \quad (r = 2, 3, \ldots). \quad (5.2.2)$$

DEFINITION 5.2.3.

$$D(x, t; \lambda) = k(x, t) - \frac{\lambda}{1!} \int_a^b k\begin{pmatrix} x, t_1 \\ t, t_1 \end{pmatrix} dt_1 + \frac{\lambda^2}{2!} \int_a^b \int_a^b k\begin{pmatrix} x, t_1, t_2 \\ t, t_1, t_2 \end{pmatrix} dt_1 dt_2$$

$$- \frac{\lambda^3}{3!} \int_a^b \int_a^b \int_a^b k\begin{pmatrix} x, t_1, t_2, t_3 \\ t, t_1, t_2, t_3 \end{pmatrix} dt_1 dt_2 dt_3 + \ldots, \quad (a \le x, t \le b), \quad (5.2.3)$$

and

$$D(\lambda) = 1 - \frac{\lambda}{1!} \int_a^b k(t_1, t_1) dt_1 + \frac{\lambda^2}{2!} \int_a^b \int_a^b k\begin{pmatrix} t_1, t_2 \\ t_1 \, t_2 \end{pmatrix} dt_1 dt_2$$

$$- \frac{\lambda^3}{3!} \int_a^b \int_a^b \int_a^b k\begin{pmatrix} t_1, t_2, t_3 \\ t_1, t_2, t_3 \end{pmatrix} dt_1 dt_2 dt_3 + \ldots. \quad (5.2.4)$$

THEOREM 5.2.1.   Let the kernel k be continuous in the square $a \le x, t \le b$, and g be integrable in $a \le x \le b$. Then the power series for $D(\lambda)$ and $D(x, t; \lambda)$ are entire functions of λ (that is to say each series converges for all λ, real or complex), and, providing that $D(\lambda) \ne 0$, the solution of the equation

$$f(x) = g(x) + \lambda \int_a^b k(x, t) f(t) \, dt \quad (a \le x \le b)$$

is integrable and given by

$$f(x) = g(x) + \lambda \int_a^b \frac{D(x, t; \lambda)}{D(\lambda)} g(t) \, dt \quad (a \le x \le b). \quad (5.2.5)$$

$D(\lambda)$ is known as Fredholm's denominator.

The solution can be written

$$f(x) = g(x) - \lambda \int_a^b h(x, t; \lambda) g(t) \, dt \quad (a \le x \le b). \quad (5.2.6)$$

The function h is known as the *resolvent kernel* of the integral equation and can be shown to satisfy

$$k(x, t) + h(x, t; \lambda) = \lambda \int_a^b k(x, s) h(s, t; \lambda) \, ds = \lambda \int_a^b h(x, s; \lambda) k(s, t) \, ds \quad (a \le x, t \le b).$$

$$(5.2.7)$$

These give rise to the following *reciprocal relationships*:

$$\left.\begin{array}{l} f(x) - \lambda \int_a^b k(x,t)f(t)\,dt = g(x) \\[3mm] g(x) - \lambda \int_a^b h(x,t;\lambda)g(t)\,dt = f(x) \end{array}\right\} \quad (a \le x \le b). \qquad (5.2.8)$$

The results known as the *Fredholm theorems* (see below) were deduced by him from this theorem.

EXAMPLE 5.2.1.   For the equation

$$f(x) = g(x) + \lambda \int_0^1 4xt^2 f(t)\,dt \quad (0 \le x \le 1),$$

we have

$$D(\lambda) = 1 - \lambda, \qquad D(x,t;\lambda) = 4xt^2, \qquad h(x,t:\lambda) = -4xt^2/(1-\lambda)$$

and the solution for $\lambda \ne 1$ is

$$f(x) = g(x) + \int_0^1 \frac{4\lambda}{(1-\lambda)} xt^2 g(t)\,dt \quad (0 \le x \le 1).$$

EXAMPLE 5.2.2.   A less trivial equation is

$$f(x) = g(x) + \lambda^2 \int_0^1 \gamma(x,t)f(t)\,dt \quad (0 \le x \le 1),$$

where the kernel γ is that of Example 5.1.1 (λ has been replaced by λ^2 for reasons which will be clear from the form of solution). The solution is

$$g(x) + \frac{\lambda}{\sin(\lambda)}\left[\sin \lambda(1-x) \int_0^x \sin(\lambda t)g(t)\,dt + \sin(\lambda x) \int_x^1 \sin \lambda(1-t)g(t)\,dt \right]$$
$$(0 \le x \le 1).$$

In this case $D(\lambda) = \sin(\lambda)/\lambda$, and the equation will not have a unique solution if $\lambda = n\pi$ for $n = 1, 2, \ldots$.

DEFINITION 5.2.4.   The kernel k is said to be a *Fredholm kernel* in $[a,b]$ if the following double integral exists

$$\int_a^b \int_a^b |k(x,t)|^2 \,dx\,dt. \qquad (5.2.9)$$

The qualifier 'in $[a,b]$' will usually be omitted when it is obvious which interval is meant. If k is a Fredholm kernel in $[a, b]$ and $f \in L^2[a, b]$ then

$$\int_a^b k(x, t) f(t) dt$$

defines a function which is in $L^2[a, b]$.

EXAMPLE 5.2.3. Kernels of the form

$$h(x, t) \log |x - t| \quad \text{and} \quad h(x, t)/|x - t|^\alpha \quad \text{for } \alpha < \tfrac{1}{2},$$

where h is a Fredholm kernel, are Fredholm kernels in any finite interval. The second type of kernel is often described as being *weakly singular*, it can be shown that the subsequent Fredholm theorems will hold for these if $\alpha < 1$. Note, however, that Theorem 2.1 is not applicable since neither of these kernels is defined in general for $t = x$.

DEFINITION 5.2.5. The *adjoint* of the kernel k is denoted by k^* and is defined by

$$k^*(x, t) = \bar{k}(t, x) \quad (a \leq x, t \leq b) \tag{5.2.10}$$

the complex conjugate of $k(t, x)$.

Clearly k is a Fredholm kernel if and only if k^* is also a Fredholm kernel.

EXAMPLE 5.2.4.

(a) If $k(x, t) = a(x) b(t)$ then $k^*(x, t) = a(t) b(x)$.
(b) If $k(x, t) = ia(x) b(t)$ then $k^*(x, t) = -ia(t) b(x)$.

DEFINITION 5.2.6.

(a) The *homogeneous Fredholm integral equation* which corresponds to (5.2.1) is

$$f(x) = \lambda \int_a^b k(x, t) f(t) dt \quad (a \leq x \leq b). \tag{5.2.11}$$

(b) If, for some value λ_0 of the scalar parameter λ, the homogeneous equation (5.2.11) has at least one non-trivial solution, then λ_0 is said to be an *eigenvalue* of the kernel (or, what is the same thing, of the homogeneous equation). The solutions are called the *eigenfunctions* of the kernel.
(c) The *adjoint equation* of (5.2.11) is

$$\varphi(x) = \lambda \int_a^b k^*(x, t) \varphi(t) dt \quad (a \leq x \leq b). \tag{5.2.12}$$

EXAMPLE 5.2.5.   The eigenvalues of the equation

$$f(x) = \lambda \int_{-1}^{1} (x - t)f(t)\, dt \quad (-1 \le x \le 1),$$

are $\pm i3^{\frac{1}{2}}/2$ and the corresponding eigenfunctions are $\pm i3^{\frac{1}{2}}x + 1$.

DEFINITION 5.2.7.   The set of eigenvalues of the homogeneous integral equation is called its *spectrum* [see I, §7.10].

   The following group of theorems is known as Fredholm's theorems. In each of them k will be assumed to be a Fredholm kernel, and g a member of $L^2[a, b]$.

THEOREM 5.2.2.   *If λ_0 is not an eigenvalue of the kernel (in other words if the homogeneous equation (5.2.11) has only the trivial, zero solution), then the inhomogeneous equation (5.2.1) has a unique solution for any function g.*

THEOREM 5.2.3.   *If λ_0 is an eigenvalue of the homogeneous equation, then $\bar{\lambda}_0$ is an eigenvalue of the adjoint equation (5.2.12) and each equation has exactly the same number of linearly independent eigenfunctions.*

EXAMPLE 5.2.6.   The eigenvalues of the following equation, which is adjoint to that of Example 5.2.5,

$$\varphi(x) = \lambda \int_{-1}^{1} (t - x)\varphi(t)\, dt \quad (-1 \le x \le 1),$$

are $\pm i3^{\frac{1}{2}}/2$ and the corresponding eigenfunctions are $i3^{\frac{1}{2}}x \pm 1$.

THEOREM 5.2.4.   *To each eigenvalue there corresponds a finite number of linearly independent eigenfunctions.*

THEOREM 5.2.5.   *If λ_0 is an eigenvalue of the homogeneous equation, then the inhomogeneous equation (5.2.1) will have a solution if and only if*

$$\int_{a}^{b} g(t)\bar{\varphi}_r(t)\, dt = 0 \quad \text{for } r = 1, 2, \ldots, n, \tag{5.2.13}$$

where $\varphi_1, \varphi_2, \ldots, \varphi_n$, are the linearly independent solutions of the homogeneous equation (5.2.12) with $\lambda = \bar{\lambda}_0$. In this case the solution will not be unique, and any two solutions will differ by a linear combination of eigenfunctions of the homogeneous equation (5.2.11).

THEOREM 5.2.6.   *The set of eigenvalues is either finite or denumerably infinite. In the latter case successive eigenvalues increase without limit.*

THEOREM 5.2.7. *Let* λ, f *and* μ, φ *be non-trivial solutions of* (5.2.11) *and* (5.2.12) *respectively, then*

$$\int_a^b f(t)\bar{\varphi}(t)dt = 0, \quad \text{if } \lambda \neq \bar{\mu}. \tag{5.2.14}$$

The next theorem is a consequence of these results and will be familiar from linear algebra.

THEOREM 5.2.8. (Fredholm's alternative) Either *the inhomogeneous equation* (5.2.1) *has a unique solution for any* $g \in L^2[a, b]$, or *the corresponding homogeneous equation* (5.2.11) *has a non-trivial solution.*

The following simple example will illustrate Fredholm's theorems.

EXAMPLE 5.2.7. The solution of the equation

$$f(x) = g(x) + 4\lambda \int_0^1 xt^2 f(t)dt \quad (0 \leq x \leq 1) \tag{5.2.15}$$

must be of the form

$$g(x) + 4\lambda x A, \tag{5.2.16}$$

where

$$A = \int_0^1 t^2 f(t)\,dt. \tag{5.2.17}$$

The substitution of (5.2.16) for $f(x)$ in (5.2.17) gives

$$(1 - \lambda)A = \int_0^1 t^2 g(t)dt. \tag{5.2.18}$$

It follows that if $\lambda \neq 1$, then A is given by

$$A = (1 - \lambda)^{-1} \int_0^1 t^2 g(t)dt,$$

and the solution of (5.2.15) is

$$g(x) + \frac{4\lambda}{(1 - \lambda)} \int_0^1 xt^2 g(t)\,dt \quad (0 \leq x \leq 1);$$

$\lambda = 1$ is easily seen to be the only eigenvalue of the homogeneous equation

$$f(x) = 4\lambda \int_0^1 xt^2 f(t)dt \quad (0 \leq x \leq 1). \tag{5.2.19}$$

For consistency in (5.2.18), when $\lambda = 1$, we must have

$$\int_0^1 t^2 g(t)\,dt = 0.$$

This verifies Theorem 5.2.3, since the only solution of the adjoint equation

$$\varphi(x) = 4\lambda \int_0^1 xt^2 \varphi(t)\,dt \quad (0 \leq x \leq 1),$$

when $\lambda = 1$ is

$$\varphi(x) = x^2.$$

The solution of (5.2.15), when $\lambda = 1$, is

$$g(x) + Cx,$$

where C is an arbitrary constant. This agrees with Theorem 5.2.5 since Cx is the eigenfunction of (5.2.19) corresponding to $\lambda = 1$.

The next example is a generalization of the last one and it is important in the theory of Fredholm equations, since it can be shown that if k is continuous then it can be approximated arbitrarily closely by a degenerate kernel. This forms the basis for Schmidt's alternative proof of Fredholm's theorems which does not depend as heavily on determinants as did the original theory. For details see Smithies (1958). It demonstrates the intimate connection between Fredholm's equation and linear algebraic equations which is reflected in Fredholm's theorems.

EXAMPLE 5.2.8.   The kernel k is said to be *degenerate* if

$$k(x,t) = \sum_{j=1}^n \alpha_j(x)\beta_j(t) \quad (a \leq x, t \leq b), \tag{5.2.20}$$

where $\{\alpha_j\}$ and $\{\beta_j\}$ are linearly independent functions. Fredholm's equation (5.2.1) becomes, in this case,

$$f(x) = g(x) + \lambda \sum_{j=1}^n \alpha_j(x) \int_a^b \beta_j(t)f(t)\,dt \quad (a \leq x, t \leq b). \tag{5.2.21}$$

It follows that f must be of the form

$$g + \lambda \sum_{j=1}^n \alpha_j A_j, \tag{5.2.22}$$

where

$$A_i = \int_a^b \beta_i(t)f(t)\,dt \quad (i = 1, 2, \ldots, n). \tag{5.2.23}$$

The substitution of (5.2.22) for f in each of the equations (5.2.23) will give the following system of linear algebraic equations:

$$A_i = \int_a^b \beta_i(t)g(t)\,dt + \lambda \sum_{j=1}^n A_j \int_a^b \alpha_j(t)\beta_i(t)\,dt \quad (i = 1, 2, \ldots, n).$$

If these have a unique solution $A_1, A_2, \ldots, A_n$, these values, when substituted in (5.2.22), will give the solution of the integral equation.

The next example shows that a homogeneous Fredholm equation need not necessarily have a non-trivial solution.

EXAMPLE 5.2.9. The only solution of

$$f(x) = \lambda \int_0^{2\pi} \cos(x)\sin(t)f(t)\,dt \quad (0 \le x < 2\pi)$$

is $f(x) = 0$. Consequently the inhomogeneous equation

$$f(x) = g(x) + \lambda \int_0^{2\pi} \cos(x)\sin(t)f(t)\,dt \quad (0 \le x < 2\pi)$$

always has a solution for any value of λ. This is

$$g(x) + \lambda \cos(x) \int_0^{2\pi} g(t)\sin(t)\,dt \quad (0 \le x < 2\pi).$$

For this to happen in Example 5.2.8, it is necessary for the α-sequence to be orthogonal to the β-sequence, that is to say

$$\int_a^b \alpha_j(t)\beta_i(t)\,dt = 0 \quad (i, j = 1, 2, \ldots, n),$$

in which case the inhomogeneous equation will have a solution for any value of λ. However, this phenomenon is unlikely to occur in a practical situation.

DEFINITION 5.2.8. The *operator K* is defined by

$$Kg(x) = \int_a^b k(x, t)g(t)\,dt \quad (a \le x \le b). \tag{5.2.24}$$

The operator which is adjoint to K is denoted by K^* and is defined by

$$K^*\varphi(x) = \int_a^b k^*(x, t)\varphi(t)\,dt \quad (a \le x \le b). \tag{5.2.25}$$

Powers of the operator K are defined for $r = 0, 1, \ldots$ by

$$K^{r+1}g(x) = \int_a^b k(x, t)K^r g(t)\,dt \quad \text{where } K^0 g(x) = g(x) \quad (a \le x \le b). \tag{5.2.26}$$

THEOREM 5.2.9.   *If k is a Fredholm kernel then both the operator K and its adjoint K* transform the space of functions which are integrable L^2 into itself.*

Equation (5.2.1) can be written as

$$f = g + \lambda K f, \qquad (5.2.27)$$

and with this notation the Liouville–Neumann series for the solution of (5.2.1) becomes formally

$$g + \sum_{r=1}^{\infty} \lambda^r K^r g. \qquad (5.2.28)$$

Although the operator notation is convenient for describing the series, we shall give a more precise description of its terms which can be used for their calculation. This is by means of the *iterated kernels* of k.

DEFINITION 5.2.9.   The *r*th *iterated kernel* k_r is defined recursively by

$$k_1(x,t) = k(x,t) \quad (a \le x, t \le b) \qquad (5.2.29)$$

$$k_{r+1}(x,t) = \int_a^b k(x,s)k_r(s,t)ds \quad (a \le x, t \le b, \quad r = 1,2,\ldots),$$

and so we can write

$$K^r g(x) = \int_a^b k_r(x,t)g(t)dt \quad (a \le x \le b).$$

The series (5.2.27) is

$$f(x) = g(x) + \sum_{r=1}^{\infty} \lambda^r \int_a^b k_r(x,t)g(t)dt \quad (a \le x \le b). \qquad (5.2.30)$$

The Liouville–Neumann series for (5.2.1) when the kernel is continuous is now seen to be the expansion of the quotient $D(x,t;\lambda)/D(\lambda)$ in ascending powers of λ. This series will converge so long as $|\lambda| < |\lambda_0|$, where λ_0 is the smallest zero in modulus of $D(\lambda)$.

Note that the *n*th iterated kernel of the adjoint kernel is the adjoint of the *n*th iterated kernel.

EXAMPLE 5.2.10.   The Liouville–Neumann series for the equation

$$f(x) = g(x) + 4\lambda \int_0^1 xt^2 f(t)\, dt \quad (0 \le x \le 1),$$

is

$$g(x) + \sum_{r=1}^{\infty} \lambda^r 4x \int_0^1 t^2 g(t)\, dt \quad (0 \le x \le 1).$$

This is a geometric series which will converge for $|\lambda| < 1$.

5.3.   HERMITIAN KERNELS, HILBERT–SCHMIDT THEORY

It is to be expected that the more restrictions that are placed on the kernel the more structure there will be in the corresponding theory. This section will be devoted to equations which have Hermitian kernels of Fredholm type. A useful guide will be found in the thory of Hermitian matrices [see I, (6.7.10)].

DEFINITION 5.3.1.   The kernel k is *Hermitian* if

$$k^*(x, t) = k(x, t) \quad (a \leq x, t \leq b). \tag{5.3.1}$$

That is to say if

$$k(x, t) = \bar{k}(t, x). \tag{5.3.2}$$

When the kernel is also real it is said to be *symmetric*; in this case the equation and its adjoint are identical.

EXAMPLE 5.3.1.

(a) The kernel

$$k(x, t) = 1 + i(x - t) + xt$$

is Hermitian.

(b) The kernel

$$k(x, t) = i(x + t)$$

is not Hermitian but $k(x, t) = (x + t)$ is symmetric.

(c) γ as defined in Example 5.1.1 is real and Hermitian and so is symmetric.

The natural setting for the Hilbert–Schmidt theory is a space of functions in which there is an inner product [see I, §§9.1 and 10.1]. If in addition the space is complete then it is called a *Hilbert space*, see IV ch. 19; for the purpose of this section we shall use a particular inner product and the norm [see I, §10.1] derived from it.

DEFINITION 5.3.2.   The *inner product* $\langle \cdot, \cdot \rangle$ is defined by

$$\langle f, g \rangle = \int_a^b f(t) \bar{g}(t) dt \tag{5.3.3}$$

for all $f, g \in L^2[a, b]$.

The *norm*, which will be denoted by $\| \cdot \|$, is defined by

$$\| f \| = [\langle f, f \rangle]^{\frac{1}{2}} = \left[\int_a^b |f(t)|^2 dt \right]^{\frac{1}{2}} \tag{5.3.4}$$

for all $f \in L^2[a, b]$.

The norm of a function is a measure of its size; it has the following properties:

(a) $\|f\| \geq 0$, and is zero is and only if f is the zero function,
(b) $\|\alpha f\| = |\alpha| \cdot \|f\|$, where α is any scalar,
(c) $\|f + g\| \leq \|f\| + \|g\|$.

Note that, in the theory of the Lebesgue integral, a function is the zero function (often called the trivial function) if it is zero at all points except those which form a set of *measure zero*. (A set of measure zero is one which can be included in a set of intervals whose total length can be made arbitrarily small—they are called *null sets* in I, Definition 4.9.2. In most, if not all, practical applications this usually means that the set consists of a finite number of points.)

DEFINITION 5.3.3.

(a) The sequence $\{e_n\}$ of (non-trivial) function form an *orthogonal sequence* if

$$\langle e_m, e_n \rangle = 0 \quad (m \neq n). \tag{5.3.5}$$

(b) If, in addition, each member of the sequence is *normalized* so that

$$\langle e_n, e_n \rangle = 1 \tag{5.3.6}$$

then it is said to be an *orthonormal sequence*. Any orthogonal sequence can be normalized to become an orthonormal sequence [see I, Definition 10.2.2].
(c) The sequence $\{e_n\}$ is *complete* if

$$\langle f, e_n \rangle = 0 \quad \text{for all } n \tag{5.3.7}$$

implies that $f = 0$, the zero function.

EXAMPLE 5.3.2.

(1) Trigonometrical series:

$$e_0(t) = 1/(2\pi)^{\frac{1}{2}}, \quad e_{2n-1} = \cos(nt)/\pi^{\frac{1}{2}}, \quad e_{2n}(t) = \sin(nt)/\pi^{\frac{1}{2}}, \quad (n = 1, 2, \ldots),$$

where

$$\langle f, g \rangle = \int_0^{2\pi} f(t)g(t)\,dt.$$

This is the orthonormal sequence which gives rise to the usual Fourier series [see IV, §20].
(2) Legendre polynomials:

$$e_n(t) = (n + \tfrac{1}{2})^{\frac{1}{2}} P_n(t) \quad (n = 0, 1, \ldots),$$

P_n is the *Legendre polynomial* of degree n [see IV, §10.3]; in this case

$$\langle f, g \rangle = \int_{-1}^{1} f(t)g(t)\,dt.$$

It can be shown that each sequence is complete.

The members of an orthonormal sequence of vectors are linearly independent [see I, §5.3] and can be thought of as providing a system of coordinate vectors each of unit length. It is therefore desirable to have a test to ensure that any element in $L^2[a, b]$ can be represented as a linear combination of them. When this is the case the sequence will provide a basis for the space in a sense which Theorem 5.3.1 will make precise.

DEFINITION 5.3.4. Let $f \in L^2[a, b]$, and $\{e_n\}$ be an orthonormal sequence. Define the sequence $\{a_n\}$ by

$$a_n = \langle f, e_n \rangle \quad (n = 0, 1, \ldots), \tag{5.3.8}$$

then the series

$$\sum_{r=0}^{\infty} a_r e_r \tag{5.3.9}$$

is the *generalized Fourier series* of f, and $\{a_n\}$ are called the *generalized Fourier coefficients* of f (relative to the sequence $\{e_n\}$) [see IV, §20.5].

THEOREM 5.3.1.   *Let f and $\{e_n\}$ be as in Definition 5.3.4, then*

$$\lim_{n \to \infty} \left\| f - \sum_{r=0}^{n} a_r e_r \right\| = 0. \tag{5.3.10}$$

This is usually written

$$f \sim \sum_{r=0}^{\infty} a_r e_r.$$

This type of convergence is called *convergence in mean* [see IV, Definition 19.2.3]. It should be emphasized that it does *not* mean that the generalized Fourier series

$$\sum_{r=0}^{\infty} a_r e_r$$

converges at each point of the interval $[a, b]$ or even almost everywhere in the interval.

The problem of the convergence of the trigonometric Fourier series has a long history culminating in Carleson's theorem which states that the trigonometric Fourier series of any function in $L^2[0, 2\pi]$ will converge to that function almost everywhere.

We have *Bessel's inequality* [see IV, §20.6.3]

$$\sum_{r=0}^{n} |a_r|^2 \le \| f \|^2 \quad \text{for all } n, \tag{5.3.11}$$

and so

$$\sum_{r=0}^{\infty} |a_r|^2 \le \| f \|^2.$$

If the sequence is complete, this becomes *Parseval's equality* [see IV, §20.6.3]

$$\sum_{r=0}^{\infty} |a_r|^2 = \| f \|^2, \tag{5.3.12}$$

which can be used to derive the following test for the completeness of an orthonormal sequence.

THEOREM 5.3.2.   *The orthonormal sequence* $\{e_n\}$ *is complete if*

$$x - a = \sum_{r=0}^{\infty} \left| \int_a^x e_r(t)\, dt \right|^2. \tag{5.3.13}$$

THEOREM 5.3.3.   *Every Hermitian Fredholm kernel has at least one (non-zero) eigenvalue and to each eigenvalue there corresponds at least one non-trivial eigenfunction.*

This theorem is in contrast to the general case where, as we saw in Example 5.2.9, a kernel may not have any eigenvalues.

THEOREM 5.3.4.   *The eigenvalues and eigenfunctions of a Hermitian Fredholm kernel and real, and the eigenfunctions which correspond to distinct eigenvalues are orthogonal.*

Each eigenvalue has at most a finite number of linearly independent eigenfunctions associated with it, and these can be arranged with the aid of the Gram–Schmidt process [see I, §7, or IV, §11] to be mutually orthogonal. We shall always suppose that this has been done, in which case the eigenfunctions will form an orthonormal sequence.
   In general there will be an infinite number of eigenfunctions.

EXAMPLE 5.3.3.   The eigenvalues of the equation

$$f(x) = \frac{1}{2\pi} \int_0^{2\pi} \frac{1 - r^2}{1 - 2r\cos(t - x) + r^2} f(t)\, dt \quad (0 \le x < 2\pi, r^2 < 1)$$

are

$$1, 1/r, 1/r^2, \ldots.$$

There is only one eigenfunction corresponding to the eigenvalue 1, namely $1/(2\pi)^{\frac{1}{2}}$, however, for $n = 1, 2, \ldots,$ there are two linearly independent (normalized) eigenfunctions corresponding to the eigenvalue $1/r^n$, namely

$$\cos(nx)/\pi^{\frac{1}{2}}, \quad \sin(nx)/\pi^{\frac{1}{2}}.$$

THEOREM 5.3.5.   *Let* $\lambda_0, \lambda_1, \ldots$ *be the eigenvalue and* $e_0, e_1, \ldots$ *the corresponding normalized eigenfunctions of the Hermitian Fredholm kernel k. Then the series*

$$\sum_{r=0}^{\infty} e_r(x)\bar{e}_r(t)/\lambda_r \tag{5.3.14}$$

converges in mean $a \leq x, t \leq b$ to $k(x, t)$. If the kernel is continuous then the convergence is uniform. Moreover

$$\sum_{r=0}^{\infty} \frac{1}{\lambda_r^2} = \int_a^b \int_a^b |k(x, t)|^2 \, dx \, dt. \tag{5.3.15}$$

EXAMPLE 5.3.4.   As a special case of Example 5.1.1, the eigenvalue problem for the differential equation

$$f''(x) + \lambda f(x) = 0 \quad (0 \leq x \leq 1), \quad \text{with } f(0) = f(1) = 0,$$

is equivalent to the homogeneous integral equation

$$f(x) = \lambda \int_0^1 \gamma(x, t) f(t) \, dt \quad (0 \leq x \leq 1).$$

It is easily verified that the non-trivial solutions of the differential equation are given by

$$\lambda_n = n^2 \pi^2, \qquad f_n(x) = \sin(n\pi x), \quad (n = 1, 2, \ldots),$$

and so these are the eigenvalues and eigenfunctions of the integral equation.

The second part of Theorem 5.3.5, equation (5.3.15), leads, after some evaluation of integrals, to the result that

$$\sum_{r=1}^{\infty} \frac{1}{r^4} = \frac{\pi^4}{90}.$$

The next theorem is a converse of this.

THEOREM 5.3.6.   *Let $\{e_n\}$ be an orthonormal sequence and $\{\lambda_n\}$ a sequence of real numbers which satisfy*

$$0 < |\lambda_n| \leq |\lambda_{n+1}| \quad \text{for } n = 0, 1, \ldots \tag{5.3.16}$$

and be such that the series

$$\sum_{r=0}^{\infty} \frac{1}{\lambda_r^2} \tag{5.3.17}$$

converges. Then the series

$$\sum_{r=0}^{\infty} \frac{e_r(x)\bar{e}_r(t)}{\lambda_r} \tag{5.3.18}$$

converges almost everywhere to a Hermitian Fredholm kernel for which $\{\lambda_n\}$ *and* $\{e_n\}$ *are the eigensystem.*

The next theorem shows that functions in the domain of the integral operator have expansions which also converge in mean.

THEOREM 5.3.7.   (Hilbert–Schmidt) *Let* $\lambda_0, \lambda_1, \ldots,$ *be the eigenvalues and* $e_0, e_1,$ *be the normalized eigenfunctions of the Hermitian Fredholm kernel k. If* $g \in L^2[a, b]$ *such that*

$$\int_a^b k(x, t) f(t) \, dt = g(x) \quad (a \leq x \leq b), \tag{5.3.19}$$

where $f \in L^2[a, b]$, *then the series*

$$\sum_{r=0}^{\infty} \langle g, e_r \rangle e_r \tag{5.3.20}$$

converges in mean to g.

If in addition k is continuous, then the series is uniformly and absolutely convergent.

COROLLARY 5.3.8.   *The series* (5.3.20) *can also be written*

$$\sum_{r=0}^{\infty} \frac{\langle f, e_r \rangle e_r}{\lambda_r}. \tag{5.3.21}$$

COROLLARY 5.3.9.   *The sequence of eigenfunctions* $\{e_n\}$ *is complete if and only if*

$$Kf = 0 \quad \text{implies that} \quad f = 0. \tag{5.3.22}$$

The orthonormal system was not assumed to be complete. This is not unreasonable since g has to lie in the range of the operator K and so is not completely arbitrary. This remark becomes pertinent in §5.5 when we consider the problem of finding f from an equation such as (5.3.19).

THEOREM 5.3.10.   *If k is a Hermitian Fredholm kernel and if the integral*

$$\int_a^b |k(x, t)|^2 \, dt \tag{5.3.23}$$

defines a bounded function, then, for $p = 2, 3, \ldots,$ *the iterated kernel* $k_p(x, t)$ *can be expanded as*

$$\sum_{r=0}^{\infty} e_r(x) \bar{e}_r(t) / \lambda_r^p, \tag{5.3.24}$$

and the series is absolutely and uniformly convergent [*see* IV, §§1.9 *and* 1.12]. *If* $f \in L^2$ *and* $p \geq 2$ *then*

$$\sum_{r=0}^{\infty} \langle f, e_r \rangle e_r / \lambda_r^p \tag{5.3.25}$$

is absolutely and uniformly convergent to $K^p f$.

The next specialization we make is to assume that the kernel is also positive definite [see I, §9.2].

DEFINITION 5.3.5.   The Hermitian k is *positive definite* if

$$\langle Kf, f \rangle = \int_a^b \int_a^b \bar{f}(x) k(x, t) f(t) \, dx \, dt > 0 \tag{5.3.26}$$

for all non-zero $f \in L^2[a, b]$.

THEOREM 5.3.11.   *A Hermitian kernel is positive definite if and only if all its eigenvalues are positive.*

THEOREM 5.3.12   (Mercer) *Let* k *be a continuous Hermitian positive definite kernel,* $\lambda_0, \lambda_1, \ldots$ *its eigenvalues, and* $e_0, e_1, \ldots$ *the corresponding orthonormalized eigenfunctions defined in* $a \leq x \leq b$, *then the series*

$$\sum_{r=0}^{\infty} \frac{e_r(x) \bar{e}_r(t)}{\lambda_r} \tag{5.3.27}$$

converges absolutely and uniformly to $k(x, t)$ *in* $a \leq x \leq b$.

COROLLARY 5.3.13.

$$\sum_{r=0}^{\infty} \frac{1}{\lambda_r} = \int_a^b k(t, t) \, dt. \tag{5.3.28}$$

EXAMPLE 5.3.5.   For the kernel γ of Example 5.1.1, this gives

$$\sum_{r=1}^{\infty} \frac{1}{r^2} = \frac{\pi^2}{6}.$$

Mercer's theorem allows us to write down (with the same notation) the general solution of the inhomogeneous equation in this case.

THEOREM 5.3.14.   (Schmidt) *Let*

$$f(x) = g(x) + \lambda \int_a^b k(x, t) f(t) \, dt \quad (a \leq x \leq b),$$

where $g \in L^2[a,b]$, k *is a continuous Hermitian positive definite kernel and* λ *is not an eigenvalue. Then*

$$f = g + \lambda \sum_{r=0}^{\infty} \frac{\langle g, e_r \rangle}{(\lambda - \lambda_r)} e_r. \tag{5.3.29}$$

This section will conclude with a description of a variational formulation for the eigenvalue problem. This is the *Rayleigh Ritz* method [see IV, §12.5] which was derived originally from energy considerations in equations of mathematical physics. The monograph by Mikhlin provides an invaluable source of material in this area.

THEOREM 5.3.15.   *Let* λ_0 *be the smallest eigenvalue in modulus of the Hermitian Fredholm kernel* k *then*

$$|\lambda_0|^{-1} = \max |\langle f, Kf \rangle| \quad \text{for all } f \in L^2 \text{ such that } \langle f, f \rangle = 1. \tag{5.3.30}$$

This gives rise to the following method of approximation which will furnish an upper bound for the smallest eigenvalue (in modulus).

Let $\{g_r\}$ be an arbitrary sequence of linearly independent members of $L_2[a,b]$, and define f_n by

$$f_n = \sum_{r=0}^{n} a_r g_r. \tag{5.3.31}$$

Then, for all choices of $\{a_r\}$, we have

$$1/|\lambda_0| \geq |\langle f_n, Kf_n \rangle| / \langle f_n, f_n \rangle. \tag{5.3.32}$$

The idea is to choose $a_0, a_1, \ldots, a_n$, so as to maximize $\langle f_n, Kf_n \rangle$ subject to the constraint that $\langle f_n, f_n \rangle = 1$.

This is easily reformulated by the use of Lagrange's method of undetermined multipliers as a linear algebraic eigenvalue problem [see IV, §§5.15 and 15.4]. The smallest eigenvalue in modulus of the resulting algebraic system will provide an upper bound for that of the integral equation.

EXAMPLE 5.3.6.   To estimate an upper bound for the smallest eigenvalue of the equation

$$f(x) = \lambda \int_0^1 \gamma(x,t) f(t) \, dt \quad (0 \leq x \leq 1),$$

where γ is the kernel of Example 5.1.1. We take

$$f_0(x) = g_0(x) = x(1 - x).$$

After the evaluation of simple integrals we have the result that

$$|\lambda_0| \leq 168/17 = 9.882,$$

the correct value of λ_0 is $\pi^2 = 9.8696\ldots$.

Unfortunately this method does not furnish lower bounds. However, these can often be found by Weinstein's method of intermediate operators (see for example Gould (1966)).

5.4.   POSITIVE KERNELS

Although it is easy to see if a kernel is symmetric, it is often not a simple matter to establish whether or not it is positive definite. There is another class of kernels for which there is a comprehensive theory; but once more it is not easy to establish when a particular kernel belongs to it. However, for this class, that of *totally positive kernels*, it is possible to approach the problem in stages. The first stage will be described here.

The theory is essentially a generalization of Perron's theorem for a square matrix with positive elements [I, Theorem 7.11.1]. This states that such a matrix has a positive eigenvalue and the components of the corresponding eigenvector have the same sign, which is usually taken to be positive.

The kernels of this chapter will be assumed to be real Fredholm kernels.

DEFINITION 5.4.1.   The kernel k is of *positive type* if

$$k(x, t) \geq 0 \quad (a \leq x, t \leq b), \tag{5.4.1}$$

and for each non-negative function h there exists a positive integer p such that

$$\int_a^b k_p(x, t) h(t) \, dt > 0 \quad (a \leq x \leq b), \tag{5.4.2}$$

where k_p is the p-th iterated kernel [Definition 5.2.9].

In particular the kernel is of positive type with $p = 1$ if

$$k(x, t) > 0 \quad (a \leq x, t \leq b), \tag{5.4.3}$$

in this case it is described as being *positive*.

The following is a generalization of Perron's theorem. It was proved first by Jentzsch in 1912 for positive kernels.

THEOREM 5.4.1.   *Let k be a kernel of positive type. Then there exists an eigenpair λ_0, f_0 with $f_0 \in C[a, b]$ such that*

$$f_0(x) = \lambda_0 \int_a^b k(x, t) f_0(t) \, dt \quad (a \leq x \leq b), \tag{5.4.4}$$

where

$$\lambda_0 > 0 \quad and \quad f_0(x) > 0 \text{ in } [a, b]. \tag{5.4.5}$$

Moreover f_0 is the only eigenfunction corresponding to λ_0, and if μ is any other *eigenvalue then*

$$\lambda_0 < |\mu|. \tag{5.4.6}$$

We shall call λ_0 the Perron eigenvalue of k.

The adjoint kernel will also be of positive type and so the same result is true for the adjoint equation (usually with a different eigenfunction of course).

The similarity between this and Perron's theorem suggests that there is some deeper result of which each is a special case. This suspicion led Krein and Rutman (1962) to develop a theory for compact linear operators which transform positive functions into positive functions; a proof of Theorem 5.4.1 will be found there [see also IV, §19.4]. A detailed exposition of the theory will be found in Krasnoselskii (1964).

There is a variational formulation for the Perron eigenvalue of a kernel of positive type. This can be traced back to a theorem due to Collatz who gave it for positive matrices.

THEOREM 5.4.2.   *Let k be a kernel of positive type with $p = 1$, then λ_0, the Perron eigenvalue of k, satisfies*

$$\min_{a \le x \le b} g(x) \Big/ \int_a^b k(x, t)g(t)\, dt \le \lambda_0 \le \max_{a \le x \le b} g(x) \Big/ \int_a^b k(x, t)g(t)\, dt \tag{5.4.7}$$

where g is any continuous positive function. When g equals f_0 then the inequalities become equalities.

This can be used to provide bounds for λ_0 as shown by the following example.

EXAMPLE 5.4.1.   The logarithmic kernel

$$k(x, t) = -\log(|t - x|) \quad (0 \le x, t \le 1),$$

satisfies the condition of Theorem 5.4.2. Then with $g(x) = 1$ it gives $0.590 < \lambda_0 < 1.0$. With

$$g(x) = \begin{cases} 1 + 4x(1 - x) & \text{for the lower bound,} \\ 1 + x(1 - x) & \text{for the upper bound,} \end{cases}$$

we get the closer inequalities $0.611 < \lambda_0 < 0.654$. The value of the eigenvalue is in fact 0.654 to three decimal places.

It is possible to relax the condition of the kernel being positive to that of being u_0-*positive*.

DEFINITION 5.4.2.   The kernel k is u_0-*positive* if there exists a non-negative function u_0 such that for each non-negative function h there are positive constants α and β (which will depend on h) such that

$$\alpha u_0(x) \le \int_a^b k(x,t)h(t)\,dt \le \beta u_0(x) \quad (a \le x \le b). \tag{5.4.8}$$

EXAMPLE 5.4.2.   The kernel γ of Example 5.1.1 is u_0-positive and the Perron eigenvalue of the equation

$$f(x) = \lambda \int_0^1 \gamma(x,t)f(t)\,dt \quad (0 \le x \le 1),$$

is

$$\pi^2 = 9.8696\ldots.$$

In this case we can take $u_0(x) = x(1-x)$. When $g(x) = x(1-x)$ we have the inequalities

$$9.615 < \lambda_0 < 11.904,$$

and with

$$g(x) = \begin{cases} x(1-x)[1 + 1.14x(1-x)] & \text{for the lower bound,} \\ x(1-x)[1 + 1.05x(1-x)] & \text{for the upper bound,} \end{cases}$$

the theorem gives

$$9.854 < \lambda_0 < 9.893.$$

There is a quantitative result which provides a bound on the relative size of the eigenvalues of a positive kernel. This is due to Hopf (1963).

THEOREM 5.4.3.   *Let λ_0 be the Perron eigenvalue of the continuous positive kernel k defined in $a \le x,\ t \le b$. If μ is any other eigenvalue, then*

$$\lambda_0 \le |\mu|(M - m)/(M + m), \tag{5.4.9}$$

where

$$M = \max_{a \le x,t \le b} k(x,t) \quad \text{and} \quad m = \min_{a \le x,t \le b} k(x,t).$$

EXAMPLE 5.4.3.   For the positive kernel

$$\frac{1}{2\pi} \frac{(1-r^2)}{1 - 2r\cos(t-x) + r^2}, \quad \text{where } 0 \le x < 2\pi \quad \text{and} \quad r^2 < 1,$$

the Perron eigenvalue is 1 and Theorem 5.4.3 gives the result that any other eigenvalue μ satisfies

$$\tfrac{1}{2}(r + r^{-1}) \le |\mu|.$$

There is a generalization of Theorem 5.4.1 for *totally positive kernels*, these are continuous kernels which satisfy

$$k\begin{pmatrix} x_1, x_2, \ldots, x_r \\ t_{11}, t_2, \ldots, t_r \end{pmatrix} > 0 \tag{5.4.10}$$

for $a \le x_1 < x_2 < \ldots < x_r \le b$ and $a \le t_1 < t_2 < \ldots < t_r \le b$ for $r = 1, 2, \ldots$.

When these inequalities hold, all the eigenvalues are positive and distinct. If the eigenvalues are ordered in increasing size, the (unique) eigenfunction which corresponds to the nth eigenvalue has exactly $n - 1$ changes of sign. A complete account will be found in the monograph by Gantmakher and Krein (1950) who were the founders of the theory. See also Karlin (1968) and Karlin and Studden (1966).

5.5.   LINEAR FREDHOLM EQUATIONS OF THE FIRST KIND

DEFINITION 5.5.1.   A *linear Fredholm integral equation of the first kind* has the form

$$\int_a^b k(x, t) f(t) \, dt = g(x); \tag{5.5.1}$$

the range of x is usually assumed to be $[a, b]$, although this is not necessary, and sometimes may not be the case.

It will be appreciated that the free term g must be in the range of the operator K otherwise there is no possibility of a reasonable solution.

EXAMPLE 5.5.1.   Let f satisfy

$$\int_0^1 \gamma(x, t) f(t) \, dt = g(x) \quad (0 \le x \le 1),$$

where γ is the kernel of Example 5.1.1. Then, since the left-hand side vanishes for $x = 0$ and 1, it follows that g must also vanish at these points. When this is so, the solution will be given by

$$f(x) = -g''(x) \quad (0 \le x \le 1).$$

This implies the additional condition, which is not obvious from the equation, that g should be twice differentiable.

Even when g is in the range of the operator K, it may be that there is more than one solution. The following is an extreme example of this situation.

EXAMPLE 5.5.2.   Suppose that f is to be found so that

$$\int_0^1 f(t) \, dt = 1,$$

one obvious solution is $f(x) = 1$, however, $1 + f_0(x)$ will also satisfy the equation when f_0 is any function such that

$$\int_0^1 f_0(t)\, dt = 0.$$

It may be possible to get a unique solution by imposing additional conditions; for example, in this case we might ask that the solution should be such that

$$\int_0^1 [f(t)]^2\, dt$$

should be a minimum. This will give the reasonable solution of $f(x) = 1$.

Considerations such as this lead to the concept of the general inverse of an operator, (see for example C. Groetsch (1977)).

We give an example to illustrate another type of problem which may arise.

EXAMPLE 5.5.3.   (Inverse Dirichlet problem) A current flows along an infinite thin superconducting strip which occupies the region $-1 \le x \le 1$, $y = 0$, $-\infty < z < \infty$. The problem is to find the distribution of current f across the breadth of the strip by measuring the horizontal force g induced by the current along the line (x, y) for $-\infty < x < \infty$ for a fixed non-zero value of y.

With appropriate units of measurement this becomes that of solving the equation

$$\frac{1}{\pi} \int_{-1}^1 \frac{(t-x)}{(t-x)^2 + y^2} f(t)\, dt = g(x) \quad (-\infty < x < \infty, \; y \neq 0).$$

It can be shown that, if f is assumed to satisfy a Hölder condition [see Definition 5.7.1 below], then

$$f(x) = -2\operatorname{Im}\left[g(x+iy)\right] - \frac{1}{\pi} \int_{-\infty}^{\infty} \frac{(t-x)}{(t-x)^2 + y^2} g(t)\, dt \quad (-1 \le x \le 1).$$

The integral is not difficult to calculate when g is known, and inaccuracies in the measurements of g will be to some extent smoothed out by the process of integration. However, the calculation of the first term involves an analytic continuation which is a notoriously unstable or ill-conditioned process in that small errors in the data may lead to large errors in the result.

The monograph by Lavrentiev (1967) is devoted to an examination of such problems. It will be found that perhaps the most difficult problem in the solution of equations of this type is that of deciding on the class of functions from which a solution is to be found. The physical background from which the problem arises may suggest the appropriate class.

For a Fredholm kernel there is a test for the existence and uniqueness of a solution of an equation of the first kind which we now give. However, the verification of these conditions may not be a simple matter in practice.

THEOREM 5.5.1 *Let k be a real symmetric positive definite Fredholm kernel with eigenvalues* $\lambda_0, \lambda_1, \ldots,$ *and the corresponding orthonormal eigenfunctions* $e_0, e_1, \ldots,$. *The equation*

$$\int_a^b k(x,t)f(t)\,dt = g(x) \quad (a \leq x \leq b) \tag{5.5.2}$$

has a unique solution if and only if the series

$$\sum_{r=0}^{\infty} |\lambda_r a_r|^2 \tag{5.5.3}$$

converges, where $\{a_n\}$ *are the generalized Fourier coefficients of g.*
  The solution is

$$\sum_{r=0}^{\infty} \lambda_r a_r e_r \tag{5.5.4}$$

which converges in mean to f.
  If, in addition, the kernel is such that

$$\int_a^b k^2(x,t)\,dt \tag{5.5.5}$$

is a bounded continuous function, then the series (5.5.4) *converges absolutely and uniformly to* f.

Although this theorem provides a complete answer to the question of solving (5.5.2) when the kernel is positive definite, it also emphasizes the difficulties that have to be overcome in the solution of a particular equation. The calculation of a complete eigensystem is a daunting task.

Note that since the series (5.5.3) cannot possibly converge for all sequences $\{a_n\}$, the equation cannot be solved for all g. The following example will illustrate this.

EXAMPLE 5.5.4. We return to the equation of Example 5.5.1. The eigensystem in this case is

$$\{n^2\pi^2, 2^{\frac{1}{2}} \sin(n\pi x)\} \quad (n = 1, 2, \ldots),$$

and so the kernel is positive definite. The generalized Fourier coefficients are the Fourier-sine coefficients of g namely

$$a_n = 2^{\frac{1}{2}} \int_0^1 g(t) \sin(n\pi t)\,dt \quad (n = 1, 2, \ldots).$$

The series (5.5.3) becomes

$$\pi^4 \sum_{r=1}^{\infty} r^4 a_r^2$$

and, for (5.5.4) to give a solution, the function g must be such that this converges. After two integration by parts we find that

$$a_n 2^{-\frac{1}{2}} = \frac{1}{n\pi}[g(0) - (-1)^n g(1)] - \frac{1}{n^2\pi^2} \int_0^1 g''(t) \sin(n\pi t) \, dt.$$

If g does not vanish at one or both of the end points, then a_n will behave like $1/n$ and the series (5.5.3) cannot converge. On the other hand, if g vanishes at both end points and g'' is continuous in $[0, 1]$, then a_n will be of order $1/n^2$ in which case the series (5.5.4) will converge to $-g''$ as required.

It is possible to extend this theory to deal with kernels which are not symmetric. This involves the notions of singular functions and values. The motivation behind this is to multiply through the equation by $K*$ to give $K*Kf = K*g$. The new integral operator has a symmetric kernel and so the above theorem can be applied. Smithies examines this problem in detail in his monograph (1958). (It is worth remarking that there is a finite dimensional analogue of this, see for example Noble and Daniel (1988).)

5.6. VOLTERRA INTEGRAL EQUATIONS

The other founder of the theory of integral equations was Volterra, and it is appropriate that a particular class of equations should bear his name. A selection of his various contributions to the theory and other writings will be found in Volterra (1959). The theory of linear Volterra equations does not have the richness of Fredholm's. On the other hand, the equations possess a flexibility which lends itself to a wide variety of numerical and manipulative techniques.

DEFINITION 5.6.1. The equation

$$f(x) = g(x) + \int_a^x k(x, t) f(t) \, dt \quad (a \le x \le b) \tag{5.6.1}$$

is a *linear Volterra equation of the second kind.*

When k is a Fredholm kernel the general Fredholm theorems are applicable, but the Hilbert–Schmidt theory is now void, since the kernel which is adjoint to that of (5.6.1) vanishes for $t \ge x$ and so equation (5.3.2) of Definition 5.3.1 is satisfied only when the kernel is identically zero. Furthermore there is one important component of Fredholm's theory which is missing and that concerns the eigenvalue problem.

THEOREM 5.6.1. *If k is a Fredholm kernel then the homogeneous equation*

$$f(x) = \lambda \int_a^b k(x, t) f(t) \, dt \quad (a \le x \le b) \tag{5.6.2}$$

has only the trivial solution $\lambda = 0$, $f(x) = 0$. In fact for (5.6.1) *Fredholm's denominator* $D(\lambda)$ *is*

$$\exp\left(-\lambda \int_a^b k(t, t)\, dt\right) \tag{5.6.3}$$

which has no zeros.

It follows from Fredholm's alternative, Theorem 5.2.8, that the inhomogeneous equation (5.6.1) will have a solution for any value of λ and any function $g \in L^2[a, b]$.

The following example is of a homogeneous equation of the second kind with a kernel which is not of Fredholm type.

EXAMPLE 5.6.1.   The equation

$$f(x) = \frac{\lambda x^{-\alpha}}{\Gamma(\alpha)} \int_0^x \frac{f(t)}{(x-t)^{1-\alpha}}\, dt \quad (0 \le x \le 1 \text{ where } 0 < \alpha < 1),$$

has the solution

$$f(x) = x^{\beta - 1} \quad \text{where } \lambda = \Gamma(\beta)/\Gamma(\alpha + \beta), \quad \text{for any } \beta > 0,$$

and so this equation has a continuum of eigenvalues.

THEOREM 5.6.2.   *The iterated Volterra kernels are given by*

$$k_{n+1}(x, t) = \begin{cases} \displaystyle\int_t^x k(x, s)k_n(s, t)\, ds & (a \le t \le x \le b), \\[2mm] 0 & (a \le x < t \le b), \end{cases} \tag{5.6.4a}$$

for $n = 1, 2, \ldots$ *with*

$$k_1(x, t) = k(x, t) \quad (a \le t \le x \le b). \tag{5.6.4b}$$

DEFINITION 5.6.2.   *The resolvent kernel for the equation* (5.6.1) *is*

$$h(x, t; \lambda) = -\sum_{r=0}^{\infty} \lambda^r k_{r+1}(x, t), \tag{5.6.5}$$

and, if k is continuous for $a \le t \le x \le b$, then the series is absolutely and uniformly convergent for all λ.

THEOREM 5.6.3.   (a) *The relationships of Theorem 5.2.7 for Volterra kernels become*

$$k(x, t) + h(x, t; \lambda) = \lambda \int_t^x k(x, s)h(s, t; \lambda)\, ds = \lambda \int_t^x h(x, s; \lambda)k(s, t)\, ds, \tag{5.6.6}$$

(b) Let k be a Fredholm kernel and $f \in L^2[a,b]$, then the unique solution of (5.6.1) is in $L^2[a,b]$ and is given by

$$g(x) - \lambda \int_a^x h(x,t;\lambda)g(t)\,dt \quad (a \le x \le b).$$
(5.6.7)

EXAMPLE 5.6.2.   Let

$$k(x,t) = \begin{cases} a(x)b(t) & (a \le t \le x \le b), \\ 0 & (a \le x < t \le b), \end{cases}$$

then

$$k_n(x,t) = \frac{a(x)b(t(}{(n-1)!}\left[\int_t^x a(s)b(s)\,ds\right]^{n-1} \quad (n = 1,2,\ldots),$$

and

$$h(x,t;\lambda) = -a(x)b(t)\exp\left\{\lambda \int_t^x a(s)b(s)\,ds\right\}.$$

DEFINITION 5.6.3.   The *linear Volterra equation of the first kind* has the form

$$g(x) = \int_a^x k(x,t)f(t)\,dt \quad (a \le x \le b).$$
(5.6.8)

This can be replaced by an equation of the second kind if

(a) g is differentiable,
(b) the kernel k is differentiable with respect to x and the resulting function is integrable,
(c) $k(x,x)$ is continuous and is never zero.

For in that case we can differentiate (5.6.8) to give

$$g'(x) = k(x,x)f(x) + \int_a^x k_x(x,t)f(t)\,dt,$$
(5.6.9)

which can be divided through by $k(x,x)$, since it does not vanish, to give a Volterra equation of the second kind. The equation will have a unique solution.

Particular equations of the first kind which can be solved explicitly will be considered in §5.8.

DEFINITION 5.6.4.   The general non-linear Volterra integral equation with a *weak algebraic singularity* has the form

$$f(x) = g(x) + \frac{1}{\Gamma(\alpha)}\int_a^x \frac{G[x,t,f(t)]}{(x-t)^{1-\alpha}}\,dt, \quad \text{where } 0 < \alpha < 1,\ a \le x \le b. \quad (5.6.10)$$

THEOREM 5.6.4. *Let G satisfy a uniform Lipschitz condition in f and be continuous in the square $a \leq x, t \leq b$. If $g \in C[a, b]$ then (5.6.10) will have a unique solution which is continuous in $[a, b]$.*

When a numerical scheme has to be constructed for a solution of such an equation, it is advisable to be aware of possible singularities in its solution. The case of $\alpha = \frac{1}{2}$, which is probably the most common, has been studied by de Hoog and Weiss (1972). They show that the equation should be regarded as a combination of two equations, each of which has a well-behaved solution.

THEOREM 5.6.5.   (de Hoog and Weiss (1972)) *Let*

$$f(x) = g_0(x) + x^{\frac{1}{2}} g_1(x) + \int_0^x \frac{G[x, t, f(t)]}{(x - t)^{\frac{1}{2}}} dt \quad (0 \leq x \leq 1), \qquad (5.6.11)$$

where

(a) *g_0, g_1 are n times continuously differentiable in $[0, 1]$,*
(b) *$G[x, t; f]$ is n times continuously differentiable with respect to x and t for $0 \leq t \leq x \leq 1$ and 2n times continuously differentiable with respect to f for all f,*
(c) *$G[x, t; f]$ satisfies a Lipschitz condition with respect to f for all f and for $0 \leq t \leq x \leq 1$.*

Then the solution of (5.6.11) is given by

$$f(x) = f_0(x) + x^{\frac{1}{2}} f_1(x) \quad (0 \leq x \leq 1), \qquad (5.6.12)$$

where f_0 and f_1 are n times continuously differentiable in $[0, 1]$, and are the unique solutions of the following coupled equations

$$f_0(x) = g_0(x) + \int_0^x t^{\frac{1}{2}} \frac{G_0[x, t; f_0(t) f_1(t)]}{(x - t)^{\frac{1}{2}}} dt \quad (0 \leq x \leq 1), \quad (5.6.13)$$

$$f_1(x) = g_1(x) + x^{-\frac{1}{2}} \int_0^x \frac{G_1[x, t; f_0(t) f_1(t)]}{(x - t)^{\frac{1}{2}}} dt$$

where the kernels are given by

$$G_0[x, t; f_0, f_1] = \{G[x, t; f_0 + f_1 t^{\frac{1}{2}}] - G[x, t; f_0 - f_1 t^{\frac{1}{2}}]\}/(2t^{\frac{1}{2}}),$$
$$G_1[x, t; f_0, f_1] = \frac{1}{2}\{G[x, t; f_0 + f_1 t^{\frac{1}{2}}] + G[x, t; f_0 - f_1 t^{\frac{1}{2}}]\}, \qquad (5.6.14)$$

for $0 \leq t \leq x \leq 1$.

EXAMPLE 5.6.3.   Let

$$f(x) = g_0(x) + x^{\frac{1}{2}} g_1(x) + \int_0^x \frac{f(t)}{(x - t)^{\frac{1}{2}}} dt \quad (0 \leq x \leq 1).$$

Then the kernels k_0 and k_1 of Theorem 5.6.5 simplify to f_1 and f_0 respectively, and the equations (5.6.13) become

$$f_0(x) = g_0(x) + \int_0^x \frac{t^{\frac{1}{2}}}{(x-t)^{\frac{1}{2}}} f_1(t)dt \quad (0 \le x \le 1),$$

$$f_1(x) = g_1(x) + \int_0^x \frac{x^{-\frac{1}{2}}}{(x-t)^{\frac{1}{2}}} f_0(t)dt \quad (0 \le x \le 1).$$

5.7. SINGULAR KERNELS

At about the same time as Fredholm and Volterra were developing their theories for reasonably well-behaved kernels, Poincaré, in his study of tides, and Hilbert, in connection with boundary-value problems, were led to investigate integral equations with singular kernels. This was continued during the 1930s by Russian mathematicians led by Mikhlin, Muskhelishvili and Vekua who used it to reformulate boundary-value problems for Laplace's equation and the biharmonic equation as integral equations. Mikhlin (1957) treats the theory of linear equations with both Fredholm and singular kernels. The monograph by Muskhelishvili (1953) is concerned solely with the theory of singular equations. Each author gives many examples of the applications of the theories. There is a useful textbook by Gakhov (1963).

It is necessary to restrict the class of functions under discussion to those which satisfy a Hölder condition.

DEFINITION 5.7.1.   The function μ satisfies a *Hölder condition* with exponent α in the region D of the complex plane if there exists a constant K such that for all $z_1, z_2 \in D$

$$|\mu(z_1) - \mu(z_2)| \le K|z_1 - z_2|^\alpha \quad \text{where } 0 < \alpha < 1. \tag{5.7.1}$$

We shall say for brevity that 'μ is in $H(\alpha)$' when it satisfies such a condition.

The theory depends on the concept of the Cauchy principal value integral [see also IV, §§9.10.2 and 9.10.4].

DEFINITION 5.7.2.   Let L be a smooth curve without double points in the complex plane, and z a point of L. If z is not an end point of the curve, the *Cauchy principal value* of the line integral

$$\int_L \frac{\mu(\zeta)}{(\zeta - z)} d\zeta \tag{5.7.2}$$

is defined to be

$$\lim_{\varepsilon \to 0+} \int_{L_\varepsilon} \frac{\mu(\zeta)}{(\zeta - z)} d\zeta, \tag{5.7.3}$$

where L_ε consists of those points ζ of L which lie outside the disc

$$|\zeta - z| < \varepsilon.$$

When the curve is closed we shall denote it by C; in this case there is no restriction on the position of the point z except that it should lie on C (when z does not lie on C the integral exists in the usual way).

THEOREM 5.7.1.   *Let μ satisfy a Hölder condition with exponent α on the smooth contour L. Then (5.7.2) defines a function which also satisfies a Hölder condition with the same exponent.*

EXAMPLE 5.7.1.   The Cauchy principal value of the integral

$$\int_a^b \frac{1}{(t-x)}\,dt \quad (a < x < b),$$

is

$$\lim_{\varepsilon \to 0+}\left[\int_a^{x-\varepsilon}\frac{1}{(t-x)}\,dt + \int_{x+\varepsilon}^b\frac{1}{(t-x)}\,dt\right] = \lim_{\varepsilon \to 0+}\{\log[\varepsilon/(x-a)] + \log[(b-x)/\varepsilon]\}$$

$$= \log[(b-x)/(x-a)].$$

We shall show how Cauchy principal value integrals can be used to reformulate the Dirichlet problem for Laplace's equation in two dimensions as a Fredholm integral equation of the second kind. This equation can be shown to have a unique solution, and so demonstrates the existence of a unique solution of the Dirichlet problem. It is an illustration of that branch of analysis now known as 'the boundary integral method' which is important in applied mathematics and numerical analysis.

The method requires the *Plemelj formulae* which give the limiting form of the Cauchy integrals occurring in complex variable theory [see IV, Theorem 9.6.1].

Before the formulae are given we make some preliminary remarks. Let L be a smooth curve without double points, z *not* be a point of L, and let μ satisfy a Hölder condition on L. Then the function w defined by

$$w(z) = \frac{1}{\pi i}\int_L \frac{\mu(\zeta)}{(\zeta - z)}\,d\zeta \tag{5.7.4}$$

is analytic in the whole of the complex plane except on L, and will vanish at infinity if L is bounded.

It is necessary to define the inside and outside of a curve. If L is closed, then this is done in the usual intuitive manner. The *inside* of L is the interior of the region which is bounded by L and which lies on the left hand when L is traversed in the direction of increasing arc length. When L is not closed, the same convention is adopted by supposing the 'immediate' interior lies to the left. The outside of L is defined in a similar manner.

Denote by $w_+(z)$ the interior limiting value of $w(z')$, given by (5.7.4), as z' tends to z, a point of L, along a path which lies entirely inside D and which is not tangential to L. Similarly let $w_-(z)$ be the exterior limit. If L is not closed then z must not be an end point.

THEOREM 5.7.2. (The Plemelj formulae)

$$w_\pm(z) = \pm \mu(z) + \frac{1}{\pi i} \int_L \frac{\mu(\zeta)}{(\zeta - z)} d\zeta \quad \text{for } z \in L. \tag{5.7.5}$$

The function μ, sometimes called the *density function*, is assumed to be in $H(\alpha)$. It can be taken to be either real or imaginary without loss of generality. The integral is a Cauchy principal value integral.

The Poincaré–Bertrand formula for the compounding, or the interchange of order, of singular integrals is a consequence of the Plemelj formulae. It shows that such integrals need to be handled with care.

THEOREM 5.7.3. (Poincaré–Bertrand formula) *Let L be a smooth arc or a closed contour, and let $k(\zeta, t)$ satisfy a Hölder condition in ζ and t on L. Then for $z \in L$,*

$$\int_L \int_L \frac{1}{(\zeta - z)(t - \zeta)} K(\zeta, t) \, dt \, d\zeta = -\pi^2 K(z, z) + \int_L \int_L \frac{1}{(t - \zeta)(\zeta - z)} K(\zeta, t) \, d\zeta \, dt. \tag{5.7.6}$$

COROLLARY 5.7.4.

$$\int_L \int_L \frac{1}{(\zeta - z)} K(\zeta, t) \, dt \, d\zeta = \int_L \int_L \frac{K(\zeta, t)}{(\zeta - z)} \, d\zeta \, dt \quad (z \in L). \tag{5.7.7}$$

EXAMPLE 5.7.2. (Interior Dirichlet problem) Let D be the interior of a simply connected domain in $\mathbb{R}^2$ (a domain is simply connected if its complement is conected and so, in particular, it will not contain holes [see also IV, Definition 13.5.1] and let C be its boundary. The boundary will be assumed to be parametrized with respect to arc length, and the function which describes this is assumed to be in $H(\alpha)$. The problem is to find a function u which satisfies Laplace's equation in D namely that

$$\frac{\partial^2 u}{\partial x^2} + \frac{\partial^2 u}{\partial y^2} = 0 \quad ((x, y) \in D)$$

and is such that

$$u(x, y) = f(z) \quad \text{for } x + iy = z \in C$$

for a given function f which is continuous on C [see IV, §§8.2 and 9.13].

Since u is harmonic in D, it is the real part of a function w which is analytic in D. We shall assume that w can be written as

$$w(z) = \frac{1}{\pi i} \int_C \frac{\mu(\zeta)}{(\zeta - z)} \, d\zeta \quad (z \in D), \tag{5.7.8}$$

where μ is a *real valued* function which is in $H(\alpha)$. Then by the first of Plemelj's formulae

$$w_+(z) = \mu(z) + \frac{1}{\pi i} \int_C \frac{\mu(\zeta)}{(\zeta - z)} \, d\zeta \quad (z \in C). \tag{5.7.9}$$

Set

$$\text{Real part } \{w_+(z)\} = f(z)$$

and take the real part of (5.7.9) to give the integral equation of the second kind

$$f(z) = \mu(z) + \frac{1}{\pi} \int_C \mu(\zeta) \, \text{Im} \, [d\zeta/(\zeta - z)], \quad \text{where } z \in C. \tag{5.7.10}$$

The kernel

$$\text{Im} \, [d\zeta/(\zeta - z)] \tag{5.7.11}$$

is a continuous function in each of its variables except at points of discontinuity of curvature of C when $\zeta = z$, where there is a step discontinuity. Consequently it is a Fredholm kernel. The equation (5.7.10) can be shown to have a unique solution, see, for example, Mikhlin (1964). When it is found, the required harmonic function can be calculated from

$$u(x, y) = \text{Re} \, \frac{1}{\pi i} \int_C \frac{\mu(\zeta)}{(\zeta - z)} \, d\zeta \quad (z \in D). \tag{5.7.12}$$

This method of solving the Dirichlet problem can be traced back to the idea of representing a harmonic function by a doublet distribution on C. For the classical approach see Kellogg (1953) or Lovitt (1950). It is probably true to say that the Dirichlet problem was the inspiration for Fredholm's theory and may be called, fancifully, the grandfather of functional analysis.

EXAMPLE 5.7.3.   (Poisson's integral formula) The problem is to calculate a function which is harmonic in a disc, in terms of its values on the circumference. This is the Dirichlet problem with the unit disc as the region D.
  Write

$$z = e^{is}, \quad \zeta = e^{i\sigma}, \quad (0 \le s, \sigma < 2\pi)$$

then (5.7.10) becomes

$$f(e^{is}) = \mu(e^{is}) + \frac{1}{2\pi} \int_0^{2\pi} \mu(e^{i\sigma}) \, d\sigma \quad (0 \le s < 2\pi). \tag{5.7.13}$$

This is an equation with a degenerate kernel and has the unique solution

$$f(e^{is}) - \frac{1}{4\pi} \int_0^{2\pi} f(e^{i\sigma}) d\sigma \quad (0 \leq s < 2\pi).$$

When this is substituted in (5.7.13) for $\mu(e^{is})$ we get, after some simplification, *Poisson's formula* [IV, §8.2]

$$u(r \cos s, r \sin s) = \frac{1}{2\pi} \int_0^{2\pi} \frac{1 - r^2}{1 - 2r \cos(s - \sigma) + r^2} f(e^{i\sigma}) d\sigma \quad (0 \leq s < 2\pi, r^2 < 1).$$

$$(5.7.14)$$

The Neumann problem for Laplace's equation, which is that of finding a function which is harmonic in a region whose normal derivative is known on the boundary of the region, can also be reformulated as the problem of solving a linear Fredholm integral equation of the second kind. For other applications in potential theory and the theory of elasticity in two dimensions see Mikhlin (1957) and Muskhelishvili (1953).

The density function μ in Example 5.7.2 could have been taken to be wholly imaginary and this would have led to an integral equation of the first kind, in which the integral would have to be interpreted as a Cauchy principal value integral.

There is a general theory for integral equations with singular kernels of the form

$$k(z, \zeta)/(\zeta - z),$$

where k satisfies a Hölder condition in ζ for each $z \in L$. This differs considerably from the Fredholm theory. We conclude this section by giving a flavour of it; for further results see Gakhov (1966), Mikhlin (1957) or Muskhelishvili (1953).

THEOREM 5.7.5.   *Let*

$$af(z) + \frac{b}{\pi i} \int_c \frac{f(\zeta)}{(\zeta - z)} d\zeta = g(z) \quad (z \in C), \qquad (5.7.15)$$

where a and b are constants, and g satisfies a Hölder condition on C, a closed smooth contour. If $a^2 \neq b^2$ then

$$f(z) = \frac{a}{a^2 - b^2} g(z) - \frac{b}{a^2 - b^2} \frac{1}{\pi i} \int_c \frac{g(\zeta)}{(\zeta - z)} d\zeta \quad (z \in C). \qquad (5.7.16)$$

When $a = \pm b$ the homogeneous equation which corresponds to (5.7.5) is

$$f(z) \pm \frac{1}{\pi i} \int_c \frac{f(\zeta)}{(\zeta - z)} d\zeta = 0 \quad (z \in C), \qquad (5.7.17)$$

which is satisfied by any function analytic outside (inside) D, the domain whose boundary is C, and so there is an unbounded number of eigenfunctions in these cases.

THEOREM 5.7.6.   *Let*

$$af(z) + \frac{b}{\pi i} \int_L \frac{f(\zeta)}{(\zeta - z)} d\zeta = g(z) \quad (z \in L),$$

$$(5.7.18)$$

where L is a smooth arc without double points, with end points α and β and $a^2 \neq b^2$. Then

$$f(z) = \frac{a}{a^2 - b^2} g(z) - \frac{b}{a^2 - b^2} \frac{1}{\pi i} \int_L \frac{\gamma(\zeta)}{\gamma(z)} \frac{g(\zeta)}{(\zeta - z)} d\zeta + A(z - \alpha)^{m-1}(z - \beta)^{-m} \quad (z \in L),$$

$$(5.7.19)$$

where $\gamma(z) = (z - \alpha)^{-m}(z - \beta)^m$, and

$$m = \frac{1}{2\pi i} \ln \left[(a + b)/(a - b) \right]$$

$$(5.7.20)$$

with $0 \leq \mathrm{Re}(m) < 1$.

The constant A is arbitrary and can be chosen so that the solution is bounded at one of the end points of the arc. (This is useful in thin aerofoil theory, where it is chosen so that the solution satisfies the Joukowski condition that the velocity at the trailing edge is finite. [IV, §9.13].)

EXAMPLE 5.7.4.

(a) Let $a = 0$, $b = 1$ in (5.7.15), then the solution of the singular integral equation of the first kind

$$\frac{1}{\pi i} \int_C \frac{f(\zeta)}{(\zeta - z)} d\zeta = g(z) \quad (z \in C)$$

is

$$f(z) = \frac{1}{\pi i} \int_C \frac{g(\zeta)}{(\zeta - z)} d\zeta \quad (z \in C).$$

If C is taken to be the real line closed by a semi-circle at infinity, these relations become, formally at least, those for the infinite Hilbert transform H which is its own inverse:

$$Hf(x) = \frac{1}{\pi i} \int_{-\infty}^{\infty} \frac{f(t)}{(t - x)} dt \quad (-\infty < x < \infty).$$

(b) Let $a = 0$, $b = i$ in (5.7.19) and let L be the interval $-1 < x < 1$, then $m = \frac{1}{2}$

and the solution of

$$\frac{1}{\pi} \int_{-1}^{1} \frac{f(t)}{(t-x)} dt = g(x) \quad (-1 < x < 1)$$

is

$$\frac{-1}{\pi} \int_{-1}^{1} \frac{(1-t^2)^{\frac{1}{2}}}{(1-x^2)^{\frac{1}{2}}} \frac{g(t)}{(t-x)} dt + A(1-x^2)^{-\frac{1}{2}} \quad (-1 < x < 1),$$

were A is a constant. This result contains, as special cases, the following well-known relations which have been useful in thin aerofoil theory:

$$\frac{1}{\pi} \int_{-1}^{1} (1-t^2)^{\frac{1}{2}} U_{n-1}(t) \frac{dt}{(t-x)} = -T_n(x) \quad (-1 < x < 1),$$

$$\frac{1}{\pi} \int_{-1}^{1} (1-t^2)^{-\frac{1}{2}} T_n(t) \frac{dt}{(t-x)} = U_{n-1}(x) \quad (-1 < x < 1),$$

(5.7.21)

where T_n and U_{n-1} are the Chebyshev polynomials of the first and second kind of degrees n and $n-1$ respectively [IV, §10.5.3 and III, §6.3.4].

Finally, we make the observation that the homogeneous equation for the unclosed contour L, namely

$$f(z) = \frac{\lambda}{\pi i} \int_{L} \frac{f(\zeta)}{(\zeta - z)} d\zeta \quad (z \in L),$$

(5.7.22)

has for its solution

$$1/[(z-\alpha)^{1-m}(z-\beta)^{m}],$$

(5.7.23)

where α and β are the end points of L and

$$m = \frac{1}{2\pi i} \log\left(\frac{1-\lambda}{1+\lambda}\right) \quad (-1 < \lambda < 1).$$

Thus there is a continuum of eigenvalues in contrast to the situation in the case of an equation which a Fredholm kernel.

5.8. DIFFERENCE KERNELS

A general treatment of the application of transform methods to the solution of certain types of integral equation will be found in IV, 13. But it is appropriate to discuss briefly a class of equations which can be solved explicitly.

Abel's integral equation (1926)

This was probably the first non-trivial integral equation to be solved explicitly. It arises as follows: suppose that a particle moves without friction under gravity on a continuous curve which lies in the vertical plane. The problem is to find the shape of the curve when the time of descent $t(y)$ from any point of the curve of vertical height y to ground level is known. It is not difficult to show that the equation of the curve satisfies the integral equation,

$$t(y) = \int_0^y \frac{f(\eta)}{[2g(y-\eta)]^{\frac{1}{2}}} \, d\eta, \qquad (5.8.1)$$

where g is the acceleration due to gravity, and the equation of the curve is given by

$$\frac{ds}{dy} = f(y).$$

Abel himself generalized the equation as

$$g(x) = \frac{1}{\Gamma(\alpha)} \int_0^x \frac{1}{(x-t)^{1-\alpha}} f(t) dt \quad \text{for } x \geq 0, \qquad (5.8.2)$$

where $0 < \alpha < 1$, and showed that

$$f(x) = \frac{1}{\Gamma(1-\alpha)} \frac{d}{dx} \int_0^x \frac{1}{(x-t)^\alpha} g(t) dt \quad \text{for } x \geq 0. \qquad (5.8.3)$$

EXAMPLE 5.8.1.   To find the shape of the curve so that the time of descent is constant, e.g. equal to 1. For this we set $t(y) = 1$. Then, with the aid of the solution (5.8.3) with $\alpha = \frac{1}{2}$, it will be found that

$$f(y) = \frac{1}{\pi} \left(\frac{2g}{y} \right)^{\frac{1}{2}}.$$

This is the equation of a cycloid with vertex at the origin (the locus of a point on the circumference of a circular wheel of radius $(2g)^{\frac{1}{2}}/\pi$ which rolls along the horizontal ground without sliding [V, §3.8.4]).

EXAMPLE 5.8.2.   (The tomography problem) One of the outstanding successes in the application of mathematics this century has been the *practical* inversion of the Radon transform [IV, §13.9]. The transform is an integral equation of the first kind and its solution forms the basis of mathematical tomography: the problem of reconstructing a three-dimensional body from two-dimensional images. We shall present here the simplest version of the problem. The body which is under examination will be assumed to have circular symmetry. For example, imagine a vertical tree trunk in the form of a right circular cylinder whose growth rings are concentric circles. Suppose that the average density $F(p)$ can be measured along the secant $r = p \cos(\theta)$, where r denotes distance from the centre of the circle. Denote by $\rho(r)$ the density on the circle of radius r, then the following relation is satisfied,

$$(1 - p^2)^{\frac{1}{2}} F(p) = \int_p^1 \frac{r}{(r^2 - p^2)^{\frac{1}{2}}} \rho(r) dr \quad \text{for } 0 < p < 1. \qquad (5.8.4)$$

(The radius of the trunk has been taken to be unity.) This can be transformed into Abel's equation in the special case of $\alpha = \frac{1}{2}$, and the solution will be found to be

$$\rho(r) = -\frac{2}{\pi r} \frac{d}{dr} \int_r^1 \frac{(1 - p^2)^{\frac{1}{2}}}{(p^2 - r^2)^{\frac{1}{2}}} F(p) dp \quad (0 < r < 1). \qquad (5.8.5)$$

The generalization from circular symmetry to a general distribution of density can be formulated as the problem of solving an infinite sequence of integral equations for the Fourier coefficients of the density function. This forms the basis of computerized tomography in which mean density measurements made by taking X-ray are transformed to give information about the actual density at any internal point. Such information is of vital importance for precise surgery. The monograph by Natterer (1986) is devoted to the mathematical aspects of this topic.

Difference Kernels

Abel's equation is a special case of the equation of the first kind

$$g(x) = \int_0^x k(x - t) f(t) dt \quad \text{for } x > 0. \qquad (5.8.6)$$

the kernel is known as a *difference kernel*. The right-hand side of (5.8.6) is usually called the *convolution (integral)* of k and f.

If g is known for $0 \le x < \infty$ and $g(0) = 0$, the equation can be solved by the use of the Laplace transform [see IV, §13 for a review of its properties]. For our purposes we need little more than its definition.

DEFINITION 5.8.1.   The *Laplace transform F of f* is

$$F(p) = \int_0^\infty e^{-pt} f(t) dt \quad \text{for } p > 0; \qquad (5.8.7)$$

f will be called the *inverse Laplace transform of F*.

(The problem of finding f from its Laplace transform is that of solving an integral equation of the first kind. The inversion formula can be found in IV, 13, where it will be seen that it involves a knowledge of values of $F(p)$ for complex values of p. This is another illustration of the remark made in §5.5 regarding the analytic continuation of the free term.)

The method is formally as follows: denote the Laplace transforms of f, g and k by F, G and K respectively. Then (5.8.6) is

$$G(p) = \int_0^\infty e^{-px} \int_0^x k(x-t)f(t)dt\,dx.$$

After the interchange of the order of integration in the double integral and the transformation $x \to x + t$, it will be seen that the double integral reduces to the product of single integrals which are the Laplace transforms of f and k, and we have

$$G(p) = F(p)K(p). \tag{5.8.8}$$

Consequently the Laplace transform of the convolution of k and f becomes the product of their Laplace transforms. This relation can be written as

$$F(p) = G(p)/K(p). \tag{5.8.9}$$

Define L as $L = 1/K$, then (5.8.9) becomes

$$F(p) = G(p)L(p). \tag{5.8.10}$$

The comparison of (5.8.8) with (5.8.10) suggests that if we can find a function l, say, whose Laplace transform is L then

$$f(x) = \int_0^x l(x-t)g(t)dt.$$

Unfortunately this is an oversimplification. This will be appreciated from the following general reasoning.

If w is a 'reasonable' function, then its Laplace transform W will vanish at infinity. Consequently $1/W$ will be unbounded at infinity and therefore cannot be the Laplace transform of a 'reasonable' function. However, it is usually possible to avoid this situation by means of some preliminary manipulations. The following examples will illustrate this.

EXAMPLE 5.8.3. Let

$$g(x) = \int_0^x (x-t)f(t)dt \quad (x \geq 0),$$

where $g(0) = 0$. The differentiation of the equation gives

$$g'(x) = \int_0^x f(t)dt$$

and so we assume that $g'(0) = 0$. A further differentiation shows that the solution is given by

$$f = g''.$$

When we use the method described above we get, since the Laplace transform of x is $1/p^2$, the equation

$$G(p) = F(p)/p^2.$$

Hence

$$F(p) = p^2 G(p).$$

But no reasonable function will satisfy

$$\int_0^\infty e^{-px} h(x) dx = p^2.$$

However, two integrations by parts will give

$$p^2 G(p) = p^2 \int_0^\infty e^{-px} g(x) dx = g(0) + pg'(0) + \int_0^\infty e^{-px} g''(x) dx.$$

And so, since $g(0) = g'(0) = 0$, we get

$$\int_0^\infty e^{-px} f(x) dx = \int_0^\infty e^{-px} g''(x) dx,$$

which gives the expected result that $f = g''$.

(It may be of interest to note that Abel's equation (5.8.2) can be regarded as the *fractional integral* of the function f, and the solution (5.8.3) as the corresponding *fractional derivative* of g. This relationship has been found to be fruitful in the investigation of certain initial-value problems for hyperbolic differential equations, see for example Baker and Copson (1939) or Duff (1956).)

The next example is rather more complicated; but the motivation behind the method of solution is the same.

EXAMPLE 5.8.4.   To find the solution of

$$g(x) = \int_0^x \log|x - t| f(t) dt \quad \text{for } x \geq 0, \text{ where } g(0) = 0. \qquad (5.8.11)$$

From Erdélyi *et al.* (1954), p. 148, equation (1), we find that the Laplace transform of the logarithmic function is

$$- \log(\gamma p)/p,$$

where

$$\log(\gamma) = \Gamma'(1) = -0.5572\ldots,$$

and so we have

$$F(p) = - G(p)p/\log(\gamma p).$$

This will be manipulated into a form suitable for inversion as follows: integrate the integral for G by parts twice and impose the condition that $g(0) = 0$. The result will be that

$$F(p) = -\left[g'(0) + \int_0^\infty e^{-px} g''(x) dx \right] \Big/ [p \log(\gamma p)]. \qquad (5.8.12)$$

From the same book of tables, p. 251, equation (11), we see that

$$1/[p \log(p)]$$

is the Laplace transform of v where

$$v(x) = \int_0^\infty \frac{x^s}{\Gamma(s+1)} ds \quad (x > 0).$$

It follows that

$$\int_0^\infty e^{-px} v\left(\frac{x}{\gamma}\right) dx = \frac{1}{p \log(\gamma p)},$$

which, when used in conjunction with equation (5.8.12), will give the solution of (5.8.11) as

$$f(x) = -v(x)g'(0) + \int_0^x g''(t) v\left(\frac{x-t}{\gamma}\right) dt.$$

The inhomogeneous equation

$$f(x) = g(x) + \int_0^x k(x-t) f(t) dt \quad (x \geq 0), \qquad (5.8.13)$$

which is usually known as the *renewal equation*, can often be solved in a similar fashion.

Take its Laplace transform, then with the above notation we have

$$F(p) = G(p) + K(p) F(p),$$

which can be solved for F to give

$$F = G/(1 - K).$$

It follows that the required function f is the inverse Laplace transform of $G/(1 - K)$, and the required solution will be the convolution of g with the inverse Laplace transform of $1/(1 - K)$. This may have to be manipulated in particular cases, as above, for the inversion formula to be applicable.

EXAMPLE 5.8.5.   The Laplace transform of the equation

$$f(x) = g(x) - \int_0^x (x-t) f(t) dt \quad (x \geq 0) \qquad (5.8.14)$$

is

$$F(p) = G(p) - p^{-2} F(p),$$

and so

$$F(p) = G(p)p^2/(1 + p^2).$$

Rearrange this as

$$F(p) = G(p) - G(p)/(1 + p^2),$$

which is now in a form suitable for inversion.

The sine function has the Laplace transform $1/(1 + p^2)$ and it is now a simple matter to deduce that the solution of the equation is given by

$$f(x) = g(x) - \int_0^x g(t)\sin(x - t)dt \quad (x \ge 0)$$

If the equation has a complicated kernel, then some ingenuity may be required to rearrange in a suitable form. But it is clearly a powerful method when supported by a collection of Laplace transforms and inverse Laplace transforms such as will be found, for example, in the Bateman collection of Erdélyi *et al.* (1954).

The technique which has been given in this section has been presented in a formal way. A useful and readable reference for a more rigorous treatment of the theory of equations with difference kernels is Bellman and Cooke (1963).

It is worth pointing out that the Laplace transform is not the only one which can be used in the solution of such equations, see, for example, Titchmarsh (1948) who bases his investigations on the Fourier transform. This leads to a more delicate analysis and requires more in the way of mathematical sophistication from the reader.

There other types of equation which can be solved by transform methods, reference should be made to IV, 13 for a description of these and for further references.

5.9. NON-LINEAR EQUATIONS

There is a particular type of non-linear Fredholm equation for which there is an attractive theory. This is Hammerstein's equation and it is based on the Hilbert–Schmidt theory of §5.3.

DEFINITION 5.9.1. An equation of the form

$$f(x) + \int_a^b k(x, t)h[t, f(t)]dt = 0 \quad (a \le x \le b) \tag{5.9.1}$$

is called a *Hammerstein equation* when k is a real symmetric positive definite Fredholm kernel and the iterated kernel k_2 is continuous in the square $a \le x, t \le b$. The function $h \equiv h[t, f]$ is assumed to be continuous in t for $a \le t \le b$, and for all f.

THEOREM 5.9.1.   *Let h satisfy*

$$|h[t, f]| < C_1|f| + C_2,$$ (5.9.2)

where C_1 and C_2 are positive constants. Then Hammerstein's equation has at least one solution which is continuous in $[a, b]$ if

$$C_1 < \lambda_0,$$ (5.9.3)

where λ_0 is the smallest eigenvalue of k (since k is positive definite, all its eigenvalues are positive). If, in addition,

$$|h[t, f_1] - h[t, f_2]| < \lambda|f_1 - f_2| \quad (a \le t \le b, -\infty < f_1, f_2 < \infty),$$ (5.9.4)

where $0 < \lambda < \lambda_0$, then the solution is unique.

EXAMPLE 5.9.1.   (Duffing's problem): The differential equation, satisfied by the angular deflection θ from the vertical of a pendulum of length l which describes forced oscillations, is

$$\frac{d^2}{dx^2}\theta(x) + \alpha^2 \sin\theta(x) = F(x), \quad \text{where } \alpha^2 = \frac{g}{l}, \ (0 \le x \le 1),$$

and F is the forcing term. The problem is to determine under what conditions the solution of the equation has period 2 when F is an odd periodic function. This can be reformulated as the problem of finding solutions of the Hammerstein integral equation

$$\theta(x) = -\int_0^1 \gamma(x, t)h[t, \theta(t)]\,dt,$$

where

$$h[t, \theta(t)] = F(t) - \alpha^2 \sin\theta(t)$$

and γ is the kernel of Example 5.1.1. This kernel satisfies the conditions for k of Definition 5.9.1. Since $|\sin\theta| \le |\theta|$ we have

$$|h[t, \theta_1] - h[t, \theta_2]| = \alpha^2|\sin\theta_1 - \sin\theta_2| \le \alpha^2|\theta_1 - \theta_2|.$$

It follows from Theorem 5.9.1 that there is at least one solution, and that if $|\alpha| < \pi$ it is unique.

A phenomenon known as *bifurcation* arises for non-linear equations which is not present for linear ones. It is not confined to integral equations and can be described in the general context of a non-linear operator equation.

Suppose that we have to solve a non-linear operator equation whose solution depends on a parameter λ, for example the equation

$$F(f, \lambda) = 0.$$ (5.9.5)

Let the pair f^*, λ^* provide a solution of the equation. Then solutions are said to *bifurcate* from f^* at λ^* if there are two or more distinct solutions which approach f^* as λ approach λ^* along different paths. In that case λ^* is said to be a *bifurcation point*. A point which is not a bifurcation point is a *regular point*.

EXAMPLE 5.9.2.   The equation

$$f(x) = x + \lambda \int_0^1 4xt[f(t)]^2 dt \quad (0 \le x \le 1), \tag{5.9.6}$$

has a degenerate kernel and its solution is of the form

$$Ax,$$

where A is a constant. The substitution of this in the equation gives the the quadratic

$$A = 1 + \lambda A^2,$$

whose zeros are

$$[1 \pm (1 - 4\lambda)^{\frac{1}{2}}]/(2\lambda).$$

It follows that the bifurcation point, λ^*, is $\frac{1}{4}$. The two solutions which bifurcate from $f^*(x) = 2x$ correspond to the two signs of the radical.

There do not seem to be necessary and sufficient conditions for a point to be a bifurcation point. We give a necessary condition for one to be present, see Berger (1977) for further details. The theorem is valid for the more general non-linear Fredholm equation.

THEOREM 5.9.2.   *Let G satisfy the following conditions:*

(a) *it is continuous in $D = \{x, t, f : a \le x, t \le b$, and $-\infty < f < \infty\}$.*
(b) *G_f exists and is continuous in D,*
(c) *G_{ff} exists in D and is bounded there.*

If λ^* is a bifurcation point for

$$f(x) = g(x) + \lambda \int_a^b G[x, t; f(t)]dt \quad (a \le x \le b), \tag{5.9.7}$$

then it is an eigenvalue of the linear homogeneous Fredholm equation

$$f_0(x) = \lambda \int_a^b G_f[x, t, f^*(t)] f_0(t)dt \quad (a \le x \le b). \tag{5.9.8}$$

$$\left(G_f = \frac{\partial G}{\partial f} \quad \text{and} \quad G_{ff} = \frac{\partial^2 G}{\partial f^2}\right).$$

This provides a test for a point to be a regular point.

COROLLARY 5.9.3.   *Let λ^*, f^* satisfy (5.9.7), then λ^* is not a bifurcation point if (5.9.8) has only the trivial solution for $\lambda = \lambda^*$.*

EXAMPLE 5.9.3.   We continue with the equation of Example 5.9.2. In this case

$$G_f = 8xtf.$$

Let $f^*(x) = -x, \lambda^* = -2$, which provide a solution for (5.9.6). The homogeneous equation (5.9.8) becomes

$$f_0(x) = -\lambda \int_0^1 8xt^2 f_0(t)dt,$$

which has a non-trivial solution only if $\lambda = -\frac{1}{2}$. Consequently $\lambda^* = -2$ is not a bifurcation point.

On the other hand, with

$$f^*(x) = 2x, \qquad \lambda^* = \tfrac{1}{4}$$

(5.9.8) becomes

$$f_0(x) = \lambda \int_0^1 16xt^2 f_0(t)dt$$

for which the eigenvalue is $\frac{1}{4}$. It follows that $\lambda^* = \frac{1}{4}$ is a bifurcation point, as we have already seen.

The general theory of non-linear integral equations depends heavily on the methods of non-linear functional analysis, and much use is made of the Fréchet derivative, which is used to provide a local linearization as in the above example. Particular equations may be susceptible to explicit solution by *ad hoc* methods. Noble, in his survey of numerical methods (in Anselone (1964)) for non-linear integral equations, quotes Chandrasekhar as saying that a particular equation was 'fortunately non-linear'.

D. K.

REFERENCES

Anselone, P. M. (1964). *Nonlinear Integral Equations*, Wisconsin.
Baker, B. B. and Copson, E. T. (1939). *The Mathematical Theory of Huyghen's Principle*, Oxford.
Bellman, R. and Cooke, K. L. (1963). *Differential-difference Equations*, Academic Press.
Berger, M. S. (1977). *Nonlinearity and Functional Analysis*, Academic Press.
Cochrane, J. A. (1972). *The Analysis of Linear Integral Equations*, McGraw-Hill.
Duff, G. F. D. (1956). *Partial Differential Equations*, Toronto University Press.
Erdélyi, A. *et al.* (1954). *Tables of Integral Transforms*, Vol I, McGraw-Hill.
Fredholm, I. (1903). Sur une classe d'équations fonctionelles, *Acta Mathematica* **27**, 365–390.
Gakhov, F. D. (1966). *Boundary Value Problems*, Pergamon.

Gantmakher, F. R. and Krein, M. G. (1950). Oscillation Matrices and Kernels and Small Vibrations of Mechanical Systems, N.B.S. trans.

Groetsch, C. W. (1977). *General Inverses of Linear Operators*, M. Dekker Inc.

Gould, S. H. (1966). *Variational Methods for Eigenvalue Problems*, Oxford.

de Hoog, F. and Weiss, R. (1972). On the Solution of Volterra Integral Equation with a Weakly Singuiar Kernal, *SIAM J. Math. Anal.* **4**, 561–573.

Hopf, E. (1963). An Inequality for Positive Linear Integral Operators, *J. Maths and Mechanics* **12**, 683–692.

Jentzsch, R. (1912). Über Integralgleichungen mit positivem Kern, *J. für reine und angewandte Math.* **141**.

Karlin, S. (1968). *Total Positivity*, Stanford.

Karlin, S. and Studden, W. (1966). *Tchebycheff Systems: with Applications in Analysis and Statistics*, Interscience.

Kellogg, O. D. (1953). *Foundations of Potential Theory*, Dover.

Krasnoselskii, M. A. (1964). *Positive Solutions of Operator Equations*, Noordhoff.

Krein, M. G. and Rutman, M. A. (1962). Linear Operators leaving Invarient a Cone in a Linear Space, *AMS Trans. Series 1* **10**,

Lavrentiev, M. M. (1967). *Some Improperly Posed Problems of Mathematical Physics*, Springer.

Lovitt, W. V. (1950). *Integral Equations*, Dover.

Mikhlin, S. G. (1957). *Integral Equations*, Pergamon.

Mikhlin, S. G. (1964). *Variational Methods in Mathematical Physics*, Pergamon.

Muskhelishvili, N. I. (1953). *Singular Integral Equations*, Noordhoff.

Natterer, F. (1986). *The Mathematics of Computerised Tomography*, Wiley.

Noble, B. and Daniel J. W. (1988). *Applied Linear Algebra*, Prentice-Hall.

Smithies, F. (1958). *Integral Equations*, Cambridge University Press.

Titchmarsh, E. C. (1948). *Theory of Fourier Integrals*, Oxford.

Tricomi, F. G. (1957). *Integral Equations*, Interscience.

Volterra, V. (1959). *Theory of Functionals and of Integral and Integro-differential Equations*, Dover (reprint).

CHAPTER 6

Dynamical Systems

6.1. INTRODUCTION

The qualitative theory of differential equations was founded by Henri Poincaré around the turn of the century, and today has grown into an extensive and vigorous field. With the rise of interest in non-linear systems and new phenomena such as 'chaos', it has become vital to many areas of application. It exists under a variety of names, of which the most appropriate is probably *dynamical systems theory*.

The qualitative theory of ordinary differential equations (ODEs) was touched upon in IV, 7.11.4 and 7.11.5. In this chapter we describe the basic ideas, methods and examples of the theory from a unified point of view.

6.1.1. Ordinary Differential Equations in the Plane

To motivate the theory we consider a system of autonomous ODEs in two real variables x, y:

$$\frac{dx}{dt} = X(x, y), \qquad \frac{dy}{dt} = Y(x, y), \qquad (6.1.1)$$

where $X, Y: \mathbb{R}^2 \to \mathbb{R}$. For simplicity we assume throughout, except in the statement of major theorems, that functions such as X and Y are *smooth*, that is, infinitely differentiable (of class C^∞ [see IV, §2.10]). This condition can usually be relaxed to r-fold differentiability (class C^r) for small r (often $r = 1$ or 2), or even to piecewise differentiability [see IV, §20.2]. However, it would unduly complicate the chapter to worry too much about the precise degree of differentiability required.

With these assumptions, solution curves $(x(t), y(t))$ to (6.1.1) with given initial conditions $x(0) = \mathbf{x}_0, y(0) = \mathbf{y}_0$, exist locally in time (t near 0), for both forward and backward time (t positive or negative), and are unique. We call such solution curves *trajectories* or *orbits* of the ODE (6.1.1).

There is a graphical method to obtain qualitative information on (6.1.1). Namely, for a grid of points (x, y) we plot the vector field $(X(x, y), Y(x, y))$, each such vector being treated as a bound vector with base point (x, y) and normalized to have length 1, unless $X(x, y) = Y(x, y) = 0$ [see V, §2.2.3]. That is, for each (x, y) in the grid we draw a line segment of unit length from (x, y) in the direction of $(X(x, y), Y(x, y))$, unless $(X(x, y)) = (0, 0)$. See also the *method of isoclines*, IV, 7.11.4.

EXAMPLE 6.1.1.   Van der Pol equation:

$$\frac{dx}{dt} = y - (x^3 - x), \qquad \frac{dy}{dt} = -x. \tag{6.1.2}$$

The vector field is plotted in Figure 1.1(a). In Figure 1.1(b) we plot selected solution trajectories $(x(t), y(t))$. The trajectories are tangent to the vector field at each point: this is the geometric content of (6.1.1). The diagram reveals the occurrence of a *limit cycle*, that is, a closed trajectory C having a neighbourhood U [see V, §5.2.3] such that all trajectories $x(t)$ with $x(0) \in C$ tend to C as $t \to \infty$. See §6.3.6.

A diagram of the trajectories of a system of ODEs is called its *phase portrait*. This forms the basis of the topological approach to dynamical systems theory.

Our main concern will be non-linear dynamical systems. However, much of the *local* behaviour of non-linear systems can be most easily described with reference to linear systems. In IV, 7.11.4, linear systems

$$\frac{d}{dt}\begin{pmatrix} x \\ y \end{pmatrix} = \begin{pmatrix} a & b \\ c & d \end{pmatrix}\begin{pmatrix} x \\ y \end{pmatrix}$$

are classified according to their phase portraits, into *nodes*, *saddles*, *foci* and *centres*, on the assumption that $ad - bc \neq 0$. (If $ad - bc = 0$, then a number of

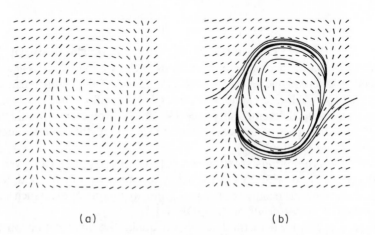

(a)   (b)

Figure 6.1.1. (a) Vector field for equation (6.1.2) (b) Some solution trajectories

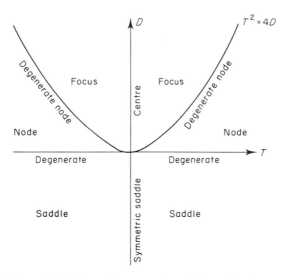

Figure 6.1.2. How the trace T and determination D
classify topological types of linear flows in the plane

degenerate cases occur, which are not important here.) One way to summarize
this information [see I, §6] is by way of the *trace*

$$T = \operatorname{tr} \mathbf{A} = a + b$$

and *determinant*

$$D = \det \mathbf{A} = ad - bc$$

of the matrix

$$\mathbf{A} = \begin{bmatrix} a & b \\ c & d \end{bmatrix},$$

see Figure 6.1.2.

If λ, μ are the eigenvalues of $\mathbf{A}$, then [see I, §7.3]

$$T = \lambda + \mu, \qquad D = \lambda\mu.$$

The origin is

a *focus* if $D > \frac{1}{4}T^2$ and $T \neq 0$,
a *node* if $D < \frac{1}{4}T^2$ and $D > 0$,
a *saddle* if $D < 0$,
a *centre* if $D > 0$, $T = 0$

and a degenerate node of various types when $D = 0$ or $D = \frac{1}{4}T^2$.

Notice that the condition for a centre, like that for a degenerate node, holds
only on a curve in (T, D)-space, whereas those for a node, focus or saddle
determine open regions. Roughly speaking, a system chosen 'at random' will
have a node, focus or saddle, but *not* a centre or a degenerate node. We will

see later that a 'typical' ODE in the plane possesses nodes, foci, saddles and limit cycles, but no other special features.

6.2.   FLOWS ON MANIFOLDS

We need a little of the language of manifolds [see V, §14.1].

An *n-dimensional manifold* M is a topological space which locally resembles an open set in $\mathbb{R}^n$. Thus there exists a *local coordinate system* or *chart* $(x_1, \ldots, x_n)$ near any given point, Figure 6.2.1. Where charts overlap we insist that each local coordinate system is *smoothly* related to the other [see V, §13.2.3].

A *dynamical system* on a manifold M is a system of n first-order ODEs in these local coordinates:

$$\frac{dx_1}{dt} = f_1(x_1, \ldots, x_n)$$

$$\frac{dx_n}{dt} = f_n(x_1, \ldots, x_n). \qquad (6.2.1)$$

The functions f_i define a *vector field* $\mathbf{f}$ on M [see V, §§13.4 and 14.3]: imagine the bound vector $\mathbf{f}$ attached to the base point x as in Figure 6.1.1(a). Solutions to (6.2.1) are curves tangent to the vector field $\mathbf{f}$, just as in the case $M = \mathbb{R}^2$ described in §6.1.1.

The manifold M is called the *phase space* of the system (6.2.1). It is necessary to use manifolds rather than Euclidean spaces $\mathbb{R}^n$, because in many applications the natural phase space is not Euclidean space. For example, consider a double pendulum (Figure 6.2.2). Its position is given by two angles (α, β), which determine a point on a 2-torus $\mathbb{T}^2$, not a plane, because angles are defined modulo 2π [see V, §14.1].

We also need the concept of a *diffeomorphism* of a manifold. This is a

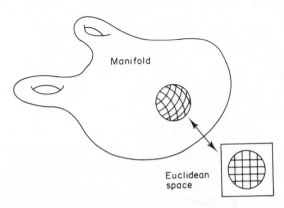

Figure 6.2.1. A chart on a manifold

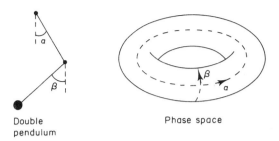

Double pendulum   Phase space

Figure 6.2.2. Phase space for a double pendulum

map $\phi: M \rightarrow M$ having an inverse ϕ^{-1} such that both ϕ and ϕ^{-1} are smooth [see V, §17.4.1]. For example, let $M = \mathbb{T}^2$ and define $\phi: \mathbb{T}^2 \rightarrow \mathbb{T}^2$ by $\phi(\alpha, \beta) = (\alpha + 1, \beta) \pmod{2\pi}$. Then $\phi^{-1}(\alpha, \beta) = (\alpha - 1, \beta) \pmod{2\pi}$, and both ϕ and ϕ^{-1} are smooth.

The basic existence and uniqueness theorem is

THEOREM 6.2.1. *Suppose that f is once differentiable and let* $x_0 \in M$. *Then near* x_0 *there exists a unique solution* $x(t)$ *to* (6.2.1) *such that* $x(0) = x_0$, *and this is defined for some time interval* $-\varepsilon < t < \varepsilon$, *where* $\varepsilon > 0$.

Solutions need not be defined for all time: the theorem is *local*. For example consider the ODE on $M = \mathbb{R}$ for which

$$\frac{dx}{dt} = 1 + x^2, \qquad x_0 = 0.$$

The solution is

$$x(t) = \tan t,$$

and this exists only for $-\pi/2 < t < \pi/2$ [see IV, §2.12 and Table 3.2.1]. Intuitively, x 'runs away to infinity' in finite time.

However, if M is compact [see V, §5.2.2], then there is a global existence theorem:

THEOREM 6.2.2. *Let f be* C^1 *and let M be compact. Then for each* $x_0 \in M$ *there exists a unique solution* $x(t)$ *with* $x(0) = x_0$, *and* $x(t)$ *is defined for all time* $t \in \mathbb{R}$.

To avoid technical problems, we assume henceforth that (6.2.1) has solutions for *all* time t for *all* initial points x_0. The solution curve $x(t)$ is the *trajectory* or *orbit* of x_0 under (6.2.1). It exists for backward time ($t < 0$) as well as forward ($t > 0$). The *phase portrait* of (6.2.1) is the system of all trajectories on M.

For example, let $M = \mathbb{T}^2 = \{(\alpha, \beta) | \alpha, \beta \in [0, 2\pi)\}$ and define

$$\frac{d\alpha}{dt} = \sin 3\beta, \qquad \frac{d\beta}{dt} = \cos 2\alpha.$$

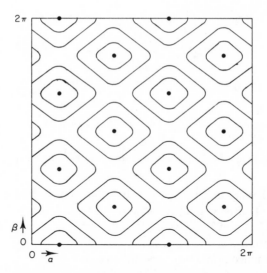

Figure 6.2.3. Phase portrait for the dynamical
system $d\alpha/dt = \sin 3\beta, d\beta/dt = \cos 2\alpha$ on the (α, β)-
torus

The phase portrait is shown in Figure 6.2.3. The dots represent fixed points
[see §6.3.1]. In general the phase portrait of a dynamical system resembles the
flow of some (non-physical) fluid through the underlying manifold. To make
this idea precise we define the *time-t flow* $\psi^t : M \to M$ by

$$\psi^t(x(0)) = x(t).$$

This map satisfies the conditions

$$\psi^0(x) = x, \tag{6.2.2}$$

$$\phi^s(\psi^t(x)) = \psi^{s+t}(x). \tag{6.2.3}$$

Further, for all t, ψ^t is a diffeomorphism of M. In general a map $\psi : \mathbb{R} \times M \to M$,
written in the form $(t, x) \mapsto \psi^t(x)$, is a *flow* if it satisfies (6.2.2), (6.2.3). Dynamical
systems theory can be reformulated as the study of flows on manifolds, because
every flow can be realized as the time t flow of a dynamical system.

6.3. BASIC CONCEPTS

Dynamical systems theory is thus the study, especially from a geometric or
topological point of view, of the qualitative properties of flows on manifolds
or their discrete analogues. Nowadays the term also covers quantitative studies
carried out by numerical methods, provided these are based upon a geometric
understanding of the nature of the flow. Because the equations for most
dynamical systems are non-linear, the computer plays an important role in the

mathematics, both as an 'experimental tool' to suggest rigorous theorems and as an aid to practical applications. However, we do not discuss numerical methods here [see III, §8].

6.3.1. Fixed Points

A point $\mathbf{x}_0 \in M$ at which the vector field $\mathbf{f}$ vanishes,

$$f_1(\mathbf{x}_0) = \ldots = f_n(\mathbf{x}_0) = 0,$$

is a *fixed point* (other terms are *critical point, equilibrium point, singular point* or *stationary point*) of $\mathbf{f}$. At a fixed point, the 'fluid' is at rest. The orbit of a fixed point is just the point itself.

The topology of the phase portrait near a fixed point is usually determined by the linearization of the vector field. By *linearization* [see IV, §5.8] we mean the linear ODE

$$\frac{dx_1}{dt} = \frac{\partial f_1}{\partial x_1}\bigg|_{\mathbf{x}_0} x_1 + \ldots + \frac{\partial f_1}{\partial x_n}\bigg|_{\mathbf{x}_0} x_n$$

$$\ldots$$

$$\frac{dx_n}{dt} = \frac{\partial f_n}{\partial x_1}\bigg|_{\mathbf{x}_0} x_1 + \ldots + \frac{\partial f_n}{\partial x_n}\bigg|_{\mathbf{x}_0} x_n.$$

More succinctly,

$$\frac{d\mathbf{x}}{dt} = \mathbf{Df}\big|_{\mathbf{x}_0} \mathbf{x}$$

where

$$\mathbf{x} = \begin{pmatrix} x_1 \\ \vdots \\ x_n \end{pmatrix}$$

and

$$\mathbf{Df} = \frac{\partial f_i}{\partial x_j}$$

is the Jacobian of $\mathbf{f}$ [see IV, §5.10].

The relation between the local structure of a non-linear system near a fixed point, and that of its linearization, is as follows.

THEOREM 6.3.1. (Hartman–Grobman theorem) *If $\mathbf{x}_0$ is a fixed point of (6.2.1) and $\mathbf{Df}\big|_{\mathbf{x}_0}$ has no purely imaginary or zero eigenvalues [see I, §7.1], then the linearized flow near $\mathbf{x}_0$ is topologically equivalent to the actual flow.*

That is, there is a homeomorphism of a neighbourhood of the point $\mathbf{x}_0$ to itself [see IV, §2.10] which carries trajectories of the linearized flow to trajectories

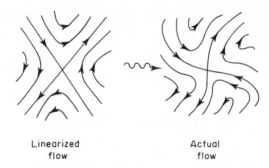

Linearized   Actual
flow   flow

Figure 6.3.1. Illustration of the Hartman–Grobma
theorem

of the actual flow, Figure 6.3.1. (This homeomorphism can also be made to preserve the parametrization of trajectories by time.)

The fixed point $\mathbf{x}_0$ is said to be *hyperbolic* or *non-degenerate* if $\mathbf{Df}|_{\mathbf{x}_0}$ has no purely imaginary or zero eigenvalues. Thus the topology of the flow near a hyperbolic critical point can be determined by linear algebra. Note that the solution to a linear system

$$\frac{d\mathbf{x}}{dt} = \mathbf{Ax},$$

with $\mathbf{A}$ an $n \times n$ matrix, is

$$\mathbf{x}(t) = e^{\mathbf{A}t}\mathbf{x}(0).$$

The Hartman–Grobman theorem *fails* in general at points $\mathbf{x}_0$ where $\mathbf{Df}|_{\mathbf{x}_0}$ is not hyperbolic. Later we will see that these are the points at which the solution may bifurcate under the variation of some additional parameter.

6.3.2. Stable and Unstable Manifolds

We now describe an important relationship between linear and non-linear systems, which associates to each fixed point of a non-linear system two manifolds: the stable manifold, along which the flow tends towards the fixed point, and the unstable manifold, along which it tends away from the fixed point. For example at a saddle point in the plane, the stable manifold is the union of the two separatrices along which the flow tends towards the saddle point, and the unstable manifold is the union of the other two separatrices, along which it tends away from the saddle point.

For a linear system on $\mathbb{R}^n$

$$\frac{d\mathbf{x}}{dt} = \mathbf{Ax},$$

with $\mathbf{A}$ an $n \times n$ matrix; we can describe the main features of the flow in the following way. Let $\lambda_1, \dots, \lambda_n$ be the eigenvalues of $\mathbf{A}$ [see I, §7.1] and divide them into three groups:

(a) $\text{Re}(\lambda_i) < 0$,
(b) $\text{Re}(\lambda_i) = 0$,
(c) $\text{Re}(\lambda_i) > 0$.

If λ is an eigenvalue of an $n \times n$ matrix $\mathbf{A}$, we define the corresponding *generalized eigenspace* to be

$$\{\mathbf{v} \in \mathbb{R}^n : (\mathbf{A} - \lambda\mathbf{I})^n \mathbf{v} = \mathbf{0}\}.$$

The sums of the generalized eigenspaces for the three types of eigenvalue λ_i above are respectively

(a) the *stable subspace* E^s
(b) the *centre subspace* E^c
(c) the *unstable subspace* E^u

of $\mathbf{A}$. The linear flow preserves these spaces, and is

towards the origin on E^s,
away from the origin on E^u.

On E^c the form of the linear flow requires a more careful analysis. For example, the flow can have a centre on E^c when dim $E^c = 2$ [see I, §5.4]—hence the name 'centre subspace'—but this is *not* the only possibility. See Figure 6.3.2.

For a general non-linear system with a fixed point $\mathbf{x}_0$ we define (local) *stable* and *unstable manifolds*

$$W^s(\mathbf{x}_0) = \{\mathbf{x} \mid \psi^t(\mathbf{x}) \to \mathbf{x}_0 \text{ as } t \to \infty\}$$
$$W^u(\mathbf{x}_0) = \{\mathbf{x} \mid \psi^t(\mathbf{x}) \to \mathbf{x}_0 \text{ as } t \to -\infty\}.$$

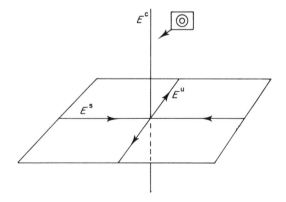

Figure 6.3.2. Stable, centre and unstable subspaces
(schematic)

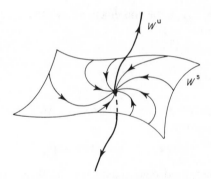

Figure 6.3.3. Stable and unstable
manifolds

See Figure 6.3.3. Intuitively, the flow ψ^t approaches $\mathbf{x}_0$ along W^s and leaves it along W^u. Sometimes W^s is called the *inset* and W^u the *outset* of $\mathbf{x}_0$.

THEOREM 6.3.2. (Stable manifold theorem) *Let* $\mathbf{x}_0$ *be a hyperbolic fixed point. Then there exist local stable and unstable manifolds* W^s, W^u *of the same dimensions as* E^s *and* E^u *for* $\mathbf{A} = \mathbf{Df}|_{\mathbf{x}_0}$, *and* W^s, W^u *are tangent to* E^s, E^u *respectively at* $\mathbf{x}_0$ *[see* V, §12.3.1].

   In other words, at a hyperbolic fixed point, the stable and unstable manifolds for the linearized flow closely approximate those for the actual flow.
   For non-hyperbolic critical points, a *centre manifold* tangent to E^c must also be introduced. But the situation here is considerably more complicated: for example, centre manifolds are not always unique, and smooth centre manifolds may not exist (see Guckenheimer and Holmes (1983)) [see also §6.3.2].

6.3.3.   Stability

   Define a critical point $\mathbf{x}_0$ to be *asymptotically stable* if there exists a neighbourhood U of $\mathbf{x}_0$ [see V, §5.2.3] such that if $\mathbf{x} \in U$ then $\psi^t(\mathbf{x}) \to \mathbf{x}_0$ as $t \to \infty$. There is a linear criterion sufficient for asymptotic stability.

THEOREM 6.3.3.   *If* $\mathbf{x}_0$ *is a hyperbolic fixed point and all the eigenvalues of* $\mathbf{Df}|_{\mathbf{x}_0}$ *have negative real parts, then* $\mathbf{x}_0$ *is asymptotically stable. If some eigenvalue has positive real part, then* $\mathbf{x}_0$ *is unstable.*

   Again, if $\mathbf{x}_0$ is not hyperbolic, stability is a more delicate question.

6.3.4. Discrete and Continuous Dynamics

For dynamics defined by a differential equation, time is a continuous variable. However, dynamics can also be defined by a difference equation

$$\mathbf{x}_{t+1} = \mathbf{f}(\mathbf{x}_t), \qquad (6.3.1)$$

where time t is descrete, $t = 0, 1, 2, 3 \dots$. Such a system is called a *discrete dynamical system*, and there is an analogous theory. Discrete systems are important for the theoretical analysis of continuous systems, as we see below.

If (as is the case for continuous dynamics) distinct initial conditions give rise to distinct motions, then the map $\mathbf{f}$ must be a diffeomorphism [see V, §17.4.1]. However, it is sometimes useful not to assume this. The analogue of the time-t flow ψ^t is the tth *iterate* $\mathbf{f}^t$, defined by

$$\mathbf{f}^0(\mathbf{x}) = \mathbf{x}$$
$$\mathbf{f}^{t+1}(\mathbf{x}) = \mathbf{f}(\mathbf{f}^t(\mathbf{x})).$$

The *orbit* of $\mathbf{x}$ is the set of points $\{\mathbf{f}^t(\mathbf{x}) | t \in \mathbb{N}\}$. If $\mathbf{f}$ is a diffeomorphism, then we define $\mathbf{f}^{-t}(\mathbf{x}) = (\mathbf{f}^{-1})^t(\mathbf{x})$ and the orbit becomes $\{\mathbf{f}^t(\mathbf{x}) | t \in \mathbb{Z}\}$.

A point $\mathbf{x}_0$ is a *fixed point* of (6.3.1) if $\mathbf{f}(\mathbf{x}_0) = \mathbf{x}_0$. It is a *periodic point of period n* if it is a fixed point of $\mathbf{f}^n$. A fixed point $\mathbf{x}_0$ of $\mathbf{f}$ is *asymptotically stable* if there exists a neighbourhood U of $\mathbf{x}_0$ such that if $\mathbf{x} \in U$ then $\mathbf{f}^t(\mathbf{x}) \to \mathbf{x}_0$ as $t \to \infty$. A periodic point of period n is asymptotically stable if its is an asymptotically stable fixed point of $\mathbf{f}^n$.

There are analogues of the stable, unstable and centre manifolds. However, comparison with behaviour of continuous and discrete *linear* systems shows that the corresponding subspaces must now be defined as follows. Let $\mathbf{L} = \mathbf{Df}|_{\mathbf{x}_0}$ be the linearization of $\mathbf{f}$ at a fixed point $\mathbf{x}_0$. Let $\lambda_1, \dots, \lambda_n$ be the eigenvalues of $\mathbf{L}$ and divide them into three groups:

(a) $|\lambda_i| < 1$,
(b) $|\lambda_i| = 1$,
(c) $|\lambda_i| > 1$.

The sums of the corresponding generalized eigenspaces are respectively

(a) the *stable subspace* E^s,
(b) the *centre subspace* E^c,
(c) the *unstable subspace* E^u.

We say $\mathbf{x}_0$ is *hyperbolic* if $E^c = \{0\}$. Most basic results, including the Hartman–Grobman theorem, generalize to discrete dynamical systems. For further details, see Guckenheimer and Holmes (1983), p. 16.

Given a diffeomorphism $\phi : X \to X$, defining a discrete dynamical system, we can 'embed' ϕ in a continuous dynamical system on a space $Y \supset X$, with $\dim Y = \dim X + 1$ [see I, §5.4]. By 'embed' we mean that if the time t flow on

Y is ψ^t, then $\psi^n(\mathbf{x}) = \phi^n(\mathbf{x})$ for all $\mathbf{x} \in X$ and for all integers n. This process, called *suspension*, is useful when constructing examples of continuous dynamical systems. To construct Y and ψ^t, first let $Z = X \times \mathbb{R}$ and let θ be the flow on Z for which $\theta^t(\mathbf{x}, 0) = (\mathbf{x}, t)$. Then introduce an equivalence relation on Z [see I, §1.3.3] by $(\mathbf{x}, s) \sim (\phi(\mathbf{x}), s + 1)$. Finally let $Y = Z/\sim$ and let ψ^t be induced by θ^t. Note that $\psi^1(\mathbf{x}, 0) \sim (\mathbf{x}, 1) \sim (\phi(\mathbf{x}), 0)$, so if we identify X with $X \times \{0\}$ then $\psi^1 = \phi$ on X, and hence $\psi^n = \phi^n$ on X for all integer n.

A converse process, the formation of a Poincaré section and its associated Poincaré mapping, is discussed in §6.3.6.

6.3.5.   Liapunov Functions

These provide a useful technique for proving stability. Suppose $\mathbf{x}_0$ is a fixed point of a flow $\mathbf{f}$, and U is a neighbourhood of $\mathbf{x}_0$. A *Liapunov function* for $\mathbf{x}_0$ is a positive definite function $L: U \to \mathbb{R}$ which decreases along trajectories. More precisely we require

(a) $L(\mathbf{x}_0) = 0$ and $L(\mathbf{x}) > 0$ for $\mathbf{x} \neq \mathbf{x}_0$,
(b) $\dot{L}(\mathbf{x}) \leq 0$ for $\mathbf{x} \in U \setminus \{\mathbf{x}_0\}$.

Here $\dot{L}(\mathbf{x})$ is determined along a trajectory, so that

$$\dot{L}(\mathbf{x}) = \sum_{j=1}^{n} \frac{\partial L}{\partial x_j} f_j(\mathbf{x}). \tag{6.3.2}$$

If a Liapunov function exists, then $\mathbf{x}_0$ must be asymptotically stable.

EXAMPLE 6.3.1.   A particle of unit mass moving in a line on a suitable non-linear spring obeys the equation

$$\dot{x} = y, \qquad \dot{y} = -x - y - x^3,$$

where x is its position, y is its velocity and dots denote t-derivatives. There is a fixed point at $(x, y) = (0, 0)$. Is it stable? To decide this, we seek a Liapunov function. In many physical systems, energy decreases as time passes, so we begin by considering the total energy. Here this is

$$E(x, y) = \tfrac{1}{2}(y^2 + x^2 + \tfrac{1}{2}x^4).$$

Is E a Liapunov function? Clearly it is positive definite. But by (6.3.2),

$$\frac{d}{dt}E(x, y) = y\dot{y} + (x + x^3)\dot{x} = -y(x + y + x^3) + (x + x^3)y = -y^2.$$

This is negative, except along the x-axis. If we modify E to

$$L(x, y) = E(x, y) + \tfrac{1}{2}xy + \tfrac{1}{4}x^2,$$

then L remains positive definite, and

$$\frac{d}{dt}L(x, y) = -\tfrac{1}{2}(x^2 + x^4 + y^2).$$

This is negative definite, so L is a Liapunov function. Thus the origin is an asymptotically stable fixed point.

There are no good general methods to find Liapunov functions, but knowledge of the dynamics (from physical intuition or numerical simulations, say) often suggests a suitable form to try.

6.3.6.   Periodic Orbits

Periodic orbits, where a trajectory forms a closed loop and the motion repeats indefinitely with some period T, are extremely important They occur around centres and on limit cycles, but they can also arise in other ways, too complicated for complete classification.

Suppose x_0 lies on a periodic orbit P. Define a *Poincaré section* Σ to be a hypersurface (a submanifold of dimension 1 less than the ambient manifold) passing through P and transverse to the flow (that is, such that no orbit near P is tangent to Σ), as in Figure 6.3.4.

Define the corresponding *Poincaré map*

$$\pi : \Sigma \to \Sigma$$

as follows:

$\pi(s) =$ the first point at which the trajectory $\{\psi^t(s)\}$ hits Σ for $t > 0$.

Then $\pi(s)$ exists for $s \in \Sigma$ close enough to x_0. Further:

THEOREM 6.3.4.   *The Poincaré map is a local diffeomorphism, and* x_0 *is a fixed point of* π.

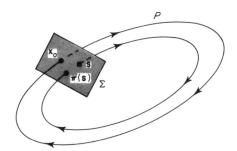

Figure 6.3.4. Poincaré section $\sum$ and the
corresponding Poincaré map π

We have defined P to be asymptotically stable if all $\mathbf{x}$ in some neighbourhood of P have orbits that approach P as $t \to \infty$. If P is asymptotically stable then the corresponding periodic orbit is a limit cycle. Thus we can detect limit cycles (including their asymptotic stability) from the *discrete* dynamics of the Poincaré map.

THEOREM 6.3.5.   *P is asymptotically stable if and only if $\mathbf{x}_0$ is an asymptotically stable fixed point of π.*

COROLLARY 6.3.6.   *If $\mathbf{x}_0$ is hyperbolic for π and all eigenvalues of $\mathbf{D}\pi|_{\mathbf{x}_0}$ have negative real part, then P is asymptotically stable.*

Classically, the stability of a periodic orbit is studied via *Floquet theory*. We describe this briefly. Let P have period T and let $\mathbf{x}(t) = \mathbf{x}(t + T)$ be the solution whose orbit is P. Linearize to get

$$\frac{d\mathbf{x}}{dt} = \mathbf{Df}|_{\mathbf{x}(t)}\mathbf{x},$$

a non-autonomous linear system with a T-periodic matrix of coefficients $\mathbf{Df}|_{\mathbf{x}(t)}$. It can be shown that such a system has a *fundamental matrix solution*

$$\mathbf{X}(t) = \mathbf{Z}(t)e^{t\mathbf{R}},$$

where $\mathbf{Z}(t)$ is T-periodic and $\mathbf{R}$ is a constant matrix whose entries may be complex. To say that $\mathbf{X}$ is a fundamental matrix solution means that the solution with initial conditions $\mathbf{x}(0) = \mathbf{x}_0$ is given by $\mathbf{x}(t) = \mathbf{X}(t)\mathbf{x}_0$.

The way solutions diverge from or converge towards P is determined by the eigenvalues of $e^{T\mathbf{R}}$. These are known as *Floquet multipliers*. One eigenvalue of $e^{T\mathbf{R}}$ is 1 (corresponding to the original periodic orbit). For asymptotic stability, all the others should lie inside the unit circle in $\mathbb{C}$. A computation shows that these other eigenvalues are in fact the eigenvalues of $\mathbf{D}\pi|_{\mathbf{x}_0}$ where π is anyPoincaré map. Thus Floquet theory can be viewed as the linearization of the concept of a Poincaré map.

6.3.7. Asymptotic Behaviour

For most purposes the important behaviour of a dynamical system is what happens as $t \to \infty$, after initial transients have died away. In this subsection we discuss various concepts associated with this *asymptotic behaviour*.

Let ψ^t be a flow on M. We say that $S \subset M$ is *invariant* (respectively *positively invariant*, *negatively invariant*) if $\psi^t(S) \subset S$ for all $t \in \mathbb{R}$ (respectively $0 < t \in \mathbb{R}$, $0 > t \in \mathbb{R}$). A point $\mathbf{x} \in M$ is *non-wandering* if, for any neighbourhood U of $\mathbf{x}$, there exist arbitrarily large times t such that

$$\psi^t(U) \cap U \neq \varnothing.$$

(Note that we do not require this condition for *all* large t. Consider a point $\mathbf{x}$ on a periodic orbit P, having period T. If U is a neighbourhood of $\mathbf{x}$ then $\psi^{nT}(U) \cap U$ contains $\mathbf{x}$ for all integers n, whence $\mathbf{x}$ is non-wandering; but if U is small then $\psi^t(U) \cap U$ is empty for times t that are not close to integer multiples of the period T.) The *non-wandering set* Ω is the set of all non-wandering points. Fixed points and periodic orbits are non-wandering.

A point $\mathbf{p} \in M$ is an *ω-limit point* of $\mathbf{x}$ if there is a sequence $\psi^{t_n}(\mathbf{x}) \to \mathbf{p}$ at $t_n \to \infty$. An *α-limit point* is defined in the same way, for $t_n \to -\infty$. Sequences (t_n) are specified here for similar reasons to those just described for the non-wandering set: we do not require $\psi^t(\mathbf{x}) \to \mathbf{p}$ as $t \to \infty$. We define the sets

$$\omega(\mathbf{x}) = \{\mathbf{p} \in M \,|\, \mathbf{p} \text{ is an } \omega\text{-limit point of } \mathbf{x}\},$$
$$\alpha(\mathbf{x}) = \{\mathbf{p} \in M \,|\, \mathbf{p} \text{ is an } \alpha\text{-limit point of } \mathbf{x}\}.$$

A closed invariant set $A \subset M$ is *attracting* if there exists a positively invariant neighbourhood U of A such that for all $\mathbf{x} \in U, \psi^t(\mathbf{x}) \to A$ as $t \to \infty$. The *basin of attraction* of A is

$$\beta(A) = \bigcup_{t \leq 0} \psi^t(U).$$

A *repelling* set s is defined in the same way using $-t$ in place of t.

Finally, as a working definition, an *attractor* for ψ^t is an attracting set that contains a dense orbit. Reversing the sign of t defines the analogous concept of a *repeller*. Intuitively, if A is an attractor, then after transients die away the system settles down to a flow on A. The topology of A, and the nature of this flow, may be very complicated. The requirement that A has a dense orbit excludes cases where A is the union of two smaller attractors.

For example, the flow in Figure 6.3.5 has a point repeller at the origin, and the limit cycle is an attractor.

Figure 6.3.5. Point repeller plus limit cycle attractor

6.3.8. Flows in the Plane

In this section we consider equations

$$\dot{x} = f(x, y), \qquad \dot{y} = g(x, y) \tag{6.3.3}$$

for $(x, y) \in \mathbb{R}^2$.

The fixed points of (6.3.3) are given by the intersection of the curves $f(x, y) = 0, g(x, y) = 0$, and have already been discussed. To find periodic orbits, two classical results are useful.

THEOREM 6.3.7.   (Poincaré–Bendixson theorem) *A non-empty compact ω- or α-limit set containing no fixed points is a periodic orbit.*

THEOREM 6.3.8.   (Bendixson's criterion) *If on a simply connected region $D \subset \mathbb{R}^2$ [see V, §5.2.6] the expression*

$$\frac{\partial f}{\partial x} + \frac{\partial g}{\partial y}$$

is not identically zero and does not change sign, then there are no periodic orbits lying entirely within D.

There is also a classification of non-wandering sets:

THEOREM 6.3.9.   (Andronov's theorem) *For flows in $\mathbb{R}^2$ the possible non-wandering sets are*

   (a) *fixed points,*
   (b) *periodic orbits,*
   (c) *unions of fixed points and the trajectories connecting them.*

Sets of type (c) are called *homoclinic orbits* if they connect a fixed point to itself, and *heteroclinic orbits* if they connect two distinct fixed points (see Figure 6.3.6).

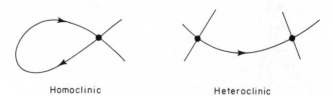

Homoclinic   Heteroclinic

Figure 6.3.6. Homoclinic and heteroclinic orbits

6.3.9.   Index Theory in the Plane

The results of this section are special to flows in the plane. Generalizations to other spaces and higher dimensions exist, but are too complicated to describe here. Given a flow in the plane, draw a loop C that does not intersect itself and runs through no fixed points. As a point c moves anticlockwise round C, the vector $(f(c), g(c))$ rotates continuously through a total angle of $2k\pi$, where k is an integer. We call k the *index* of C, Figure 6.3.7, and write $k = \mathrm{ind}(C)$. If C encloses a unique fixed point $\mathbf{x}_0$ then k is the *index of* $\mathbf{x}_0$. For the standard types of fixed point we can tabulate the index:

Fixed point	Index
sink	$+1$
source	$+1$
centre	$+1$
saddle	-1

Clearly if C is a periodic orbit its index is 1. Indices are additive in the following sense:

THEOREM 6.3.10.   (Index theorem) *The index of a loop is equal to the sum of the indices of the fixed points within it.*

There is also a formula

$$\mathrm{ind}(C) = \frac{1}{2\pi} \int_C \frac{f\,dg - g\,df}{f^2 + g^2}. \tag{6.3.4}$$

Index = −1

Figure 6.3.7. The index of a saddle

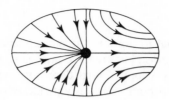

Figure 6.3.8. A saddle-node

COROLLARY 6.3.11.

(*a*) *Every periodic orbit encloses at least one fixed point.*

(*b*) *If all fixed points inside C are hyperbolic then C contains $n + 1$ sinks or sources and n saddles, for some n.*

(*c*) *If there is a unique fixed point inside C then it must be a source or a sink.*

None of the above results apply without modification to flows on 2-manifolds other than $\mathbb{R}^2$.

An important degenerate fixed point is the *saddle-node*, of index 0, which can be thought of as a saddle and a node merged together: see Figure 6.3.8.

6.3.10.   Special Systems

We mention three important special types of flow.

Gradient Systems

Let $V:\mathbb{R}^n \to \mathbb{R}$ and call V the *potential*. The associated *gradient flow* is

$$\frac{dx_i}{dt} = -\frac{\partial V}{\partial x_i}.$$

We can regard V as a Liapunov function. Solution trajectories cut the level sets

$$\{x \mid V(x) = \text{const.}\}$$

orthogonally, and flow 'downhill' in V. There can thus be no limit cycles or homoclinic orbits.

EXAMPLE 6.3.2.   Let the potential be $V:\mathbb{R}^2 \to \mathbb{R}$, where $V(x, y) = \frac{1}{2}(x^2 + y^2)$. Then the gradient flow is defined by

$$\dot{x} = -x, \qquad \dot{y} = -y.$$

Thus $x(t) = e^{-t}x_0, y(t) = e^{-t}y_0$, where (x_0, y_0) is the initial point. All orbits tend towards the origin as $t \to \infty$: this accords with the fact that the origin is the unique minimum of V. To see that V is a Liapunov function, we compute $\dot{V}$ along orbits:

$$\dot{V} = x\dot{x} + y\dot{y} = -(x^2 + y^2) < 0 \quad \text{unless } x = y = 0.$$

Hamiltonian Systems

These are highly important in applications, arising, for example, from dynamics without friction. In classical form we have n position coordinates $q_1, \ldots, q_n$ and n corresponding momentum coordinates $p_1, \ldots, p_n$. Let $\mathcal{H}$ denote a real-valued function $\mathcal{H}(q_1, \ldots, q_n, p_1, \ldots, p_n)$ called the *Hamiltonian*. The corresponding *Hamiltonian flow* or *Hamiltonian system* is given by

$$\frac{dq_i}{dt} = \frac{\partial \mathcal{H}}{\partial p_i}, \qquad \frac{dp_i}{dt} = -\frac{\partial \mathcal{H}}{\partial q_i}. \tag{6.3.5}$$

EXAMPLE 6.3.3. The equations for a planar pendulum may be written in Hamiltonian form, using the Hamiltonian

$$\mathcal{H}(p, q) = \tfrac{1}{2}p^2 + \cos q,$$

where q is its angle to the vertical and p is the corresponding angular momentum. Explicitly we have

$$\frac{dq}{dt} = \frac{\partial \mathcal{H}}{\partial p} = p, \qquad \frac{dp}{dt} = -\frac{\partial \mathcal{H}}{\partial q} = \sin q.$$

Explicit solution of these equations requires complicated transcendental functions (elliptic functions). However, the qualitative features of the motion can easily be deduced by observing that $\mathcal{H}(p, q)$ is a conserved quantity. That is, along orbits we have

$$\frac{d}{dt}\mathcal{H}(p, q) = p\dot{p} - \dot{q}\sin q = p\sin q - p\sin q = 0,$$

so $\mathcal{H}(p, q) = \text{constant}$ on orbits. Thus the orbits are level curves of $\mathcal{H}(p, q)$ and can easily be plotted and analysed.

Hamiltonian systems have many special properties. For example, a Hamiltonian system in the plane has no sources or sinks, only foci and saddles. The quantity $\mathcal{H}$ is constant on orbits, as in the example above. Hamiltonian systems are best treated as a separate context for dynamics, paralleling but differing from the theory of general dynamical systems. See Arnold (1978), MacKay and Meiss (1987), and Guckenheimer and Holmes (1983) for further remarks. We draw attention to two topics that are extremely important in applications: the *KAM theorem*, which describes perturbations near quasiperiodic orbits, and the *Melnikov method*, used to detect homoclinic orbits.

Non-autonomous Systems

A *non-autonomous system*, for example a forced one, takes the form

$$\frac{d\mathbf{x}}{dt} = \mathbf{f}(\mathbf{x}, t).$$

Now the vector field f varies with time.

EXAMPLE 6.3.4.   The Duffing oscillator is a forced non-linear oscillator, and has an equation of the form

$$\ddot{x} + x - x^3 = \cos(\omega t),$$

where ω is the forcing frequency. If we let $y = \dot{x}$, then the equation becomes

$$\dot{x} = y, \qquad \dot{y} = x^3 - x - \cos(\omega t),$$

which is a non-autonomous dynamical system because of the explicit t-dependence.

The solutions are indescribably complicated!

A non-autonomous system can be considered as an ordinary dynamical system if we introduce a new variable x_{n+1} which plays the role of time, and write

$$\frac{d\mathbf{x}}{dt} = f(\mathbf{x}, x_{n+1}), \qquad \frac{dx_{n+1}}{dt} = 1,$$

with $x_{n+1} = 0$ at $t = 0$ as initial condition. Thus there is no real loss of generality in considering only autonomous systems. However, the *interpretation* of results may be different in the autonomous and non-autonomous cases.

6.3.11.   Structural Stability

The concept of structural stability arose from attempts to describe the 'typical' behaviour of a dynamical system—behaviour that persists when *the system itself* is perturbed. It must be distinguished from the stability of an individual solution—behaviour that persists when the *initial conditions of a fixed system* are perturbed. The precise notion of structural stability presented here is not entirely satisfactory for this purpose, but no effective substitute has yet been defined and the concept still plays an influential role.

For simplicity we consider dynamical systems defined on a *compact manifold* K. We begin with three definitions.

DEFINITION 6.3.1.   A vector field $\mathbf{f}$ [see V, §13.4] is an *ε-perturbation* of a vector field $\mathbf{g}$ if $\mathbf{f}$ and $\mathbf{g}$, and also the first partial derivatives of $\mathbf{f}$ and $\mathbf{g}$, differ in norm by at most ε.

Here the norm of a vector is $\|\mathbf{x}\| \sqrt{(x_1^2 + \ldots + x_n^2)}$ as usual [see I, §10.1].

DEFINITION 6.3.2. Two vector fields $\mathbf{f}$ and $\mathbf{g}$ on K are *topologically equivalent* if there is a homeomorphism $\phi : K \to K$ which sends orbits of $\mathbf{f}$ to orbits of $\mathbf{g}$ [see V, §5.1].

In other words, $\mathbf{f}$ and $\mathbf{g}$ have topologically equivalent phase portraits.

DEFINITION 6.3.3. A vector field $\mathbf{f}$ on K is *structurally stable* if there exists $\varepsilon > 0$ such that every ε-perturbation of $\mathbf{f}$ is topologically equivalent to $\mathbf{f}$.

In other words, the phase portrait of a structurally stable vector field retains its qualitative form under sufficiently small perturbations.

EXAMPLE 6.3.5. For simplicity we here relax the condition that K be compact [see V, §5.2.2]: the flows described below can easily be extended to the sphere by adding a point at infinity.

(a) The vector field $f(x, y) = (-y, x)$ on $\mathbb{R}^2$, whose associated dynamical system is

$$\dot{x} = -y, \qquad \dot{y} = x,$$

is structurally unstable. To see this, let

$$g(x, y) = f(x, y) - \varepsilon(x^2 + y^2)(x, y).$$

Then for all $\varepsilon \neq 0$ the phase portraits of $\mathbf{f}$ and $\mathbf{g}$ (Figure 6.3.9) are topologically inequivalent.

(b) The vector field $f(x, y) = (-y, x) - (x^2 + y^2)(x, y)$ on $\mathbb{R}$ is structurally stable. Although we cannot prove it here, every sufficiently small perturbation of $\mathbf{f}$ has a topologically equivalent phase portrait, namely, a stable limit cycle with a source inside it (see Guckenheimer and Holmes (1983)).

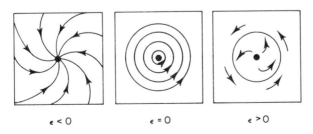

$\varepsilon < 0 \qquad\qquad \varepsilon = 0 \qquad\qquad \varepsilon > 0$

Figure 6.3.9. An example of structural instability

Structural stability is a formal way to capture the idea of 'behaviour that is not destroyed by small changes to the system'. It was originally hoped that it would be a *generic* property, that is, be possessed by 'almost all' dynamical systems. That hope turned out to be misplaced, but the concept remains important.

For flows on a compact surface (2-manifold) [see V, §14.1] there is a complete characterization of structural stability.

THEOREM 6.3.12.   (Peixoto's theorem) *A vector field on a compact surface is structurally stable if and only if*

(a) *the number of periodic orbits and fixed points is finite,*
(b) *each periodic orbit and fixed point is hyperbolic,*
(c) *there are no heteroclinic or homoclinic orbits,*
(d) *the non-wandering set Ω consists only of fixed points and periodic orbits.*

For orientable compact surfaces, the structurally stable dynamical systems *are* generic.

This result motivates the following definition, which applies to manifolds of arbitrary dimension.

DEFINITION 6.3.4.   A dynamical system is *Morse–Smale* if

(a) the number of periodic orbits and fixed points is finite,
(b) all stable and unstable manifolds intersect transversely,
(c) the non-wandering set Ω consists only of fixed points and periodic orbits.

Morse–Smale systems are always structurally stable, but the converse is false.

6.4.   ATTRACTORS, BIFURCATIONS AND CHAOS

In this section we adopt a more zoological approach. The aim is to amass a collection of specimens that display some of the more important kinds of non-linear dynamic behaviour. The main topics covered are attractors, which we have defined already, bifurcations—changes in the qualitative behaviour of a dynamical system as a parameter is varied—and chaos, the occurrence of apparently random behaviour in a deterministic system.

6.4.1.   Attractors

We take up the concept of an attractor from §6.3.7. Recall that this is a closed attracting invariant set that contains a dense orbit. Intuitively, it captures the 'long-term behaviour' of the system. The presence of a dense orbit is a technical requirement: it means that the attractor is not the union of two smaller attractors, so that all regions of the attractor contribute to the long-term behaviour observed.

The simplest attractors are the classical ones:

(a) sinks (point attractors),
(b) stable limit cycles (topological circle attractors).

Both of these types of attractor are structurally stable (at least, if they are hyperbolic). It is relatively easy to define dynamical systems that have more complicated attractors, but most such attempts lead to structurally unstable systems. That is, more complicated attractors tend to be fragile, in the sense that they break up into classical ones if the vector field is suitably perturbed.

Can structurally stable attractors, other than the classical ones, exist? This question is the real starting point of modern dynamical systems theory, because the answer, surprisingly, is 'yes'. They are known as *strange attractors*.

Before describing some examples, we must recall a fundamental construction from point-set topology, known as the *Cantor set*. Begin with the unit interval

$$I_0 = [0, 1].$$

Remove the open middle third to obtain

$$I_1 = [0, \tfrac{1}{3}] \cup [\tfrac{2}{3}, 1].$$

Remove their middle thirds to obtain

$$I_2 = [0, \tfrac{1}{9}] \cup [\tfrac{2}{9}, \tfrac{1}{3}] \cup [\tfrac{2}{3}, \tfrac{7}{9}] \cup [\tfrac{8}{9}, 1]$$

and so on. Then the Cantor set is

$$K = \bigcap_{n=0}^{\infty} I_n.$$

It is an uncountable compact totally disconnected topological space [see V, §5.2]. The phrase 'Cantor set' is also used to denote any topological space that is homeomorphic to the above.

EXAMPLE 6.4.1.   (The solenoid) This example is constructed by first defining a discrete dynamical system on a solid torus and then suspending it [§6.3.4] to obtain a flow. Let T be a solid torus [see V, §14.1] and define a map $\phi: T \to T$

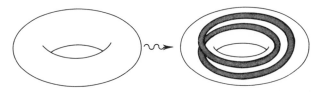

Figure 6.4.1. Wrapping a solid torus twice inside itself

which maps T inside itself by wrapping it twice round the hole in the middle (Figure 6.4.1). Under iteration of ϕ, all initial points tend to an attractor which is the intersection of all iterates,

$$A = \bigcap \phi^n(T).$$

This attractor A has an intricate geometry, and topologically its local structure is the cartesian product of an interval and a Cantor set.

We may now suspend ϕ (strictly speaking, we must extend the domain of definition of ϕ to obtain a diffeomorphism, but this is easy) to get a flow ψ on a space $Y \supset T$ in which ϕ 'embeds'. Then dim $Y = 3$ and ψ has an attractor B which locally look like $A \times \mathbb{R}$. Thus B is not even a manifold, so it certainly is not a classical point or circle attractor.

It is clear that ϕ is structurally stable—any small perturbation also wraps the torus T twice inside itself and hence leads to the same geometry. It can then be proved that ψ is also structurally stable, at least if the construction is specified a little more carefully.

EXAMPLE 6.4.2.   (The Smale horseshoe) This is a strange saddle rather than a strange attractor, but it is instructive and important.

Let X be the unit square $[0, 1] \times [0, 1] \subset \mathbb{R}^2$ and define a map $\phi : X \to \mathbb{R}^2$ as in Figure 6.4.2. Then ϕ folds X into a U-shape and replaces it overlapping itself. We require ϕ to be a composition of dilations in the horizontal and vertical

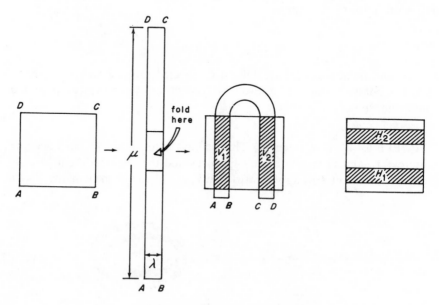

Figure 6.4.2. The Smale horsehoe

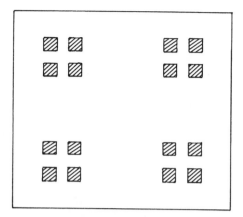

Figure 6.4.3. Strange saddle in the Smale
horsehoe: early stage of construction

directions on the two shaded regions H_1 and H_2, which map to V_1 and V_2. Let

$$\Lambda = \{\mathbf{x} \,|\, \varphi^n(\mathbf{x}) \in X \text{ for all } n \in \mathbb{Z}\}.$$

Then the form of Λ is roughly as shown in Figure 6.4.3. This suggests, correctly, that Λ is the cartesian product of two Cantor sets, one for the vertical direction and one for the horizontal.

Let Σ be the set of bi-infinite sequences $\{a_i \,|\, i \in \mathbb{Z}\}$ on two symbols $\{0, 1\}$. That is, each a_i is either 0 or 1. Define a metric d on Σ [see I, §10.1] by

$$d(a, b) = \sum_{i = -\infty}^{\infty} \varepsilon_i 2^{-|i|},$$

where $\varepsilon_i = 0$ if $a_i = b_i$, 1 if $a_i \neq b_i$. Define the *shift map* $\sigma : \Sigma \to \Sigma$ by

$$(\sigma a)_i = a_{i+1}.$$

So σa is a shifted one place to the left.

By analysing the geometry of Λ and ϕ it can be shown that there is a homeomorphism

$$f : \Lambda \to \Sigma$$

such that

$$f(\phi(x)) = \sigma(f(x)).$$

In other words, the structure of ϕ acting on Λ is topologically the same as that of σ acting on Σ.

Now the action of σ on Σ is 'obvious'. For example, an element $a \in \Sigma$ is a periodic point for σ if and only if it is periodic as a sequence of 0s and 1s. For example, the fixed points are

$$\ldots 0000000 \ldots$$
$$\ldots 1111111 \ldots$$

and the points of period 2 are

$$\ldots 010\mathbf{1}010\ldots$$
$$\ldots 101\mathbf{0}101\ldots$$

where the bold symbol indicates a_0.

It is not hard to find a point $a\in\Sigma$ whose orbit is dense in Σ. Arrange all finite sequences of 0s and 1s in a countable list L. Construct a by starting with some sequence stretching from a finite position to $-\infty$, and adding the sequences in the list L successively on to the right-hand end. Then a defines a dense orbit for σ on Σ, so Λ contains a dense orbit for ϕ.

Dynamically, Λ is not an attractor—it behaves rather like a saddle. Its geometric form is essentially unchanged if small changes are made to ϕ, that is, it is structurally stable.

The correspondence with symbol sequences is an example of *symbolic dynamics*, which is an important technique in the subject (see Guckenheimer and Holmes (1983), chapter 5).

EXAMPLE 6.4.3.   (Lorenz attractor) Edward Lorenz (1963) introduced this attractor in the course of an investigation of atmoshperic dynamics, with implications for weather prediction.

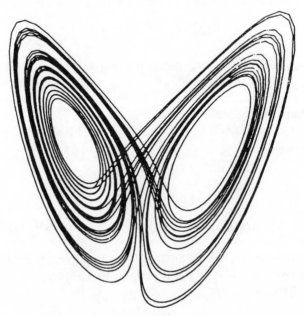

Figure 6.4.4. The Lorenz attractor (courtesy of Colin Sparrow, Cambridge)

The *Lorenz equations* determine a flow on $\mathbb{R}^3$, given by

$$\frac{dx}{dt} = \sigma(y - x), \qquad \frac{dy}{dt} = \rho x - y - xz, \qquad \frac{dz}{dt} = -\beta z + xy,$$

where $\sigma, \rho, \beta \in \mathbb{R}$ are parameters. The dynamic behaviour depends upon the values of σ, ρ and β. The case $\sigma = 10, \beta = 8/3$ and $\rho = 28$ is celebrated, because here the dynamical behaviour settles down to Figure 4.4, known as the *Lorenz attractor*.

The Lorenz attractor has two lobes, each approximately a surface, which meet in the middle. Trajectories wind round the left-hand lobe several times, then round the right, then round the left again. The number of turns on each side varies in an apparently random fashion.

The 'surfaces' involved are actually composed of infinitely many very close layers, like a Cantor set.

While it has not yet been proved rigorously that the Lorenz system has a strange attractor, the numerical evidence certainly points that way. Various geometric approximations to the Lorenz system definitely do have strange attractors.

6.4.2.   Bifurcations

The equations for dynamical systems often contain parameters other than the variables $\mathbf{x}$ that represent the state of the system. If these parameters are denoted by $\lambda = (\lambda_1, \dots, \lambda_k)$ then we have equations of the form

$$\frac{dx_1}{dt} = f_1(x_1, \dots, x_n; \lambda_1, \dots, \lambda_k)$$

$$\dots$$

$$\frac{dx_n}{dt} = f_n(x_1, \dots, x_n; \lambda_1, \dots, \lambda_k)$$

or more simply

$$\frac{d\mathbf{x}}{dt} = \mathbf{f}(\mathbf{x}, \lambda).$$

As λ varies, the phase portrait of the corresponding flow usually changes 'gradually', so that, in particular, it remains the same up to topological equivalence. But at certain 'critical' values of λ, the topology of the phase portrait can change. (Classically, changes in the number of fixed points or periodic orbits are paramount: these, in particular, change the topology of the phase portrait.) Such a value of λ is said to be a *bifurcation point* of the dynamical system. More precisely:

DEFINITION 6.4.1.   Let $\mathbf{f}(\mathbf{x}, \lambda)$ be a family of vector fields parametrized by λ. A parameter value $\lambda = \lambda_0$ is a *bifurcation point* if there exists λ_1 arbitrarily close to λ_0 such that $\mathbf{f}(\mathbf{x}, \lambda_0)$ and $\mathbf{f}(\mathbf{x}, \lambda_1)$ are not topologically equivalent.

In other words, λ_0 is a point at which the dynamical system $\mathbf{f}(\mathbf{x}, \lambda)$ becomes structurally unstable.

Here we shall mostly restrict attention to the case $k = 1$, so that the *bifurcation parameter* $\lambda \in \mathbb{R}$. Until recently very little was known about multi-parameter bifurcation, $k > 1$, but there has been an explosion of research in the area: see Golubitsky and Schaeffer (1985), Golubitsky, Stewart and Schaeffer (1988), Golubitsky and Guckenheimer (1986), Guckenheimer (1984) and Guckenheimer and Holmes (1983), for example.

There are many different kinds of bifurcation, and it is the task of bifurcation theory to provide methods for detecting, analysing, and classifying them. Here we concentrate on the simplest cases:

(1) *Steady state bifurcation*, where the number of fixed points changes.
(2) *Hopf bifurcation*, where a periodic trajectory is created around a fixed point.

EXAMPLE 6.4.4.   Consider the equation

$$dx/dt = f(x, \lambda) \equiv \lambda x - x^3$$

where $x, \lambda \in \mathbb{R}$. The steady states are given by $0 = \lambda x - x^3 = x(\lambda - x^2)$. If $\lambda < 0$ then the only steady state is $x = 0$, if $\lambda > 0$ there are three steady states $x = -\sqrt{\lambda}, 0, \sqrt{\lambda}$. The corresponding phase portraits are shown in Figure 6.4.5. There is a bifurcation point $\lambda = 0$ where the topology of the phase portrait (in particular, the *number* of steady states) changes. For $\lambda < 0$ there is a single sink at $x = 0$: when $\lambda > 0$ this sink becomes a source and 'throws off' two more sinks at $\pm \sqrt{\lambda}$. The behaviour of the system can be summarized in a *bifurcation diagram* (Figure 6.4.6). This shows, for each value of λ, the corresponding steady states x. It takes the form of a straight line plus a parabola.

There is a simple necessary condition for bifurcation.

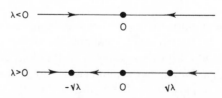

Figure 6.4.5. Steady-state bifurcation of a
sink into two sinks and a source

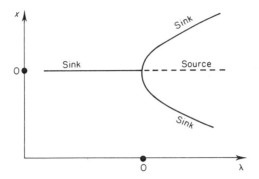

Figure 6.4.6. Bifurcation diagram corresponding to Figure 6.4.5: the pitchfork

THEOREM 6.4.1. *Let* $\mathbf{f}: \mathbb{R}^{n+k} \to \mathbb{R}^n$ *be smooth and suppose that* $\mathbf{f}(\mathbf{x}_0, \lambda_0) = \mathbf{0}$. *In order for the dynamical system*

$$dx/dt = \mathbf{f}(\mathbf{x}, \lambda)$$

to have a steady-state bifurcation at $(\mathbf{x}_0, \lambda_0)$, *it is necessary that the Jacobian matrix* $[\partial f_i / \partial x_j]$ *be singular at* $(\mathbf{x}_0, \lambda_0)$ *[see I, §6.4].*

The proof is simple: if the Jacobian is non-singular, then the implicit function theorem [see IV, §5.13] implies that near $(\mathbf{x}_0, \lambda_0)$ the equation $\mathbf{f}(\mathbf{x}, \lambda) = \mathbf{0}$ has a unique solution. Therefore no bifurcation can occur.

The condition is also 'morally' sufficient for steady-state bifurcation, in the following sense. If the Jacobian is singular, then there exist small perturbations of $\mathbf{f}$ for which a steady-state bifurcation must occur. In practice, workers in bifurcation theory often take the singularity of the Jacobian as the definition of a steady-state bifurcation.

Although a bifurcation involves the appearance of a structurally unstable flow—at the bifurcation point—it is possible for the bifurcation to take place 'in a structurally stable fashion'. This is the case if all sufficiently small perturbations of the system $\mathbf{f}(\mathbf{x}, \lambda)$ also possess a bifurcation of the same topological type. (In effect, the structurally stable flows that appear before and after the bifurcation can 'stabilize' the occurrence of the structurally unstable flow that provides a transition between them.) The more bifurcation parameters there are, the more complicated is the type of bifurcation that can occur in a structurally stable way. The number of bifurcation parameters needed for this is known as the *codimension* of the bifurcation.

There is thus a hierarchy of types of bifurcation, ordered by codimension. Golubitsky and Schaeffer (1985) have proved a classification theorem for local steady-state bifurcations, using methods from singularity theory. They use a different convention from ours above, taking the dynamical system in the form

$$dx/dt + \mathbf{f}(\mathbf{x}, \lambda) = \mathbf{0}$$

Codim	Type	Normal form $\varepsilon, \delta = \pm 1$	Defining conditions $= 0$ always include f, f_x	Non-degeneracy conditions $\neq 0$
1	limit point	$\varepsilon x^2 + \delta \lambda$	—	$\varepsilon = \operatorname{sgn} f_{xx}, \delta = \operatorname{sgn} f_\lambda$
2	transcritical bifurcation	$\varepsilon(x^2 - \lambda^2)$	f_λ	$\varepsilon = \operatorname{sgn} f_{xx}, \operatorname{sgn} \det \mathbf{D}^2 \mathbf{f} < 0$
3	isola centre	$\varepsilon(x^2 + \lambda^2)$	f_λ	$\varepsilon = \operatorname{sgn} f_{xx}, \operatorname{sgn} \det \mathbf{D}^2 \mathbf{f} > 0$
2	hysteresis	$\varepsilon x^3 + \delta \lambda$	f_{xx}	$\varepsilon = \operatorname{sgn} f_{xxx}, \delta = \operatorname{sgn} f_\lambda$
3	asymmetric cusp	$\varepsilon x^2 + \delta \lambda^3$	$f_\lambda, \det \mathbf{D}^2 \mathbf{f}$ choose $y, f_{yy} = 0$	$\varepsilon = \operatorname{sgn} f_{xx}, \delta = \operatorname{sgn} f_{yyy}$
3	pitchfork	$\varepsilon x^3 + \delta \lambda x$	f_{xx}, f_λ	$\varepsilon = \operatorname{sgn} f_{xxx}, \delta = \operatorname{sgn} f_{x\lambda}$
3	quartic fold	$\varepsilon x^4 + \delta \lambda$	f_{xx}, f_{xxx}	$\varepsilon = \operatorname{sgn} f_{xxxx}, \delta = \operatorname{sgn} f_\lambda$

Table 6.4.1. Recognition conditions for steady-state bifurcation

rather than

$$dx/dt = \mathbf{f}(\mathbf{x}, \lambda).$$

Effectively this changes $\mathbf{f}$ to $-\mathbf{f}$. We employ their convention in Table 6.4.1. (Note that they also define codimension in a slightly different way: despite this, Table 6.4.1 follows our definition above in this respect.)

Without loss of generality we may translate coordinates so that the bifurcation points is $(\mathbf{x}_0, \lambda_0) = (\mathbf{0}, \mathbf{0})$. In order for a steady-state bifurcation to occur at $(\mathbf{0}, \mathbf{0})$ we must have $\det [\partial f_i / \partial x_j](\mathbf{0}, \mathbf{0}) = 0$ by Theorem 6.4.1. Each type of bifurcation has an associated *normal form* which determines the topology of the set of steady states. For each normal form there is a system of *defining conditions*, expressions in partial derivatives of $\mathbf{f}$ which must vanish at $(\mathbf{0}, \mathbf{0})$. There is also a system of *non-degeneracy conditions*, expressions which must *not* vanish—otherwise the bifurcation has still higher codimension. The combination of the defining and non-degeneracy conditions solves the *recognition problem* for the type of bifurcation.

EXAMPLE 6.4.5. Bifurcation problems of the same type as Example 6.4.4, where $f(x, \lambda) = x^3 - \lambda x$, are recognized as follows. The defining conditions are

$$f = f_x = f_{xx} = f_\lambda = 0$$

and the non-degeneracy conditions are

$$f_{xxx}, f_{x\lambda} \neq 0,$$

all derivatives being evaluated at $(x, \lambda) = (0, 0)$. Any steady-state bifurcation satisfying these conditions will have the characteristic 'pitchfork' shape of Figure 6.4.6.

We now move on to Hopf bifurcation, in which periodic orbits are created by the loss of stability of a steady state.

EXAMPLE 6.4.6. Consider the dynamical system given in cartesian coordinates by

$$dx/dt = \lambda x + y - x(x^2 + y^2)$$
$$dy/dt = -x + \lambda y - y(x^2 + y^2).$$

In polar coordinates [see V, §1.2] this takes a more tractable form

$$d\theta/dt = -1$$
$$dr/dt = \lambda r - r^3.$$

Then $\theta(t) = -t$, so there is a clockwise angular motion at constant speed. Meanwhile r obeys the same equation as in Example 6.4.4. However, $r \geq 0$ in polar coordinates, so as far as the radial motion is concerned we see a sink at $r = 0$ for $\lambda < 0$, but a source at $r = 0$ and a sink at $r = \sqrt{\lambda}$ when $\lambda > 0$. The corresponding phase portraits are thus as in Figure 6.4.7.

The sink for $\lambda = 0$ becomes a source, and throws off a *stable limit cycle*. This is the simplest example of Hopf bifurcation (Figure 6.4.7), named after Eberhard Hopf who (following work of A. A. Andronov) obtained a simple sufficient condition for it to occur.

THEOREM 6.4.2. (Hopf bifurcation theorem) *Suppose that the dynamical system*

$$dx/dt = \mathbf{f}(\mathbf{x}, \lambda)$$

has a fixed point $(\mathbf{x}_0, \lambda_0)$ *at which the following hold:*

(a) *The Jacobian* $\mathbf{Df}_{(\mathbf{x}_0, \lambda)}$ *of* $\mathbf{f}$ *at* $(\mathbf{x}_0, \lambda)$ *has a pair of eigenvalues* $\sigma(\lambda) \pm i\omega(\lambda)$ *for* λ *near* λ_0.
(b) *When* $\lambda = \lambda_0$, $\sigma(\lambda) = 0$ *and* $\omega(\lambda) = \omega \neq 0$.
(c) $\mathbf{Df}(\mathbf{x}_0, \lambda_0)$ *has no other purely imaginary eigenvalues that are integer multiples of* $i\omega$.
(d) *(Eigenvalue crossing condition)* $d\sigma(\lambda)/d\lambda \neq 0$ *at* $\lambda = \lambda_0$.

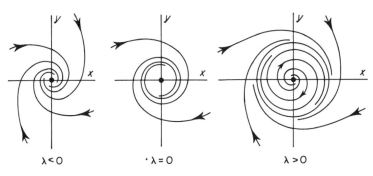

$\lambda < 0$ $\cdot\lambda = 0$ $\lambda > 0$

Figure 6.4.7. Hopf bifurcation

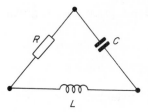

Figure 6.4.8. An electronic circuit which can undergo Hopf bifurcation

Then a continuous branch of limit cycles bifurcates from $(\mathbf{x}_0, \lambda_0)$, *with period tending to* $2\pi/\omega$ *as* $\lambda \to \lambda_0$.

Thus the Hopf theorem says that, subject to technicalities, a non-linear system whose Jacobian has an imaginary eigenvalue possesses a periodic solution. Many generalizations and extensions of the Hopf theorem are now known, but this one often suffices and is easy to use.

EXAMPLE 6.4.7.   Consider the electronic circuit of Figure 6.4.8. The circuit equations are

$$i_C = C dv_C/dt, \qquad v_L = L di_L/dt, \qquad v_R = \phi(i_R),$$
$$i_R = i_L = -i_C, \qquad v_R + v_L = v_C,$$

where the i are currents, the v voltages, and $v_R = \phi(i_R)$ is the non-linear characteristic of the resistor R. To normalize the equations, set

$$x = i_L, \qquad y = -(C/L)^{1/2} v_C, \qquad \tau = (LC)^{1/2} t,$$

yielding

$$dx/d\tau = -y - f(x), \qquad dy/d\tau = x$$

for $f(x) = (L/C)^{1/2} \phi(x)$. The simplest non-linear characteristic is

$$f(x) = x^3 - \lambda x,$$

where λ controls the amount of gain in the device R.
  The Jacobian matrix is

$$\mathbf{Df} = \begin{pmatrix} \lambda & -1 \\ 1 & 0 \end{pmatrix},$$

whose eigenvalues are

$$\tfrac{1}{2}(\lambda \pm \sqrt{\lambda^2 - 4}).$$

If λ is between -2 and 2 we have complex eigenvalues $\sigma(\lambda) \pm i\omega(\lambda)$, where

$$\sigma(\lambda) = \tfrac{1}{2}\lambda, \qquad \omega(\lambda) = \tfrac{1}{2}\sqrt{4 - \lambda^2}.$$

The conditions of Theorem 6.4.2 hold at $\lambda = 0, (x, y) = (0, 0)$. Thus there exists a bifurcating branch of periodic solutions, corresponding to oscillations of the circuit.

If condition (d) of Theorem 6.4.2 is violated, we have *degenerate Hopf bifurcation*. A complete theory, analogous to steady-state bifurcation, has been obtained by Golubitsky and Langford (1981) (see also Golubitsky and Schaeffer (1985)). It the vector field **f** has symmetry, a version of the Hopf theorem remains true (see Golubitsky and Stewart (1985), or Golubitsky, Stewart and Schaeffer (1988)).

Other kinds of bifurcation of small codimension are studied in Guckenheimer (1984) and Guckenheimer and Holmes (1983), Chapter 7. We mention in particular the Takens–Bogdanov bifurcation (Guckenheimer and Holmes (1983), §7.3), in which a homoclinic orbit may be created. Steady-state and Hopf bifurcation for discrete dynamical systems are studied in Guckenheimer and Holmes (1983), §3.5. Other texts on bifurcation theory include Chow and Hale (1982) and Iooss and Joseph (1981).

6.4.3. Chaos

In §6.4.1 we described 'apparently random' behaviour in the Lorenz system. However, the equations that define this system are fully deterministic: given initial conditions lead to uniquely specified behaviour. The sense in which a deterministic model can produce 'random' effects is a major discovery of the last few decades, giving rise to a central branch of dynamical systems theory— chaotic dynamics.

The word *chaos*—more accurately, *deterministic chaos*—refers to any instance of such behaviour. Example 6.4.2, the Smale horseshoe, provides an excellent example. Recall that the dynamic there is equivalent to the shift map on the space Σ of bi-infinite binary sequences. Generate an initial sequence $a = (a_i)$ by the following rule. If $i < 0$ let $a_i = 0$. If $i \geq 0$, then, for each i, determine the value of a_i by tossing a coin. Let $a_i = 0$ if the coin lands 'heads', 1 if it lands 'tails'. Then a is a 'random' sequence in the sense that there is no specific pattern to the a_i.

Now take this sequence a (or the corresponding point $\alpha = f^{-1}(a)$ in the strange saddle Λ) as initial conditions and iterate the dynamics. The successive values of $\phi(\alpha)$ correspond to the sequences $\sigma(a)$, $\sigma^2(a)$, and so on. Focus only on the 0th entry of these sequences: these run in turn through the values $a_0, a_1, a_2, a_3, \ldots$ and so on. But these are our 'random' sequence. In other words, Λ contains an initial point α such that the iterates of α under ϕ mimic the terms a_i of a random sequence. It is in this sense that ϕ acting on Λ contains apparently random dynamics. Suitable initial conditions lead to random dynamical evolution. Indeed almost all initial conditions do.

EXAMPLE 6.4.8. (Logistic mapping) This example is a discrete dynamical system (using a mapping that is not a diffeomorphism). It has been widely studied because of its simple defining equations and surprisingly complex behaviour.

The logistic mapping $f:[0, 1] \rightarrow [0, 1]$ has the form

$$f(x) = kx(1 - x),$$

where k is a constant between 0 and 4. Iterating the mapping we get the discrete dynamical system $x_{t+1} = kx_t(1 - x_t)$.

Geometrically, the logistic mapping stretches or compresses the line segment in a non-uniform manner, and then folds it in half. For given k, the mapping folds the interval up and lays it down on top of the interval between 0 and $k/4$.

The range of k values between 0 and 3 is the *steady-state regime*, the least interesting from the point of view of dynamics. Pick k in this range, say $k = 2$, and iterate the mapping. The values converge to a unique point. In other words, there is a point attractor, a stable steady state.

When k is exactly 3, the fixed point is 'marginally stable': convergence to it is *extremely* slow. This is a sign that something dramatic is about to happen. Indeed, when $k > 3$, the fixed point becomes unstable.

What happens when $k > 3$, say $k = 3.2$? Now the value of x_t flips alternately between two distinct numbers. This is a *period two cycle*. So the steady state loses stability and becomes periodic.

If k is increased to about 3.5 the period two attractor also goes unstable, and a period four cycle appears. By $k = 3.56$ the period has doubled again to eight; by $k = 3.567$ it has reached 16, and thereafter we observe a rapid sequence of doublings to periods of 32, 64, 128, ... So rapid is this *period-doubling cascade* that by $k = 3.58$ the period has doubled infinitely often. At that point, the logistic mapping becomes chaotic. At the maximum value, $k = 4$, a given trajectory passes arbitrarily close to every point of the interval. The entire interval has become an attractor.

However, the onset of chaos is not as regular as the above may suggest. For example, when $k = 3.835$, there is a period 3 cycle:

$$0.152\,0744 \rightarrow 0.494\,5148 \rightarrow 0.958\,6346.$$

If k is now increased *very* gently the periods then go 6, 12, 24, 48, 96, ... in a new period-doubling cascade.

Similarly, when $k = 3.739$, there is a cycle of period *five*

$$0.841\,1372 \rightarrow 0.499\,6253 \rightarrow 0.934\,7495 \rightarrow 0.228\,0524 \rightarrow 0.658\,2304.$$

The parameter k is not just a simple 'chaos generator'. Increasing k does not always makes the dynamics more complicated. On the contrary, buried within the chaotic regime are 'windows' of regular behaviour.

This sequence of events is illustrated in Figure 6.4.9. To get an overview of the entire dynamic behaviour of the logistic mapping for all values of k in one go, we draw its *bifurcation diagram*. This is a graph with k running horizontally and x vertically. Above each value of k, those x-values that lie on the attractor

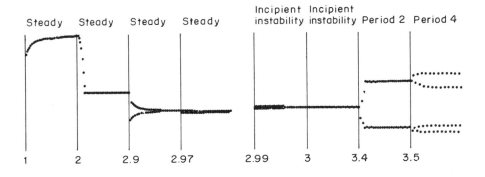

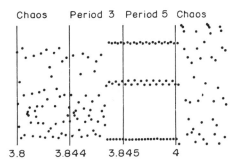

Figure 6.4.9. Sample orbits of the logistic mapping; the vertical bars indicate changes
in the value of k, after which the map is iterated 50 times

for that k are marked. Each vertical slice gives a picture, in the interval from
0 to 1, of the corresponding attractor. The result is shown in Figure 6.4.10. For
$k < 3$ there is a single smooth curve, representing a point attractor. At $k = 3$
this curve split into two, splitting again and again as k runs through the period-
doubling regime. Around $k = 3.58$ the bifurcation diagram culminates in
infinitely many branches and the system goes chaotic. The branches broaden

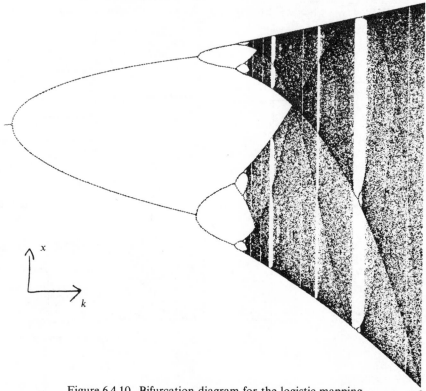

Figure 6.4.10. Bifurcation diagram for the logistic mapping

into bands of chaotic attractors, and the bifurcation diagram is pocked with random dots.

The structure of this bifurcation diagram is quite remarkable. For example, the manner in which the successive period-doublings take place has certain *universal* characteristics. In particular, suppose that successive bifurcations take place at k-values $k_1, k_2, \ldots$. Then the quotient $(k_{n+1} - k_n)/(k_{n+2} - k_{n+1})$ tends to a limit $\delta = 4.669\,201\,609\ldots$. The universal property is that the corresponding quotients tend to the *same* value in *any* similar period-doubling cascade, for functions other than $f(x) = kx(1 - x)$. This celebrated result, first conjectured by Feigenbaum (1978), forms the basis of some of the deepest work on chaotic dynamics. Its proof uses a method known as *renormalization.*

Chaotic dynamics is now a broad subject with its own range of methods. For further details, consult in particular Berry, Percival and Weiss (1978), Bedford and Swift (1988), Holden (1986) and Thompson and Stewart (1986). A large number of seminal papers in the subject are collected together in Hao

Bai-Lin (1984), including those of Lorenz and Feigenbaum. Methods for reconstructing chaotic attractors from experimental observations, a highly important topic, are surveyed by Broomhead and King (1986). The history, mathematical content, and applications of chaos are described in non-technical language in Stewart (1989).

I.S.

REFERENCES

Arnold, V. I. (1978). *Mathematical Methods of Classical Mechanics*, Springer.
Bedford, T. and Swift, J. W. (eds.) (1988). *New Directions in Dynamical Systems*, Cambridge University Press.
Berry, M. V., Percival, I. O. and Weiss, N. (eds.) (1987). *Dynamical Chaos*, Royal Society.
Broomhead, G. S. and King, G. P. (1986). On the Qualitative Analysis of Experimental Dynamical Systems, in *Nonlinear Phenomena and Chaos*, Ed. S. Sarkar, Adam Hilger, 113–144.
Chow, S. N. and Hale, J. K. (1982). *Methods of Bifurcation Theory*, Springer.
Feigenbaum, M. J. (1978). Quantitative Universality for a Class of Nonlinear Transformations, *J. Stat. Phys.* **19**, 25–52 (Reprinted in Hao Bai-Lin (1984)).
Golubitsky, M. and Guckenheimer, J. (eds.) (1986). Multiparameter Bifurcation Theory, *Contemporary Math.* **56**, Amer. Math. Soc.
Golubitsky, M. and Langford, W. F. (1981). Classification and Unfoldings of Degenerate Hopf Bifurcations, *J. Diff. Eq.* **41**, 375–415.
Golubitsky, M. and Schaeffer, D. G. (1985). *Singularities and Groups in Bifurcation Theory*, Vol. I, Springer.
Golubitsky, M. and Stewart, I. N. (1985). Hopf Bifurcation in the Presence of Symmetry, *Arch. Rational Mech. Anal.* **87**, 107–165.
Golubitsky, M., Stewart, I. N. and Schaeffer, D. G. (1988). *Singularities and Groups in Bifurcation Theory*, Vol. II, Springer.
Guckenheimer, J. (1984). Multiple Bifurcation Problems of Codimension Two, *SIAM J. Math. Anal.* **15**, 1–49.
Guckenheimer, J. and Holmes, P. (1983). *Nonlinear Oscillations, Dynamical Systems, and Bifurcations of Vector Fields*, Springer.
Hao Bai-Lin (ed.) (1984). *Chaos*, World Scientific.
Holden, A. (ed.) (1986). *Chaos*, Manchester University Press.
Iooss, G. and Joseph, D. D. (1981). *Elementary Stability and Bifurcation Theory*, Springer.
Lorenz, E. N. (1963). Deterministic Non-periodic Flow, *J. Atoms. Sci.* **20**, 130–141 (Reprinted in Hao Bai-Lin (1984)).
MacKay, R. S. and Meiss, J. D. (eds.) (1987). *Hamiltonian Dynamical Systems*, Adam Hilger.
Stewart, I. N. (1989). *Does God Play Dice? The Mathematics of Chaos*. Basil Blackwell.
Thompson, J. M. T. and Stewart, H. B. (1986). *Nonlinear Dynamics and Chaos*, Wiley.

CHAPTER 7
Control Theory

7.1. INTRODUCTION

The early development of control theory was based upon a continuous, linear, time-invariant, single-input/single-output model. In this approach a system is considered as a black box into which is fed an input signal and out of which a response, or output, is observed and measured. The analysis of such a model is carried out using the frequency approach in which the Laplace transform of input and output gives rise to a transfer function [see IV, §13.4]. This line of investigation led to design methods characterized by Bode and Nichols charts, Nyquist diagrams and root-locus [see Engineering Guide, §14.10]. More recently these frequency domain methods have been generalized to multi-input/multi-output systems (Rosenbrook, (1970); MacFarlane, (1980)). An algebraic theory for multivariable discrete systems has also been developed by Kalman, Falb and Arbib (1969).

A state space approach to the investigation of dynamical control systems arose from intense interest in aerospace systems following the end of the Second World War. This alternative to the frequency domain methods incorporates the state of the system as well as input and output variables. Based upon the state space model of a system, a geometric approach to the structural synthesis (Feedback control) of multivariable systems was developed and is described by Wonham (1985).

Geometric methods using the tools of differential geometry [see V, §12], Lie groups [see V, §14.5] and Lie algebras [see V, §14.6] have provided most of the results for non-linear control systems. Contributions to this area have been collected together in the books by Isidori (1989) and Banks (1988). Very recently a differential algebraic approach to non-linear system theory has been suggested by Fliess (1986). Here the methods of differential Galois theory, as described by Pommaret (1983), are used to investigate problems such as the invertibility and decoupling of non-linear control systems.

Much of the work described above has been concerned with concepts such as controllability, observability and realization. However, another important area of considerable activity has been that of optimal control. State space has been the main setting for this area of control theory which has much in common with the calculus of variations [see IV, §12]. Pontryagin *et al.* (1962) with his minimum principle and Bellman (1957) with dynamic programming [see IV, §16] have made two of the important contributions to the problem of finding optimal control trajectories with respect to some measure of system performance. Special features such as singular controls and junction conditions (Bell and Jacobson (1975)) have arisen from certain classes of optimal control problems.

The present chapter attempts to give an insight into some of the main contributions in the above-mentioned areas of control theory.

7.2.   LINEAR SYSTEMS

7.2.1.   Equivalent Systems

Consider a linear, continuous, time-invariant dynamical system (hereafter called a linear system—see Chapter 6) described by the equations

$$\dot{\mathbf{x}}(t) = \mathbf{A}\mathbf{x}(t) + \mathbf{B}\mathbf{u}(t), \tag{7.2.1}$$

$$\mathbf{y}(t) = \mathbf{C}\mathbf{x}(t), \tag{7.2.2}$$

where $\mathbf{x}(t) \in \mathbb{R}^n$, $\mathbf{u}(t) \in \mathbb{R}^m$, $\mathbf{y}(t) = \mathbb{R}^p$, and $\mathbf{A}, \mathbf{B}$ and $\mathbf{C}$ are real $n \times n, n \times m$ and $p \times n$ matrices respectively [see Engineering Guide, §14.10].

Taking Laplace transforms with zero initial conditions [see IV, §13.4] in (7.2.1) gives

$$s\bar{\mathbf{x}}(s) = \mathbf{A}\bar{\mathbf{x}}(s) + \mathbf{B}\bar{\mathbf{u}}(s),$$

i.e.

$$(s\mathbf{I} - \mathbf{A})\bar{\mathbf{x}}(s) = \mathbf{B}\bar{\mathbf{u}}(s),$$

whence [see I, §6.4]

$$\bar{\mathbf{x}}(s) = (s\mathbf{I} - \mathbf{A})^{-1}\mathbf{B}\bar{\mathbf{u}}(s).$$

Taking Laplace transforms in (7.2.2) and substituting for $\bar{\mathbf{x}}$ yields

$$\bar{\mathbf{y}}(s) = \mathbf{C}(s\mathbf{I} - \mathbf{A})^{-1}\mathbf{B}\bar{\mathbf{u}}(s) = \mathbf{G}(s)\bar{\mathbf{u}}(s),$$

where

$$\mathbf{G}(s) = \mathbf{C}(s\mathbf{I} - \mathbf{A})^{-1}\mathbf{B} \tag{7.2.3}$$

is the *transfer function matrix* for system (7.2.1–2) and relates the input and output transforms of that system.

The concepts of *controllability* and *observability* of linear systems are closely related to the effects of a linear transformation of coordinates in the state space [see I, §5.13]. Consider a non-singular [see I, §6.4] coordinate transformation (the independent time variable t is omitted for convenience)

$$\mathbf{z} = \mathbf{T}\mathbf{x} \qquad (|\mathbf{T}| \neq 0).$$

Equations (7.2.1) and (7.2.2) in the new state vector $\mathbf{z}$ become

$$\dot{\mathbf{z}} = \mathbf{T}\mathbf{A}\mathbf{T}^{-1}\mathbf{z} + \mathbf{T}\mathbf{B}\mathbf{u} \tag{7.2.4}$$

$$\mathbf{y} = \mathbf{C}\mathbf{T}^{-1}\mathbf{z}. \tag{7.2.5}$$

In order to find the transfer function of this new system (7.2.4) and (7.2.5), take Laplace transforms again with zero initial conditions to give

$$\bar{\mathbf{y}} = \mathbf{CT}^{-1}(s\mathbf{I} - \mathbf{TAT}^{-1})^{-1}\mathbf{TB}\bar{\mathbf{u}} = \mathbf{G}_1(s)\bar{\mathbf{u}},$$

where

$$\mathbf{G}_1(s) = \mathbf{CT}^{-1}(s\mathbf{I} - \mathbf{TAT}^{-1})^{-1}\mathbf{TB} \qquad (7.2.6)$$

is the transfer function matrix for system (7.2.4–5). But

$$(\mathbf{T}^{-1}(s\mathbf{I} - \mathbf{TAT}^{-1})^{-1}\mathbf{T})^{-1} = \mathbf{T}^{-1}(s\mathbf{I} - \mathbf{TAT}^{-1})\mathbf{T} = s\mathbf{I} - \mathbf{A}$$

so that

$$\mathbf{T}^{-1}(s\mathbf{I} - \mathbf{TAT}^{-1})^{-1}\mathbf{T} = (s\mathbf{I} - \mathbf{A})^{-1}.$$

It follows that (7.2.6) can be written as

$$\mathbf{G}_1(s) = \mathbf{C}(s\mathbf{I} - \mathbf{A})^{-1}\mathbf{B}$$
$$= \mathbf{G}(s)$$

by (7.2.3).

The two systems (7.2.1–2) and (7.2.4–5) related by the transformation **T** are said to be *algebraically equivalent*. From the above analysis it is seen that algebraically equivalent systems have the same transfer function matrix. This means that all algebraically equivalent systems having a given transfer function would be equally valid representations for a system with that transfer function.

7.2.2. Controllability

If the initial state is given by $\mathbf{x}(t_0) = \mathbf{x}_0$, then the solution of (7.2.1) is given by [see Engineering Guide, §14.10]

$$\mathbf{x}(t) = \Phi(t, t_0)\left[\mathbf{x}_0 + \int_{t_0}^{t} \Phi(t_0, \tau)\mathbf{Bu}(\tau)d\tau \right],$$

where

$$\Phi(t, t_0) = \exp[\mathbf{A}(t - t_0)].$$

By choosing $t_0 = 0$ and $\mathbf{x}(t_1) = \mathbf{0}$ for a given finite time t_1, this solution reduces to

$$-\mathbf{x}_0 = \int_0^{t_1} e^{-\mathbf{A}\tau}\mathbf{Bu}(\tau)d\tau. \qquad (7.2.7)$$

DEFINITION 7.2.1. The representation (7.2.1) is said to be *completely state controllable*, or the pair $(\mathbf{A}, \mathbf{B})$ is *controllable*, if for any initial state $\mathbf{x}(0) = \mathbf{x}_0$ there exists a finite time t_1 and an input $\mathbf{u}:[0, t_1] \to \mathbb{R}^m$ such that $\mathbf{x}(t_1) = \mathbf{0}$.

In Definition 7.2.1 the initial state $\mathbf{x}_0$ is also termed a *controllable state*. An alternative definition for complete state controllability is given next.

DEFINITION 7.2.2.   The linear system (7.2.1–2) is said to be completely state controllable if it is not algebraically equivalent to any system of the form

$$\begin{pmatrix} \dot{\bar{\mathbf{x}}} \\ \dot{\tilde{\mathbf{x}}} \end{pmatrix} = \begin{pmatrix} \mathbf{A}_{11} & \mathbf{A}_{12} \\ \mathbf{0} & \mathbf{A}_{22} \end{pmatrix} \begin{pmatrix} \bar{\mathbf{x}} \\ \tilde{\mathbf{x}} \end{pmatrix} + \begin{pmatrix} \mathbf{B}_1 \\ \mathbf{0} \end{pmatrix} \mathbf{u},$$

$$\mathbf{y} = (\mathbf{C}_1, \mathbf{C}_2) \begin{pmatrix} \bar{\mathbf{x}} \\ \tilde{\mathbf{x}} \end{pmatrix}, \tag{7.2.8}$$

where $\bar{\mathbf{x}}$ and $\tilde{\mathbf{x}}$ are vectors of dimension $\bar{n}$ and $\tilde{n} = n - \bar{n}$ respectively and $\mathbf{A}_{11}, \ldots, \mathbf{C}_2$ are matrices of appropriate dimensions obtained by partitioning $\mathbf{A}, \mathbf{B}$ and $\mathbf{C}$.

Notice that a system of the form (7.2.8) can be illustrated diagrammatically as shown in Figure 7.2.1. Clearly, controllability is a property which depends only on the matrices $\mathbf{A}$ and $\mathbf{B}$ and is independent of the output matrix $\mathbf{C}$. It can be proved (Barnett and Cameron, (1985)) that system (7.2.1) is completely state controllable if and only if the $n \times mn$ controllability matrix

$$\mathscr{C}(\mathbf{A}, \mathbf{B}) = (\mathbf{B}, \mathbf{A}\mathbf{B}, \mathbf{A}^2\mathbf{B}, \ldots, \mathbf{A}^{n-1}\mathbf{B}) \tag{7.2.9}$$

is of full rank, i.e. rank $(\mathscr{C}) = n$ [see I, §5.6].

EXAMPLE 7.2.1.

$$\dot{x}_1 = 2x_1 + x_2 + u, \qquad \dot{x}_2 = -x_1 - u.$$

For this system

$$\mathbf{A} = \begin{pmatrix} 2 & 1 \\ -1 & 0 \end{pmatrix}, \qquad \mathbf{B} = \mathbf{b} = \begin{pmatrix} 1 \\ -1 \end{pmatrix}.$$

The non-singular coordinate transformation $\mathbf{z} = \mathbf{T}\mathbf{x}$ where

$$\mathbf{T} = \begin{pmatrix} 4 & 3 \\ 1 & 1 \end{pmatrix}$$

yields new state equations

$$\dot{z}_1 = z_1 + z_2 + u, \qquad \dot{z}_2 = z_2.$$

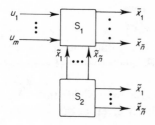

Figure 7.2.1. A system which is not completely state controllable

The second equation is independent of z_1 and u and so the transformed system is in the form of (7.2.8) and Figure 7.2.1. The original system is therefore not completely state controllable. This result can also be obtained from the fact that

$$\mathscr{C}(\mathbf{A}, \mathbf{b}) = \begin{pmatrix} 1 & 1 \\ -1 & -1 \end{pmatrix} \quad \text{with rank}(\mathscr{C}) = 1.$$

7.2.3. Observability

The solution of system (7.2.1–2) with initial condition $\mathbf{x}(0) = \mathbf{x}_0$ can be expressed as

$$\mathbf{y}(t) - \mathbf{C} \int_0^t e^{\mathbf{A}(t-\tau)} \mathbf{B} \mathbf{u}(\tau) d\tau = \mathbf{C} e^{\mathbf{A}t} \mathbf{x}_0.$$

DEFINITION 7.2.3. The representation (7.2.1–2) is said to be *completely observable*, or the pair $(\mathbf{A}, \mathbf{C})$ is *observable*, if any initial state $\mathbf{x}_0$ is uniquely determined from the knowledge of the output $\mathbf{y}(t)$ over the interval $[0, t_1]$ and when the input is completely known.

In Definition 7.2.3 the input $\mathbf{u}(\cdot)$ may be assumed null since if system (7.2.1–2) is completely observable for $\mathbf{u}(\cdot) = 0$, then it is completely observable for any $\mathbf{u}(\cdot)$. The initial state $\mathbf{x}_0$ in Definition 7.2.3 is called an *observable state*.

An alternative definition of complete observability is as follows.

DEFINITION 7.2.4. The linear system (7.2.1–2) is said to be completely observable if it is not algebraically equivalent to any system of the form

$$\begin{pmatrix} \dot{\bar{\mathbf{x}}} \\ \dot{\tilde{\mathbf{x}}} \end{pmatrix} = \begin{pmatrix} \mathbf{A}_{11} & \mathbf{0} \\ \mathbf{A}_{21} & \mathbf{A}_{22} \end{pmatrix} \begin{pmatrix} \bar{\mathbf{x}} \\ \tilde{\mathbf{x}} \end{pmatrix} + \begin{pmatrix} \mathbf{B}_1 \\ \mathbf{B}_2 \end{pmatrix} \mathbf{u},$$

$$\mathbf{y} = (\mathbf{C}_1, \mathbf{0}) \begin{pmatrix} \bar{\mathbf{x}} \\ \tilde{\mathbf{x}} \end{pmatrix}, \tag{7.2.10}$$

where $\bar{\mathbf{x}}$ and $\tilde{\mathbf{x}}$ are vectors of dimension $\bar{n}$ and $\tilde{n} = n - \bar{n}$ respectively ($\bar{n}$ and $\tilde{n}$

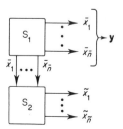

Figure 7.2.2. A system which is not completely observable

not necessarily the same dimensions as in Definition 7.2.2) and $\mathbf{A}_{11}, \ldots, \mathbf{C}_1$ are matrices (in general different from those in (7.2.8)) of appropriate dimensions obtained by partitioning $\mathbf{A}, \mathbf{B}$ and $\mathbf{C}$.

As for (7.2.8), so a system of the form (7.2.10) can be illustrated diagrammatically as in Figure 7.2.2.

It can be proved (Barnett and Cameron (1985)) that system (7.2.1–2) is completely observable if and only if the $n \times pn$ matrix

$$\mathcal{O}(\mathbf{A}, \mathbf{C}) = (\mathbf{C}^T, \mathbf{A}^T \mathbf{C}^T, (\mathbf{A}^2)^T \mathbf{C}^T, \ldots, (\mathbf{A}^{n-1})^T \mathbf{C}^T) \tag{7.2.11}$$

is of full rank, i.e. rank $(\mathcal{O}) = n$ [see I, §5.6].

EXAMPLE 7.2.2.

$$\dot{x}_1 = x_2, \qquad \dot{x}_2 = x_3,$$
$$\dot{x}_3 = -10x_1 - 9x_2 - 4x_3 + u, \qquad y = x_1 + 2x_2. \tag{7.2.12}$$

Here

$$\mathbf{A} = \begin{pmatrix} 0 & 1 & 0 \\ 0 & 0 & 1 \\ -10 & -9 & -4 \end{pmatrix}, \qquad \mathbf{C} = (1, 2, 0).$$

Then

$$\mathcal{O}(\mathbf{A}, \mathbf{C}) = \begin{pmatrix} 1 & 0 & -20 \\ 2 & 1 & -18 \\ 0 & 2 & -7 \end{pmatrix} \quad \text{with rank}(\mathcal{O}) = 3.$$

Thus, system (7.2.12) is completely observable.

7.2.4. Controllable and Unobservable Subspaces

Systems which are not completely state controllable and not completely observable have state space representations given by (7.2.8) and (7.2.10). It turns out that the pair $(\mathbf{A}_{11}, \mathbf{B}_1)$ in (7.2.8) is controllable and the pair $(\mathbf{A}_{11}, \mathbf{C}_1)$ in (7.2.10) is observable. So, in general, there do exist some controllable states and some observable states even if the full system is neither completely controllable nor completely observable. Kalman's *canonical decomposition theorem* (Furuta, Sano and Atherton (1988)) enables the state space to be decomposed into four subspaces. One of these is the *controllable subspace* $\mathcal{X}_c$ which is the set of all controllable states in the state space $\mathcal{X}$. Through (7.2.7) the controllable subspace can be defined as

$$\mathcal{X}_c \triangleq \left\{ \mathbf{x}_0 : \forall \mathbf{u}, \mathbf{x}_0 = -\int_0^t e^{-\mathbf{A}\tau} \mathbf{B} \mathbf{u}(\tau) \, d\tau \right\}$$

and is clearly a real vector space [see I, §5.2]. It can be shown that $\mathcal{X}_c$ is actually the range space $\mathscr{R}(\mathscr{C})$ of the controllability matrix (7.2.9). Figure 7.2.1 illustrates the decomposition of system (7.2.1–2) into the controllable subsystem S_1 and the uncontrollable subsystem S_2.

Now the state space $\mathbb{R}^n$ may be written as [see I, §5.11]

$$\mathbb{R}^n = \mathscr{R}(\mathcal{O}) \oplus \ker \mathcal{O}^T, \tag{7.2.13}$$

where $\mathcal{O}$ is the observability matrix (7.2.11). For a system of the form (7.2.10), since the dimension of $\mathscr{R}(\mathcal{O})$ is equal to the rank of $\mathcal{O}$, the latter being less than n, say n_0, it follows from (7.2.13) that the nullity [see I, §5.11] of $\mathcal{O}^T$ is $n - n_0$. The subspace $\ker \mathcal{O}^T$ is called the *unobservable subspace*, the set of all unobservable states.

The canonical decomposition theorem of Kalman then states that system (7.2.1–2) is equivalent to a system of the form

$$A = \begin{pmatrix} A_{11} & A_{12} & A_{13} & A_{14} \\ 0 & A_{22} & 0 & A_{24} \\ 0 & 0 & A_{33} & A_{34} \\ 0 & 0 & 0 & A_{44} \end{pmatrix}, \quad B = \begin{pmatrix} B_1 \\ B_2 \\ 0 \\ 0 \end{pmatrix},$$

$$C = (0, C_2, 0, C_4),$$

where A_{ii} is a square matrix of order n_i ($i = 1, \ldots, 4$), and the other matrices are of appropriate dimensions. Here

$$n_1 = \dim(\mathscr{R}(\mathcal{O}) \cap \ker \mathcal{O}^T),$$
$$n_2 = \dim \mathscr{R}(\mathcal{O}) - n_1,$$
$$n_3 = \text{nullity of } \mathcal{O}^T - n_1,$$
$$n_4 = n - n_1 - n_2 - n_3.$$

The outcome of this theorem is that the system (7.2.1–2) can be decomposed into four subsystems described under the following four headings:

(i) controllable and unobservable,
(ii) controllable and observable,
(iii) uncontrollable and unobservable,
(iv) uncontrollable and observable.

Examples and the proof of the theorem can be found in Furuta, Sano and Atherton (1988).

7.2.5. Module Formulation

Consider a linear, discrete, time-invariant, dynamical system [see §6.3.4] described by the equations

$$\mathbf{x}(t + 1) = \mathbf{Ax}(t) + \mathbf{Bu}(t), \quad \mathbf{y}(t) = \mathbf{Cx}(t) \quad (t \in \mathbb{Z}). \tag{7.2.14}$$

In order to simplify the presentation, suppose a single-input/single-output system is considered first.

The *input space* Ω contains all real, infinite sequences

$$u = (\ldots 0, 0, u(-p), u(-p+1), \ldots, u(-1), u(0); 0, 0, \ldots), \qquad (7.2.15)$$

$u(t) \in \mathbb{R}$, with all but a finite number of terms zero and where $\forall t > 0 \; u(t) = 0$. The *output space* Γ is the set of all real, infinite sequences

$$y = (\ldots, 0, 0; y(1), y(2), \ldots),$$

$y(t) \in \mathbb{R}$, where $\forall t \leq 0$, $y(t) = 0$. A map $f : \Omega \to \Gamma$ can then be defined as a linear, zero-state, input/output map.

A binary operation [see I, §1.5] $\circ$ on Ω may be chosen as that of *concatenation*. This is defined by taking any two sequences $u, v \in \Omega$, with

$$u = (\ldots, 0, 0, u(-p), u(-p+1), \ldots, u(-1), u(0); 0, 0, \ldots),$$
$$v = (\ldots, 0, 0, v(-q), v(-q+1), \ldots, v(-1), v(0); 0, 0, \ldots),$$

and then

$$u \circ v = (\ldots, 0, 0, u(-p), \ldots, u(0), v(-q), \ldots, v(0); 0, 0, \ldots).$$

Clearly, concatenation is an associative operation [see I, §8.2.1] and with the zero sequence $(\ldots, 0, 0, \ldots, 0, 0; 0, 0, \ldots)$ as identity, $(\Omega, \circ)$ is a monoid [see I, §8.4]. This monoid is the basis for several different equivalence relations [see I, §1.3.3], including the *Nerode equivalence relation* $\sim$: for two inputs $u, v \in \Omega$,

$$u \sim v \quad \text{if and only if} \quad f(u \circ w) = f(v \circ w) \quad \forall w \in \Omega,$$

and the *Myhill equivalence relation* $\sim$: for two inputs $u, v \in \Omega$,

$$u \sim v \quad \text{if and only if} \quad f(w_1 \circ u \circ w_2) = f(w_1 \circ v \circ w_2) \quad \forall w_1, w_2 \in \Omega.$$

The significance of these equivalence relations in control and automata theory can be found in Kalman, Falb and Arbib (1969).

For a multi-input system, each term $u(t)$ of sequence u (7.2.15) in the input space Ω is an m-vector $(u_1(t), \ldots, u_m(t))^T$ and may be represented by a polynomial obtained from the z-transform [see IV, §13.5] of the infinite sequence $(\ldots, 0, 0, \mathbf{u}(-p), \ldots, \mathbf{u}(0); 0, 0, \ldots)$, namely

$$\bar{\mathbf{u}}(z) = \sum_{t=-p}^{0} \mathbf{u}(t) z^{-t}$$

$$= \begin{pmatrix} u_1(-p) \\ \vdots \\ u_m(-p) \end{pmatrix} z^p + \begin{pmatrix} u_1(-p+1) \\ \vdots \\ u_m(-p+1) \end{pmatrix} z^{p-1} + \ldots + \begin{pmatrix} u_1(-1) \\ \vdots \\ u_m(-1) \end{pmatrix} z + \begin{pmatrix} u_1(0) \\ \vdots \\ u_m(0) \end{pmatrix} = \begin{pmatrix} \bar{u}_1(z) \\ \vdots \\ \bar{u}_m(z) \end{pmatrix},$$

where

$$\bar{u}_j(z) = u_j(-p) z^p + u_j(-p+1) z^{p-1} + \ldots + u_j(-1) z + u_j(0) \in \mathbb{R}[z].$$

Thus, elements of Ω may be represented by m-vectors such as $\bar{\mathbf{u}}(z)$ with components from the ring $\mathbb{R}[z]$ [see I, §2.3], i.e. $\bar{\mathbf{u}}(z) \in \mathbb{R}^m[z]$. In fact, $\Omega \cong \mathbb{R}^m[z]$.

Since scalar multiplication of $\mathbb{R}^m[z]$ over $\mathbb{R}[z]$ is defined in the obvious way, $\mathbb{R}^m[z]$ is readily made into an $\mathbb{R}[z]$-module. Further details on the $\mathbb{R}[z]$-module structure on Ω (and on the output space Γ) can be found in Kalman, Falb and Arbib (1969).

7.3   NON-LINEAR SYSTEMS

A general non-linear control system may be expressed in the form

$$\dot{\mathbf{x}}(t) = \mathbf{X}(\mathbf{x}(t), \mathbf{u}(t)) \tag{7.3.1}$$

$$\mathbf{y}(t) = \mathbf{Y}(\mathbf{x}(t)) \tag{7.3.2}$$

where the state function $\mathbf{x}: [0, \infty) \to M$ has values belonging to a smooth real analytic manifold M [see V, §14]. A special case worthy of note is when M is a Lie group [see V, §14.5], for example when $\mathbf{x}(t) \in GL(n, \mathbb{R})$ [see V, §14.4]. On the other hand, in many applications M is simply $\mathbb{R}^n$ as has been assumed in §2. The control function $\mathbf{u}: [0, \infty) \to \Omega$ has values belonging to some open or closed subset Ω of $\mathbb{R}^m$. The function $\mathbf{u}$ belongs to some class of admissible controls $\mathcal{U}$; for example, the class of all piecewise continuous functions defined on $[0, \infty)$ [see IV, §§2.1 and 20.2]. The vector field $X: M \times \Omega \to TM$ maps elements from the product space $M \times \Omega$ to the tangent bundle TM of the manifold M [see V, §§14.3 and 13.4].

7.3.1.   Linear-analytic Systems

One of the approaches to the mathematical theory of non-linear systems has been to use the ideas and techniques of differential geometry (see V, §§12 and 14). The most common type of mathematical model considered in this approach has been a special case of (7.3.1), namely

$$\dot{\mathbf{x}}(t) = \mathbf{f}(\mathbf{x}(t)) + \sum_{j=1}^{m} u_j(t)\mathbf{g}_j(\mathbf{x}(t)), \qquad \mathbf{x}(0) = \mathbf{x}_0, \tag{7.3.3}$$

called a linear-analytic system and which is described in terms of local coordinates.

Let $V(M)$ be the set of all analytic vector fields on M and regard $V(M)$ as a Lie algebra over $\mathbb{R}$ under the Lie bracket operation defined as follows.

DEFINITION 7.3.1.   For any $\mathbf{f}, \mathbf{g} \in V(M)$ the Lie bracket $[\mathbf{f}, \mathbf{g}]$ of $\mathbf{f}$ and $\mathbf{g}$ is given by

$$[\mathbf{f}, \mathbf{g}] = \frac{\partial \mathbf{g}}{\partial \mathbf{x}} \mathbf{f} - \frac{\partial \mathbf{f}}{\partial \mathbf{x}} \mathbf{g}.$$

EXAMPLE 7.3.1.

$$\dot{x}_1 = ux_3, \qquad \dot{x}_2 = x_1, \qquad \dot{x}_3 = x_1 x_2.$$

It follows that

$$\mathbf{f}(\mathbf{x}) = \begin{pmatrix} 0 \\ x_1 \\ x_1 x_2 \end{pmatrix}, \qquad \mathbf{g}(\mathbf{x}) = \begin{pmatrix} x_3 \\ 0 \\ 0 \end{pmatrix}$$

$$[\mathbf{f},\mathbf{g}](\mathbf{x}) = \begin{pmatrix} 0 & 0 & 1 \\ 0 & 0 & 0 \\ 0 & 0 & 0 \end{pmatrix} \begin{pmatrix} 0 \\ x_1 \\ x_1 x_2 \end{pmatrix} - \begin{pmatrix} 0 & 0 & 0 \\ 1 & 0 & 0 \\ x_2 & x_1 & 0 \end{pmatrix} \begin{pmatrix} x_3 \\ 0 \\ 0 \end{pmatrix}$$

$$= \begin{pmatrix} x_1 x_2 \\ 0 \\ 0 \end{pmatrix} - \begin{pmatrix} 0 \\ x_3 \\ x_2 x_3 \end{pmatrix} = \begin{pmatrix} x_1 x_2 \\ -x_3 \\ -x_2 x_3 \end{pmatrix}.$$

The set $V(M)$ can also be considered as a module over the ring $C(M)$ of C^∞ real-valued functions on M.

DEFINITION 7.3.2.   A subset D of $V(M)$ is said to be *involutive* if, $\forall \mathbf{f}, \mathbf{g} \in D$, $[\mathbf{f},\mathbf{g}] \in D$.

EXAMPLE 7.3.2.

$$\mathbf{f}(\mathbf{x}) = \begin{pmatrix} 0 \\ x_3 \\ -x_2 \end{pmatrix}, \qquad \mathbf{g}(\mathbf{x}) = \begin{pmatrix} -x_3 \\ 0 \\ x_1 \end{pmatrix}, \qquad \mathbf{h}(\mathbf{x}) = \begin{pmatrix} x_2 \\ -x_1 \\ 0 \end{pmatrix}.$$

It is easy to show that $[\mathbf{f},\mathbf{g}] = \mathbf{h}$, $[\mathbf{g},\mathbf{h}] = \mathbf{f}$ and $[\mathbf{h},\mathbf{f}] = \mathbf{g}$ so that $\{\mathbf{f},\mathbf{g},\mathbf{h}\}$ is an involutive set of $V(\mathbb{R}^3)$.

Given the vector fields $\mathbf{f}, \mathbf{g}_1, \ldots, \mathbf{g}_m$ of system (7.3.3), let $L(\mathbf{f}, \mathbf{g}_1, \ldots, \mathbf{g}_m)$ be the Lie algebra [see V, §14.6] generated by these vector fields. Furthermore, denote by $L(\mathbf{f}, \mathbf{g}_1, \ldots, \mathbf{g}_m)(\mathbf{x})$ the subspace of $\mathbb{R}^n$ [see I, §5.5] defined by

$$L(\mathbf{f}, \mathbf{g}_1, \ldots, \mathbf{g}_m)(\mathbf{x}) \overset{\Delta}{=} \{h(\mathbf{x}) : h \in L(\mathbf{f}, \mathbf{g}_1, \ldots, \mathbf{g}_m)\}.$$

This subspace is clearly a finite-dimensional vector space of dimension not greater than n [see I, §5.4].

7.3.2.   Accessibility

For simplicity assume that $M = \mathbb{R}^n$.

DEFINITION 7.3.3.   The *reachable set at time* t_1 for system (7.3.3) is the set of all states $x_1 \in \mathbb{R}^n$ such that there exists a control function $u:[0, t_1] \to \Omega$ belonging to $\mathscr{U}$ which yields a solution to (7.3.3) satisfying $x(t_1) = x_1$.

If the reachable set of Definition 7.3.3 is denoted by $R(x_0, t_1)$ then the set

$$R(x_0) = \bigcup_{t_1 \geqslant 0} R(x_0, t_1)$$

is called the *reachable set*.

DEFINITION 7.3.4.   System (7.3.3) is *accessible at* x_0 if $R(x_0)$ contains a neighbourhood of some state $x \in \mathbb{R}^n$, i.e. if $R(x_0)$ has a non-empty interior [see V, §5.2].

System (7.3.3) is *accessible* if it is accessible from each $x_0 \in \mathbb{R}^n$ and is *controllable* if $\forall x_0 \in \mathbb{R}^n$, $R(x_0) = \mathbb{R}^n$.

THEOREM 7.3.1.   *System* (7.3.3) *is accessible at* x_0 *if and only if* $L(\mathbf{f}, \mathbf{g}_1, \ldots, \mathbf{g}_m)(x_0) = \mathbb{R}^n$ *and is accessible if*, $\forall x \in \mathbb{R}^n$, $L(\mathbf{f}, \mathbf{g}_1, \ldots, \mathbf{g}_m)(x) = \mathbb{R}^n$.

EXAMPLE 7.3.3.

$$\dot{x}_1 = x_1 + u, \qquad x_1(0) = \sigma_1, \tag{7.3.4}$$
$$\dot{x}_2 = x_1^2, \qquad\quad x_2(0) = \sigma_2.$$

By inspection,

$$\mathbf{f}(x) = \begin{pmatrix} x_1 \\ x_1^2 \end{pmatrix}, \qquad \mathbf{g}(x) = \begin{pmatrix} 1 \\ 0 \end{pmatrix},$$

and direct calculation yields

$$[\mathbf{f}, \mathbf{g}](x) = \begin{pmatrix} -1 \\ -2x_1 \end{pmatrix}, \qquad [\mathbf{f}, [\mathbf{f}, \mathbf{g}]](x) = \begin{pmatrix} 1 \\ 0 \end{pmatrix}, \qquad [\mathbf{g}, [\mathbf{f}, \mathbf{g}]](x) = \begin{pmatrix} 0 \\ -2 \end{pmatrix}.$$

All other brackets generated by $\mathbf{f}$ and $\mathbf{g}$ are null. Then $L(\mathbf{f}, \mathbf{g})$ contains the vector fields $\mathbf{f}, \mathbf{g}, [\mathbf{f}, \mathbf{g}], [\mathbf{f}, [\mathbf{f}, \mathbf{g}]]$ and $[\mathbf{g}, [\mathbf{f}, \mathbf{g}]]$. It follows that the subspace $L(\mathbf{f}, \mathbf{g})(x)$ includes, for all $x \in \mathbb{R}^2$, the vectors $(1, 0)^T$ and $(0, -2)^T$ so that $L(\mathbf{f}, \mathbf{g})(x) = \mathbb{R}^2$. Thus, system (7.3.4) is accessible by Theorem 7.3.1.

Although system (7.3.4) is accessible, it is not controllable. This follows from the second equation in (7.3.4) which implies that

$$x_2(t) = \sigma_2 + \int_0^t x_1^2(\tau) d\tau \geq \sigma_2.$$

Thus, no matter what control u is used, the system cannot be controlled to any point $(x_1(t), x_2(t))$ for which $x_2(t) < \sigma_2$.

7.4. OPTIMAL CONTROL

7.4.1. Pontryagin's Minimum Principle

A fairly general statement of an optimal control problem is to find, in a class of state variables $x_i(t)$ $(i = 1, \ldots, n)$ and control variables $u_j(t)$ $(j = 1, \ldots, m)$ satisfying differential equations

$$\dot{x}_i = f_i(\mathbf{x}, \mathbf{u}, t) \tag{7.4.1}$$

and end conditions

$$\mathbf{x}(t_0) = 0, \quad t_0 \text{ specified,}$$

$$\psi_q[\mathbf{x}(t_1), t_1] = 0 \quad (q = 1, \ldots, k \leq n + 1),$$

(in which t_1 may or may not be specified), those variables $x_i(t)$, $u_j(t)$ which minimize a performance index

$$J[\mathbf{u}(\cdot)] = G[\mathbf{x}(t_1), t_1] + \int_{t_0}^{t_1} L(\mathbf{x}(t), \mathbf{u}(t), t) dt. \tag{7.4.2}$$

In many practical problems the control vector $\mathbf{u}(\cdot)$ must lie in a closed, bounded region Ω [see V, §5.2]. There is no loss of generality in posing the problem above as a minimization problem since to maximize J one may always minimize $-J$.

Such an optimal control problem is clearly associated with problems in the calculus of variations [see IV, §12] although the appearance of bounded control variables makes the former problem differ from the classical formulation. An optimal control problem with performance index in the form of (7.4.2) is known as a *Bolza* problem. With the integral omitted from (7.4.2) the problem is termed a *Mayer* problem whereas with the function G omitted it is called a *Lagrange* problem. All three types of problem are equivalent to one another and transformations exist to take one type of problem to another. For example, a Bolza problem can be transformed to a Lagrange problem by introducing a new variable x_{n+1} defined by

$$\dot{x}_{n+1} = 0, \qquad x_{n+1}(t_0) = \frac{G[\mathbf{x}(t_1), t_1]}{t_1 - t_0}.$$

The performance index in (7.4.2) can then be written as

$$J = \int_{t_0}^{t_1} (L + x_{n+1}) dt.$$

On the other hand, a Bolza problem can be transformed to a Mayer problem by introducing a new variable x_{n+1} such that

$$\dot{x}_{n+1} = L(\mathbf{x}, \mathbf{u}, t), \qquad x_{n+1}(t_0) = 0,$$

The performance index in (7.4.2) can then be written in the form

$$J = G[\mathbf{x}(t_1), t_1] + x_{n+1}(t_1).$$

Necessary conditions for optimality of a control $\mathbf{u}(\cdot)$ in the Bolza problem described above are based upon a Hamiltonian H [see IV, §17.4.6] defined in terms of adjoint variables $\lambda_i(t)$:

$$H(\mathbf{x}, \mathbf{u}, \lambda, t) \overset{\Delta}{=} L(\mathbf{x}, \mathbf{u}, t) + \sum_{i=1}^{n} \lambda_i(t) f_i(\mathbf{x}, \mathbf{u}, t). \tag{7.4.3}$$

The necessary conditions are then given by the equations

$$\dot{\lambda} = -\frac{\partial H}{\partial \mathbf{x}}, \tag{7.4.4}$$

$$\left[H(t_1) + \frac{\partial G}{\partial t_1} \right] dt_1 + \left(\frac{\partial G}{\partial \mathbf{x}(t_1)} - \lambda(t_1) \right)^T dx(t_1) = 0, \tag{7.4.5}$$

subject to

$$\frac{\partial \psi_q}{\partial t_1} dt_1 + \frac{\partial \psi_q}{\partial \mathbf{x}(t_1)} dx(t_1) = 0 \quad (q = 1, \ldots, k), \tag{7.4.6}$$

$$H(\mathbf{x}, \mathbf{u}^0, \lambda, t) \leq H(\mathbf{x}, \mathbf{u}, \lambda, t) \quad (t_0 \leq t \leq t_1), \tag{7.4.7}$$

where $\mathbf{u}^0$ is the optimal control which minimizes the performance index J in (7.4.2). Condition (7.4.7) simply states that the optimal control $\mathbf{u}^0$ minimizes the Hamiltonian H at every instant of time $t \in [t_0, t_1]$. This condition is known as *Pontryagin's minimum principle*. Generally, this result leads to a bang-bang solution in which the controls $u_j(t)$ are switched between their bounds at discrete times. In the special case where $\mathbf{u}$ is unbounded, equation (7.4.7) implies

$$\frac{\partial H}{\partial \mathbf{u}} = \mathbf{0}. \tag{7.4.8}$$

An important special case of the *transversality* conditions (7.4.5) and (7.4.6) is when none of the final states $x_i(t_1)$ are specified. The conditions then reduce to

$$H(t_1) = -\frac{\partial G}{\partial t_1}, \tag{7.4.9}$$

$$\lambda(t_1) = \frac{\partial G}{\partial \mathbf{x}(t_1)} \quad (i = 1, \ldots, n). \tag{7.4.10}$$

If the final time t_1 is also specified then (7.4.9) is not valid.

In the case where the Hamiltonian H does not contain the time t explicitly, there exists a first integral of the set of differential equations (7.4.1) and (7.4.4) which is

$$H(\mathbf{x}, \mathbf{u}, \lambda) = \text{constant}.$$

That is to say, under the conditions states, the Hamiltonian H remains constant throughout the time interval $[t_0, t_1]$.

A special case of the above optimal control problem is to minimize a quadratic performance index when the system equations are linear, the so-called LQP problem. This is the subject of the next example.

EXAMPLE 4.4.1.   [see Engineering Guide, §14.11]

$$J = \tfrac{1}{2}\mathbf{x}^{\mathsf{T}}(t_1)\mathbf{M}\mathbf{x}(t_1) + \frac{1}{2}\int_{t_0}^{t_1} (\mathbf{x}^{\mathsf{T}}\mathbf{Q}\mathbf{x} + \mathbf{u}^{\mathsf{T}}\mathbf{R}\mathbf{u})dt,$$

subject to the linear system

$$\dot{\mathbf{x}}(t) = \mathbf{A}\mathbf{x}(t) + \mathbf{B}\mathbf{u}(t),$$

where $\mathbf{M}$ and $\mathbf{R}$ are symmetric positive-definite [see I, §9.2] and $\mathbf{Q}$ is symmetric. The control $\mathbf{u}$ is assumed to be unbounded.

The Hamiltonian is given by

$$H(\mathbf{x}, \mathbf{u}, \lambda) = \tfrac{1}{2}\mathbf{x}^{\mathsf{T}}\mathbf{Q}\mathbf{x} + \tfrac{1}{2}\mathbf{u}^{\mathsf{T}}\mathbf{R}\mathbf{u} + \lambda^{\mathsf{T}}(\mathbf{A}\mathbf{x} + \mathbf{B}\mathbf{u}).$$

Then (7.4.4) and (7.4.8) yield

$$\dot{\lambda} = - H_{\mathbf{x}} = - \mathbf{Q}\mathbf{x} - \mathbf{A}^{\mathsf{T}}\lambda, \qquad (7.4.11)$$

$$H_{\mathbf{u}} = \mathbf{R}\mathbf{u} + \mathbf{B}^{\mathsf{T}}\lambda = 0 \quad \Rightarrow \quad \mathbf{u} = - \mathbf{R}^{-1}\mathbf{B}^{\mathsf{T}}\lambda. \qquad (7.4.12)$$

Try a solution $\lambda = \mathbf{P}(t)\mathbf{x}$, where $\mathbf{P}(\cdot)$ is a time-varying symmetric matrix but otherwise arbitrary. Then

$$\dot{\lambda} = \dot{\mathbf{P}}\mathbf{x} + \mathbf{P}\dot{\mathbf{x}} = \dot{\mathbf{P}}\mathbf{x} + \mathbf{P}(\mathbf{A}\mathbf{x} + \mathbf{B}\mathbf{u}) = \dot{\mathbf{P}}\mathbf{x} + \mathbf{P}\mathbf{A}\mathbf{x} - \mathbf{P}\mathbf{B}\mathbf{R}^{-1}\mathbf{B}^{\mathsf{T}}\mathbf{P}\mathbf{x}$$

by (7.4.12). This result, together with (7.4.11) leads to

$$(\dot{\mathbf{P}} + \mathbf{P}\mathbf{A} - \mathbf{P}\mathbf{B}\mathbf{R}^{-1}\mathbf{B}^{\mathsf{T}}\mathbf{P} + \mathbf{Q} + \mathbf{A}^{\mathsf{T}}\mathbf{P})\mathbf{x} = 0$$

which in turn implies that matrix $\mathbf{P}$ satisfies the matrix Riccati differential equation [see IV, §7.10]

$$\dot{\mathbf{P}} = \mathbf{P}\mathbf{B}\mathbf{R}^{-1}\mathbf{B}^{\mathsf{T}}\mathbf{P} - \mathbf{A}^{\mathsf{T}}\mathbf{P} - \mathbf{P}\mathbf{A} - \mathbf{Q}. \qquad (7.4.13)$$

From (7.4.12) the control is linear state feedback in the form

$$\mathbf{u} = - \mathbf{R}^{-1}\mathbf{B}^{\mathsf{T}}\mathbf{P}\mathbf{x}.$$

The transversality condition (7.4.10) with $G = \tfrac{1}{2}\mathbf{x}^{\mathsf{T}}(t_1)\mathbf{M}\mathbf{x}(t_1)$ yields

$$\lambda(t_1) = \mathbf{M}\mathbf{x}(t_1)$$

so that $\mathbf{P}(t_1) = \mathbf{M}$. With this end condition on matrix $\mathbf{P}$ the Riccati equation (7.4.13) may be solved backwards in time to give $\mathbf{P}(\cdot)$.

In the special case where the final time is infinite, $\mathbf{P}(t)$ reduces to a constant matrix $\mathbf{P}_0$. The matrix Riccati equation then reduces to a set of algebraic equations and the matrix $\mathbf{R}^{-1}\mathbf{B}^{\mathsf{T}}\mathbf{P}_0$ is a constant matrix $\mathbf{K}$. It follows that in this case $\mathbf{u} = - \mathbf{K}\mathbf{x}$, a constant feedback control law.

The next example illustrates the application of Pontryagin's principle to a problem in which the control is bounded.

EXAMPLE 7.4.2.

$$\dot{x}_1 = x_2 - u, \qquad x_1(0) = 0$$
$$\dot{x}_2 = u, \qquad x_2(0) = 0, \qquad |u| \le 1.$$

It is required to maximize the performance index $x_1(2)$. Expressing this as a minimization problem, it is of Mayer type with $G[x(t_1), t_1] = -x_1(2)$ since minimizing $-x_1(2)$ will maximize $x_1(2)$ as required.

The Hamiltonian H is given by (7.4.3) and is

$$H = \lambda_1(x_2 - u) + \lambda_2 u.$$

(7.4.4) yields

$$\dot{\lambda}_1 = 0, \qquad \dot{\lambda}_2 = -\lambda_1.$$

$t_1 = 2$ is specified and so (7.4.9) is not valid. (7.4.10) yields

$$\lambda_1(2) = -1, \qquad \lambda_2(2) = 0.$$

Then, $\forall t \in [0, 2]$, $\lambda_1(t) = -1$, $\lambda_2(t) = t - 2$.

The coefficient of u in H is $\lambda_2 - \lambda_1 = t - 1$ and so

$$\lambda_2 - \lambda_1 < 0 \quad \text{for } t < 1,$$
$$\lambda_2 - \lambda_1 > 0 \quad \text{for } t > 1.$$

Then by Pontryagin's principle (7.4.7) the coefficient of u, namely H_u called the *switching function*, determines the control u as

$$u = -\operatorname{sgn}(\lambda_2 - \lambda_1),$$

i.e.

$$u = 1 \quad \text{when } t < 1,$$
$$u = -1 \quad \text{when } t > 1.$$

When $u = 1$,

$$\dot{x}_1 = x_2 - 1, \qquad x_1(0) = 0,$$
$$\dot{x}_2 = 1, \qquad x_2(0) = 0,$$

which yields

$$x_1(t) = \tfrac{1}{2}(t - 2)t, \qquad x_2(t) = t.$$

The state trajectory in the (x_1, x_2)-space is then given by the parabola [see V, §1.3]

$$2x_1 + 1 = (x_2 - 1)^2 \quad \text{for } 0 \le t < 1.$$

At $t = 1$ the control is switched to the value $u = -1$. Then

$$\dot{x}_1 = x_2 + 1, \qquad x_1(1) = -\tfrac{1}{2},$$
$$\dot{x}_2 = -1, \qquad x_2(1) = 1.$$

On integrating these equations with respect to time t,

$$x_1(t) = -3 + 3t - \tfrac{1}{2}t^2,$$
$$x_2(t) = 2 - t,$$

and the state trajectory for $1 < t \leq 2$ is again a parabola

$$2x_1 - 3 = -(x_2 + 1)^2.$$

Since the Hamiltonian does not contain the time t explicitly, H remains constant throughout the trajectory. To verify this, note that, $\forall t \in [0, 1), u = 1$ and

$$H = -(x_2 - 1) + (t - 2) = -(t - 1) + (t - 2) = -1.$$

For $t \in (1, 2]$, when $u = -1$,

$$H = -(x_2 + 1) - (t - 2) = -(2 - t + 1) - (t - 2) = -1.$$

Thus, the Hamiltonian stays constants at the value -1.

7.4.2. Singular Control

Suppose for the time being that an optimal control problem contains only one control variable u. When the Hamiltonian of such a problem is linear in u the sign of the switching function H_u determines the possible values for u by Pontryagin's minimum principle (as in Example 7.4.2). However, if the switching function H_u vanishes identically over the time interval $[t_0, t_1]$, then Pontryagin's principle gives no information about the possible optimal control. Such a problem is called a *singular control problem* (Bell and Jacobson (1975)). Since H_u vanishes identically in such a problem, it is sometimes possible to obtain the control by differentiating H_u with respect to time a sufficient number of times until the (singular) control appears. It is known that an even number of differentiations ($2q$ say) are required before u can appear explicitly. The parameter q is called the *order* of the singular problem.

There are two necessary conditions for optimality of a singular control. The *generalized Legendre–Clebsch* (GLC) condition states that, $\forall t \in [t_0, t_1]$,

$$(-1)^q \frac{\partial}{\partial u}\left(\frac{d^{2q}H_u}{dt^{2q}}\right) \geq 0, \tag{7.4.14}$$

where q is the order of the singular problem as defined above. This result remains true for a vector control but there is then an additional necessary condition which states that, $\forall t \in [t_0, t_1]$,

$$\frac{\partial}{\partial \mathbf{u}}\left(\frac{d^p H_u}{dt^p}\right) = 0 \quad \text{for } p \text{ odd.} \tag{7.4.15}$$

The second necessary condition for optimality of a singular vector control is the *Jacobson condition*. This states that, $\forall t \in [t_0, t_1], \exists$ a time-varying, $n \times n$

symmetric matrix $\mathbf{Q}$ such that

$$\dot{\mathbf{Q}} = -H_{xx} - f_x^T\mathbf{Q} - \mathbf{Q}f_x, \qquad \mathbf{Q}(t_1) = G_{xx}[\mathbf{x}(t_1), t_1] \qquad (7.4.16)$$

and

$$(H_{ux} + f_u^T\mathbf{Q})f_u \geq 0. \qquad (7.4.17)$$

EXAMPLE 7.4.3. Consider the optimal control problem:

$$\text{minimize} \quad J = \frac{1}{2}\int_0^{3\pi/2}(-x_1^2 + x_2^2)dt$$

subject to

$$\dot{x}_1 = x_2, \qquad x_1(0) = 0, \qquad\qquad (7.4.18)$$
$$\dot{x}_2 = u, \qquad x_2(0) = 1, \qquad |u| \leq 1. \qquad (7.4.19)$$

The Hamiltonian for this problem is

$$H(x_1, x_2, u) = \tfrac{1}{2}(-x_1^2 + x_2^2) + \lambda_1 x_2 + \lambda_2 u.$$

The switching function H_u is given by

$$H_u = \lambda_2.$$

The adjoint equations (7.4.4) and transversality conditions (7.4.10) are

$$\dot{\lambda}_1 = x_1, \qquad \lambda_1(3\pi/2) = 0, \qquad (7.4.20)$$
$$\dot{\lambda}_2 = -x_2 - \lambda_1, \qquad \lambda_2(3\pi/2) = 0. \qquad (7.4.21)$$

If $\lambda_2 \equiv 0$, then $\dot{\lambda}_2 \equiv 0$ and it follows from (7.4.21) that

$$\lambda_1 = -x_2.$$

Thus

$$\dot{\lambda}_1 = -\dot{x}_2 = -u$$

and then (7.4.20) yields $x_1 = -u$. Since from (7.4.18) and (7.4.19) $\ddot{x}_1 = u$, the state x_1 satisfies the differential equation $\ddot{x}_1 = -x_1$. With initial conditions on x_1 and $\dot{x}_1$ from (7.4.18) and (7.4.19), the solution for state x_1 is given by [see IV, §2.12]

$$x_1(t) = \sin t.$$

From (7.4.18) and (7.4.19), it is clear that

$$x_2(t) = \cos t \quad \text{and} \quad u(t) = -\sin t.$$

Also, $\lambda_1(t) = -\cos t$ and $\lambda_2(t) = 0$ satisfy (7.4.20) and (7.4.21). The control $u(t) = -\sin t$ is a singular control for the problem posed.

Differentiating the switching function H_u with respect to time:

$$H_u = \lambda_2,$$
$$\dot{H}_u = \dot{\lambda}_2 = -x_2 - \lambda_1 \quad \text{from (7.4.21)},$$
$$\ddot{H}_u = -\dot{x}_2 - \dot{\lambda}_1 = -u - x_1 \quad \text{from (7.4.19) and (7.4.20)}.$$

Thus

$$(-1)\frac{\partial}{\partial u}\ddot{H}_u = 1 > 0$$

and the GLC condition is satisfied.

To test the Jacobson condition write the matrix $\mathbf{Q}$ in (7.4.16) as

$$\mathbf{Q} = \begin{pmatrix} q_{11} & q_{12} \\ q_{12} & q_{22} \end{pmatrix}.$$

Then since

$$\mathbf{f}_u = \begin{pmatrix} 0 \\ 1 \end{pmatrix} \quad \text{and} \quad H_{xu} = \begin{pmatrix} H_{x_1 u} \\ H_{x_2 u} \end{pmatrix} = \begin{pmatrix} 0 \\ 0 \end{pmatrix},$$

it follows that

$$(H_{ux} + \mathbf{f}_u^T \mathbf{Q})\mathbf{f}_u = q_{22} \tag{7.4.22}$$

$$H_{xx} = \begin{pmatrix} -1 & 0 \\ 0 & 1 \end{pmatrix}, \quad \mathbf{f}_x = \begin{pmatrix} 0 & 1 \\ 0 & 0 \end{pmatrix}, \quad G_{xx} = 0.$$

Also, by writing $\tau = 3\pi/2 - t$ (reverse time), (7.4.16) becomes

$$\dot{q}_{11} = -1, \qquad q_{11}(0) = 0,$$
$$\dot{q}_{12} = q_{11}, \qquad q_{12}(0) = 0,$$
$$\dot{q}_{22} = 1 + 2q_{12}, \qquad q_{22}(0) = 0,$$

where here a dot over a variable denotes differentiation with respect to τ. The solution to these equations is

$$q_{11} = -\tau, \qquad q_{12} = -\tfrac{1}{2}\tau^2, \qquad q_{22} = \tau - \tfrac{1}{3}\tau^3,$$

so that (7.4.17) and (7.4.22) together give

$$\tau - \tfrac{1}{3}\tau^3 \geq 0.$$

This is satisfied provided that $\tau \leq \sqrt{3}$.

Thus, both necessary conditions are satisfied for $\tau \in [0, \sqrt{3}]$. However, it can be shown that the solution above is non-optimal for $\pi/2 < \tau < \sqrt{3}$ and so the two necessary conditions are not sufficient for optimality. Further details can be found in Bell and Jacobson (1975).

7.4.3. Junction Conditions

Many trajectories arising from optimal control problems contain both bang-bang (non-singular) arcs and singular arcs. Such trajectories are called *partially singular* trajectories and a point at which a non-singular arc joins a singular one is called a *junction*. A necessary condition for optimality at a junction is as follows:

Let t_c be a point at which singular and non-singular subarcs of an optimal scalar control u are joined and let q be the order of the singular problem. Suppose the strengthened ((7.4.14) with strict inequality) GLC condition is satisfied at t_c and assume that the control is piecewise analytic in a neighbourhood of t_c [see IV, §2.10]. Let $u^{(r)}$ $(r \geq 0)$ be the lowest order derivative of u which is discontinuous at t_c. Then $q + r$ is an odd integer.

EXAMPLE 7.4.4.   Consider the optimal control problem

$$\text{minimize} \quad \int_0^2 x^2 \, dt,$$

subject to

$$\dot{x} = u, \qquad x(0) = 1, \qquad |u| \leq 1.$$

The Hamiltonian H and the switching function H_u are given by

$$H = x^2 + \lambda u, \qquad H_u = \lambda.$$

The adjoint equation and transversality condition give

$$\dot{\lambda} = -2x, \qquad \lambda(2) = 0.$$

By Pontryagin's principle $u = -\operatorname{sgn} \lambda$ so that subarcs are either bang-bang $(u = \pm 1)$ or singular $(\lambda \equiv 0)$.

If u is chosen initially with value 1 then $x(t) = 1 + t$, whereas with a value $-1, x(t) = 1 - t$. Clearly, the latter choice reduces the value of x and thus reduces the integrand of the performance index. In fact, it is not difficult to show that the optimal control is as follows:

$$\forall t \in [0, 1), \quad u(t) = -1 \quad \text{(nonsingular arc),}$$
$$\forall t \in (1, 2], \quad u(t) = 0 \quad \text{(singular arc),}$$

with a junction at $t = 1$. The order of the singular arc is obtained from

$$H_u = \lambda, \qquad \dot{H}_u = \dot{\lambda} = -2x, \qquad \ddot{H}_u = -2\dot{x} = -2u,$$

so that $q = 1$.

The control u is discontinuous at $t = 1$ so that $r = 0$. Thus $q + r = 1$ which is odd and the junction theorem is satisfied.

7.4.4   Dynamic Programming [see IV, §16]

Consider a discrete system governed by the non-linear difference equation

$$\mathbf{x}(i) = \mathbf{f}[\mathbf{x}(i-1), \mathbf{u}(i)] \quad (i = 1, \ldots, N), \tag{7.4.23}$$

where $\mathbf{x}(i) \in \mathbb{R}^n$, $\mathbf{u}(i) \in \mathbb{R}^m$, $\mathbf{f}: \mathbb{R}^n \times \mathbb{R}^m \to \mathbb{R}^n$. It is assumed that $\mathbf{x}(0)$ is known. An optimal control problem is to find a control sequence $\mathbf{u}(1), \ldots, \mathbf{u}(N)$ which produces a state sequence $\mathbf{x}(1), \ldots, \mathbf{x}(N)$ from (7.4.23) such that control and state together yield a minimum value of a performance index

$$J = G[\mathbf{x}(N)].$$

Bellman (1957) proposed a method for solving such an optimal control problem based upon a *principle of optimality*, a method known as dynamic programming which is essentially a multi-stage decision process. The principle of optimality states that an optimal policy has the property that whatever the current state $\mathbf{x}(i-1)$ and current decision $\mathbf{u}(i)$, the remaining decisions must constitute an optimal policy with regard to the state $\mathbf{x}(i)$ resulting from the decision $\mathbf{u}(i)$. The proof of this principle is by contradiction and is almost self-evident. The principle simply says that any part of an optimal trajectory is itself optimal.

Dynamic programming is a computational algorithm based on Bellman's principle. The algorithm commences at the last decision stage (Figure 7.4.1). For simplicity, assume state x and control u are scalars and suppose that state $x(N-1)$ has already been determined. In order to determine the optimal decision $u(N)$, a search over a suitable discrete set of values $u^1(N), \ldots, u^q(N)$ must be carried out to determine which one minimizes $G[x(N)]$ (Figure 7.4.2). If $x(N)$ is generated by the difference equation

$$x(N) = f[x(N-1), u(N)],$$

then the optimal $u(N)$ is determined from the minimization

$$\min_{u(N)} G\{f[x(N-1), u(N)]\}.$$

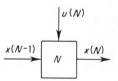

Figure 7.4.1. The last decision stage

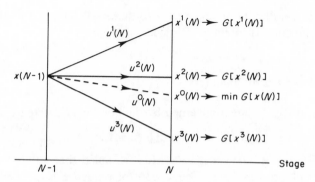

Figure 7.4.2. Decision search for optimal $u(N)$

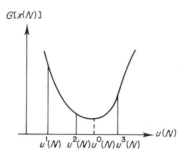

Figure 7.4.3. Optimal $u(N)$ and
minimum $G[x(N)]$

Some linear interpolation will generally be sufficient to obtain an approximate value for the optimal decision (or control) $u^0(N)$ (Figure 7.4.3). Let the minimum value of $G[x(N)]$ associated with state $x(N-1)$ be denoted by $S[x(N-1)]$ so that

$$S[x(N-1)] \overset{\Delta}{=} \min_{u(N)} G\{f[x(N-1), u(N)]\},$$

where S is called the *optimality function*. Thus, the function value $S[x(N-1)]$ at state $x(N-1)$ is the minimum value of G obtainable from that state. Notice that the minimization of G is carried out first over $u(N)$ and then having found $u(N)$ as a function of $x(N-1)$, treat S as a function of $x(N-1)$ only. This procedure may be repeated for a range of states $x^1(N-1),\ldots,x^p(N-1)$ and the optimality function value found for each of these states. If the required optimal trajectory then happens to pass through one of these states at the $(N-1)$-stage, then the value of the optimality function at that stage gives the minimum value of the performance index from that state.

One may then move to the $(N-2)$-stage and calculate the values of the optimality function from a range of states $x^1(N-2),\ldots,x^p(N-2)$ by again using a set of control values $u^1(N-1),\ldots,u^q(N-1)$ for each state and choosing the optimal control for that state. Linear interpolation at the $(N-1)$-stage will generally again be necessary to find the values of the optimality function

$$S[x(N-2)] = \min_{u(N-1)} S[x(N-1)],$$

which, by (7.4.23), can be written as

$$S[x(N-2)] = \min_{u(N-1)} S\{f[x(N-2), u(N-1)]\}.$$

Continuing in this manner it is not difficult to see that at stage k $(k = 1,\ldots,N-1)$ the value of the optimality function from state $x(k)$ is given by the general recurrence relation [see I, §14.13]

$$S[x(k)] = \min_{u(k+1)} S\{f[x(k), u(k+1)]\}. \tag{7.4.24}$$

Clearly, a boundary condition for this relation is

$$S[x(N)] = G[x(N)]. \tag{7.4.25}$$

Working backwards through each stage, the first stage is reached where $x(0)$ is given. It is then possible to find the optimal control $u(1)$ from $x(0)$ and, by working forward, to patch together the individual optimal trajectories from one stage to the next until the final optimal control $u(N)$ is obtained.

By definition, the optimal state and control must satisfy the recurrence relation (7.4.24) and, furthermore, having found the optimality function S from (7.4.24) the optimal solution is known. The relation (7.4.24) together with the boundary condition (7.4.25) therefore form a necessary and sufficient conditon for an optimum. The case of vector state and vector control can be handled by a simple extension of the above scalar case; in fact (7.4.24) simply becomes

$$S[\mathbf{x}(k)] = \min_{\mathbf{u}(k+1)} S\{\mathbf{f}[\mathbf{x}(k), \mathbf{u}(k+1)]\} \tag{7.4.26}$$

and boundary condition (7.4.25) becomes

$$S[\mathbf{x}(N)] = G[\mathbf{x}(N)].$$

The recurrence relation (7.4.26) is called the Hamilton–Jacobi–Bellman (HJB) equation since in its continuous form it is closely related to the Hamilton–Jacobi equation of classical mechanics [IV, §17.4.8]. The continuous form of the HJB equation can be derived from the discrete case given above in the following way.

Consider the optimality function S as a function of state and stage, where stage k represents discrete time kh, h being the time interval between consecutive stages. Then $S[\mathbf{x}(k)]$ may be written as

$$S[\mathbf{x}(x), kh] = \min_{\mathbf{u}(k+1)} S\{\mathbf{f}[\mathbf{x}(k), \mathbf{u}(k+1)], (k+1)h\}$$

in place of (7.4.26). As the interval h is reduced, so this recurrence relation approaches the equation

$$S(\mathbf{x}, t) = \min_{\mathbf{u}} \left[S(\mathbf{x}, t) + \sum_{i=1}^{n} \frac{\partial S}{\partial x_i} h f_i(\mathbf{x}, \mathbf{u}) + h \frac{\partial S}{\partial t} \right], \tag{7.4.27}$$

where $f_i (i = 1, \ldots, n)$ are the components of the vector function $\mathbf{f}$. Since $S(\mathbf{x}, t)$ is not a function of $\mathbf{u}$ (7.4.27) leads to

$$\frac{\partial S}{\partial t} + \min_{\mathbf{u}} \sum_{i=1}^{n} \frac{\partial S}{\partial x_i} f_i(\mathbf{x}, \mathbf{u}) = 0. \tag{7.4.28}$$

This equation is the continuous form of the HJB equation. The control $\mathbf{u}$ must be chosen to minimize

$$\sum_{i=1}^{n} \frac{\partial S}{\partial x_i} f_i(\mathbf{x}, \mathbf{u}),$$

resulting in a control law $\mathbf{u} = \mathbf{u}(\mathbf{x}, t)$. The partial differential equation (7.4.28) with this control law must then be solved for $S(\mathbf{x}, t)$ subject to the boundary condition $S[\mathbf{x}(t_1), t_1)] = G[\mathbf{x}(t_1)]$, where $t_1 = Nh$ [see IV, §8].

EXAMPLE 7.4.5. Minimize

$$J = x_2(t_1),$$

subject to

$$\dot{x}_1 = u, \qquad \dot{x}_2 = \tfrac{1}{2}x_1^2 + \tfrac{1}{2}u^2.$$

The HJB equation (7.4.28) for this problem is

$$S_t + \min_u \, [S_{x_1} u + \tfrac{1}{2}S_{x_2}(x_1^2 + u^2)] = 0, \tag{7.4.29}$$

where $S_t = \partial S/\partial t$, etc. Minimization over u yields

$$S_{x_1} + uS_{x_2} = 0,$$

i.e.

$$u = - S_{x_1}/S_{x_2}. \tag{7.4.30}$$

Substitution into (7.4.29) gives a non-linear partial differential equation [see IV, §8.6]

$$S_t - \tfrac{1}{2}(S_{x_1})^2/S_{x_2} + \tfrac{1}{2}S_{x_2}x_1^2 = 0, \tag{7.4.31}$$

with boundary condition $S[x(t_1), t_1] = x_2(t_1)$.

To solve this equation with the given boundary condition, a trial solution

$$S(x, t) = x_2 + \tfrac{1}{2}c(t)x_1^2, \qquad c(t_1) = 0 \tag{7.4.32}$$

is suggested. Substituting this trial solution into (7.4.31) leads to

$$\dot{c} = c^2 - 1, \qquad c(t_1) = 0,$$

which can easily be solved to give

$$c(t) = \tanh (t_1 - t)$$

[see IV, §2.13]. It follows from (7.4.32) that

$$S(x, t) = x_2 + \tfrac{1}{2}x_1^2 \tanh (t_1 - t)$$

whence, from (7.4.30),

$$u = - x_1 \tanh (t_1 - t).$$

This is the optimal control which yields the minimum value of $x_2(t_1)$.

D. J. B.

REFERENCES

Banks, S. P. (1988). *Mathematical Theories of Nonlinear Systems*, Prentice Hall.
Barnett, S. and Cameron, R. G. (1985). *Introduction to Mathematical Control Theory*, Clarendon Press.
Bell, D. J. and Jacobson, D. H. (1975). *Singular Optimal Control Problems*, Academic Press.
Bellman, R. (1957). *Dynamic Programming*, Princeton University Press.
Fliess, M. (1986). Some Remarks on Nonlinear Invertibility and Dynamic State-feedback, *Theory and Applications of Nonlinear Control Systems*, Eds. C. I. Byrnes and A. Lindquist, North-Holland, 115–121.
Furuta, K., Sano, A. and Atherton, D. (1988). *State Variable Methods in Automatic Control*, Wiley.
Isidori, A. (1989). *Nonlinear Control Systems: An Introduction* Communications and Control Engineering Series, Springer-Verlag.
Kalman, R. E., Falb, P. L. and Arbib, M. A. (1969). *Topics in Mathematical System Theory*, McGraw-Hill.
MacFarlane, A. G. J. (ed.) (1980). *Complex Variable Methods for Linear Multivariable Feedback Systems*, Taylor and Francis.
Pommaret, J. F. (1983). *Differential Galois Theory*, Gordon and Breach.
Pontryagin, L. S., Boltyanskii, V. G., Gamkrelidze, R. V. and Mishchenko, E. F. (1962). *The Mathematical Theory of Optimal Processes*, Interscience.
Rosenbrock, H. H. (1970). *State-space and Multivariable Theory*, Nelson.
Wonham, W. M. (1985). *Linear Multivariable Control: A Geometric Approach*, Springer-Verlag.

CHAPTER 8

The Finite Element Method

8.1. INTRODUCTION

The finite element method is a powerful and flexible tool for the approximate solution of physical problems modelled by systems of differential equations. It first appeared, at least in its modern form, in the engineering literature of the 1950s as a technique for solving structural analysis problems. However, it quickly became obvious that the method was equally applicable to many other areas, and finite elements have flourished in the physical sciences.

A great practical advantage of the finite element method is that the construction of the approximation scheme is an intrinsic part of the process. One consequence of this is that, while a significant programming effort is required to produce a code, when the code is ready it can cope with a very wide range of problems. This is in contrast with the finite difference method where, for example, boundary conditions and irregular boundaries often defeat attempts at a general computer program [see, e.g. II, §9]. The problem of programming effort has itself been ameliorated in recent years with the appearance of reliable commercial software. The problem description (geometry, physical parameters, loads, fluxes, etc.) can be supplied as data. This flexibility has led to the production of finite element libraries, which can often be used with little or no knowledge of the method. It is still, however, the case that there are areas in which these packages are inapplicable, and, even when they are appropriate, an understanding of the finite element method greatly enhances their intelligent use.

In this chapter we shall concentrate our efforts on the techniques and analysis which are particular to the finite element method. Therefore, while numerical integration schemes are essential for the construction of the approximation, and finite difference methods are equally essential for tackling time-dependent problems, we shall refer the reader to other sources. In addition, we shall not discuss matrix solution techniques at length.

The structure of the remainder of this chapter is as follows: The basic ideas underpinning the method are introduced in a simple example in §8.2. In §8.3, we shall show how to take a differential equation and pose it in a form suitable for finite element analysis. We shall also use this opportunity to introduce some useful notation. §8.4 introduces the basic building blocks for the finite element method, the basis (or shape) functions in two dimensions, and outlines the conditions they must satisfy for various problems.

The assembly of the finite element system is typically carried out in a sequence of small calculations related to individual elements (this is one of the reasons that the method enjoys great flexibility), and the organization of these calculations is described in §8.5. §8.6 deals with the accuracy of the finite element approximation and, in particular, shows how this depends on the smoothness of the solution of the exact equation. Extensions to time-dependent problems are discussed in §8.7.

It is neither possible nor desirable to explain the depth or subtlety of the finite element method in one short chapter. We shall concentrate on explaining the basic principles, through use of simple examples, in order to equip the reader to tackle the many more advanced texts.

8.2.   FIRST EXAMPLE

As a first step let us consider the following example from linear elasticity (for those readers unfamiliar with elasticity, the model is unimportant, the approximation technique is).

We have a one-dimensional beam of unit length, held fixed at both ends and subject to a force uniformly distributed along its length. Naturally the beam undergoes a deflection. In order to calculate this deflection, we can minimize the potential energy of the beam. Taking u to be the displacement, and after suitable scaling, we can write this as

$$E(u) = \int_0^1 \left\{ \frac{1}{2}\left(\frac{du}{dx}\right)^2 - u \right\} dx, \qquad (8.2.1)$$

and we can (in principle) minimize expression (8.2.1) over all 'suitable' functions u such that $u(0) = u(1) = 0$ (what is meant by 'suitable' will be made clearer later in the chapter). In this case we can solve the problem explicitly—a simple calculus of variations argument [see IV, §12] leads to $u(x) = \frac{1}{2}x(1 - x)$—but in general we can find no explicit expression for the solution of even quite simple problems. This type of problem is often referred to as *infinite dimensional*, since we have an infinity of functions u from which to choose the one which minimizes $E(u)$.

We can reduce our range of functions to those which can be described with the aid of a finite number of parameters and obtain a (*finite dimensional*) problem which is much more simple to solve than the original, but the cost we must pay is that this will only give us an approximation to the solution we seek. The aim therefore is to produce an approximate problem which is easy to solve but where the approximation remains 'close' to the true solution u. The finite element method achieves these twin goals.

An idea, which should be familiar to those with a knowledge of interpolation [see III, §2.3], is to take our approximation to u to be a piecewise polynomial [see III, §2.3] denoted by (say) $u_{h,p}$. We can divide the interval $[0, 1]$, over which we need to solve the problem, into smaller intervals of size h, and in each of these subintervals the function $u_{h,p}$ is a polynomial of degree p. Of course we usually, though not invariably, assume that the function $u_{h,p}$ is continuous across the boundaries of the subintervals. To give a concrete, and rather crude, example: let $h = \frac{1}{2}$ and $p = 1$, so we have divided in interval into two parts, and assumed $u_{\frac{1}{2},1}$ to-be linear in each (and continuous). Therefore if we write

$$\phi(x) = \begin{cases} 2x & \text{if } 0 \le x \le \frac{1}{2}, \\ 2 - 2x & \text{if } \frac{1}{2} \le x \le 1, \end{cases}$$

then $u_{\frac{1}{2},1} = a\phi(x)$ for some a. We only need to substitute this into expression (2.1) and calculate the minimum value of $E(u_{\frac{1}{2},1})$ to obtain $a = \frac{1}{8}$ and our approximation is

$$u_{\frac{1}{2},1}(x) = \begin{cases} \frac{1}{4}x & 0 \le x \le \frac{1}{2}, \\ \frac{1}{4}(1 - x) & \frac{1}{2} \le x \le 1. \end{cases}$$

Of course this approximation is rather coarse and we might suspect that if we break the interval into n subintervals (elements) of width $h = 1/n$ then we could attempt to use a function which is linear in each element. Fourtunately there is a simple way to describe such functions which is a generalization of the above idea. For each consecutive pair of elements we can define a function $\phi_i(x)$, such that

$$\phi_i(x) = \begin{cases} \dfrac{1}{h}[x - (i - 1)h] & (i - 1)h \le x \le ih \\ \dfrac{1}{h}[(i + 1)h - x] & ih \le x \le (i + 1)h \\ 0 & \text{elsewhere} \end{cases} \quad (i = 1, \ldots, n - 1).$$

It is easily seen that any function which is piecewise linear over these elements (and zero at the ends of the interval) can be written as

$$u_{h,1} = \sum_{i=1}^{n-1} u_i \phi_i(x),$$

and again all have to do is minimize $E(u_{h,1})$ over the values $u_1, \ldots, u_n$.

This example, whilst illustrating the basic idea of the finite element method, does not illuminate its many advantages. In particular, as each of the basis functions ϕ_i have the same shape, shifted along the axis, many of the calculations involved in determining $E(u_{h,p})$ are identical. Also, it is not necessary to be able to write the problem in terms of a minimization problem, in order to apply the finite element method. Lastly, as we shall see, we end with a matrix system to solve and, since each of the ϕ_i are zero over most of the solution interval, the resulting matrix can be made sparse and banded. These points will be made clearer in the next few sections.

8.3. STEADY-STATE (ELLIPTIC) PROBLEMS

Let us consider, as an example, the case of a two-dimensional body, which we shall call Ω, subject to a source of heat described by some function $f(x, y)$ defined on Ω. Further, we shall assume that the body has boundary Γ, and that on one part of the boundary, say Γ_D, the temperature u is held constant at $u = 0$ and on the remainder, Γ_N, the body is insulated, therefore $\partial u/\partial n = 0$, where $\partial/\partial n$ denotes the derivative taken in the sense of the outward normal to the boundary. The subscripts D and N refer to 'Dirichlet' and 'Neumann', which is how these types of boundary conditions are usually referred to in the theory of partial differential equations.

The standard mathematical model for such a problem is:

Problem (E_1).

$$-\nabla^2 u = f \quad \text{in } \Omega,$$

$$u = 0 \qquad \text{on } \Gamma_D,$$

$$\frac{\partial u}{\partial n} = 0 \qquad \text{on } \Gamma_N,$$

where

$$\nabla^2 \equiv \frac{\partial^2}{\partial x^2} + \frac{\partial^2}{\partial y^2}.$$

In order to treat this problem, we shall rewrite it in a different form. Firstly, if $-\nabla^2 u = f$ then, taking any 'suitable' function v, such that $v = 0$ on Γ_D, we have

$$\int_\Omega -\nabla^2 u \cdot v\, d\Omega = \int_\Omega f \cdot v\, d\Omega.$$

Now integrating this by parts leads to

$$-\int_\Gamma \frac{\partial u}{\partial n} \cdot v\, d\Gamma + \int_\Omega \nabla u \cdot \nabla v\, d\Omega = \int_\Omega f \cdot v\, d\Omega. \tag{8.3.1}$$

However, $v = 0$ on Γ_D so (8.3.1) becomes

$$-\int_{\Gamma_N} \frac{\partial u}{\partial n} \cdot v\, d\Gamma + \int_\Omega \nabla u \cdot \nabla v\, d\Omega = \int_\Omega f \cdot v\, d\Omega,$$

and $\partial u/\partial n = 0$ on Γ_N, so finally the problem becomes:

Problem (E_2). Find u where $u = 0$ on Γ_D such that

$$\int_\Omega \nabla u \cdot \nabla v\, d\Omega = \int_\Omega f \cdot v\, d\Omega, \tag{8.3.2}$$

for all v such that $v = 0$ on Γ_D.

Note that we have included the Neuman boundary condition in the equation, therefore it need no longer be considered explicitly. For this reason, this type of Neumann boundary condition is often referred to as *natural*.

We must consider now what is meant by 'suitable' functions. The question is simply answered, firstly the function u must satisfy the Dirichlet boundary condition and the function v must be zero where we impose a Dirichlet boundary condition on u. Secondly the functions u and v must be such that the left-hand side of (8.3.2) is well defined, that is

$$\int_\Omega \nabla v \cdot \nabla v \, d\Omega < \infty \tag{8.3.3}$$

(this definition is rather lax, but it will serve the purposes of this chapter). One immediate thing we should note is that for a function u to be a candidate for a solution of problem (E_1) then at least we must be able to differentiate it twice with respect to both x and y. Such a function is often called a *strong* solution of the problem. It is obvious from the above manipulation that any solution of (E_1) is also a solution of (E_2). However, for a function u to be a candidate for a solution of problem (E_2) it need only satisfy (8.3.3) (plus Dirichlet boundary conditions), that is, it needn't even be once differentiable. It is easily seen, for example, that the two-dimensional equivalents of the functions ϕ_i of the previous section satisfy condition (8.3.3). In fact this is critical in finite element error analysis. Therefore, problem (E_2) has many more candidates for a solution than the original problem (E_1).

The form of the problem (E_2) is variously referred to as the *weak* form by mathematicians, or the *principle of virtual work* by engineers, and the solution of problem (E_2) is called the weak solution of the differential equation. This term deserves some remarks, since it is often assumed by non-mathematicians that weak is equivalent to poor.

Firstly, from a physical point of view, the differential equation may be overly restrictive as a model. Take, for instance, the example of §8.2. We can derive a differential equation from the minimization principle (8.2.1), but if we look at the form (8.2.1) we see that it only involves integrals of first derivatives. To proceed to the differential equation $(-u_{xx} = 1)$, we first have to *assume* that the function u is twice differentiable [see IV, §3.4]. This is not a physical assumption, it is only a mathematical convenience. This sort of operation also takes place in, for example, fluid dynamics. There, the natural conservation laws are often expressed in integral form and further assumptions must be included, in order to derive a differential equation.

Secondly, from a mathematical point of view, we have said that any strong solution is also a weak solution. In addition to this, it is often the case that we can prove there is exactly one weak solution (this can be done for the two examples above, and also in much greater generality). So the situation becomes: either there is a strong solution, in which case it is also the weak solution and can be found by solving problem (E_2), or there isn't strong solution anyway and we can still solve the weak problem.

Thirdly, any differential equation or minimization problem of the types which are of interest to us here can be cast in weak form, so if we learn to deal with the weak form we can cope with most problems for which the finite element method is suited.

To end this section, we shall look at a slightly more general problem which will serve as the model problem in §8.4, and §8.5. This is a second-order scalar problem in two dimensions. For vector problems (e.g. two- or three-dimensional elasticity) the reader is referred to Zienkiewicz and Taylor (1989) and for fourth-order problems to Ciarlet (1978). The extensions to these cases are, however, relatively straightforward and the reader is encouraged to attempt them.

Consider the following differential equation.

Problem (E_3). Find u such that

$$-\nabla\cdot(b\nabla u) + cu = f \quad \text{on } \Omega,$$

$$u = g_1 \quad \text{on } \Gamma_D,$$

$$b\frac{\partial u}{\partial n} = g_2 \quad \text{on } \Gamma_N,$$

where $b = b(x, y)$ and has minimum value strictly greater than zero, and $c = c(x, y)$ is greater than or equal to zero. $g_i (i = 1, 2)$ are functions defined on the boundary.

By following the same reasoning as before (left as an exercise), we achieve the weak form of this problem.

Problem (E_4). Find $u \in V$ such that

$$\int_\Omega \{b\nabla u\cdot\nabla v + cu\cdot v\}d\Omega = \int_\Omega f\cdot v\,d\Omega + \int_{\Gamma_N} g_2\cdot v\,d\Gamma$$

for all $v \in V^0$.

Here V refers to the set of functions satisfying $v = g_1$ on Γ_D and which satisfy condition (8.3.3) above, and V^0 refers to the set of functions satisfying $v = 0$ on Γ_D and satisfying condition (8.3.3). In order to simplify notation, from now on we shall refer to the generic problem.

Problem (E_5). Find $u \in V$ such that

$$a(u, v) = f(v) \quad \text{for all } v \in V^0.$$

So in the example above

$$a(u, v) = \int_\Omega \{b\nabla u\cdot\nabla v + cu\cdot v\}\,d\Omega,$$

$$f(v) = \int_\Omega f\cdot v\,d\Omega + \int_{\Gamma_N} g_2\cdot v\,d\Gamma.$$

Note that the form $f(v)$ is linear in v and that the form $a(u,v)$ is bilinear (i.e. linear in u and v separately) and symmetric ($a(u,v) = a(v,u)$). The linearity of the forms is inherited from the linearity of the differential equation, non-linear differential equations lead to non-linear finite-dimensional problems. The finite element procedure follows by analogy in the non-linear case, but we are left with the additional concern of solving a system of simultaneous non-linear equations (with all of the attendant concerns over uniqueness, etc.), for more details see Zienkiewicz and Taylor (1989). The symmetry of the bilinear form follows from the fact that the above differential equation is *self-adjoint* [see IV, §7.6.3], in fact it is used as a common definition of self-adjointness. Non-self-adjoint problems lead to asymmetric matrices, using the standard finite element procedure (the *Galerkin* procedure), where the functions v are chosen from the same set as the solutions u. For non-self-adjoint problems the *Petrov–Galerkin* procedure is often adopted, where the functions v have a different form from those which describe the solution u. This technique is very similar to *upwinding* in finite differences. We shall implicitly assume these properties of symmetry and bilinearity in what follows, for extensions see Strang and Fix (1973) and Zienkiewicz and Taylor (1989).

In the next section we shall see how to produce a finite-dimensional approximation to this problem.

8.4. FINITE ELEMENT BASIS FUNCTIONS

Our aim, as was suggested in §8.2, is to produce a set of linearly independent piecewise polynomial functions over a grid in such a way that we can easily reproduce any piecewise polynomial function as a linear combination of these. Such a set of functions is called a *basis* or *base*. For the moment, we shall assume that the region Ω is rectangular with axes parallel to the x and y axes and that it has been subdivided into a union [see I, §1.2.1] of rectangles also with the same property. The case for triangular subdivisions can be worked through similarly, and the reader is encouraged to do this. A full treatment for triangles can be found in Zienckiewicz and Taylor (1989) or Strang and Fix (1973).

There are many different ways to create finite element basis functions, but the two most common categories are referred to as *standard* and *hierarchical*. The terms are somewhat misleading as, at least at the time of publication, the hierarchical bases seem to be more popular. However, it is only the most up to date of commercial codes which presently include hierarchical bases. For piecewise bilinear functions the two are identical and we shall describe this first.

In a similar fashion, as in §8.2, we shall produce basis functions ϕ_i which are centred on a node and are only non-zero in the surrounding elements. Assume, for ease of exposition, that in the mesh described above there is a patch of four elements as shown in Figure 8.4.1. It is a simple matter to create a function which is equal to one at the centre node, zero at the outside edges, and bilinear in

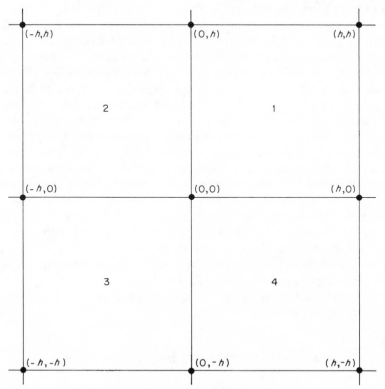

Figure 8.4.1. Subsection of mesh

each element, such a function is

$$\phi(x, y) = \begin{cases} (1 - x/h)(1 - y/h) & \text{in element 1,} \\ (1 + x/h)(1 - y/h) & \text{in element 2,} \\ (1 + x/h)(1 + y/h) & \text{in element 3,} \\ (1 - x/h)(1 + y/h) & \text{in element 4,} \\ 0 & \text{elsewhere in the mesh,} \end{cases}$$

(see Figure 8.4.2). One immediate property of this function is that its definition is very similar in each element. For that reason, no matter what the shape or size of the original elements, we shall be able to perform most of our calculations on a *reference* element, and by simple change of variables map back onto the original element.

Following the above example, we can number the nodes in the mesh $i = 1, \ldots, n$ and for each node i produce a similar function ϕ_i. It is clear that any piecewise bilinear function on the mesh can then be written as a linear combination of these ϕ_i, so that, for example,

$$u_{h,1} = \sum_{i=1}^{n} u_i \phi_i.$$

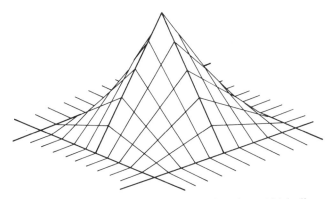

Figure 8.4.2. Piecewise bilinear basis functions (thick lines
denote element boundaries)

One other convenient property is that the coefficients u_i above have a direct physical interpretation; at node j, $u_{h,1} = u_j$. This is due to the fact that, at any node j, *only* the basis function ϕ_j is non-zero. As we move to higher-order polynomials and consider the difference between the standard and hierarchical formulations we shall see that the standard form retains this property whilst the hierarchical form does not.

Strictly, we should have said that we define these functions at any node in the mesh where the solution is unknown (i.e. not at Dirichlet boundary points); however, in practice we initially ignore this condition when constructing the finite element problem and impose the boundary conditions later.

When we consider higher-order polynomials, we must make some choices. In order to describe the quadratic basis functions let us first return to the one-dimensional case. We have, for example, an interval split up into elements of length h. We want to produce simple basis functions so that we can easily describe any continuous piecewise quadratic function as a linear combination of these. Now, any quadratic function has three degrees of freedom (for example a, b and c in $ax^2 + bx + c$), so we need three different functions in each element to be able to describe it. Suppose that one of the elements is the interval $[0, h]$. Let us put *three* nodes in the element, one each at $x = 0$, $x = h/2$ and $x = 1$. Now we shall create three quadratic functions ψ_i which are equal to 1 at each of the nodes respectively, and zero at the other two. After a little algebra

$$\psi_1 = (1 - 2*x/h)*(1 - x/h),$$
$$\psi_2 = 4*x/h*(1 - x/h),$$
$$\psi_3 = -x/h*(1 - 2*x/h).$$

Now, all we need do is match these functions across the element boundaries so as to make them continuous. One of the functions ψ_2 is zero on the boundaries, so we need do nothing more with it. For ψ_1 we need only match it up with the equivalent of ψ_3 in the element adjacent to the left, so we obtain, essentially, two different kinds of basis functions. For internal nodes, say the node at $x = h/2$ in the element over $[0, h]$ we have

$$\phi(x) = 4*x/h*(1 - x/h), \tag{8.4.1}$$

and for nodes on element boundaries the definition must extend over two elements, so for two adjacent elements on $[-h, 0]$ and $[0, h]$ we have

$$\phi(x) = \begin{cases} (1 + 2*x/h)*(1 + x/h) & \text{in } [-h, 0] \\ (1 - 2*x/h)*(1 - x/h) & \text{in } [0, h], \end{cases}$$

(see Figure 8.4.3).

The form just described is the *standard* form. The *hierarchical* form has the same basis functions already described for the linear case together with the functions described as in (8.4.1) (see Figure 8.4.4). Again, some simple algebra shows that any continuous piecewise quadratic function on the mesh can be written as a linear combination of *either* the standard basis *or* the hierarchical basis. The coefficients will, of course, be different. The accuracy of the finite element approximation does not depend on the basis chosen for our piecewise polynomials, so theoretically the standard and hierarchical bases are

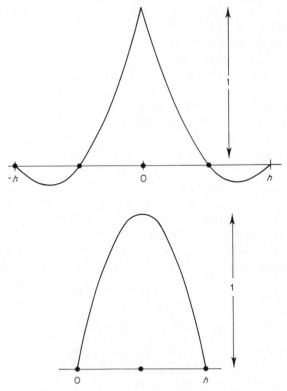

Figure 8.4.3. Standard one-dimensional quadratic basis
functions

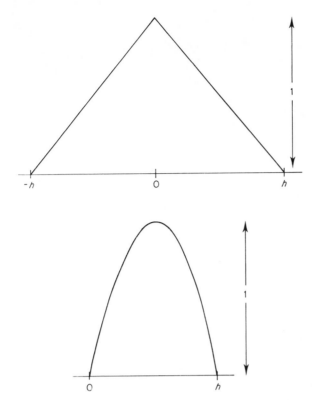

Figure 8.4.4. Hierarchical one-dimensional basis functions

interchangeable. Computational considerations may, however, lead us to prefer one above the other (Babuška, Griebel and Pitkäranta (1989)).

To return to the two-dimensional case, we can use these one-dimensional quadratic functions to produce biquadratic functions. On the reference element of Figure 8.4.5 we superimpose nine nodes and produce the portions of the basis functions which are non-zero in this element by taking products of the one-dimensional functions. So for the standard basis

$$\phi_1(x, y) = (1 - 2*x/h)*(1 - x/h)*(1 - 2*y/h)*(1 - y/h),$$
$$\phi_2(x, y) = 4*x/h*(1 - x/h)*(1 - 2*y/h)*(1 - y/h),$$
$$\phi_5(x, y) = 16*x/h*(1 - x/h)*y/h*(1 - y/h),$$

the remainder of the functions are formed in the obvious symmetric fashion. For the hierarchical basis the equivalent functions are

$$\phi_1(x, y) = (1 - x/h)*(1 - y/h),$$
$$\phi_2(x, y) = 4*x/h*(1 - x/h)*(1 - y/h),$$
$$\phi_5(x, y) = 16*x/h*(1 - x/h)*y/h*(1 - y/h).$$

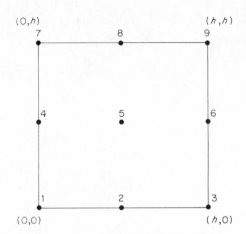

Figure 8.4.5. Biquadratic reference element

These functions are drawn in Figure 8.4.6. In order to construct continuous functions, we then match this element with its neighbours, as we did in the one-dimensional case. So the entire basis functions for nodes 1, 3, 7 and 9 are defined over four elements in the obvious way. Those for nodes 2, 4, 6 and 8 over two elements and the remaining one at node 5 is internal to one element.

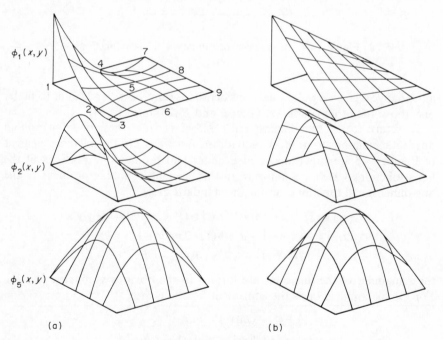

(a)   (b)

Figure 8.4.6. Two-dimensional biquadratic basis functions (a) Standard
(b) Hierarchical

Often the internal node is dropped, leaving only the eight boundary nodes (in either the standard or hierarchical form). Such an element is called a *serendipity* element. The reason for the name is that the early finite element theory suggested that it might be necessary to have nine nodes for the approximation to be good and it was happily found that this need not be the case (Zienkiewicz and Taylor (1989)).

One advantage of the hierarchical family is seen by, for example, discarding the basis functions corresponding to nodes 4, 5 and 6. Then any linear combination of the remaining functions describes a function which is linear in the y direction but quadratic in the x direction. This has the obvious application that if we know the solution to be smooth in one direction and oscillatory in the other, then we can approximate this well, without using a large number of basis functions. This is possible using standard bases, but is more difficult. Also, the hierarchical form allows us to match up elements of different degree very easily. If the above element has a linear element to the north and west and a quadratic element to the south and east, then we need only drop the basis functions corresponding to nodes 4 and 8 and we are automatically guaranteed continuity.

The extensions to cubic and higher-order elements may be found by repeating the steps above. We have still more choices to make, for example one family of elements with good approximation properties has serendipity type elements for levels 1, 2 and 3 (roughly corresponding to linear, quadratic, cubic) and then introduces an internal biquadratic function at the next level. For descriptions of these finite element families and discussion of their relative merits we refer to Zienkiewicz and Taylor (1989), Strang and Fix (1973) and Babuška, Griebel and Pitkäranta (1989).

There are many other finite element families. For example for fourth-order problems it is sometimes necessary that the derivatives of the finite element functions have to be continuous (although this condition can often be dropped). This is due to the fact that *second* derivatives appear in the weak form. This is very difficult to arrange in two dimensions (see for example the comments in Strang and Fix (1973) relating to basis functions with continuous derivatives on a triangle). However, it is possible, and one way to do this is to have two types of basis function representing the function and its derivative respectively. Such basis functions are referred to as the *Hermite* family after Hermite interpolation [see III, §6.5.3], in which case the ones described in detail above are *Lagrange* basis functions. At the end of §8.6 we shall discuss the *patch test*, a computational test which allows us to decide when we can use elements which do not satisfy continuity requirements. It is also a useful procedure which can be used to test the correctness of a finite element code.

We occasionally relax the continuity requirement entirely and produce basis functions which are polynomials (often, but not invariably, constant) in each element, and which do not match at element boundaries. These are used in so-called *mixed* methods (see Oden and Reddy (1976)) or the more exotic H^{-1} methods (see Oden and Reddy (1976) or Rachford and Wheeler (1975)). They are also widely used in the related boundary element method. For the purposes of this chapter we shall not concern ourselves with such extensions, leaving the interested reader to explore the above references.

In the next section we shall show how to use these basis functions to assemble the finite element system and therefore allow us to obtain the coefficients in the approximation.

8.5.   THE ASSEMBLY OF THE FINITE ELEMENT SYSTEM

The finite element aproximation to problem (E) can be stated quite simply by replacing V by the set of linear combinations of the basis functions (which satisfy the Dirichlet boundary conditions). For this we shall choose the obvious notation $V_{h,p}$ and so the approximate problem is:

Problem ($E_{h,p}$).   Find $u_{h,p}$ such that

$$a(u_{h,p}, v_{h,p}) = f(v_{h,p}) \quad \text{for all } v_{h,p} \in V^0_{h,p}.$$

Now remember that the finite element approximation is of the form

$$u_{h,p} = \sum_{i=1}^{n} u_i \phi_i,$$

(some of the u_i may already have been fixed to force the approximation to satisfy boundary conditions), so substituting this into the above equation, we obtain

$$\sum_{i=1}^{n} u_i a(\phi_i, v_{h,p}) = f(v_{h,p}) \quad \text{for all } v_{h,p} \in V^0_{h,p}. \tag{8.5.1}$$

Finally, if we make equation (8.5.1) true for any of the functions ϕ_j in the set $V^0_{h,p}$ then it must be true for an arbitrary linear combination of them, and hence for all functions in $V^0_{h,p}$, so we obtain

$$\sum_{i=1}^{n} u_i a(\phi_i, \phi_j) = f(\phi_j) \quad (j = 1, \ldots, n). \tag{8.5.2}$$

This is simply a matrix equation

$$\mathbf{Au} = \mathbf{f}.$$

where the elements of $\mathbf{A}$ and $\mathbf{f}$ are given by $a_{ij} = a(\phi_i, \phi_j), f_j = f(\phi_j)$. Therefore the task we now face is that of calculating the elements of the matrix $\mathbf{A}$ (often called the *stiffness* matrix, from elasticity) and the vector $\mathbf{f}$ (the *load* vector).

Consider again the one-dimensional case with piecewise linear basis functions and with

$$a(u, v) = \int_0^1 \frac{du}{dx} \frac{dv}{dx} dx.$$

Now, each basis function ϕ_i is only non-zero over two elements, therefore $a(\phi_i, \phi_j) \neq 0$ only if $j = i - 1$, $i, i + 1$. Therefore, in this case, the matrix is tridiagonal [see III, §4.9]. A moment's reflection will convince the reader that for the general problem, in arbitrary number of dimensions, the stiffness matrix is at least sparse (i.e. most of the entries are zero) and if the nodes are numbered properly then it is banded (the non-zero entries are clustered around the main diagonal). These properties are very desirable, both to cut down computer storage requirements and also to construct efficient matrix solvers.

Now, as the basis functions have different definitions over each element, it makes sense to calculate $a(\phi_i, \phi_j)$ element by element. For example:

$$a(\phi_i, \phi_i) = \int_{(i-1)h}^{(i+1)h} \frac{d\phi_i}{dx} \frac{d\phi_i}{dx} dx$$

$$= \int_{(i-1)h}^{ih} \frac{d\phi_i}{dx} \frac{d\phi_i}{dx} dx + \int_{ih}^{(i+1)h} \frac{d\phi_i}{dx} \frac{d\phi_i}{dx} dx.$$

Each of the integrals in the right-hand side of the above equation take place over one element. For the other two cases:

$$a(\phi_{i-1}, \phi_i) = \int_{(i-1)h}^{ih} \frac{d\phi_{i-1}}{dx} \frac{d\phi_i}{dx} dx = a(\phi_i, \phi_{i-1}).$$

In order to perform the necessary calculations we shall transform all the integrals into integrals over $[0, 1]$. The generic case is

$$\int_{ih}^{(i+1)h} \frac{d\phi_j}{dx} \frac{d\phi_k}{dx} dx, \tag{8.5.3}$$

where ϕ_j and ϕ_k are linear functions over the interval $[ih, (i+1)h]$, equal to zero at one of the ends and equal to 1 at the other. The appropriate transformation is

$$x = ih * (1 - \xi) + (i + 1)h * \xi.$$

Note that the functions $(1 - \xi)$ and ξ appearing in the transformation happen to be the basis functions defined on the reference element $[0, 1]$, and the other terms ih and $(i + 1)h$ are the global coordinates of the element nodes. This property is very useful and will carry over to two dimensions. To continue, then $dx = h \, d\xi$ and

$$\frac{d}{dx} = \frac{1}{h} \frac{d}{d\xi}.$$

So expression (8.5.3) becomes

$$\int_{ih}^{(i+1)h} \frac{d\phi_j}{dx} \frac{d\phi_k}{dx} dx = \frac{1}{h} \int_0^1 \frac{d\Phi_j}{d\xi} \frac{d\Phi_k}{d\xi} d\xi. \tag{8.5.4}$$

Usually we refer to x as the global coordinate and ξ as the local coordinate.

The functions Φ_j and Φ_k are just $1 - \xi$ or ξ depending on the orientation of the functions ϕ_j and ϕ_k. Therefore the calculation, in the above example, turns out to be *independent* of the element. This is not necessarily the case for a general problem, but the calculations are similar enough to be carried out element by element. If we define

$$\Phi_1 = 1 - \xi, \qquad \Phi_2 = \xi,$$

then we have, essentially, four calculations to perform:

$$\frac{1}{h}\int_0^1 \frac{d\Phi_i}{d\xi}\frac{d\Phi_j}{d\xi}d\xi \quad (i,j=1,2).$$

These are stored in a 2×2 *element matrix*. In the above example, it is easy to calculate that this matrix is

$$\mathbf{A}_{\mathrm{el}} = \begin{pmatrix} \dfrac{1}{h} & -\dfrac{1}{h} \\ -\dfrac{1}{h} & \dfrac{1}{h} \end{pmatrix}.$$

It is obvious that, for example, if the elements were of different size, h_i, then the above matrix would be different for every element with h replaced by h_i. We have to assemble these element matrices $\mathbf{A}_{\mathrm{el}}$, calculated for each element, into the global matrix $\mathbf{A}$. In order to do this we introduce *local* and *global* node numbering. In this example the node numbering is illustrated in Figure 8.5.1. This must be done with a mapping from element numbers and local node numbers into global node numbers. In the computational procedure we have a two-dimensional array $NN(\cdot,\cdot)$ whose (i,j)th entry is the global number corresponding to element i, local number j.

Returning to the evaluation of the global elements in equation (8.5.3), we see that the (j,k)th entry of the element stiffness matrix corresponding to element i is added into the global matrix in position $(NN(i,j), NN(i,k))$. Starting from a matrix of zeros we *assemble* the global stiffness matrix by taking each element matrix in turn and *adding* the entries into the appropriate position in $\mathbf{A}$. If we do this in the present case we obtain

$$\mathbf{A} = \begin{pmatrix} 1/h & -1/h & 0 & \cdots & 0 \\ -1/h & 2/h & -1/h & \cdots & 0 \\ 0 & -1/h & 2/h & \cdots & 0 \\ \vdots & \vdots & \vdots & \ddots & \vdots \\ 0 & 0 & 0 & \cdots & 1/h \end{pmatrix}.$$

The load vector can be calculated and assembled in a similar way, taking care when we transform variables. We have a 2×1 element vector with generic term

$$\int_{(i-1)h}^{ih} f(x)\cdot\phi_j(x)dx,$$

Figure 8.5.1. Local and global node numbering

and performing the same change of variables as above we obtain

$$h \int_0^1 f(ih*(1-\xi)+(i+1)h*\xi)\cdot\Phi_j(\xi)d\xi,$$

where Φ_j is as above. Of course the calculation of these element matrices and load vectors would not ordinarily be done analytically, but rather numerically and we return to this point later in this section.

Note that, so far, we have not forced the finite element approximation to satisfy the boundary conditions. If we assume that the boundary conditions are $u(0)=0=u(1)$, then this implies that the coefficients of the first and last basis functions should be zero. If we examine our assembled equation, and put the first and last coefficients of the solution vector $\mathbf{u}$ equal to zero, then we see that this is equivalent to striking out the first row and first column, and last row and last column of the matrix $\mathbf{A}$ and the first and last entries of the vector $\mathbf{f}$. Suppose we had more general Dirichlet boundary conditions, say $u(0)=g$ and $u(1)=0$. Now we have already dealt with the second type of condition, if $u(0)=g$, then we have to impose this on the finite element approximation, i.e. make $u_{h,p}(0)=g$ and this implies that the coefficient of the first basis function (corresponding to the node at $x=0$) must be g, so the first entry in the solution vector $\mathbf{u}$ is g. Eliminating the first row and rearranging the remaining ones, it becomes clear that this is equivalent to

(a) replacing f_j in the load vector by $f_j - a_{1,j}*g$, $j=1,\ldots,n$, and
(b) eliminating the first row of $\mathbf{A}$ and $\mathbf{f}$ and eliminating the first column of $\mathbf{A}$.

In practice this method of imposing Dirichlet boundary conditions leads to some computational inconveniences. Particularly in higher dimensions, this procedure may lead us into the position of eliminating internal rows and columns of the global stiffness matrix, which in turn forces us to 'close up' the matrix. For this reason other methods are often preferred; however, they all are equivalent to the procedure above and for further details the reader is referred to Zienkiewicz and Taylor (1989).

Now, for any arbitrary Neumann boundary condition, say $u_x(0)=g$, then, as we have seen before, the general term of the right-hand side of the finite element equations will be

$$f_i = \int_0^1 f\cdot\phi_i dx + g\cdot\phi_i(0),$$

so if $g = 0$, which is often the case, the right-hand side is just the one we have already calculated. If $g \neq 0$ then all the terms in **f** will be unchanged, except the first. This is because $\phi_i(0) = 0$, for $i \neq 0$. Also $\phi_0(0) = 1$, so

$$f_0 = \int_0^1 f \cdot \phi_0 dx + g,$$

$$f_i = \int_0^1 f \cdot \phi_i dx \quad (i = 1, \ldots, n).$$

In practice, we would normally calculate the load vector ignoring the Neumann condition and then simply adjust the appropriate term(s).

We have gone into some detail over the one-dimensional case as it contains the essential elements of the assembly procedure. In higher dimensions the process is exactly the same: we give the nodes local and global numbers, evaluate element matrices and vectors and assemble them into their global equivalents, imposing boundary conditions on the appropriate nodes at the end of the operation. One step does, however, call for special mention, the transformation to the standard element. Again taking the case of quadrilateral elements with piecewise bilinear basis functions (triangles or higher order bases can be worked through as an example), we must transform an arbitrary quadrilateral into a reference square. We shall take this to be the square $\{-1 \leq \xi \leq 1, -1 \leq \eta \leq 1\}$ (many authors take the square $\{0 \leq \xi \leq 1, 0 \leq \eta \leq 1\}$, this is simply a matter of personal preference and will have no effect on the approximation). Now if our element has vertices with coordinates (x_i, y_i), $i = 1, \ldots, 4$ (see Figure 8.5.2), then the transformation takes the form

$$x = x_1 * \tfrac{1}{4}(1 + \xi)(1 + \eta) + x_2 * \tfrac{1}{4}(1 - \xi)(1 + \eta)$$
$$+ x_3 * \tfrac{1}{4}(1 - \xi)(1 - \eta) + x_4 * \tfrac{1}{4}(1 + \xi)(1 - \eta),$$
$$y = y_1 * \tfrac{1}{4}(1 + \xi)(1 + \eta) + y_2 * \tfrac{1}{4}(1 - \xi)(1 + \eta)$$
$$+ x_3 * \tfrac{1}{4}(1 - \xi)(1 - \eta) + y_4 * \tfrac{1}{4}(1 + \xi)(1 - \eta).$$

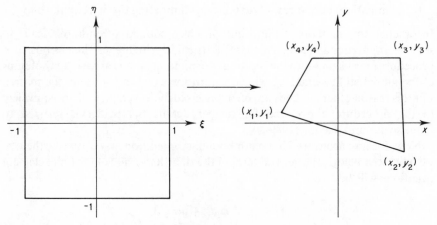

Figure 8.5.2. Local to global coordinate mapping

(See the remark following equation (8.5.3).) Now taking the particular example of calculating the (i,j)th member of this element matrix of a problem like (E_2), then we must calculate

$$\int_{\Omega_{el}} \nabla_{x,y}\phi_i \cdot \nabla_{x,y}\phi_j \, dx\, dy$$

(with the obvious notation Ω_{el} for the element and $\nabla_{x,y}$ for the gradient operator in the global coordinates). Now, by the chain rule we have

$$dx = \frac{\partial x}{\partial \xi}d\xi + \frac{\partial x}{\partial \eta}d\eta = (x_\xi d\xi + x_\eta d\eta),$$

$$dy = \frac{\partial y}{\partial \xi}d\xi + \frac{\partial v}{\partial \eta}d\eta = (y_\xi d\xi + y_\eta d\eta),$$

so denoting by $\mathbf{J}$ the Jacobian matrix of the transformation, [see IV, §5.12]

$$\mathbf{J} = \begin{pmatrix} x_\xi & x_\eta \\ y_\xi & y_\eta \end{pmatrix},$$

we have

$$dx\, dy = |\mathbf{J}|\, d\xi\, d\eta.$$

Similarly

$$\frac{\partial}{\partial \xi} = x_\xi \frac{\partial}{\partial x} + y_\xi \frac{\partial}{\partial y} \quad \text{and} \quad \frac{\partial}{\partial \eta} = x_\eta \frac{\partial}{\partial x} + y_\eta \frac{\partial}{\partial y}.$$

Or in more compact notation

$$\nabla_{\xi,\eta} \equiv \mathbf{J}^T \nabla_{x,y}.$$

The term then becomes

$$\int_{-1}^{1}\int_{-1}^{1} (\mathbf{J}^T)^{-1}\nabla_{\xi,\eta}\Phi_i \cdot (\mathbf{J}^T)^{-1}\nabla_{\xi,\eta}\Phi_j |\mathbf{J}|\, d\xi\, d\eta,$$

or in full

$$\int_{-1}^{1}\int_{-1}^{1} \left(y_\eta \frac{\partial \Phi_i}{\partial \xi} - y_\xi \frac{\partial \Phi_i}{\partial \eta}\right) \cdot \left(y_\eta \frac{\partial \Phi_j}{\partial \xi} - y_\xi \frac{\partial \Phi_j}{\partial \eta}\right) + \left(-x_\eta \frac{\partial \Phi_i}{\partial \xi} + x_\xi \frac{\partial \Phi_i}{\partial \eta}\right)$$

$$\cdot \left(-x_\eta \frac{\partial \Phi_j}{\partial \xi} + x_\xi \frac{\partial \Phi_j}{\partial \eta}\right)\frac{d\xi\, d\eta}{|\mathbf{J}|}, \tag{8.5.5}$$

where the Φ are the basis functions on the reference element.

In the more general case of problem (E_4), for the sake of example with $b = c = 1$, there are two terms in the right-hand side:

$$\int_\Omega \dot{\nabla}_u \cdot \nabla_v \, d\Omega \quad \text{and} \quad \int_\Omega u \cdot v \, d\Omega.$$

The first term produces the stiffness matrix as before. The second produces a *mass* matrix. The general term of the element mass matrix is easily seen to be

$$\int_{\Omega_{el}} \phi_i \cdot \phi_j \, d\Omega,$$

which transforms into

$$\int_{-1}^{1} \int_{-1}^{1} \Phi_i \cdot \Phi_j |\mathbf{J}| \, d\xi \, d\eta.$$

The mass matrix will appear again when we consider time-dependent problems.

A moderate amount of care must be taken when designing the finite element mesh. If the elements are unreasonably distorted, or if, for example, midside nodes are placed too near to vertices, then the transformation above can become non-unique, that is different local coordinates can be mapped into the same global coordinates. Fortunately, there is an easy check for this problem: if the determinant of the Jacobian matrix $|\mathbf{J}|$, remains the same sign over the reference element then the mapping is unique (one can see, from expression (8.5.5) that we have problems if $|\mathbf{J}| = 0$). For low-order elements the situation becomes simple, a necessary condition for bilinear elements is that they are *convex*, that is no internal angle should be greater than π. For quadratic elements we have the additional restriction that the midside nodes should not be adjacent to the vertices (see Zienkiewicz and Taylor (1989) and Strang and Fix (1973) for exact details). In general, however, we must check the sign of $|\mathbf{J}|$, and most commercial software performs this check automatically.

We have only considered bilinear basis functions, the procedure is very similar for higher-order functions. One flexibility we are afforded when using higher-order functions is that we can use elements with curved sides (see Zienkiewicz and Taylor (1989)), and map these into the reference element using the obvious combination of local basis functions and nodal coordinates as we did in the bilinear case. This sort of mapping is called *isoparametric*, we can also have *subparametric* and *superparametric* mappings but these are less common.

The description above necessarily refers to an idealization of the true situation. In practice we are going to depart from this procedure and there are three common ways of this occurring. Firstly, we may, for some reason, decide to use basis functions which do not satisfy the stated inter-element continuity requirements. This is uncommon, but not unknown, for second-order problems, but is, however, almost the rule for fourth-order problems (see the remarks above). Secondly, we have assumed that it is possible to force the finite element solution to satisfy the Dirichlet boundary conditions. This may be difficult in two respects: even with curved elements it may not be possible to follow the boundary of a physical region exactly, and anyway the basis functions might not be able to match the conditions exactly. Lastly, we have assumed that we have calculated the terms in the stiffness matrix and the load vector exactly. It will be seen in §8.6 that the simple error estimates which we produce rely on these three properties not being violated. Although it is possible to generalize these estimates, it is equally possible to produce finite element approximations which fail to be even close to a true solution. The problem is very similar to that of *consistency* in finite difference methods. We mention again the *patch test* which will be described at the end of §8.6 which allows us to answer some of the questions above.

The penultimate question is that of actually calculating the entries of the element matrices and vectors. As was said above, this is seldom done analytically. The most common numerical integration technique employed in *Gaussian* integration (see Kopal (1961)), which is easy here since we are always calculating on squares (or cubes in three dimensions). The program then need only be able to evaluate the basis functions and their derivatives at the Gauss points, together with the values of any other physical parameters involved. Numerical integration has been treated at length in the above reference and the reader is referred there for further information. One choice we must make is the order of the Gauss rule—we wish to keep it as low as possible without sacrificing accuracy. It is unnecessary, and anyhow too expensive, that the numerical integration rule should calculate the integrals of the polynomials which appear exactly. Remember, the terms in the stiffness matrix are integrals of *squares* of polynomials and a high-order rule would be required for exact integration. It is straightforward to show (see Strang and Fix, (1973)) that, for a 2*m*th order problem, it is sufficient that the integration rule integrates the *m*th derivatives of the basis functions exactly. This guarantees convergence to the exact solution as the mesh is refined. However, we must always be aware of the possible lack of accuracy, particularly with rapidly changing (in space) loads, and therefore be prepared to use higher-order quadrature than is implied above. A common example would be a load which is concentrated over a small area.

One surprising effect of numerical integration is that it can *improve* the accuracy of an approximation. In effect (see Zienkiewicz and Taylor (1989)), numerical integration makes the elements less stiff allowing them to follow the deformation (in mechanics) or solution more closely.

To obtain the finite element approximation it now only remains to solve the resulting matrix system. This is by no means a trivial matter, but the sparse banded nature of the equations allows us to exploit an abundance of efficient software which is available. Many of these packages are based on classical techniques which are well known. We shall, however, mention in passing two methods: firstly the *frontal method* of Irons (1970) which is closely linked with the assembly process. This method resembles Gaussian elimination which eliminates each variable as soon as the relevant equation has been assembled. It is usually described in terms of a *front* sweeping through the grid during the assembly procedure. Only the variables currently on the front need be kept in storage and thus most work is carried out to minimize the front width. This method is probably the most popular for solving finite element equations. The second (class of) method(s) which deserve special mention are *multigrid* methods (see Hackbusch (1985) for a full treatment and Briggs (1987) for a particularly elegant introduction). These methods are iterative methods which exploit the fact that standard iterative techniques rapidly smooth the component of the equation residual which has wavelength similar to the mesh size, once the residual has become smooth it can be represented accurately on a much coarser grid. If the computations are then continued on this coarse grid we can save a great deal of work. Many multigrid methods can be shown to solve the matrix equation to within acceptable accuracy in a number of operations proportional to the number of unknowns. This is clearly an optimal estimate; however, it is often the case that the proportionality constant is large and therefore we may have to consider very large problems before multigrid methods come into their own.

8.6.   ACCURACY OF THE FINITE ELEMENT APPROXIMATION

In this section we shall describe some of the techniques used for analysing the discretization error in the finite element method, and also present some other results whose derivation is too involved for this chapter but which are none the less useful to know when applying the method.

The natural way to measure the error in finite element approximation is using the *energy norm* $\|u\|$, which is related to the problem in the following way:

$$\|u\|^2 = a(u, u).$$

In elasticity problems this is precisely the energy, and we shall show that the energy norm of the error is minimized by the finite element method. It is a simple matter to show (Ciarlet (1978)) that the energy norm is ≥ 0 for all functions and that it is only equal to 0 for the zero function.

The most powerful result from finite element analysis is that (at least for symmetric forms) the method gives us the best approximation possible from all linear combinations of the chosen basis functions. We shall consider problem (E) and its approximation, problem ($E_{h,p}$). The two statements give us

$$a(u, v) = f(v) \qquad \text{for all } v \in V^0,$$
$$a(u_{h,p}, v_{h,p}) = f(v_{h,p}) \quad \text{for all } v_{h,p} \in V^0_{h,p}.$$

We are interested in the energy norm of the error, $e = u - u_{h,p}$. The critical point is to note that if the mesh follows the boundary exactly and the basis functions are continuous, then $v_{h,p} \in V^0$, therefore we can substitute $v_{h,p}$ for v in the first of the above pair of equations obtaining

$$a(u, v_{h,p}) = f(v_{h,p}) \quad \text{for all } v_{h,p} \in V^0_{h,p}$$
$$a(u_{h,p}, v_{h,p}) = f(v_{h,p}) \quad \text{for all } v_{h,p} \in V^0_{h,p}.$$

Subtracting the second equation from the first, we obtain

$$a(e, v_{h,p}) = 0 \quad \text{for all } v_{h,p} \in V^0_{h,p}.$$

Now, if we consider the energy norm of e, then

$$a(e, e) = a(e, u - u_{h,p}),$$

however, $u_{h,p} \in V_{h,p}^0$, therefore

$$a(e, e) = a(e, u)$$
$$= a(u - u_{h,p}, u)$$
$$= a(u, u) - a(u_{h,p}, u)$$
$$= a(u, u) - a(u_{h,p}, u_{h,p}),$$

a result which is often expressed as the *energy in the error* is equal to the *error in the energy*.

Starting from the energy norm of the error and proceeding in a different direction, we obtain

$$a(e, e) = a(e, u) = a(e, u - v_{h,p}) \quad \text{for all } v_{h,p} \in V_{h,p}^0. \tag{8.6.1}$$

It is easy to show, at least for the bilinear forms which we have been considering, that the *Cauchy–Schwarz* inequality holds (Ciarlet (1978)), that is

$$a(u, v) \le \sqrt{a(u, u)} \cdot \sqrt{a(v, v)},$$

so applying this to equation (8.6.1) we obtain

$$a(e, e) \le \sqrt{a(e, e)} \cdot \sqrt{a(u - v_{h,p}, u - v_{h,p})}, \tag{8.6.2}$$

or, rearranging inequality (8.6.2),

$$\|e\| = \|u - u_{h,p}\| \le \|u - v_{h,p}\| \quad \text{for all } v_{h,p} \in V_{h,p}^0. \tag{8.6.3}$$

Statement (8.6.3) says simply that amongst all possible choices for our approximation to u from our finite element discretization we have chosen the one which minimizes the error. An immediate consequence of this is that if the exact solution u can be expressed in terms of our basis functions then this will be our approximation (since in this case $e = 0$ and we can do no better).

This result also forms the foundation for the *patch test*. We have implicitly assumed above that the basis functions have the appropriate degree of continuity, if they don't then we don't necessarily expect (8.6.3) to hold. The patch test tells us that we can use elements without the necessary continuity requirements if they reproduce the exact solution when they are capable of doing so (Zienkiewicz and Taylor (1989)). To apply the patch test, for example when the elements are piecewise linear, we construct an arbitrary patch of elements and supply the program with data which are consistent with a piecewise linear (or often simply linear) solution. This should then be reproduced exactly by the code. Even when we do have the correct continuity requirements, the above procedure is a simple check for the existence of bugs in the code.

The result (8.6.3) is the starting point for most finite element error estimates, since if we bound the last term, we also bound the first. The standard procedure, if for example we are considering what happens when we reduce the mesh size h, would be to replace $v_{h,p}$ above by the *interpolant* of u obtained by interpolating using our finite element basis functions. Inequality (8.6.3) guarantees that the actual approximation will be better or equal to the interpolant.

The results we obtain, using this technique and others, show us that the quality of the approximation depends heavily on the presence of singularities in the exact solution, and the stronger the singularity the less good the approximation. We state two results below, both for *quasi uniform* meshes. These meshes are such that as $h \to 0$ or $p \to \infty$, then the inequalities

$$\bar{h}/\underline{h} < K_1, \qquad \bar{p}/\underline{p} < K_2,$$

hold, where $\bar{h}$(respectively $\bar{p}$) is the maximum element diameter (respectively polynomial degree) and $\underline{h}$ (respectively $\underline{p}$) is the minimum element diameter (respectively polynomial degree) and K_i are constants independent of the mesh. If m is a measure of the strength of the singularity then

$$\|e\| \le Ch^{\min(p, m-1)} \tag{8.6.4}$$

for h-refinement and

$$\|e\| \le Cp^{1-m}$$

for p-refinement. It is not an easy problem to determine m, but values for examples can be found in various texts (e.g. Ciarlet (1978)). The important point is that, for many physical problems, $m \in [1, 2]$. To see the effect of this, say $m = \frac{3}{2}$, then inequality (8.6.4) shows us that, no matter what the degree of approximation used (i.e. $p = 1, 2, \ldots$), if we subdivide the elements we can expect no better than $h^{\frac{1}{2}}$ convergence. More simply, in two dimensions, subdividing each quadrilateral into four, with a consequential fourfold increase in the size of the problem, only reduces the error by a factor of approximately 0.7. There are many other estimates, in other norms, but the following is a summary of the consequences of these.

(a) The p-method can be expected to be more accurate than the h-method, if any singularity in the solution lies on an interelement boundary. Therefore, if we are unhappy with the quality of our results we gain more by increasing the order of the polynomial rather than subdividing the mesh.

(b) In general we do well to stay away from elements which are too distorted, for example, long thin elements. This is not an absolute requirement and sometimes we can do better with these elements, but their thoughtless use degrades the quality of the approximation.

(c) It is possible, by using adaptive analysis (see Zienkiewicz and Taylor (1989)), to improve greatly on the estimates above. We can obtain *optimal meshes*, that is ones where the error is the smallest possible for the number of basis functions used. However, the finite element user is well advised to design what might be termed as an *obvious* mesh. So if we expect the solution to vary rapidly in one region, we use small elements or high-degree polynomials there.

8.7. TIME-DEPENDENT PROBELMS

We shall begin by discussing *parabolic* [see IV, §8.5] problems, for example the distribution of temperature in a body through the course of time. *Hyperbolic* problems present greater difficulties. The model problem we shall choose (in one space and one time dimension) is:

Problem (P_1).   Find $u(x,t) \in [0,1] \times [0,T]$ such that

$$\frac{\partial u}{\partial t} - \frac{\partial^2 u}{\partial x^2} = f(x,t),$$

$$u(x,0) = u_0(x) \quad \text{(initial condition)},$$
$$u(0,t) = u(1,t) = 0 \quad \text{(boundary conditions)}.$$

This problem is the typical mathematical model for the temperature distribution in a rod of unit length, which has some initial temperature $u_0(x)$, is held at a fixed temperature at each end and is subject to a source of heating described by the function $f(x,t)$. A first inclination might be to discretize in both time and space using finite elements. However, this is generally unsuccessful, perhaps because using finite elements in time links all of the time levels in both the positive and retrograde sense. That is, the approximate solution at time t_2 can affect that at t_1 when $t_1 < t_2$. This destroys the natural evolutionary character of the physical problem. If we discretize in space using finite elements we obtain a system of coupled ordinary differential equations to solve. Therefore the usual approach is to use finite elements in space and *finite differences* in time.

Following the procedure adopted in §8.3, we obtain a weak form of problem (P_1) by multiplying the equation by an appropriate function v (the conditions to be satisfied by v are the same as those for elliptic problems) and integrating in space, after some integration by parts [see IV, §4.3] we obtain

Problem (P).   Find $u(x,t)$, such that $u(x,t) = u_0(x)$ and

$$\int_0^1 \frac{\partial u}{\partial t} v \, dx + \int_0^1 \frac{\partial u}{\partial x}\frac{\partial v}{\partial x} \, dx = \int_0^1 fv \, dx,$$

for all v such that $v(0) = v(1) = 0$.

We adopt the notation

$$(u,v) = \int_0^1 uv \, dx \quad \text{and} \quad a(u,v) = \int_0^1 \frac{\partial u}{\partial x}\frac{\partial v}{\partial x} \, dx,$$

(with the obvious changes for different problems as in §8.3).

Now we create our finite element basis functions in exactly the same way as before. In particular this means, at least for standard methods, that the basis functions are *independent of time*. (Some new techniques, such as the *moving finite element method* have time-dependent basis functions, but these are by no means standard.) We then allow the coefficients in the finite element approximation to vary with time, that is, given a choice of basis functions $\phi_i(x)$ our approximation takes the form

$$u_{h,p}(x,t) = \sum_{i=1}^{n} u_i(t)\phi_i(x).$$

We substitute this into Problem (P) to obtain (letting $v = \phi_i, i = 1, \ldots, n$ as before)

Problem $(P_{h,p})$.   Find $u_{h,p}$ such that

$$\sum_{i=1}^{n} \frac{du_i}{dt}(\phi_i, \phi_j) + \sum_{i=1}^{n} u_i a(\phi_i, \phi_j) = f(\phi_j) \quad (j = 1, 2, \ldots, n).$$

We see the mass matrix appearing in the first term on the left side of the above equation, and the stiffness matrix as before. Therefore we have obtained a coupled system of n ordinary differential equations of the form $\mathbf{M\dot{u}} + \mathbf{Au} = \mathbf{f}$, where $\mathbf{u}$ is the vector of unknown coefficients, $\mathbf{\dot{u}}$ its time derivative, $\mathbf{M}$ is the mass matrix and $\mathbf{A}$ is the stiffness matrix. The mass matrix can be calculated and assembled in the same way as the stiffness matrix, and will, in general, have the same sparsity pattern.

We can now apply standard finite difference algorithms to this system (see Richtmeyer and Morton (1967)). The algorithm should be chosen with the usual concerns over accuracy and stability. To illustrate some of the particular problems which arise when we apply finite differences to a system like the one above, we shall take a brief look at the forward Euler algorithm. This is probably the simplest and best known of finite difference algorithms. Discretizing the system $\mathbf{M\dot{u}} + \mathbf{Au} = \mathbf{f}$ we obtain $\mathbf{Mu}_{n+1} = \mathbf{Mu}_n - \Delta t(\mathbf{Au}_n - \mathbf{f}_n)$, where the subscript n refers to the solution at time $n\Delta t$. Therefore, at each time step we have to solve a matrix problem. The inverse of $\mathbf{M}$ is a full matrix, so even though we might calculate the inverse (or some decomposition, for example) we have to pay a storage penalty. One common approach is called *lumping* (Zienkiewicz and Taylor (1989)). Here, the idea is to replace the mass matrix by a diagonal matrix and thus simplify the system. A common approach would be to replace the diagonal term in each row by the sum of all of the terms in that row, and to set the off-diagonal terms to zero. There are several interpretations of this procedure (Zienkiewicz and Taylor (1989)). For simple linear triangular elements it amounts to using shape functions which are piecewise constant in an area surrounding each node when calculating the mass matrix (retaining the standard basis for the stiffness matrix and the load vectors). However, this procedure can lead us into trouble with higher-order elements (we can, for example, end up with zeros on the diagonal) and the second interpretation proves to be more fruitful. If, in the standard basis, we construct a numerical quadrature scheme whose sampling points are at the element nodes, then we are automatically guaranteed that the off-diagonal terms in the mass matrix are zero. This is because if we take two different basis functions, then if one is non-zero at any given node the other must be zero at the same node, therefore their product is zero at all nodes. Having chosen the nodes as sampling points we then choose the weights, usually to make the integration rule as high an order as possible. It can be shown that this gives us a sufficiently good numerical integration scheme to guarantee a good approximation. Unfortunately, at the time of writing there is no accepted technique for lumping the hierarchical mass matrix.

Another popular method is to use an iterative scheme for solving the problem. Often this is done by first obtaining a reasonable approximation via a lumped matrix as an initial iterate and then iterating on the full mass matrix. Typically, only a handful of iterations are required for sufficient accuracy.

The forward Euler scheme suffers from the constraint of being only conditionally stable, therefore we must restrict the length of the time step to obtain even a reasonable approximation. Further, the finer the mesh, the harsher this restriction. If we move to a more stable difference scheme, for example Crank–Nicolson, we obtain a system of the form

$$\frac{2\mathbf{M} + \mathbf{A}\Delta t}{2}\mathbf{u}_{n+1} = \frac{2\mathbf{M} - \mathbf{A}\Delta t}{2}\mathbf{u}_n + \frac{(\mathbf{f}_{n+1} + \mathbf{f}_n)\Delta t}{2}.$$

The lumping technique is no longer available to us as we must invert the matrix $\mathbf{M} + \mathbf{A}\Delta t$. If we lumped both $\mathbf{M}$ and $\mathbf{A}$ we would decouple all of the grid points and hence we would no longer have an acceptable approximation. In this case we are left with the two other choices, compute and store an inverse (or decomposition) of $2\mathbf{M} + \mathbf{A}\Delta t$, or use an iterative technique. If we choose to use an iterative technique it would be standard practice to choose $\mathbf{u}_n$ as our initial iterate for $\mathbf{u}_{n+1}$.

As an example of a *hyperbolic* problem we shall look at the second-order *wave* equation:

Problem (H). Find $u(x,t)$ such that

$$u_{tt} - u_{xx} = 0$$

with appropriate initial and boundary conditions.

Following exactly the same steps as above we obtain a system

$$\mathbf{M\ddot{u}} + \mathbf{Au} = 0,$$

and again we have a coupled system of ordinary differential equations to solve. A useful exercise would be to derive the appropriate system for the equation $u_t + u_x = 0$. In principle, we can obtain high accuracy for many hyperbolic equations using finite element methods; however, the approximate solution may lack precisely the features of the exact solution which interest us. In particular, the simplest non-linear hyperbolic conservation law $u_t + (\frac{1}{2}u^2)_x = 0$ develops shocks (discontinuities in the solution) even from smooth initial data. The position of these shocks is of great physical interest; however, the very nature of the finite element approximation prohibits any discontinuities from occurring in the approximation. Various techniques have been developed to overcome this problem, with varying degrees of success, for example introducing artificial viscosity as we do with finite difference methods. However, there is no consensus as to a best method and those interested in pursuing this type of problem would do well to consult the research literature.

A. W. C.

REFERENCES

Babuška, I., Griebel, M. and Pitkäranta, J. (1989). *The problem of selecting shape functions for a p-type finite element. Int. J. Num. Meth. Engng.*, **28**, 1891–1908.

Briggs, W. L. (1987). *A Multigrid Tutorial*, SIAM.

Ciarlet, P. G. (1978). *The Finite Element Method for Elliptic Problems*, North-Holland.

Hackbusch, W. (1985). *Multi-Grid Methods and Applications*, Springer-Verlag.

Irons, B. M. (1970). A frontal solution program for finite element analysis, *Int. J. Num. Meth. Engng.* **2**, 5–32.

Kopal, Z. (1961). *Numerical Analysis*, Chapman and Hall.

Oden, J. T. and Reddy, J. N. (1976). *An Introduction to the Mathematical Theory of Finite Elements*, Wiley-Interscience.

Rachford, H. H. and Wheeler, M. F. (1975). *An H^{-1} Galerkin procedure for the two-point boundary value problem*, in *Mathematical Aspects of Finite Element Methods in Partial Differential Equations*, Ed. C. de Boor, Academic Press.

Richtmeyer, R. and Morton, K. W. (1967). *Difference Methods for Initial Value Problems*, Wiley-Interscience.

Strang, G. and Fix, G. J. (1973). *An Analysis of the Finite Element Method*, Prentice-Hall.

Zienkiewicz, O. C. and Taylor, R. L. (1989). *The Finite Element Method*, 4th edn., McGraw-Hill.

CHAPTER 9

Computational Complexity

9.1. INTRODUCTION

Computational complexity addresses itself to the quantitative aspects of the solution of computational problems.

Let us start with a simple example. Two decimal numbers of length m and n digits respectively are to be added together. The sequence of computational steps that will produce their sum involves a repetitive use of one operation, namely that of adding together two one-digit numbers, one such number from each of the two decimal numbers. We shall call this operation a basic operation for this elementary summation method.

The operation of adding together two one-digit numbers produces a single digit and, possibly, a carry. Repeatedly applying this operation to the two numbers in *from-right-to-left* order, we shall arrive after a certain number of steps at the value of their sum. In general, the lengths of the two numbers are not equal, that is $m \neq n$, and so the total number of steps, or basic operations, that will be required to compute the sum equals the larger of the two lengths, $l = \max \{m, n\}$. We then say that addition of two numbers with m and n digits *costs* l basic operations of addition. This number is also called the time complexity of addition of two decimal numbers, and is denoted by $T(m, n)$, a function of m and n,

$$T(m, n) = \max \{m, n\}.$$

If the two decimal numbers are of the same length of n digits, then the time complexity will, of course, be equal to

$$T(n) = n.$$

The numbers m and n define the *size* of the problem to compute the sum of two decimal numbers, and the *time complexity function* $T(m, n)$, or $T(n)$, is a *function of the size of the problem*.

Consider the following further examples and note the role of the parameter n.

(i) Find the largest in a sequence of n integers.
(ii) Solve a set of n linear algebraic equations

$$a_{11}x_1 + a_{12}x_2 + \ldots + a_{1n}x_n = b_1$$
$$a_{21}x_1 + a_{22}x_2 + \ldots + a_{2n}x_n = b_2$$
$$\vdots$$
$$a_{n1}x_1 + a_{n2}x_2 + \ldots + a_{nn}x_n = b_n$$

(iii) Sort a sequence of n distinct integers into descending order.

(iv) Evaluate a polynomial $P_n(x) = \sum_{k=0}^{n} a_{n-k}x^k$ at $x = x_0$.

In each of these problems the parameter n provides a measure of the size of the problem in the sense that the time required to solve the problem will increase with n.

9.1.1. Definition of an Algorithm

In the summation of two numbers the sequence of steps to compute the sum is called a summation algorithm.

Addition of two numbers is an example of a very simple algorithm where it is convenient to estimate the cost of computation in terms of just one basic operation, the addition of two one-digit numbers. However, more frequently a computational algorithm involves several types of basic operations. Such operations are used by the algorithm in a precise and systematic way to compute the result. Let us take another simple example. We are required to compute the product of two decimal numbers of length m and n respectively. Here we can easily see that the elementary school algorithm involves the use of two basic operations, the operation of multiplying two one-digit numbers (which produces one digit and, possibly, a carry) and the operation of adding two one-digit numbers (which produces one digit and, possibly, a carry).

We shall now estimate the cost of this computation. Since the decimal numbers are of lengths m and n digits respectively, the algorithm will require mn basic operations of one-digit multiplication. To estimate the total number of basic addition operations required, we note that the multiplication stage produces at most $s = \max \{m, n\}$ decimal numbers to be added together, each of length at most $l = \max \{2m, 2n\}$ digits. This gives

$$2mn < 2s^2$$

basic addition operations.

We can thus say that the elementary algorithm to compute the product of two decimal numbers costs

$$T(m, n) = 3mn < 3s^2$$

operations of addition and of multiplication of one-digit numbers.

In order to estimate meaningfully computational cost, or time complexity, of a computational process, we need to define properly an algorithm, which is a name for a computational process that can be carried out in practice.

DEFINITION 9.1.1. An *algorithm* is a step-by-step procedure for solving a problem. It consists of a finite set of unambiguous rules which specify a finite sequence of operations that provide the solution to a problem, or to a specific class of problems.

Several important features of this definition must be emphasized.

Feature One. Each step of an algorithm must be unambiguous and precisely defined. The actions to be carried out must be rigorously specified.

Feature Two. An algorithm must always arrive at a problem solution after a finite number of steps. Indeed, a useful algorithm must require a reasonable number of steps.

Feature Three. Every meaningful algorithm possesses zero or more inputs and provides one or more outputs. Inputs are defined as quantities which are given to the algorithm initially before it is executed, and the outputs as quantities which have a specified relation to the inputs and which are delivered at the completion of the algorithm's execution. In the simple algorithms for summation of two numbers and for the product evaluation of two numbers, the inputs are the two numbers to be added or multiplied together respectively, and the outputs are their sum or their product respectively.

Feature Four. It is preferable that the algorithm should be applicable to any member of a class of problems rather than to only a single problem. The concept of an algorithm can be further formalized in a number of ways, for example, in terms of Turing machines or of computer programs in some pseudo-code or a proper programming language. Examples of algorithms are given later in the text, when solutions to various problems are presented.

With regards to an algorithm, a problem can be viewed as consisting of a *domain* of the problem's *instances* and of a *question* that can be asked about any of the instances. An algorithm is then said to *solve* a problem P if, given any instance I of P as input data, it will generate the answer of P's question for I. One classical example in the theory of algorithms is that there are well-defined mathematical problems for which no algorithms *can* exist. For these problems, given a particular instance, one may be able to come up with an answer for that instance, while there is no general procedure that will apply to *all* instances.

9.1.2. Measures of Complexity

Much of computer science research consists of designing and analysing enormous numbers of algorithms. A theory of computational complexity uses rigorous mathematical techniques to evaluate algorithms in a manner independent of specific implementations, computers or data.

The fundamental concept in this respect is that of a *measure* of the algorithm's performance. Efficiencies of various algorithms are then compared using their computed measures. If the measure is machine independent, one observes, for example, that good algorithms tend to remain good if they are expressed in different programming languages or run on different machines.

The two most useful measures are the *time* required to execute the algorithm and the *memory* needed by the algorithm. We have already introduced the time function of an algorithm in the earlier section. The memory function is the space requirement of the algorithm. This may be for the storage of matrices, intermediate data, etc. The memory function represents the peak amount required.

On the program level, other measures, such as the program length, which is indicative of computation time, and the depth of the program, i.e. the numbers of layers of concurrent steps into which the problem can be decomposed, are useful. Depth corresponds to the time that the program would require under parallel computation.

The time of the algorithm and storage space are important measures and are particularly appropriate if the algorithm/program is to be run often. The time of the algorithm is the factor restricting the size of problems that can be solved by the computer and the program length in some sense measures the simplicity of an algorithm. The program length as a measure of the algorithm's efficiency is most appropriate if programming time is important or if the program is to be run frequently.

9.2.   ARITHMETIC COMPLEXITY OF COMPUTATIONS

The arithmetic complexity of computations is used to estimate the cost of the algorithms, where the set of basic operations required to compute the solution consists of the arithmetic operations, $+$, $-$, $*$, $/$, perhaps extended to include the square root operation and the operations of max and min.

The major classes of arithmetic problems include algebraic processes, such as the solution of linear systems, matrix multiplication and inversion, the evaluation of a determinant, the evaluation of a polynomial, at a point or at several points, and the evaluation of polynomial derivatives, iterative computations, e.g. the root-finding problem, the solution of linear systems and the multiplication of two integers.

9.2.1.   Matrix Multiplication

$$\begin{pmatrix} a_{11} & a_{12} \\ a_{21} & a_{22} \end{pmatrix} \begin{pmatrix} b_{11} & b_{12} \\ b_{21} & b_{22} \end{pmatrix} = \begin{pmatrix} c_{11} & c_{12} \\ c_{21} & c_{22} \end{pmatrix}$$

The usual method for computing the product of two $n \times n$ matrices requires n^3 basic multiplications and $n^2(n-1)$ basic additions:

$$c_{11} = a_{11}b_{11} + a_{12}b_{21},$$

$$c_{12} = a_{11}b_{12} + a_{12}b_{22},$$
$$c_{21} = a_{21}b_{11} + a_{22}b_{21},$$
$$c_{22} = a_{21}b_{12} + a_{22}b_{22}.$$

Here by the basic operation of multiplication we understand an operation to compute the product of two entries of the input matrices, e.g. $a_{21}b_{11}$, and by the basic operation of addition, the addition of two basic products, e.g. $a_{21}b_{11} + a_{22}b_{21}$. So to compute the product of two 2×2 matrices we need eight basic multiplications and four basic additions.

Thus the time complexity of the usual matrix multiplication algorithm is given as

$$T(n) = n^3 + n^2(n-1).$$

The time complexity is important for large-size problems, that is, for large values of n. It also follows that, for large n, the time-function value $T(n)$, as in the matrix product example, will be dominated by the value of the leading term n^3. This term determines the magnitude of the time function and this is why this term only is of importance. This fact is recorded using the following mathematical notation

$$T(n) = O(n^3),$$

which means that *there is some constant c such that the algorithm's running time on all inputs of size n is bounded by cn^3*. The precise value of c would depend on the computer used. The time complexity of the usual matrix multiplication algorithm is cubic: when n is a hundred, the time function value is a million; whenever n doubles, the time function increases eightfold. It is in the case of problems with such *explosive* algorithms that one attempts to design more efficient algorithms, with lower magnitude of the time function. For the matrix multiplication problem, when a new algorithm due to Strassen (1969) was published it was greeted as a nice surprise, because it showed a way to compute the matrix product in $O(n^{2.81})$ operations.

9.2.2. Matrix Product by Strassen's Algorithm

Let $\mathbf{A}$ and $\mathbf{B}$ be two square matrices of dimension $n = 2m$, an even number, and their product be denoted by $\mathbf{C}$.

$$\begin{pmatrix} A_{11} & A_{12} \\ A_{21} & A_{22} \end{pmatrix} \begin{pmatrix} B_{11} & B_{12} \\ B_{21} & B_{22} \end{pmatrix} = \begin{pmatrix} C_{11} & C_{12} \\ C_{21} & C_{22} \end{pmatrix}$$

To evaluate $\mathbf{C} = \mathbf{AB}$ using Strassen's method we compute

$$Q_1 = (A_{11} + A_{22})(B_{11} + B_{22}),$$
$$Q_2 = (A_{21} + A_{22})B_{11},$$
$$Q_3 = A_{11}(B_{12} - B_{22}),$$

$$Q_4 = A_{22}(-B_{11} + B_{21}),$$
$$Q_5 = (A_{11} + A_{12})B_{22},$$
$$Q_6 = (-A_{11} + A_{21})(B_{11} + B_{12}),$$
$$Q_7 = (A_{12} - A_{22})(B_{21} + B_{22}),$$

then

$$C_{11} = Q_1 + Q_4 - Q_5 + Q_7,$$
$$C_{21} = Q_2 + Q_4,$$
$$C_{12} = Q_3 + Q_5,$$
$$C_{22} = Q_1 + Q_3 - Q_2 + Q_6.$$

Here seven multiplications of $m \times m$ matrices and 18 additions of $m \times m$ matrices are required.

This gives the time complexity of the algorithm as

$$T(n) = 7m^3 + 7m^2(m-1) + 18m^2,$$

where $n = 2m$, since the product of two $m \times m$ matrices uses m^3 multiplications and $m^2(m-1)$ additions, and the addition of two $m \times m$ matrices uses m^2 additions. Substituting $m = n/2$ into the time complexity expression we obtain

$$T(n) = (\tfrac{7}{8})n^3 + (\tfrac{7}{8})n^2(n + \tfrac{22}{7}).$$

Whenever $n > 30$ we have both

$$(\tfrac{7}{8})n^3 < n^3 \quad \text{and} \quad (\tfrac{7}{8})n^2(n + \tfrac{22}{7}) < n^2(n-1).$$

The key to the economy lies in the fact that the Strassen algorithm calls for only seven multiplications of $m \times m$ matrices rather than eight.

In general, whenever m itself is an even number, the number of arithmetic operations can be reduced by using the same algorithm to compute each of the seven products of $m \times m$ matrices.

Since the advent of the computer the most important results in computational complexity relate to the computation processes on binary numbers. Taking $n = 2^s$ to be a power of 2 and denoting by $m(s)$ the number of multiplications needed to multiply two $n \times n$ matrices and by $a(s)$ the number of additions, we obtain

$$m(s + 1) = 7m(s),$$
$$a(s + 1) = 7a(s) + 18(4^s).$$

Using the initial conditions $m(0) = 1$ and $a(0) = 0$ we obtain

$$m(s) = 7^s = (2^s)^{\log_2 7} = n^{\log_2 7} = n^{2 \cdot 807},$$
$$a(s) = 6(7^s - 4^s) = 6((2^s)^{\log_2 7} - (2^s)^2) = 6(n^{\log_2 7} - n^2).$$

Extensions of the algorithm include cases where n is not the power of 2. An appropriate 'padding' of the two matrices by zeros is possible for such cases, so as to make their dimension a power of 2. The estimate then becomes $M(n) \leq 7n^{\log_2 7} - 42n^2$, where now $M(n)$ denotes the number of multiplications needed to multiply two $n \times n$ matrices. A similar result can be derived for the total number of additions $A(n)$, to show that the Strassen algorithm requires only $4.7n^{2.807}$ operations, in total, compared with $2n^3 - n^2$ operations of the usual method.

9.2.3. Lower Bounds on Matrix Multiplication

One important question that the theory of computational complexity attempts to answer is the following: What is the minimum number of basic operations needed to solve a given problem? This number is called the *lower bound* on the computational complexity of the problem. The other important question in practical computations is whether the lower bound is an attainable lower bound, that is, whether an algorithm exists or, if not, whether it is possible to design an algorithm that would require no more than the minimum number of basic operations to obtain the solution.

Strassen's algorithm, by expressing the formulae for the multiplication of two 2×2 matrices with seven basic multiplications, enabled one to construct an $O(n^{2.807})$ algorithm. If we knew how to multiply 2×2 matrices in six multiplications, we would have an $O(n^{\log 6})$, $O(n^{2.59})$, algorithm. However, it has been proved that seven basic multiplications is the minimum number required to find the product of two 2×2 matrices.

More generally, multiplying two small $k \times k$ matrices in only m multiplications could yield a Strassen-like recursive algorithm of complexity $O(n^m)$. An algorithm due to Pan (1978, 1984) is one such algorithm. It multiplies two 70×70 matrices in $O(n^{2.795})$ basic multiplications. In Pan's algorithm the product of $n \times p$ and $p \times n$ matrices is evaluated using the trace of the product of three matrices of dimensions $n \times p, p \times m$ and $m \times n$. We shall first illustrate the relationship between the product of two matrices and the trace of the product of three matrices, and how it can be used to compute the product matrix, on two small examples.

EXAMPLE 9.2.1. Consider the product matrix $\mathbf{C}$ of a 3×2 matrix $\mathbf{A}$ and a 2×3 matrix $\mathbf{B}$:

$$\begin{pmatrix} a_{11}a_{12}a_{13} \\ a_{21}a_{22}a_{23} \end{pmatrix} \times \begin{pmatrix} b_{11}b_{12} \\ b_{21}b_{22} \\ b_{31}b_{32} \end{pmatrix} = \begin{pmatrix} c_{11}c_{12} \\ c_{21}c_{22} \end{pmatrix}.$$

The bilinear form of the traditional algorithm gives

$$c_{ij} = \sum_{k=1}^{3} a_{ik}b_{kj} \quad (i,j = 1,2),$$

or

$$c_{11} = a_{11}b_{11} + a_{12}b_{21} + a_{13}b_{31},$$
$$c_{12} = a_{11}b_{12} + a_{12}b_{22} + a_{13}b_{32},$$
$$c_{21} = a_{21}b_{11} + a_{22}b_{21} + a_{23}b_{31},$$
$$c_{22} = a_{21}b_{12} + a_{22}b_{22} + a_{23}b_{32}.$$

The trilinear form of the product of the three matrices **A**, **B** and **C** is, on the other hand, given as

$$\sum_{i,j,k} a_{ij}b_{jk}c_{ki}$$

$$= a_{11}b_{11}c_{11} + a_{11}b_{12}c_{21} + a_{12}b_{21}c_{11} + a_{12}b_{22}c_{21} + a_{13}b_{31}c_{11}$$
$$+ a_{13}b_{32}c_{21} + a_{21}b_{11}c_{12} + a_{21}b_{12}c_{22} + a_{22}b_{21}c_{12} + a_{22}b_{22}c_{22}$$
$$+ a_{23}b_{31}c_{12} + a_{23}b_{32}c_{22}$$
$$= (a_{11}b_{11} + a_{12}b_{21} + a_{13}b_{31})c_{11} + (a_{21}b_{11} + a_{22}b_{21} + a_{23}b_{31})c_{12}$$
$$+ (a_{11}b_{12} + a_{12}b_{22} + a_{13}b_{32})c_{21} + (a_{21}b_{12} + a_{22}b_{22} + a_{23}b_{32})c_{22}.$$

In the trilinear form equation the coefficient of c_{ij} is the (j, i) element of the product matrix **C**. This fact can be used to actually compute the product matrix **C** of the two matrices **A** and **B**. Example 9.2.2 shows how this is done for two 4×4 matrices.

EXAMPLE 9.2.2.   Given are two 4×4 matrices

$$\mathbf{A} = \begin{pmatrix} 5 & 3 & -1 & 4 \\ 2 & -5 & 7 & 9 \\ -7 & 4 & 0 & 5 \\ 8 & -3 & -4 & 1 \end{pmatrix} \quad \text{and} \quad \mathbf{B} = \begin{pmatrix} 2 & 1 & -4 & -6 \\ 4 & 3 & -9 & 1 \\ 6 & 7 & 7 & 5 \\ -8 & 0 & 2 & 1 \end{pmatrix}.$$

Find their product **C**.

Solution.   There is more than one way in which the trilinear form can be expressed. In our example we shall use the following representation of the trilinear form:

$$\sum_{i,j,k} a_{ij}b_{jk}c_{ki}$$

$$= \sum_{\substack{i,j,k:\underbrace{i+j+k}_{\text{is even}}}} (a_{ij} + a_{k+1,i+1})(b_{jk} + b_{i+1,j+1})(c_{ki} + c_{j+1,k+1})$$

$$- \sum_{i,k} a_{k+1,i+1} \sum_{\substack{j:i+j+k \\ \text{is even}}} (b_{jk} + b_{i+1,j+1})c_{ki}$$

$$- \sum_{i,j} a_{ij} b_{i+1,j+1} \sum_{\substack{k:i+j+k \\ \text{is even}}} (c_{ki} + c_{j+1,k+1})$$

$$- \sum_{j,k} \sum_{\substack{i:i+j+k \\ \text{is even}}} (a_{ij} + a_{k+1,j+1}) b_{jk} c_{j+1,k+1}.$$

Substituting for the a_{ij} and b_{ij} and collecting the coefficients of c_{ij} we get

$$\sum_{i,j,k} a_{ij} b_{jk} c_{ki}$$

$$= -16c_{11} - 46c_{12} - 38c_{13} - 28c_{14} + 7c_{21}$$
$$+ 36c_{22} + 5c_{23} - 29c_{24} - 46c_{31} + 104c_{32} + 2c_{33}$$
$$- 31c_{34} - 28c_{41} + 27c_{42} + 51c_{43} - 70c_{44}.$$

Since the coefficient of c_{ij} is the (j, i)th entry in the matrix $\mathbf{C}$, we have

$$\mathbf{C} = \begin{pmatrix} -16 & 7 & -46 & -28 \\ -46 & 36 & 104 & 27 \\ -38 & 5 & 2 & 51 \\ -28 & -29 & -31 & -70 \end{pmatrix}$$

In this example we have used a particular trilinear form representation to compute the product matrix $\mathbf{C}$. It requires, as can be seen, 80 multiplications of various terms involving a_{ij} and b_{ij}. This is, of course, inferior to both Strassen's method (with 49 multiplications) and the traditional method (with 64 multiplications). However, by clever choice of the trilinear form representations so as to minimize its computational time, Pan was able to suggest an algorithm, which yields superior results to Strassen's method, for the matrices of size $n > 6$.

So, what is the overall minimum number of basic operations (or *arithmetics*) needed to evaluate the product of two $n \times n$ matrices? At present it can only be said that since the problem has $2n^2$ inputs and n^2 outputs, a lower bound on the number of arithmetics must be $O(n^2)$. However, the fastest known matrix multiplication algorithm uses $O(n^{2 \cdot 496})$ arithmetics and is due to Pan.

9.2.4.   The Fast Fourier Transform (FFT)

Assume a given set of n points

$$a_0, a_1, a_2, \ldots, a_{n-1}.$$

The discrete Fourier transform (DFT) [for Fourier Transform, see IV, §13.2], applied to these points yields a set of n points

$$A_0, A_1, A_2, \ldots, A_{n-1},$$

where A_p is computed by the formula

$$A_p = \sum_{q=0}^{n-1} w^{pq} a_q \quad (p = 0, 1, \ldots, n-1, w = e^{2\pi i/n}).$$

Straightforward computation of A_p uses $n-1$ complex multiplications and $n-1$ complex additions, thus yielding a computational procedure of complexity $O(n^2)$.

Cooley and Tukey (1965) observed that, whenever $n = rs$ is a composite number, a more efficient method of computation exists.

Each p in $0 \le p \le n-1$ can be written uniquely as

$$p = p_1 + p_2 r \quad (0 \le p_1 < r, 0 \le p_2 < s),$$

and each q in $0 \le q \le n-1$ as

$$q = q_1 s + q_2 \quad (0 \le q_1 < r, 0 \le q_2 < s).$$

Therefore

$$A_{p_1 + p_2 r} = \sum_{q_1=0}^{r-1} \sum_{q_2=0}^{s-1} w^{(p_1 + p_2 r)(q_1 s + q_2)} a_{q_1 s + q_2}$$

$$= \sum_{q_1=0}^{r-1} \sum_{q_2=0}^{s-1} w^{p_1 q_1 s} w^{p_1 q_2} w^{p_2 q_1 rs} w^{p_2 r q_2} a_{q_1 s + q_2}$$

$$= \sum_{q_1=0}^{r-1} \sum_{q_2=0}^{s-1} (w^s)^{p_1 q_1} w^{p_1 q_2} (w^r)^{p_2 q_2} a_{q_1 s + q_2},$$

since

$$w^{rs} = w^n = 1.$$

Thus

$$A_{p_1 + p_2 r} = \sum_{q_2=0}^{s-1} (w^r)^{p_2 q_2} \left[w^{p_1 q_2} = \sum_{q_1=0}^{r-1} (w^s)^{p_1 q_1} a_{q_1 s + q_2} \right].$$

Hence to compute each A_p we need to

(i) Compute the DFT of r points, for each q_2,

$$b_{p_1 q_2} = \sum_{q_1=0}^{r-1} (w^s)^{p_1 q_1} a_{q_1 s + q_2}.$$

(ii) Compute the factor

$$c_{p_1 q_2} = w^{p_1 q_2} b_{p_1 q_2}.$$

(iii) Compute the DFT of s points,

$$A_p = \sum_{q_2=0}^{s-1} (w^r)^{p_2 q_2} c_{p_1 q_2}.$$

This requires $(r + s - 1)$ multiplications and $(r + s - 2)$ additions. In order to compute n values A_p we shall need $n(r + s - 1)$ multiplications and $n(r + s - 2)$ additions.

If r and s are themselves composite numbers, the same idea can be used to compute the DFT of r or s points. In particular, when $n = 2^s$ is a power of 2, this computation process uses $2^s s = n\log_2 n$ multiplications and $2^s(s - 2) = n(\log_2 n - 2)$ additions. Thus the new algorithm, known as the FFT, is of complexity $O(n\log n)$, a staggering reduction compared with $O(n^2)$.

The FFT takes advantage of the periodic nature of the *sine* and *cosine* functions to greatly reduce the number of multiplications required in evaluating the DFT.

The tremendous reduction in computing time of the DFT which the FFT algorithm offers makes the DFT one of the most powerful mathematical tools used in the solution of many important practical problems.

9.2.5. Product of Two Integers

In the introduction we have already discussed the usual algorithm for computing the product of two n-digit decimal numbers. As has been shown, the algorithm's time cost is

$$T(n) \le 3n^2 = O(n^2)$$

of operations of single-digit multiplications and single-digit additions.

Another algorithm for computing the product of two numbers was developed in 1962 by two mathematicians, Karatsuba and Ofman (1963). This new algorithm computes the product in $O(n^{\log 3})$ time and is based on the following idea.

Let a and b be two n-digit numbers, where $n = 2m$, an even number. We can represent the two numbers uniquely as

$$a = a_0 + a_1 10^m \quad \text{and} \quad b = b_0 + b_1 10^m,$$

where a_0, a_1, b_0 and b_1 are m-digit numbers. The product p is then expressed as

$$p = ab = (a_0 + a_1 10^m)(b_0 + b_1 10^m)$$
$$= a_0 b_0 + (a_1 b_0 + a_0 b_1)10^m + a_1 b_1 10^{2m}.$$

The computation of p can be viewed as consisting of four steps:

Step 1. Compute $a_0 b_0$.
Step 2. Compute $a_1 b_0 + a_0 b_1$.
Step 3. Compute $a_1 b_1$.
Step 4. Carry out $n = 2m$ single-digit additions.

In this algorithm the operation of multiplication by the power of 10 is ignored since it represents a simple shift operation. Thus the algorithm requires four multiplications of m-digit numbers and two additions of $2m$-digit numbers.

This is the same as the estimate on the number of operations in the usual method. The key to the new algorithm is the way in which $a_0 b_0, a_1 b_0 + a_0 b_1$ and $a_1 b_1$ are computed. Their computation is based on the identities

$$a_0 b_0 = a_0 b_0,$$
$$a_0 b_1 + a_1 b_0 = (a_0 - a_1)(b_1 - b_0) + a_0 b_0 + a_1 b_1,$$
$$a_1 b_1 = a_1 b_1,$$

from which it follows that to obtain the left-hand side values of the identities one needs to find three products of two m-digit numbers and carry out some additions and subtractions. If the addition or subtraction of two single-digit numbers with the possible carry or borrow is taken as a unit of addition then the product of two n-digit numbers uses three multiplications of m-digit numbers and $4n$ units of additions.

Further savings are due to the fact that the suggested computation mechanism can be applied recursively when computing each of $a_0 b_0$, $(a_0 - a_1)(b_0 - b_1)$ and $a_1 b_1$.

In this problem, if we now assume that the two numbers are represented as binary numbers, then the new algorithm yields the result that for $n = 2^s$, a power of 2, the product of two $n = 2^s$ digit numbers is obtained using 3^s multiplications of single-digit numbers and $8(3^s - 2^s)$ units of addition, under the initial condition that for $n = 2^0 = 1$ only one single multiplication and no units of addition are needed.

In summary, this method to compute the product of two n-digit numbers uses at most $3 \times 3^{\log_2 n} = 3n^{\log_2 3} = 3n^{1.59}$ single-digit multiplications and $8 \times 3 \times 3^{\log_2 n} = 24n^{\log_2 3} = 24n^{1.59}$ units of addition.

9.2.6. Worst-case and Average-case Analysis

The algorithms that have been discussed in this chapter are all very well behaved in the sense that one can predict quite accurately the number of steps required for a problem instance. With the same size n, the prediction does not vary from one instance to another.

However, there are many algorithms where the number of steps varies tremendously among instances of the same size. Since a complexity function takes on a single value for each size, two approaches are open for the complexity analysis. In the first approach, one distinguishes between the counts of the *average* and the *maximum* number of steps, over all inputs of size n. Some examples of this analysis are given later in the text. In the second approach, the complexity analysis distinguishes between the *average-case* problem instance and the *worst-case* problem instance analysis.

The worst-case analysis provides the algorithm's performance *guarantee*: the problem will always require no more time or space than the worst-case estimate, but its drawback is that 'pathological' cases are allowed to determine the complexity, even though they may be extremely rare in practice. For example, if a problem is such that its solution requires 10^{12} operations 1 per cent of the time but only 1000 operations 99 per cent of the time, then, by the worst-case bound judgement, the problem is very slow. Yet in many applications one is concerned with the algorithms which are run many times and so one is concerned with the *average* or *typical* efficiency of solving the problem instance. For example, sorting and searching algorithms, discussed in the next section are in this category, because of their frequent use in various applications. For such algorithms, an average-case analysis is almost always more realistic than the worst-case analysis.

On the other hand, the average-case analysis also has drawbacks, in particular, since it involves a decision on a probability distribution for the problem instances. It may happen that to find a mathematically tractable distribution, that also models the problem instances encountered in practice, is not easy or even possible. Often the average-case analysis provides a prediction that one can have confidence in only if one has to solve a large number of instances.

9.3.   COMPLEXITY OF DATA PROCESSING PROBLEMS

Data processing problems include such processes as *sorting* the data elements in some order, *searching* through the ordered set of data to establish whether or not an element is in the set, *updating* the data set by *inserting* a new element into the set or *deleting* an element from the set. The nature of these problems implies that the elements of a set are defined so that the notions of *greater than*, *equal to* and *less than* are meaningful within the set.

9.3.1.   Search Problem

Suppose we are given a sequence of n numbers. Such a sequence is normally denoted by

number$[1,...,n]$ = number$[1]$, number$[2]$, number$[3]$,..., number$[n]$.

We wish to find the largest value among the numbers of the sequence.

A simple method is to scan the sequence sequentially, comparing each number with the largest one found to that point. This *algorithm* can be described as follows.

Algorithm findmaximum

```
if n > 0 then
    index := 1
    max := number[index]
    while index < n do
        index := index + 1
```

```
   if max < number[index]
   then max: = number[index]
   endif
  enddo
endif
```

There are two *basic operations* involved in the algorithm. These are the *comparison* of the 'max' value with the current element value of array 'number' and the *assignment* of a new value to the max. The **while**-loop must be repeated $n-1$ times, and this sets *lower bounds* on the execution time of the algorithm; the first max assignment statement in the algorithm is always executed, and so if the first element happens to be the largest one, then the max assignment statement within the **while**-loop is never executed, which means that the assignment is done only once. On the other hand, if the array is an increasing sequence of numbers then max is assigned a new value in total n times. Hence 1 and n are the extreme values. As for the average, there are $n!$ possible arrangements of the numbers.

In order to obtain the average number of assignment statement executions we could examine every one of $n!$ possible arrangements of n numbers, counting the assignment executions for each arrangement, then computing the total sum of the executions and dividing it by the number of arrangements, $n!$. This is too many arrangements to examine even for small values of n. Instead we can arrive at the result analytically.

We have noted that the initial assignment of the max value is always executed once, so the count of executions begins at 1. Assuming all permutations are equally likely, the probability that the second element is larger than the first is $\frac{1}{2}$. The probability that the third element is larger than either of the first two elements is $\frac{1}{3}$. Continuing in this way, we obtain the average of assignments equal to the sum

$$1 + \frac{1}{2} + \frac{1}{3} + \dots + \frac{1}{n} = H_n,$$

which for large n is equal to

$$\log_e n + 0.577.$$

Since the $\log n$ grows much slower than n itself, the time complexity of the algorithm is determined by the **while**-loop time and thus is proportional to n.

Some very efficient searching and sorting algorithms have been designed. For example it is known that searching on a sequence of n numbers requires about $\log n$ comparisons, and sorting of n numbers requires about $n \log n$ comparisons.

9.3.2. A Binary Search Algorithm

When a quick retrievel of information is important, a data structure known as the binary search tree is often used.

DEFINITION 9.3.1. A binary search tree for an ordered sequence $S = s_1, s_2, \ldots, s_n$ is a labelled binary tree in which each node i is labelled by an element $s_i \in S$ such that

(i) for each node j in the left subtree of $i, s_j < s_i$,
(ii) for each node j in the right subtree of $i, s_j > s_i$, and
(iii) for each element s which is in S, there is exactly one node i such that $s_i = s$.

The levels in a binary search tree are numbered from 1 (at the root level) upwards, where k is the last level.

The binary search tree has two notable properties: each node can have at most two children (subnodes), and the elements that identify the nodes are arranged so that at any node the smaller element is in the left subtree. In such a set-up searching for a particular element can be very efficient; a comparison at each node indicates whether to take the left branch or the right, and so the number of remaining possibilities is halved at each node. The maximum efficiency is achieved when the tree is perfectly balanced, i.e. when every node has exactly two children, with the exception perhaps of the penultimate level nodes which may have one or no children. The average number of element comparisons for a successful search, i.e. when the element is found to be in the

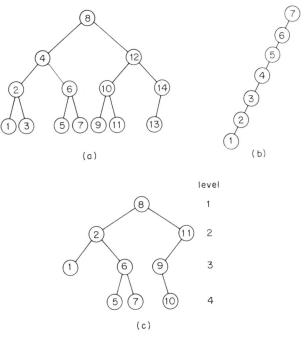

Figure 9.3.1. Binary search tree (a) Perfectly balanced search tree for 14 elements (b) Degenerate tree for 7 elements (c) General tree for 9 elements

tree, is then equal to $\log_2 n$. In the worst case, where the tree has degenerated to just one child linked to each node, the average number of comparisons is one-half the number of nodes, $n/2$. Figure 9.3.1 illustrates three instances of the binary search tree.

Let us now study the following search problem. Suppose n element values are read in random sequence and inserted into a tree, which is initially empty. The elements are placed in the left-to-right ordering. A recursive algorithm can be designed both to add a new element and to search for one that is already present.

Algorihm treesearch (treenode)

(The algorithm searches for element KEY in a tree defined by treenode. If KEY is not found, the algorithm inserts it.)
if treenode = nil   //the tree is empty//
then insert new node with element KEY
else
    case KEY **of**
        : KEY < treenode.element: *treesearch*(*treenode.left*);
        : KEY > treenode.element: *treesearch*(*treenode.right*);
        : KEY = treenode.element: process node   //found//
    endcase
endif

Assume that the elements are the integers 1 to n and that all permutations of elements are equally likely. Let some element i arrive first and so become the root of the tree. When the rest of the elements have been inserted there will be $i - 1$ nodes to the left of the root and $n - i$ nodes to the right. If i happens to be the midpoint of the range, the tree is perfectly balanced at this highest level; if i is equal to 1 or n, the first branches of the tree are completely unbalanced. Continuing the argument recursively we note that if the second element to arrive is j and it is less than i, then when the tree is filled there will be $j - 1$ nodes in the branch to the left of j and $i - j - 1$ nodes in the branch to the right.

We shall now evaluate the average number of comparisons using the notion of the tree path length.

DEFINITION 9.3.2.   The *path length* of the tree is the sum of the distances of all nodes from the root, and the distance of a node from the root is one less than its level. The average path length of the tree is the path length of the tree, divided by the number of nodes in the tree.

For example, in Figure 9.3.1(c),

(i) the distance of node 9 from the root is equal to 2,
(ii) the path length of the tree is equal to 17, and
(iii) the average path length of the tree is equal to $\frac{17}{9} = 1.89$.

To calculate the average path length of the tree in the general case we put forward the following argument.

At the root, the path length can be viewed as follows:

(i) 1 (for the root node versus the root nodes of the two subtrees),
(ii) plus the path length of a subtree with $i - 1$ nodes,
(iii) plus the path length of a subtree with $n - i$ nodes.

These lengths are not known, but they can be calculated by applying the same procedure at the next level in the tree. Ultimately the end of each branch is reached, where every node has a path length of either 0 or 1. Denote the path length of the tree on n elements, and with the element i designated as the root, by $P_n(i)$; if the root is not emphasized, then the path length is simply denoted by P_n. The recursive definition of the path length must be averaged for all possible values of i from 1 to n. We have

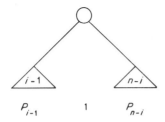

$$P_n(i) = [(i - 1)(P_{i-1} + 1) + 1 + (n - i)(P_{n-i} + 1)]/n$$

and then

$$P_n = \frac{1}{n} \sum_{i=1}^{n} P_n(i) = 2\left(1 + \frac{1}{n}\right)H_n - 3$$

$$= 2\log_e n - 1.845$$

for large n. That is, the average path length of a tree is $O(\log n)$.

9.3.3. Comparison Sort Problem

A computational model for sorting four distinct numbers, $K1, K2, K3, K4$, is shown in Figure 9.3.2. It is a binary decision tree, where a node represents an operation of comparison of two numbers, Km and Kn, and the left and right arcs represent the decision paths, one of which is taken upon the comparison outcome. After Km is compared with Kn, if Km is less than Kn, then one follows the left arc, otherwise one follows the right arc. It is very important to preserve consistency throughout the process in terms of 'if the left-hand-side number in the relation $Km < Kn$? is less than the right-hand-side number, then follow the left arc, otherwise follow the right arc'.

Each path from the root of the tree to its terminal node, which is denoted

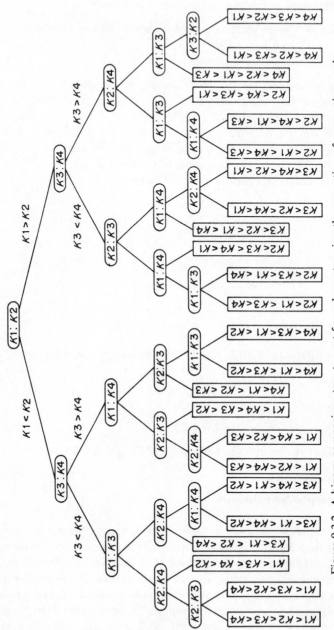

Figure 9.3.2. A binary comparison tree to sort four elements using the operation of comparison only

by□, is meaningful, that is no redundant comparisons are allowed in the tree. If $K1 < K2$ and $K1 > K3$ then it is 'known' that the ordered sequence is $K3 < K1 < K2$, without comparing $K2$ and $K3$.

If the root of the tree is on level 1, then at the lowest level, k, in the tree there are at most 2^k nodes. On the other hand, there are $n!$ permutations of n numbers. This gives $2^k \geq n!$ or $k \geq \log_2 n!$, which for large n is approximately equal to $n\log n$. This is the lowest bound on the number of comparisons required to sort n numbers. There are many comparison sort algorithms for which this lower bound is the worst-case time bound. At the same time, for many other comparative sorts, $O(n\log n)$ is the average-case time, while the algorithm's worst-case time is $O(n^2)$.

9.4. COMPLEXITY OF COMBINATORIAL PROBLEMS

The field of combinatorial algorithms concerns the problems of performing computations, e.g. grouping, arrangement, ordering, selection, on discrete finite mathematical structures. Operations on matrices, the FFT algorithms, sorting of a sequence of elements or searching for an element in a sequence are examples of combinatorial problems. As a rule, several algorithms are available to solve these problems, and for some problems one can even point out the 'best' or an optimal algorithm. An important question is that of a relationship between performance of an algorithm and whether it is termed as 'good' algorithm.

Intuitively, we would like to call an algorithm 'good' when it is sufficiently efficient to be usable in practice, but this is a rather vague notion. A better idea is to consider *an algorithm 'good' if its worst-case complexity is bounded by a polynomial function of the problem size, n.* Algorithms for matrix manipulation and searching and sorting algorithms are all 'polynomial time algorithms' and hence 'good'.

9.4.1. Polynomial and Exponential Time Algorithms

In general, a polynomial complexity algorithm is defined as an algorithm whose complexity function is $O(p(n))$ for some polynomial p and problem size n. If an algorithm is found to be polynomial bounded when implemented on one type of computer, it will be polynomial bounded, though perhaps by a polynomial of a different degree, when implemented on virtually any other computer.

An algorithm that cannot be bounded by a polynomial time function is termed 'exponentially bounded'. For example, if the algorithm's time grows at least as fast as c^n for some $c > 1$, it is an 'exponential time' algorithm. There are 'subexponential' functions such as $n^{\log n}$, which are neither polynomially bounded nor exponential; for simplicity they are also referred to as 'exponential' functions.

The difference between polynomials and exponentials becomes clear if we

Time complexity function	Problem size		
	20	40	60
n	0.00002 s	0.00004 s	0.00006 s
$n\log n$	0.00009 s	0.0002 s	0.0004 s
n^3	0.008 s	0.064 s	0.216 s
n^5	3.2 s	1.7 min	13.0 min
$n^{\log n}$	0.4 s	5.6 min	9 h
2^n	1.0 s	12.7 days	366 centuries
n^n	3×10^{10} centuries		

Table 9.4.1.   Comparison of several complexity functions (we assume that the problem of size 1 takes 10^{-n} s to compute)

look at the functions from an 'asymptotic' point of view. A given exponential function may initially yield smaller values than a given polynomial function, but there always is an N such that for all $n > N$ the polynomial grows much slower than the exponential. In Table 9.4.1, growths of some polynomial functions are compared to certain exponential functions.

In practical terms, the difference between the algorithms with polynomial and those with exponential complexity lies in the simple fact that when using a polynomial time algorithm one would normally obtain the computed solution, while with an exponential time algorithm one's whole lifetime would not be long enough to see the computed solution. Exceptions are the extreme cases, where an algorithm with the polynomial time of very high degree will be impractical for applications whereas an exponential algorithm would solve small-size instances. However, both extreme cases are of no practical importance.

Another point about the polynomial-time algorithms is that they are able to take advantage of the technological progress in the speed of computers. Suppose we have two algorithms, one with the run time $O(n^3)$ and another with $O(2^n)$, both of which can solve a problem with $n = 100$ in one hour. If a new computer becomes available with the speed increased by a factor of 1000, the polynomial time algorithm can now solve in one hour instances of the size $N = (1000)^{1/3} n = 10n$, a *multiplicative factor* of 10, whereas the exponential algorithm improves on the problem size by an *additive factor* of 10, since here we have $N = \log_2 1000 + n = 9.97 + n$.

Many combinatorial problems which obviously admit some algorithms, seem to have no 'good' algorithms for their solution, in spite of strenuous efforts to discover such algorithms.

One example of such a problem is
The maximum stable set problem. Given a graph, find in it a maximum number of vertices, no two of which are adjacent.

A graph with n vertices has 2^n subsets of vertices. An algorithm for solving the problem is based on an exhaustive search. This gives the complexity function $O(2^n)$, an exponential function. Since this function indicates the number of 'basic computational steps' required by the algorithm, obtaining a solution for large-size instances of the problem quickly becomes impracticable. However, no one has yet discovered a substantially faster algorithm.

Other examples of problems in the same category are:

The travelling salesman problem (TSP). Find the shortest route for a salesman who must visit n cities. [see Definition 9.4.1 and V, Part A, §6.6.1]

The k-clique problem. Does an undirected graph [see V, 6] on n nodes contain a complete subgraph on k nodes?

The assignment problem. Find a minimum sum subset of the elements in an $n \times n$ matrix, with exactly one element in each row and in each column.

The problem of factoring a large number. Find all the primes [see I, §4.2] that divide the given large number exactly.

These and other problems of a similar nature have been studied over a long time and the only algorithms for their solution that are known are of exponential computing time. It has been conjectured that certain problems might not be at all solvable by 'reasonable' polynomial time algorithms.

For practical applications it is important to resolve fully whether the problem is solvable by a polynomial time algorithm.

One way to do this is to establish lower bounds on the complexity of the problem. Some important results have been proved, but there still remain many open questions. For problems known to be of polynomial time, e.g. sorting, matrix operations, the 'good' algorithms are generally made possible because of the gain of some deeper insight into the structure of the problem. Important research is going on in the area of understanding crucial properties of difficult combinatorial problems.

On the other hand, years of efforts to design efficient algorithms brought about a number of design techniques and strategies, which are sufficiently general to be used in developing algorithms for many different classes of problems. Design methods such as recursion, optimization techniques, dynamic programming, graph search,branch-and-bound, backtracking, data-updating methods and graph mapping often yield effective algorithms in solving large classes of problems. The next section gives a concise account of these methods.

9.4.2.   Efficient Algorithm Design Techniques

An important aspect of the algorithm design is working out various trade-offs to achieve an overall performance of the algorithm which is as efficient as possible, for example, trading off time against storage, or trading off time against accuracy in a numerical solution.

Recursion

One of the oldest and most widely used algorithm design techniques is known as the divide-and-conquer approach. The most general formulation of an algorithm using this approach is as follows.

Given problem instance I.

if I is divisible into smaller instances, $I_1, I_2, \ldots, I_k$, **then**

    Solve I_1;

    Solve I_2;

    $\vdots$

    Solve I_k;

    Combine the k partial solutions into a solution for I

else Solve I directly

Recursion is a special case of the divide-and-conquer approach, where the problem instance is divided into a small number of *smaller-size instances of the same type as the original instance* and the smaller instances are divided in the same way. Eventually the instances become small enough to be solved directly. The solutions to the smaller instances are then combined to give solutions to the bigger instances, until the solution to the original instance is computed.

Recursion is important in many ways. A number of specific design techniques are inherently recursive and so recursion is a natural way to describe algorithms obtained by these techniques. In its own right, as a divide-and-conquer method, recursion is an important algorithm design technique. Whenever possible, balancing the competing costs is a useful rule in achieving an efficient design, and this is why the recursive algorithms which require division of the problem into smaller instances of approximately equal size have, in general, a better overall performance. On the other hand, as a warning, recursion may lead to exponential time algorithms because though only a small number of smaller instances is created, some of them are resolved many times, giving exponential time complexity.

Optimization

A large class of problems, whatever their physical details may be, can be mathematically formulated as a problem where we have n inputs and are required to obtain a subset of the inputs that satisfies certain constraints.

Any subset that satisfies these constraints is called a *feasible* solution. A feasible solution that either maximizes or minimizes a certain function defined on the problem, the *objective function*, is an *optimal* solution. We are often required to find an optimal solution. There is normally an obvious way to determine a feasible solution but not necessarily an optimal solution.

Efficient algorithms available for the solution of such problems are often based on one of the two techniques, greed and augmentation. The greedy method can be demonstrated in an example.

EXAMPLE 9.4.1. (A postage stamp problem) Consider a postage stamp problem, where there is a set of stamps of six different denominations, 25p, 20p, 10p, 3p, 2p and 1p. A particular letter we are mailing requires the total postage of 80p. We need to find a valid subset of the stamps. Almost without thinking we start with the largest denomination stamp, since all denominations are less than the total postage required, and build a subset element by element. We would get, say, $25p + 25p + 25p + 3p + 2p$. This method of making up the postage is known as a greedy algorithm.

The greedy method suggests that at any individual stage a 'locally optimal' option, in some particular sense, is selected. However, it is easy to see that the postage solution in Example 9.4.1 can be 'improved' in the sense of obtaining a smaller number of stamps yielding the needed postage total, e.g. four 20p stamps yield the same total as the five stamps chosen in the first solution; another four-stamp combination is two 25p, one 20p and one 10p. The example shows that not every greedy approach succeeds in producing the best result overall. It may produce a good result for a while, yet the overall result may be poor.

Still, there are many problems where greedy algorithms can be relied upon to produce 'good' solutions with high probability. For example, if the problem is such that an exhaustive search is the only way to obtain an optimal solution, then the greedy method can be the only real, and wise, choice.

A classical example that illustrates this point is the travelling salesman problem (TSP) (Lawler et al. (1985)). This problem is formulated using the notion of a graph.

DEFINITION 9.4.1. A *graph* is a collection of *vertices* (*nodes*) and *edges*, and is usually denoted by $G = (V, E)$. Vertices are simple objects that can have names, e.g. 'cities' and other properties; an edge is a connection between two vertices, e.g. a connection between two 'cities'.

A *path* from vertex v to u in a graph is a list of vertices in which successive vertices are connected by edges in the graph; the first vertex in the list is v and the last is u.

A *simple path* is a path in which no vertex is repeated.

A *cycle* is a path that is simple except that the first and last vertex are the same (a path from a point back to itself).

A *simple cycle* is a cycle where no edge is repeated.

A graph with no cycles is called a tree.

A graph is *connected* if there is a path from every vertex to every other vertex in the graph.

A graph with all edges present is called a *complete* graph; a graph with relatively few edges is called *sparse*; a graph with relatively few of the possible edges missing is called *dense*.

An edge is said to be *incident* with the vertices it joins.

A vertex is said to have *degree k* if there are k edges incident with it.

In a *weighted* graph, numbers (weights) are assigned to each edge to represent, say, distances or costs.

In a *directed* graph edges are 'one-way': an edge may go from v to u or from u to v but not both. If no such directions are assigned to the edges, the graph is an *undirected* graph.

EXAMPLE 9.4.2. (The travelling salesman problem (TSP)) A salesman, starting from his home city, is to visit, exactly once, each city on a given list of

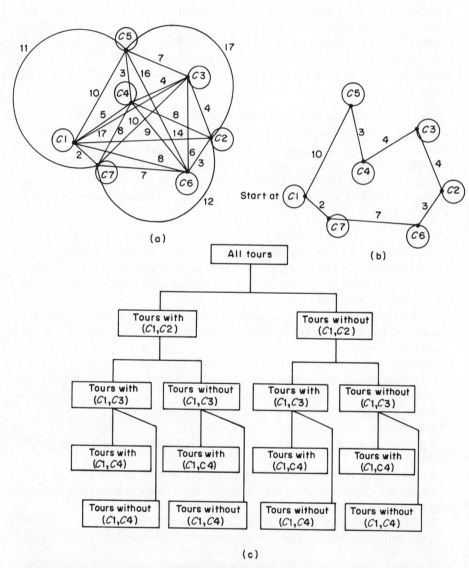

Figure 9.4.1. The TSP for an instance with seven cities (a) Seven cities and distances between them (b) A greedy solution to (a) (c) The start of a solution tree for the TSP instance in (a)

n cities and then return home. It is advantageous for him to select the order in which to visit the cities so that the total distance travelled in the tour would be as small as possible. It may be assumed that the salesman knows, for each pair of cities, the distance between the two cities. However, it is not at all obvious how to select the minimum-distance tour.

With a view to its solution, the problem can be formulated as a graph problem: we have an undirected graph with weights on the edges. It is reasonable to assume that all edges exist, that is, that the graph is complete. A tour is a simple cycle that includes all the vertices. We wish to find a *tour* which minimizes the sum of these edge weights. In Figure 9.4.1(a) an example is shown of the TSP, a graph with seven cities (vertices).

The only known algorithms that produce optimal solutions to this problem are of the 'try-all-possibilities' variety. They all have extremely expensive running times. On the other hand, the TSP has a number of practical applications and therefore efficient algorithms are essential.

A Greedy Algorithm for TSP

A greedy algorithm for the TSP considers the shortest edges first and accepts an edge into the solution subset if it, together with the edges already in the subset, does not cause a vertex to have degree three or more, and does not form a cycle, unless the number of edges in the subset equals the number of vertices in the problem instance; otherwise the edge is rejected. The collection of edges selected under these criteria will form a collection of unconnected paths, until the last step, when the single, remaining path is closed to form a tour.

An example of a greedy solution for a seven 'cities' graph is shown in Fig. 9.4.1(b).

The greedy algorithms are easy to implement and they are fast. In situations where the greedy method does not work, an approach of *iterative improvements* can sometimes help. This general technique assumes a start with any solution to the constraints and then looks for a way to *augment* the weight of the solution by making local changes. The new solution is then improved again and in the same way. The process is continued until no improvement is possible. Such a locally optimal solution, under appropriate conditions, is also globally optimal. Often the augmentation method is a good heuristic.

Dynamic Programming

This technique is a special kind of recursion applied to the problems for which there is no obvious way to divide a problem into a small number of subproblems whose solutions can be combined to solve the original problem. In such cases the problem's instance is partitioned into as many smaller instances as necessary and a track is kept of the smaller instances generated so as to make sure that the same instance is never solved twice. This latter condition normally ensures a *polynomial time* algorithm.

The most efficient way to achieve an overall control over the development of smaller instances is to create a table of all the smaller instances one might ever have to solve. Sometimes one can discard the solutions for some smaller instances as the computation proceeds and reuse the space for larger instances. The filling-in of the table of instances to obtain a solution for the original instance has been termed *dynamic programming.*

After the table has been filled in, only certain solutions from the table are used to build up a sequence of solutions leading to an optimal solution to the original instance. This sequence of solutions, known as an optimal solution sequence, is arrived at by employing the *property of optimality*, which requires that *for any intermediate state of the problem and the corresponding intermediate solution for this state, the solutions of the subsequent subproblems must constitute an optimal solution sequence with regard to this intermediate solution.*

A dynamic programming algorithm may vary in form from problem to problem, but the filling-in of a table and the order in which this is done remain a common theme in all applications of the approach.

9.4.3. Graph Searching

A fundamental problem concerning graphs is the path problem. In its simplest form it requires one to determine whether or not there exists a path in the given graph $G = (V, E)$ starting at vertex v and ending at vertex u. The TSP is one example of the path problem. A more general problem would be to determine for a given starting vertex $v \in V$ all vertices u such that there is a path from v to u. This problem can be solved by starting at vertex v and systematically searching the graph G for vertices that can be reached from v.

The search is an examination of the edges of a graph using the following basic procedure.

```
Mark all edges and vertices of the graph 'new'
while 'new' vertex do
        Choose a 'new' vertex and mark it 'old'
        while 'new' edges lead away from the 'old' vertex do
                if the other endpoint of the 'new' edge is a 'new' vertex
                then
                        Mark the edge 'old'
                        Mark the endpoint-vertex 'old'
                endif
        enddo
    enddo
```

If G is *not connected* the algorithm ensures a complete traversal of the graph. If G is *connected*, the search generates a *spanning tree*, i.e. a subset of edges of the graph that connects all nodes. The root of the spanning tree is the start vertex. The edges of the spanning tree are the edges which lead to new vertices

when visited. The properties of the spanning tree depend upon the criteria used
to select the starting vertex and the edges to explore.

For some simple graph problems, such as finding connected components,
any order of exploration is satisfactory. For more difficult graph problems,
e.g. TSP, the exploration order is crucial. There are two basic search methods.

In a *breadth-first* approach one starts at a vertex v. All unvisited vertices
adjacent to v are visited next, i.e. the edge selected in the innermost **while**-loop

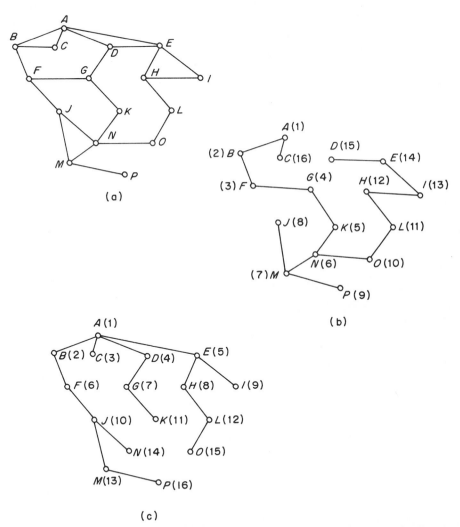

Figure 9.4.2. An example of a graph search (a) An example of a connected undirected
graph of 16 vertices (b) A depth-first search of graph (a) (c) A breadth-first search of
graph (a)

is an edge out of vertex with candidate edges. Such a search partitions the vertices into levels depending on their distance from the start vertex. At level 1 we have the start vertex v; at level 2 we have all vertices adjacent to v; at level 3, all vertices adjacent to the vertices of level 2, and so on.

In a *depth-first* search the exploration of a vertex is suspended as soon as a new vertex is reached. At this time the exploration of the new vertex u begins, i.e. the edge selected in the innermost **while**-loop is an edge out of the vertex u. When this new vertex has been explored, the exploration of v is continued. A depth-first search can be implemented as a recursive algorithm.

With respect to the two search methods the spanning trees are distinguished as the *breadth-first* and *depth-first spanning trees*. Figure 9.4.2 shows two such trees. Both breadth-first and depth-first searches, if efficiently implemented, require $O(n + e)$ time to explore an n-vertex, e-edge graph.

Other useful searches are the *topological search*, which labels the vertices of an n-vertex acyclic directed graph with integers from the set $1, \ldots, n$, such that the presence of the edge $\langle i, j \rangle$ in the graph implies that $i < j$, and the *lexicographic search*, which searches for the lexicographically maximum spanning tree.

DEFINITION 9.4.2.   Given two spanning trees which connect two vertices, v and u, of a weighted graph,

$$T_1 = \langle v_1, v_2 \rangle, \ldots, \langle v_{m-1}, v_m \rangle$$

and

$$T_2 = \langle u_1, u_2 \rangle, \ldots, \langle u_{m-1}, u_m \rangle,$$

where

$$v_1 = u_1 = v \quad \text{and} \quad v_m = u_m = u.$$

T_1 is *lexicographically greater* than T_2 if there is some k $1 < k < m$, such that

$$\text{weight}(\langle v_{i-1}, v_i \rangle) = \text{weight}(\langle u_{i-1}, u_i \rangle) \quad \text{for } 1 < i < k,$$

and

$$\text{weight}(\langle v_{k-1}, v_k \rangle) < \text{weight}(\langle u_{k-1}, u_k \rangle),$$

or else

$$\text{weight}(\langle v_{i-1}, v_i \rangle) = \text{weight}(\langle u_{i-1}, u_i \rangle) \quad \text{for } 1 < i < m.$$

A spanning tree connecting two vertices, v and u, which is not lexicographically less than any other tree connecting the same two vertices, is said to be *lexicographically maximum* among the trees connecting v and u.

Backtracking and Branch-and-bound

It is a fact that the exhaustive search algorithms are not adequate for solving a difficult combinatorial problem of a practically interesting size. And still the concept of enumeration is at the heart of many of to-day's algorithms. Such algorithms have some usefulness because they employ various means of limiting the amount of enumeration that has to be done. They differ from one another only in how they go about it. Dynamic programming is one technique of this type. Another approach based on the same idea is known as branch-and-bound (BB) technique. A typical BB algorithm contains three elements: partitioning into subproblems, selection of subproblems (branching) and bounding. Let us turn again to the TSP. For the TSP, separation is done by dividing a given set of tours into two subsets; in one subset, tours must pass over the link joining a certain pair of cities, and, in the other subset, no tour may use that link. As the enumeration proceeds, the subsets decrease in size until it finally becomes possible to solve the subproblems. If this occurs before their number becomes enormous, the problem will be solved.

The most important possibility for limiting the number of generated subsets is the discovery of a sufficiently high lower bound for the subsets under investigation. If it can be shown, by some means, that all subsets of the current set can yield no lower value for the objective function that some value already obtained elsewhere, then that set need no longer be pursued.

When suitable strategies are used for the crucial choice of which subsets to explore first, so that high bounds are obtained soon and large areas of the problem are ruled out of the search, the technique is very powerful.

The systematic search for the feasible solution set for an optimum solution to the problem is organized as a tree. Many tree organizations may be possible. Figure 9.4.1(c) shows an example of the TSP and its tree organization for the BB search.

A backtracking algorithm uses the same ideas as the BB—it partitions the problem into a set of subproblems by systematically generating the nodes of a search tree with the solutions of the subproblems. Boundary functions are then used to reduce, by as many as possible, the number of nodes examined for their status as a feasible solution to the problem. Eventually the problem solution is obtained. The difference between the two methods lies in the way in which the tree of the subproblem solutions is generated.

Both methods begin with the root node. The remaining nodes, however, are generated by applying the depth-first approach in backtracking and the breadth-first approach in BB. Boundary functions are used carefully enough so that, at the conclusion of the process, at least one answer node is always generated if the problem requires us to find all solutions.

9.5.   THEOREM PROVING BY MACHINE

Ever since the advent of computers, one of the strongest and proudest ambitions of scientists has been to endow the machine with genuine powers of reasoning. Considerable efforts are continually undertaken in this direction. In particular, attempts are made to enable the computer to carry out logical and mathematical reasoning by proving theorems of pure logic or of mathematics (Chang and Lee (1973); Bundy (1983); Ramsey (1988)).

We shall illustrate one problem from the area of theorem proving by machine. Results obtained for this problem illuminate some important points about complexity theory in general.

A few introductory definitions will be helpful. The simplest system which reflects something of mathematical reasoning is the propositional calculus, the formal language in which separate logical statements, which individually may be either true or false, are joined together by the lexical elements NOT, AND, OR and IMPLIES. A logical system that has greater expressive power than the propositional calculus is the first-order predicate calculus. Basic statements in this language are formed from symbols representing *individual elements* and *predicates(properties)*, and *functions* on them; and compound statements are formed with the quantifiers ∀('for all') and ∃('there exists').

There is a precise notion of a *proof* of a statement of the predicate calculus, such that a statement is provable if and only if it is valid. Using the first-order predicate calculus it is possible to formulate a great deal of mathematics.

9.5.1. The First-order Integer Addition Problem

The system $\mathcal{N} = \langle N + \rangle$ is given which consists of the natural numbers $N = 0, 1, \ldots$, and the operation '+' of addition.

The formal language $\mathcal{L}$ employed for discussing properties of the system, the so-called first-order predicate language, has

(i) variables $x, y, z, \ldots$ ranging over natural numbers;
(ii) the operation symbol +, equality =;
(iii) the logical connectives AND, OR and IMPLIES;
(iv) the quantifiers ∀ and ∃, and parentheses.

A sentence s is a formula in which every variable is bound by a quantifier. For example, a sentence such as

$$\exists x \forall y (x + y = y)$$

is a formal transcription of 'there exists a number x so that for all numbers y, $x + y = y$.' This sentence is true in $\mathcal{N}$ (there is the null element in $\mathcal{N}$).
Other examples are

$\forall x \forall y (x + y = y + x)$ is true, i.e. addition is commutative;
$\forall x \exists y (x = y + y)$ is false, i.e. not all numbers are even;
$\forall x \forall y [\exists a (x + a = y)$ AND $\exists a (y + a = x)$ IMPLIES $x = y]$ is true, i.e. if $x \le y$ and $y \le x$ then $x = y$.

The set of true sentences in $\mathcal{N}$ is denoted by PA (*Pressburger arithmetic*). In 1929 Pressburger showed that for the PA there exists an algorithm that, on input of any given sentence s of the language $\mathcal{L}$, gives output YES or NO to the question whether this particular sentence is in PA. The PA is thus a problem which can be solved; such problems are called *decidable*.

9.5.2. Decidable and Undecidable Problems

A problem for which an algorithm exists is called decidable. An *undecidable* problem is a problem that cannot be solved by any algorithm. The concepts of decidability and undecidability are fundamental in mathematical logic, where it is a common task to decide whether a certain problem (or statement or theory) is true or false. The undecidable problems establish a formal framework in logic for questions that cannot, in principle, be resolved by computational means as we understand them.

In 1936 Turing demonstrated that certain problems are so hard that they are *undecidable in the sense that no algorithm at all can be given to solve them.* Turing further proved that it is impossible to specify any algorithm which, given an arbitrary computer program and an arbitrary input to that program, can decide whether or not the program will eventually halt when applied to that input. This is the famous *halting theorem* (Cutland (1980)).

In the same year Church showed that provability, and hence validity, in the predicate calculus is undecidable (Church 1936). A variety of other problems have since been proved undecidable. One of the most significant results among these problems is due to Matijasevich. It concerns the following problem.

Let $P(x_1,\ldots,x_n)$ be a polynomial in the variables $x_1,\ldots,x_n$ with integer coefficients. Then the equation

$$P(x_1,\ldots,x_n) = 0$$

for which integer solutions are sought is called a *diophantine* equation. Diophantine equations do not always have solutions. For instance, no integer value of x would satisfy the equation $x^3 - 2 = 0$.

In 1900 Hilbert posed the problem, which since then has been known as Hilbert's tenth problem, of *whether there is an effective algorithm that will determine whether any given diophantine equation has a solution.*

Matijasevich has shown that there is no such procedure, that is, that Hilbert's tenth problem is undecidable.

9.5.3. Algorithms for Decidable Problems

Proving the decidability of a problem is the first step in solving the problem: PA has been proved decidable, the propositional calculus is decidable, and many other problems have been found to be decidable.

The next step is actually to construct an algorithm for a decidable problem, to obtain the solution. As it turns out, there are a number of important problems that are decidable as far as the principle of the matter is concerned, but every algorithm that may be constructed for their solution takes such a vast amount of computing time that the problem remains 'undecidable in practical terms', or 'uncomputable in practice'. This fact first became apparent in the 1960s and early 1970s. The notion of an *intractable* problem was born then (Garey and Johnson (1979); Lewis and Papadimitriou (1981)).

DEFINITION 9.5.1.   A problem that is so hard that no polynomial time algorithm can solve it is called an *intractable* problem. This, in general, means that an *intractable* problem can be either solvable by an exponential time algorithm or undecidable.

The intractable undecidable problems have a spirit of finality about them: once the problem is proved to be undecidable, that is the end of the matter. It is the intractable decidable problems that carry with them a real challenge in complexity theory.

The first examples of intractable decidable problems were obtained in the early 1960s, though these problems are 'artificial', that is, specially constructed to have the appropriate properties. Only in the early 1970s were the first 'natural intractable decidable' problems discovered. Today there is a wealth of such problems proved, notably, in automata theory, formal language theory and mathematical logic.

The PA problem is one such problem. Many attempts were made to devise an efficient algorithm to solve PA. When such programs were run on a computer, the computation terminated only on the simplest instances of the problem. Then, in 1974, Fisher and Rabin showed that the reason for this is the fact that the inherent lower bound on the complexity of algorithms for PA is a double exponential function of the input n, $2^{2^{cn}}$, where c is a constant (Fisher and Rabin (1974)).

Their result shows that the fact that PA is decidable is of little use in developing practical algorithms for this problem as the explosive growth of the double exponential function suggests that any algorithm for PA will use hopelessly huge amounts of time on relatively short sentences thus rendering the problem practically unsolvable.

9.6.   COMPUTATIONAL COMPLEXITY AND CLASS *NP*

We know that

  (i) a polynomial time algorithm is an algorithm whose time complexity is $O(p(n))$ for some polynomial function p in the size of the problem instance, n; for example the time function can be $T(n) = n^2 + n + 1$ that is $O(n^2)$;

 (ii) an algorithm whose time complexity cannot be bounded by a polynomial function is termed an exponential time algorithm;

(iii) a problem which is so hard that no polynomial time algorithm can solve it is termed an intractable problem; an intractable problem, for some problem instances, may turn out to be so difficult that it is unsolvable not in principle, but in practice.

Another large group of problems consists of problems for which no efficient algorithms are known, except for exponentially increasing time, yet no one has been able to prove that these problems do not have polynomial time solutions. Problems of this kind are termed *apparently intractable*.

One of the major tasks of complexity theory is to develop means that would help to establish which natural problems are intractable and which are solvable in polynomial time.

One approach used to this end is proving upper and lower bounds on the time and space complexity of specific problems. The bounds, for example, can be studied in terms of a worst-case performance measure of a given algorithm. And though, for some problems, a worst-case bound may be too pessimistic, close upper and lower bounds, if obtained, provide a realistic performance guarantee of the algorithm.

9.6.1. Non-deterministic Algorithms

Intractable and apparently intractable problems are much harder to understand in terms of their crucial properties, particularly in the usual environment of *deterministic* algorithms: the notion of such an algorithm presupposes the property that the result of every operation is uniquely defined. Deterministic algorithms agree with the way programs are executed on a computer.

Suppose now that the usual restriction on the outcome of every operation is removed. An algorithm will now be allowed to contain operations whose outcome is not uniquely defined but limited to a specified set of possibilities. The machine executing such operations is allowed to choose any one of these outcomes *subject to a certain termination condition*. As an example, let us assume the following termination condition: whenever there is a set of choices that leads to a successful completion of the solution, then one such set of choices is always made and the algorithm terminates successfully. This leads to the concept of a *non-deterministic* algorithm.

A machine capable of executing a non-deterministic algorithm is called a *non-deterministic machine*. The fundamental assumption about the non-deterministic machine is that it has the ability to select a 'correct' element from the set of allowed choices—if such an element exists—every time a choice is to be made. Whenever successful termination is possible, a non-deterministic machine makes a sequence of choices which is a shortest sequence leading to a successful termination.

For uniformity the theory formally restricts itself to *decision problems*, that is, problem whose question requires only a YES or NO answer. This does not restrict the applicability of the theory as many problems are naturally expressed as decision problems. For example, the *TS decision problem* on n 'cities' is formulated as follows: *Is there a tour over n cities of total length not greater than D?*

A non-deterministic algorithm may be interpreted as consisting of two stages, a *guessing* and a *checking* stage. The concept of a non-deterministic machine handles the guessing stage: it picks out at random an input data set. The algorithm then examines the input for a YES answer to the problem's question and outputs a YES/NO result accordingly. If the *checking stage* of the algorithm requires polynomial time for its completion, the algorithm is termed a *polynomial time non-deterministic* algorithm.

9.6.2.   The *NP*-space

Failing to find close explicit upper and lower bounds on the complexity of the given problem, one may seek a proof that the complexity of the problem is related to that of some other problem; in other words, one attempts to classify the problem as being 'complete' in some larger class of problems. Such a result relates the complexity of the particular problem to that of the larger class as a whole.

One important class of problems, known as the *NP-space problems*, groups together the problems that are *decidable in principle*, that is, problems that

   (i) are *solvable* in principle, by computing means as we know them;
  (ii) *can be formulated* as decision problems;
 (iii) are *soluble in polynomial time* by a non-deterministic machine.

The class *NP* includes an enormous number of practical problems that occur in business and industry.

The term '*NP*-space problem' can be interpreted as 'non-deterministic polynomial time' problem, and, in turn, the class of problems solvable by a deterministic algorithm in polynomial time has been given the name '*P-space*'.

9.6.3.   Relationship between *P* and *NP* Spaces

The relation between the classes *P* and *NP* is fundamental.

Every decision problem solvable by a polynomial time deterministic algorithm is also solvable by a polynomial time non-deterministic algorithm, that is $P \subseteq NP$. To see this, one simply needs to observe that any deterministic algorithm can be used as the checking stage of a non-deterministic algorithm. If $R \in P$, and AL is any polynomial time deterministic algorithm for R, we can obtain a polynomial time non-deterministic algorithm for R merely by using AL as the checking stage and ignoring the guess. Thus $R \in P$ implies $R \in NP$.

However, what we would like to know is *whether or not $P = NP$*. Researchers have for years been attempting without success to find polynomial time algorithms for certain problems in NP, such as, for example, TSP. This lack of success gave rise to a conjecture that $P \neq NP$. Yet, no proof of this conjecture has been found either. This problem is considered to be the most important open problem in computer science and one of the most important open problems in mathematics to-day.

There are currently two major lines of research into this dilemma.

(i) Proving that the problems of the *NP*-class are intractable, i.e. that the problems are so difficult that no polynomial complexity algorithms can possibly solve them. Unfortunately, at present, proving $P \neq NP$ seems just as hard as proving $P = NP$.

(ii) Examining the relationship between various *NP* problems so as to gain further insight into inherent properties of the problems within the class. For example, within the class *NP* a further subclass of problems is identified as *NP-complete* problems. Complete problems have the property that all problems in the *NP* class, including other *NP*-complete problems, are *polynomially reducible* to any one of them, in the following sense.

9.6.4. *NP*-complete Problems

DEFINITION 9.6.1. In an *NP*-space problem *R1* is called *polynomially transformable* or *polynomially reducible* to problem *R2* if there is a polynomial time deterministic algorithm which transforms or 'reduces' *R1* to *R2*. This is denoted '$R1 \propto R2$'. Two problems *R1* and *R2* are called *polynomially equivalent* if $R1 \propto R2$ and $R2 \propto R1$. Problem *R* is called *NP*-complete if $R \in NP$ and every problem in *NP* is polynomially reducible to *R*.

A proof that an *NP* problem is *NP*-complete is a proof that the problem is not in *P* (does not have a deterministic polynomial time algorithm) unless every *NP* problem is in *P*. In other words *all NP-complete problems are of equivalent difficulty: if one NP-complete problem can be solved in polynomial time by a deterministic algorithm then all other problems in NP can be solved in polynomial time; if one NP-complete problem can be shown to require exponential time then all the other NP-complete problems in NP require exponential time.*

The notion of *NP*-complete problem was introduced in 1970 by Cook who established the first-ever *NP*-complete problem, the so-called *satisfiability problem*. He proved thereby that *if the satisfiability problem can be solved in polynomial time by a deterministic algorithm then P = NP* (Cook (1971, 1973)).

9.6.5. Open Problems

By definition, a polynomial time non-deterministic algorithm checks in polynomial time a proposed string of symbols for a YES solution. It may, however, not be able as 'quickly' to check for a NO answer. The class *NP* is defined by the existence of a polynomial time non-deterministic algorithm for checking for a YES solution only.

A problem for which a non-deterministic algorithm can be constructed such that it will check for a NO-solution in polynomial time, is said to belong to the class co*NP*. The class co*NP* consist of all problems that are the complement of some problems in *NP*. Intuitively, the problems in *NP* are of the form 'determine whether a solution exists', whereas the complementary problems in co*NP* are of the form 'show that there are no solutions'. It is not known whether *NP* = co*NP*, but there are problems that fall in the intersection *NP* ∩ co*NP*. An example of such a problem is *the composite numbers problem*: given an integer *n*, determine whether *n* is composite, i.e. whether there exist factors *p* and *q* such that *n* = *pq*. *Note that the problem of finding factors may be harder than showing their existence.*

The problems that belong to both *NP* and co*NP* classes are termed *open*. Recently three important open problems have been shown to be in the class *P*. These are the linear programming problem, the problem of determining whether two graphs of degree at most *d* are isomorphic, and the problem of factoring polynomials in one variable.

9.6.6.   Beyond the *NP*-class

There are many important problems which do not seem to be solvable in polynomial time by a non-deterministic machine. For example, the problem of proving that a given Boolean expression is always 1 (or 0) does not seem to be possible to solve in polynomial time with a non-deterministic algorithm. Another problem in this category is the so-called *placement problem*, i.e. placing a set of

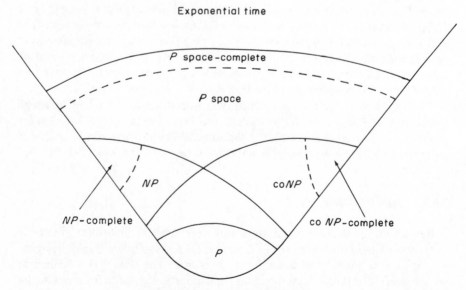

Figure 9.6.1. Complexity classes

circuits such that the total length of wires is less than some specified value L, or proving that there exists no placement with a wire length less than L. This problem does not seem to be possible to solve in a non-deterministically polynomial time.

A further dimension is added to computational complexity theory if the emphasis is centred on another major resource required by a computation. The amount of computer memory required by the computation is often just as important as the time it takes to complete the computation. The class of all problems solvable in polynomial space is termed the P-space. All problems solvable non-deterministically in polynomial time can be solved in polynomial space, so the P-space includes NP and coNP.

However, the question of whether there exist problems solvable in polynomial space that cannot be solved in polynomial time remains unresolved. It is conjectured that P-space may contain problems that are thought by some to be harder than the problems in NP and coNP.

An even more powerful conjecture suggests that there exists a different set of problems complete in P-space. The P-space-complete problems are the problems in P-space such that if any one of them is in NP, then P-space $= NP$, or if any one of the P-space-complete problems is in P, then P-space $= P$.

Figure 9.6.1 shows some important complexity classes and their possible relationships.

L. K.

REFERENCES

Aho, A. V., Hopcroft, J. E. and Ullman, J. D. (1983). *Data Structures and Algorithms,* Addison-Wesley.

Bundy, A. (1983). *The Computer Modelling of Mathematical Reasoning,* Academic Press.

Chang Ch-L. and Lee R. Ch-T. (1973). *Symbolic Logic and Mechanical Theorem Proving,* Academic Press.

Church, A. (1936). An Unsolvable Problem of Elementary Number Theory, *Am. J. Math.* **58**, 345–363.

Cook, S. A. (1971). The Complexity of Theorem-proving Procedures, *Proc. 3rd Annual ACM Symp. Theory of Computing,* Shaker Hts., Ohio, 151–158.

Cook, S. A. (1973). A Hierarchy for Non-deterministic Time Complexity, *J. Comput. System Sci.* **7**, 343–353.

Cooley, J. W. and Tukey, J. W. (1965). An Algorithm for the Machine Calculation of Complex Fourier Series, *Math. Comput.* **19**(90), 297–301.

Cooley, J. W., Lewis, P. A. W. and Welch, P. D. (1977). The Fast Fourier Transform and its Application to Time Series Analysis, in *Statistical Methods for Digital Computers,* Eds. K. Enslien, A. Ralston and H. S. Wilf, John Wiley.

Cutland, N. L. (1980). *Computability. An Introduction to Recursive Function Theory,* Cambridge University Press.

Fisher, M. J. and Rabin, M. O. (1974). Super-exponential Complexity of Pressburger Arithmetic, in *Complexity of Computation,* SIAM-AMS Proc. 7, Ed. R. Karp, 27–42.

Garey, M. R. and Johnson, D. S. (1979). *Computers and Intractability. A Guide to the Theory of NP-completeness,* W. H. Freeman and Co.

Karatsuba, A. and Ofman, Yu. (1963). Multiplication of Multidigit Numbers on Automata, *Soviet Physics Dokl.* **7**, 595–596.

Kronsjo, L. (1985). *Computational Complexity of Sequential and Parallel Algorithms*, John Wiley.

Lawler, E. L., Lenstra, J. K., Rinnooy Kan, A. H. C. and Shmoys, D. B. (Eds.) (1985). *The Travelling Salesman Problem: A Guided Tour of Combinatorial Optimization*, John Wiley.

Lewis, H. R. and Papadimitriou, C. H. (1981). *Elements of the Theory of Computation*, Prentice-Hall.

Pan, V. (1978). Strassen Algorithm is not Optimal, Trilinear technique of Aggregating, Uniting and Cancelling for Constructing Fast Algorithms for Matrix Multiplication, *Proc. 19th Annual Symposium on the Foundations of Computer Science*, Ann Arbor, MI, 166–176.

Pan, V. (1984) How can we Speed up Matrix Multiplication? *SIAM Rev.* **26**(3), 393–415.

Ramsey, A. (1988). *Formal Methods in Artificial Intelligence*, CUP.

Strassen, V. (1969). Gaussian Elimination is not Optimal, *Num. Math.* **13**, 354–356.

Winograd, S. (1980). *Arithmetic Complexity of Computations*, CBMS-NSF Regional Conference Series in Applied Mathematics, Society for Industrial and Applied Mathematics, Philadelphia, Pennsylvania, 19103.

CHAPTER 10

Non-cooperative Finite Games

10.1. INTRODUCTION

Von Neumann's original article on 'The Theory of Parlour Games' was published before the Second World War (von Neumann (1928)) but the theory of games lacked sufficient credibility to be used for strategic defence considerations during the war. Since then the subject has found more constructive applications in a variety of disciplines, notably economics and biology. Following the war most of our knowledge was represented in von Neumann and Morgenstern's book, 'The Theory of Games and Economic Behaviour' (1953). Then three seminal papers by John Nash published in the early 1950's aroused great academic interest in the subject. The 'Nash program' stressed the need to provide behavioural descriptions based on non-cooperative games, but very little was actually achieved in this area until the present decade. Most research in the 1950s and 1960s was in the fields of co-operative games and infinite games, and for an overview of this the reader should consult the four volumes 'Contributions to the Theory of Games' (Kuhn and Tucker (1953)) and the sequel 'Advances in Game Theory' (Dresher, Shapley and Tucker (1964)). During the 1970s significant advances were made in the fields of incomplete information, repeated games and evolutionary games—these are discussed in §§10.3–10.6. In the early part of this decade Ariel Rubinstein introduced a non-cooperative bargaining game (Rubinstein (1982)) that was to provide the basis for the completion of the Nash program by Binmore, Rubinstein and Wolinsky in 1986—an overview of non-cooperative bargaining theory is given in §10.7. The other main area of recent research, described in §10.8, started with the theory of equilibrium selection in non-cooperative games having a multiplicity of Nash equilibrium points. The most successful selection criterion—in that it always identifies a unique solution in finite games of perfect recall—is the 'tracing procedure' of Harsanyi and Selten (1988). However, there is much to be said for Aumann's correlated equilibria (Aumann (1974, 1987)). There are indeed many different methods of selecting a solution in such games and many game theorists are now dissatisfied with this whole approach. Instead of tailoring the solution concept to suit the game, an alternative is to augment the non-cooperative model of the game to incorporate information about the 'intelligence' of one's opponent, i.e. to model players as finite automata. The

293

chapter ends with a taste of some of these most recent advances in the theory of games.

In §10.2 we introduce the basic tools used in analysing non-cooperative games, using as little abstraction as possible. For this we must first introduce the reader to some fundamental concepts such as the distinctions between finite and infinite games, and non-cooperative and cooperative games. So let us now give an informal definition of a *game*. Loosely speaking this is a situation in which one or more *players* have to choose between various *actions* which are not necessarily the same for all players. When these choices have been executed this determines an *outcome*, and then each player receives a *pay-off* depending on the outcome. This description of a game encompasses many economic, social and biological situations as well as standard games such as poker and chess. The players may be people, animals or economic agents and one of the players may be 'chance'. A *finite game* is one with a finite set of players, each having a finite set of possible actions (the reader should note the distinction between 'actions' and 'strategies', drawn in §10.2.1).

The definition of a *pay-off* depends on the nature of the players. If they are animals, then the pay-off is measured in terms of the increase in *Darwinian fitness* accruing to the animals as a result of the actions taken [see §10.6]. But in most of this chapter we assume that players are social or economic agents, and their pay-offs are measured by increments in the utility representation of their preferences. So each player is endowed with a preference relation '$\succsim$' over the set of possible outcomes Φ, and assuming this is total, reflexive, transitive, continuous and monotonic [see I, §1.3 and IV, §§2.1 and 2.7] then there exists a continuous function $u:\Phi \to \mathbb{R}_+$ such that $u(x) \geq u(y)$ if and only if $x \succsim y$. This function is called the *utility function* which represents $\succsim$. In general they have no cardinal value but under certain conditions on $\succsim$ (concerning preferences over *lotteries* over Φ—see Luce and Raiffa (1957), §2.5) it can be represented by a *Von Neumann and Morgenstern utility function* (von Neumann and Morgenstern (1953)) which is unique up to affine transformations [see I, 5.13]. So if one sets the zero and scale, one *can* associate a cardinal value with the values of a von Neumann and Morgenstern utility function. However, this still does not allow interpersonal comparison of utility values (see Luce and Raiffa (1957), §2.7). Representation of preferences by such von Neumann and Morgenstern utility functions is only really applicable when there is some uncertainty in the environment—hence the need for $\succsim$ to satisfy conditions over lotteries on Φ. But if the friction in the model is due to time rather than uncertainty we can still associate pay-offs with utility values, of their *time preference* representation. More details are given in §10.7.2.

We can now draw the distinction between a cooperative and a non-cooperative game. In a *cooperative game* we have no need to analyse players strategies: all we require is knowledge of the *pay-off region X*, i.e. the subset of $\mathbb{R}^n_+$ consisting of pay-off vectors associated with different outcomes in Φ. An example is given in Figure 10.1.1. We assume that players communicate before the game for as long as is necessary for them to reach agreement on an outcome,

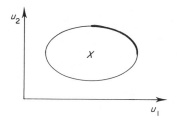

Figure 10.1.1

which will then be enforced by an independent arbitrator. The particular outcome that is agreed is the *solution* to the cooperative game (or 'solution set' if there is more than one possible outcome), and we identify this by assuming that it satisfies various *axioms*, depending on the solution concept employed. One common axiom is that of *Pareto optimality*, i.e. no player can increase his pay-off without decreasing some other player's pay-off. In Figure 10.1.1 the Pareto optimal set is the bold part of the boundary. For examples of some axiomatic schemes which define various solution concepts for cooperative games, see §10.7.1 and I, 13.3.

In a *non-cooperative game* the 'solution' is not an outcome but a set of strategies, one for each player (i.e. a particular *stragety profile*) and we assume that players are 'rational' in that they choose their strategy in order to maximize their own pay-off. No pre-play communication is allowed and there is no need for an independent arbitrator. The players are in a *contest* situation if they have opposing interests (i.e. prefer different outcomes in Φ), but this does not preclude the possibility of cooperation! Experimental studies of human and animal behaviour have provided much empirical evidence of cooperative behaviour in contest situations—see Axelrod (1984) for a review of literature in this area. So the theory of non-cooperative games provides the tools for a strategic analysis of cooperative behaviour, whilst cooperative game theory is the axiomatic approach to cooperation. More details and an account of the important results of the strategic approach to cooperation are given in §10.4 and §10.6.3.

10.2. ELEMENTARY THEORY OF NON-COOPERATIVE FINITE GAMES

10.2.1. Finite Games in Extensive Form

In I, 13.5 there is a brief introduction to extensive games. Here we describe, mainly by examples, the formal language of finite games in extensive form (also called *game trees*) and analyse the optimal behaviour of the players.

EXAMPLE 10.2.1. In the simple game of Figure 10.2.1, player *I* has pure strategies *t* and *b* and player *II* has pure strategies *l* and *r* [see I, 13.2.1]. Both

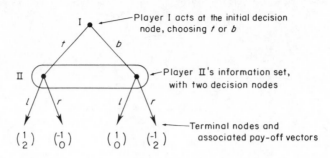

Figure 10.2.1

players move simultaneously, which is captured in the game tree by placing player *II*'s decision nodes in an *information set*. This means that she does not know for certain which node the play is at, although she may have *beliefs* (a probability distribution over the set of nodes in an information set) as in Example 10.2.2. Clearly this game could be drawn with player *II* at the initial node and player *I* governing moves *t* and *b* at the subsequent information set, because it is a static game (though not in 'static form', a term used by some authors for the normal form). In this game the set of *actions* at each decision node *are* the pure strategies of one player, but in games where players have more than one move at different points in time this is not the case [see Example 10.2.2]. Notice that the set of actions at each decision node in an information set must be the same, for otherwise the player might know which decision node has been reached. Associated with every possible play of the game is a pay-off vector, being the von Neumann and Morgenstern utility values or Darwinian fitnesses associated with that outcome: for example, if *I* plays *b* and *II* plays *r* then their pay-offs are -1 and 2 respectively.

The extensive form is particularly useful in dynamic games, providing a natural representation of the sequential nature of players actions.

EXAMPLE 10.2.2.   The game of Figure 10.2.2 begins with a chance move '*o*' and has only one player, who believes that chance chooses *a* with probability $\frac{1}{4}$ and *b* with probability $\frac{3}{4}$. His actions at information set h_1 are *l* and *r* and at information set h_2 are *L* and *R*, so he has four pure strategies *lL, lR, rL* and *rR*. Notice that a strategy requires specification of his move at h_2 even though that information set will never be reached if he plays *r* at h_1. It is the determination of strategies at information sets that are never reached under optimal play that makes games like this so difficult to analyse [see §10.3].

DEFINITION 10.2.1.   A *behavioural strategy* for player *i* is a set of probability distributions, one for every information set belonging to him, over the set of actions available at the information set.

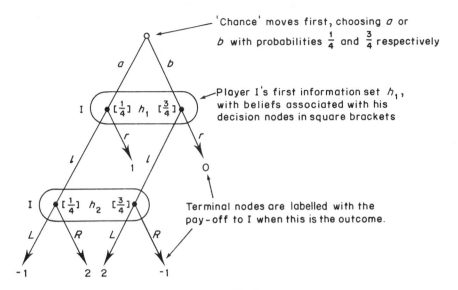

Figure 10.2.2

This concept should be distinguished from that of a *mixed strategy*, which is a probability distribution over the set of pure strategies for player i [see I, 13.2.2]. A *properly mixed strategy* is one which assigns strictly positive probability to each pure strategy.

DEFINITION 10.2.2. Two strategy profiles $\sigma = \{\sigma_i\}$ and $\tau = \{\tau_i\}$, where σ_i and τ_i are (pure, mixed or behavioural) strategies for player i, are *equivalent* if all nodes are reached with the same probabilities under both σ and τ.

In games of *perfect recall* (where no player forgets any past moves) behavioural and mixed strategies are equivalent, a general result due to Kuhn and Tucker (1953), p. 213: (see also Robert Aumann's paper in Dresher, Shepley and Tucker (1964) for the generalization of the theorem to infinite games).

THEOREM 10.2.1. *In finite games of perfect recall, to every mixed strategy profile there exists an equivalent behavioural strategy profile, and vice versa.*

EXAMPLE 10.2.3. In Figure 10.2.3 I has pure strategies lL, lR, rL and rR; II has pure strategies aA, aB, bA and bB. The behavioural strategies indicated are equivalent to the mixed strategy profile $\{(\frac{1}{4}, \frac{1}{2}, \frac{1}{4}, 0), (\frac{1}{2}, 0, \frac{1}{2}, 0)\}$.

Now consider the solution of games in extensive form. The notion of *dominance* was introduced in I, §13.2.5, and there applied to games in normal

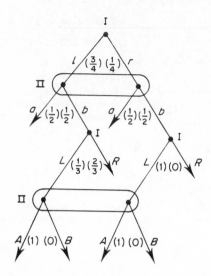

Figure 10.2.3

form. In extensive form games, *Zermelo's algorithm* consists of deleting all
dominated actions at penultimate decision nodes and thence proceeding by
backwards induction [see IV, §16.2.1].

EXAMPLE 10.2.4.   Application of Zermelo's algorithm to the game in
Figure 10.2.4 produces the sequence of reductions shown in Figure 10.2.5. A
faster algorithm which sometimes also leads to a greater reduction, due to Kuhn
(1950), is to construct a sequence of games by deleting all dominated actions
together with the subgames they determine. Applying this to the game in
Figure 10.2.4 results in the same solution in just two steps (Figure 10.2.6). In
general neither Zermelo's nor Kuhn's algorithm reduce the game to a unique

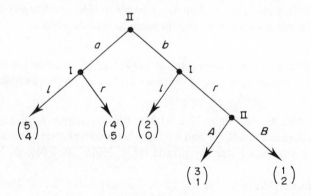

Figure 10.2.4

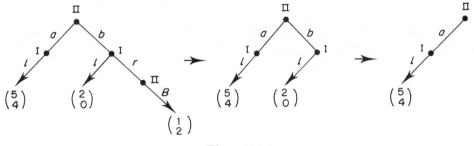

Figure 10.2.5

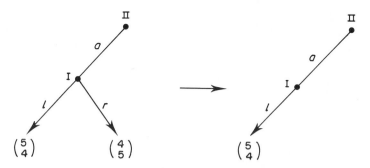

Figure 10.2.6

play. Indeed the problem of multiplicity of such 'solutions' casts much doubt on our standard methods of analysis of non-zero sum games [see §10.8].

10.2.2. Normal Form of Two-person Finite Games

The normal form of the game in Example 10.2.1 is the bi-matrix representation in Figure 10.2.7.

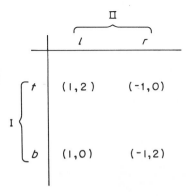

Figure 10.2.7

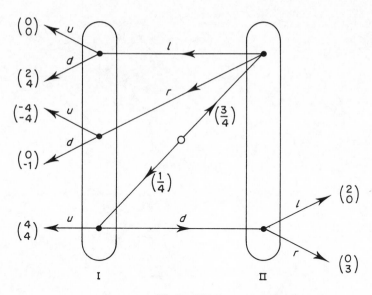

Figure 10.2.8

EXAMPLE 10.2.5.   In the game of Figure 10.2.8 chance moves first with the probabilities shown. In such games of chance the normal form contains the *expected* pay-off matrices [see II, §8.1], and the expected pay-off vectors in this game are

$$E(u, l) = \tfrac{3}{4}(0, 0) + \tfrac{1}{4}(4, 4) = (1, 1),$$

and similarly

$$E(u, r) = (-2, -2), \qquad E(d, l) = (2, 3), \qquad E(d, r) = (0, 0)$$

as shown in Figure 10.2.9.

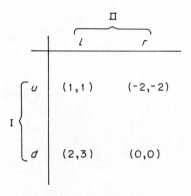

Figure 10.2.9

For a general finite two-person game in normal form we specify two $n \times m$ pay-off matrices $\mathbf{A}$ and $\mathbf{B}$ [see I, §6.2] when player I has n pure strategies and player II has m pure strategies, the elements a_{ij} and b_{ij} being the pay-offs to players I and II respectively, when I employs his ith pure strategy and II uses his jth pure strategy. If the game involves chance moves, then $\mathbf{A}$ and $\mathbf{B}$ will be expected pay-off matrices. The game is *zero-sum* if $\mathbf{A} + \mathbf{B} = \mathbf{0}$ or more generally if $\mathbf{A} + \mathbf{B} = \mathbf{C}$ is a matrix of constants (i.e. every element is the same), since adding a constant to *all* pay-offs to a player does not alter either player's optimal behaviour. Zero-sum games have been discussed in I, §13.2 and here, unless otherwise stated, we assume the game to be non-zero-sum.

Let $\mathbf{p}$ and $\mathbf{q}$ be mixed strategies for player I and player II respectively, so $\mathbf{p}$ is an $n \times 1$ probability vector and $\mathbf{q}$ is an $m \times 1$ probability vector, i.e. $\mathbf{p} = (p_1, p_2, \ldots, p_n)^{\mathrm{T}}$ and $p_1 + p_2 + \ldots + p_n = 1$; similarly $\mathbf{q} = (q_1, q_2, \ldots, q_m)^{\mathrm{T}}$ and $q_1 + q_2 + \ldots + q_m = 1$. Then the *expected pay-offs* to I and II respectively are

$$E_1(\mathbf{p}, \mathbf{q}) = \mathbf{p}^{\mathrm{T}} \mathbf{A} \mathbf{q} \quad \text{and} \quad E_2(\mathbf{p}, \mathbf{q}) = \mathbf{p}^{\mathrm{T}} \mathbf{B} \mathbf{q}.$$

The set (E_1, E_2) in $\mathbb{R}^2$, as $\mathbf{p}$ as $\mathbf{q}$ range over all possible mixed strategies, is called the *pay-off region* of the game.

EXAMPLE 10.2.6.   Each player in the game of Example 10.2.1 has two pure strategies, and so mixed strategies are $\mathbf{p} = (p, 1 - p)^{\mathrm{T}}$ and $\mathbf{q} = (q, 1 - q)^{\mathrm{T}}$ where $0 \le p, q \le 1$. The expected pay-offs are

$$E_1(p, q) = 2q - 1 \quad \text{and} \quad E_2(p, q) = 2(1 - p) + 2q(2p - 1),$$

and the pay-off region is as shown in Figure 10.2.10.

DEFINITION 10.2.3.   A strategy profile $(\hat{\mathbf{p}}, \hat{\mathbf{q}})$ is called a *Nash equilibrium* if it maximizes a players expected pay-off given that the other player is employing their own Nash equilibrium stategy. That is, for all $(\mathbf{p}, \mathbf{q})$

$$\hat{\mathbf{p}}^{\mathrm{T}} \mathbf{A} \hat{\mathbf{q}} \ge \mathbf{p}^{\mathrm{T}} \mathbf{A} \hat{\mathbf{q}} \quad \text{and} \quad \hat{\mathbf{p}}^{\mathrm{T}} \mathbf{B} \hat{\mathbf{q}} \ge \hat{\mathbf{p}}^{\mathrm{T}} \mathbf{B} \mathbf{q}.$$

The obvious generalization to more-person games in normal form holds (see Nash (1951)).

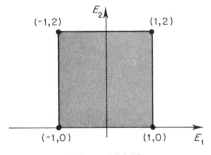

Figure 10.2.10

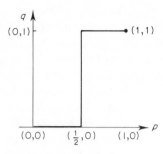

Figure 10.2.11

EXAMPLE 10.2.7.   Pure Nash equilibria are easy to identify in two-person games. For example, the game in Example 10.2.1 has two pure Nash equilibrium (t, l) and (b, r). For, suppose that t is a Nash equilibrium strategy for player I, then the corresponding Nash equilibrium strategy for player II would be l (she gets 2 instead of 0). Now suppose that l is a Nash equilibrium strategy for II, then I would play any mixture of t and b, including the pure strategy t. This type of circular argument also identifies (b, r) as a pure Nash equilibrium, as the reader can easily verify.

To find any mixed Nash equilibria of this game, consider the expected pay-offs in Example 10.2.6: since E_1 is independent of p *any* mixed strategy for I could be part of a Nash equilibrium, but suppose that $p = \hat{p}$ and $q = \hat{q}$ defines a mixed Nash equilibrium, so that $\hat{q}$ must maximize $E_2(\hat{p}, q)$. If $\hat{p} < \frac{1}{2}$ then $\hat{q} = 0$ and if $\hat{p} > \frac{1}{2}$ player II should play $\hat{q} = 1$. Finally if $p = \frac{1}{2}$ player II gets 0 whatever she plays and so $\hat{q}$ can be anything in $[0, 1]$. The full set of Nash equilibria are conveniently expressed in Figure 10.2.11.

THEOREM 10.2.2.   *Every finite non-cooperative game has at least one Nash equilibrium in mixed strategies.*

This is *Nash's theorem* (Nash (1951)): assuming mixed strategies are allowed, it tells us that existence is not a problem. A result of Harsanyi (1973b) states that when a game has a finite number of Nash equilibria this number is usually odd [as in Example 10.2.8]. However, there may be infinitely many Nash equilibria even in very simple games, as we have seen in Example 10.2.7. How then can we identify a unique solution of the game? Do we *need* to identify a unique solution? Consider the normal form game in Figure 10.2.7. If I thinks that II will play l as her Nash equilibrium strategy then I will play his corresponding Nash equilibrium strategy, t. But if I guessed wrong and II intends to play r, the play will be (t, r) which is not a Nash equilibrium. So in this game the multiplicity of Nash equilibria is a problem because players may have different Nash equilibria in mind when they play.

The problem of identifying a unique solution in non-zero-sum finite games has been the subject of much recent research and is returned to in §10.8.2 of this exposition. For the moment we only mention one case when the multiplicity of Nash equilibria is not a problem and that is when all Nash equilibria are *equivalent* (i.e. all Nash equilibrium pay-offs are the same) and *interchangeable* (i.e. if $(\hat{\mathbf{p}}_1, \hat{\mathbf{q}}_1)$ and $(\hat{\mathbf{p}}_2, \hat{\mathbf{q}}_2)$ are any two Nash equilibria then so are $(\hat{\mathbf{p}}_1, \hat{\mathbf{q}}_2)$ and $(\hat{\mathbf{p}}_2, \hat{\mathbf{q}}_1)$). Clearly in this case the play will still be optimal whatever Nash equilibrium players have in mind.

Two-person zero-sum games are the only class of non-cooperative game where all Nash equilibria are interchangeable and equilivalent (see Owen (1982)). Moreover in two-person zero-sum games the set of Nash equilibria coincides with the set of maximin strategies (called 'solutions' in I, 13.2) so, whatever solution players have in mind, the pay-offs are always the same. It is in this sense that two-person zero-sum games can be said to be *solvable*.

A 'solution' of a non-zero-sum game is much more difficult to identify except in certain very special cases. For not only may there be many Nash equilibria, which are not interchangeable and equivalent in general, but the Nash concept is not the only definition of optimal behaviour: the idea of *maximin strategies* can be extended to the non-zero-sum case, being that strategy which maximizes a players minimum pay-off and these are not usually the same as Nash equilibria. The maximum of minimum pay-offs is called player i's *value*—the pay-off he can guarantee against any strategy choice of his opponent provided he himself plays his maximin strategy. In general the sum of all players values is not zero except in zero-sum games.

EXAMPLE 10.2.8.   Consider the problem of a couple that have different interests: he wants to play football (F) and she wants to play cards (C). The normal form of this game is shown in Figure 10.2.12. This is called the *economic battle of the sexes* (not to be confused with the biological battle of the sexes in §10.6). The circular argument used in Example 10.2.7 here identifies two pure Nash equilibria: $\{C, C\}$ and $\{F, F\}$. Let $\mathbf{p} = (p, 1 - p)^{\mathrm{T}}$ and $\mathbf{q} = (q, 1 - q)^{\mathrm{T}}$ be

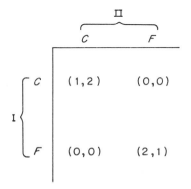

Figure 10.2.12

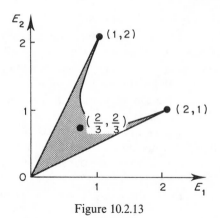

Figure 10.2.13

mixed strategies for I and II respectively. Then the pay-off functions are

$$E_1(p,q) = p(3q - 2) + 2(1 - q),$$
$$E_2(p,q) = q(3p - 1) + (1 - p).$$

Consider a Nash equilibrium where $p = \hat{p}$ and $q = \hat{q}$. If $3\hat{q} - 2 > 0$ then $\hat{p}$ must be 1 (to maximize E_1) and so $\hat{q} = 1$ (to maximize E_2), which gives the Nash equilibrium $\{C, C\}$. Similarly $3q - 2 < 0$ implies $\hat{p} = 0$ which implies $\hat{q}$ must be 0 and the Nash equilibrium $\{F, F\}$ is obtained. Clearly the only properly mixed Nash equilibrium is when $3\hat{q} - 2 = 0$ and $3\hat{p} - 1 = 0$, i.e. $\{(\frac{1}{3}, \frac{2}{3})^T$ and $(\frac{2}{3}, \frac{1}{3})^T\}$, and the mixed Nash equilibrium pay-off is $\frac{2}{3}$ to each player. The pay-off region is shown in Figure 10.2.13 with the three Nash equilibria pay-offs indicated. Note that the (unique) maximin strategy pair is $\{(\frac{2}{3}, \frac{1}{3})^T$ and $(\frac{1}{3}, \frac{2}{3})^T\}$, which may be found by the methods described in I, §13.2, and that this guarantees each player $\frac{2}{3}$.

Although each player's value is the same as their pay-offs when they play the mixed Nash equilibrium, the maximin strategy is not a Nash equilibrium. Also no Nash equilibria are interchangeable or equivalent, so it is impossible to identify a solution for this game using standard analytical tools (Figure 10.2.13).

Often we would want, at least, our 'solution' to be a Nash equilibrium, and so further constraints on the equilibrium definition are required if we are to reduce the number of possible solutions. Most of the following sections address this problem, §10.8 in particular. But before we think about refining the equilibrium concept we mention some useful tools for finding Nash equilibria in games larger than 2×2 (as most of our examples have been). The first is

THEOREM 10.2.3.   *If* $\hat{\mathbf{p}}, \hat{\mathbf{q}}$ *is a mixed strategy Nash equilibrium pair with* $\sigma_1, \sigma_2, \ldots, \sigma_k$ *being the pure strategies to which* $\hat{\mathbf{p}}$ *assigns positive probability, and* $\tau_1, \tau_2, \ldots, \tau_l$ *being the pure strategies to which* $\hat{\mathbf{q}}$ *assigns positive probability, then*

$$E_1(\sigma_1, \hat{\mathbf{q}}) = E_1(\sigma_2, \hat{\mathbf{q}}) = \ldots = E_1(\sigma_k, \hat{\mathbf{q}}) = E_1(\hat{\mathbf{p}}, \hat{\mathbf{q}})$$

and

$$E_2(\hat{\mathbf{p}}, \tau_1) = E_2(\hat{\mathbf{p}}, \tau_2) = \ldots = E_2(\hat{\mathbf{p}}, \tau_l) = E_2(\hat{\mathbf{p}}, \hat{\mathbf{q}})$$

This is intuitively obvious, for if $E_1(\sigma_1, \hat{\mathbf{q}}) > E_1(\sigma_2, \hat{\mathbf{q}})$ say, then player I can do better against $\hat{\mathbf{q}}$ by playing a mixed strategy other than $\hat{\mathbf{p}}$, which assigns zero probability to σ_2 and greater probability to σ_1.

Secondly, we recall the concept of *domination* [see I, §13.2.5] and notice that, by definition, no Nash equilibrium will assign positive probability to a strictly dominated strategy.

EXAMPLE 10.2.9.   In the game of Example 10.2.5, d strictly dominates u for I and l strictly dominates r for II. Hence $\{d, l\}$ is the unique Nash equilibrium of this game.

10.2.3.   Subgame-perfect Equilibria

We end this section with the first refinement of Nash equilibria, due to Selten (1965, 1973), who calls them *subgame perfect equilibria* to distinguish them from

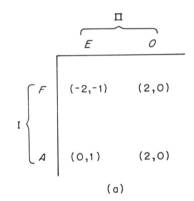

(a)

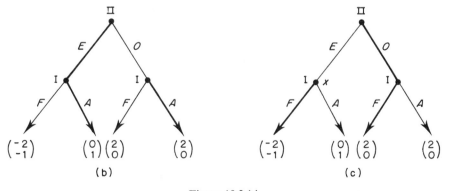

Figure 10.2.14

his 'trembling hand perfect equilibria' for games with incomplete information, discussed in the next section.

EXAMPLE 10.2.10.   The game in Figure 10.2.14 illustrates the *chain-store paradox* (Selten (1978)), so called because player *I* may be thought of as a monopolist and player *II* as a potential entrant to the market. *I* has two pure strategies, *F*, to fight the potential entrant with economic barriers to entry (such as predatory pricing etc.), and *A*, to aquiesce and allow entry into the market. Player *II* also has two pure strategies, *E*, to attempt to enter the market, and *O*, to stay out. From the normal form it is clear that there are two pure Nash equilibria $\{A, E\}$ and $\{F, O\}$ and these are marked on the extensive form in Figure 10.2.14(b) and (c). (There are also mixed Nash equilibria where *II* plays *O* and *I* plays *F* with probability between $\frac{1}{2}$ and 1, but these are not necessary for explaining the subgame perfect equilibrium concept.) Now the monopolist can never lose by aquiescing, but it only costs nothing for the monopolist to fight *provided entry does not occur*. So $\{F, O\}$ is not a plausible equilibrium—it is Nash but not subgame perfect—and $\{A, E\}$ is the 'rational' expected play. Using the definition below, the equilibrium $\{F, O\}$ is not a subgame perfect equilibrium because *F* is not a Nash equilibrium at the subgame beginning at node *x* in Figure 10.2.14(c). On the other hand $\{A, E\}$ is a subgame perfect equilibrium because it is a Nash equilibrium at every proper subgame. (A *proper subgame* of an extensive form game is a proper subtree of the game tree—having terminal nodes, a unique initial node, etc. [see I, §13.5].)

DEFINITION 10.2.4.   A *subgame-perfect equilibrium* of a game *G* is a Nash equilibrium of *G* which is also a Nash equilibrium of every proper subgame of *G*.

Subgame-perfect equilibria can be identified using the backwards induction principle (c.f. Zermelo's algorithm in §10.2.1). An example of this is the solution of *Rubinstein's game* [see §10.7.2]. Finally we mention that although some Nash equilibria may have weakly dominated pure strategies in their support, subgame-perfect Nash equilibria never assign positive probability to weakly dominated strategies. The relationship between dominated strategies and equilibria is returned to in §10.8.2.

10.3.   GAMES WITH INCOMPLETE INFORMATION

10.3.1.   Bayesian Equilibria

Games with *incomplete information*, which refers to some aspect of the rules of the game such as not knowing other players pay-offs, their strategy sets or the information other players have about the game, should be distinguished from games with *imperfect information*, which means that players do not know all the actions available to other players, including chance. Games with imperfect information have been discussed extensively in the literature (for example see Kuhn and Tucker (1953), pp. 193–216). In this section we show how to model games with incomplete information and define three important solution concepts, beginning with *Bayesian equilibria* which were introduced in three seminal papers by Harsanyi (1967–68).

First we describe Harsanyi's *sequential expectations* model for games with incomplete information. Consider a finite two-person game in normal form with pay-off matrices **A** and **B** but where player *I* only knows his own pay-offs, **A**, and player *II* only knows her own pay-offs, **B**. The two-player restriction is for ease of exposition only and the reader should have no difficulty with extending the model to the *n*-player case. Similarly the assumption that information about pay-offs is incomplete is just for simplification and the extension of the model to other forms of incomplete information is straightforward. Now *I*'s strategy choice will depend on his *first-order expectations* of **B**, a subjective probability distribution over all possible **B**, denoted $P_1^1(\mathbf{B})$, and similarly *II*'s strategy choice depends on $P_2^1(\mathbf{A})$, her first-order expectations of **A**. But since *I*'s optimal strategy choice depends on the strategy employed by *II*, player *I*'s optimal strategy also depends on his expectations (or 'beliefs') about player *II*'s first-order expectations of **A**. Harsanyi calls these player *I*'s *second-order expectations*, a subjective probability distribution $P_1^2(P_2^1)$ over all possible $P_2^1(\mathbf{A})$, and player *II* holds similar second-order expectations. By induction each player's optimal choice of strategy depends on his beliefs about the other player's beliefs about his beliefs... about the pay-off matrices. The set of both player's *n*th-order expectations $\{P_1^n(P_2^{n-1}),\ P_2^n(P_1^{n-1})$ for $n = 2, 3, \ldots\}$ is called the *universal belief space*.

Harsanyi used this model to analyse such games as games with *complete* information, provided that player's beliefs satisfy certain *consistency* requirements. That is, if every player's subjective probability distributions over the alternative possibilities for the opponent's pay-offs can be regarded as conditional probability distributions [see II, 3.9] derived from one basic probability distribution over the unknown parameters, then the infinite regress of his sequential expectations model can be by-passed by analysing the game as one of complete information where chance moves first over the unknown parameters according to this basic probability distribution. This assumption of common priors is known as the *Harsanyi doctrine*. Harsanyi calls this complete information game the *Bayes equivalent* to the original incomplete information game, and the Nash equilibria of the Bayes equivalent game (which always exist, by Nash's theorem) are called *Bayesian equilibria* of the original game. If players have consistent beliefs, then the basic probability distribution *P* from which they are derived is unique (Harsanyi (1967–68)) and *common knowledge*, in that every player knows *P* and knows that every other player knows *P* and knows that every player knows that every other player knows *P*, etc. (Aumann (1976)).

EXAMPLE 10.3.1.   Suppose that player I and player II can each be of two types: I_1 or I_2 and II_1 or II_2, and that players only know their own types. Each player's beliefs about the type of the other player are shown in Figure 10.3.1. These beliefs are consistent with types being chosen according to the basic probability distribution shown in Figure 10.3.2 since they coincide with the conditional probabilities under this distribution. For example,

$$P(II_1|I_1) = \tfrac{1}{8}/(\tfrac{1}{8} + \tfrac{1}{4}) = \tfrac{1}{3},$$

which is I_1's subjective probability assigned to II being II_1 in Figure 10.3.1 [see II, §3.9]. On the other hand the beliefs shown in Figure 10.3.3 are not consistent with any basic probability distribution over player's types.

   Assuming that the two players hold the consistent beliefs of Figure 10.3.1, suppose that they play one of four possible 2×2 zero-sum games with pay-off matrices A_{ij}, being the pay-off matrix for I if I_i plays II_j (and so $-A_{ij}$ is II's pay-off matrix). Let

$$A_{11} = \begin{pmatrix} 0 & 1 \\ 0 & 0 \end{pmatrix}, \quad A_{12} = \begin{pmatrix} 1 & 0 \\ 0 & 0 \end{pmatrix}, \quad A_{21} = \begin{pmatrix} 0 & 0 \\ 1 & 0 \end{pmatrix}, \quad A_{22} = \begin{pmatrix} 0 & 0 \\ 0 & 1 \end{pmatrix}.$$

In the Bayes equivalent game, chance moves first by choosing these matrices with probabilities $\tfrac{1}{8}, \tfrac{1}{4}, \tfrac{1}{4}$ and $\tfrac{3}{8}$ respectively. In this game each player has four pure strategies as shown in Figure 10.3.4: TB for I means 'play T if I_1 and B if I_2', RL for II means 'play R if II_1 and L if II_2' etc. The expected pay-off to

	II_1	II_2
I_1	$(\tfrac{1}{3},$	$\tfrac{2}{3})$
I_2	$(\tfrac{2}{5},$	$\tfrac{3}{5})$

I about II

	II_1	II_2
I_1	$\begin{pmatrix} \tfrac{1}{3} \end{pmatrix}$	$\begin{pmatrix} \tfrac{2}{5} \end{pmatrix}$
I_2	$\begin{pmatrix} \tfrac{2}{3} \end{pmatrix}$	$\begin{pmatrix} \tfrac{3}{5} \end{pmatrix}$

II about I

Figure 10.3.1

	II_1	II_2
I_1	$\tfrac{1}{8}$	$\tfrac{1}{4}$
I_2	$\tfrac{1}{4}$	$\tfrac{3}{8}$

Figure 10.3.2

	II_1	II_2
I_1	$(\tfrac{1}{2},$	$\tfrac{1}{2})$
I_2	$(\tfrac{1}{4},$	$\tfrac{3}{4})$

I about II

	II_1	II_2
I_1	$\begin{pmatrix} \tfrac{1}{3} \end{pmatrix}$	$\begin{pmatrix} \tfrac{1}{5} \end{pmatrix}$
I_2	$\begin{pmatrix} \tfrac{2}{3} \end{pmatrix}$	$\begin{pmatrix} \tfrac{4}{5} \end{pmatrix}$

II about I

Figure 10.3.3

		II			
		LL	*LR*	*RL*	*RR*
I	*TT*	$\frac{1}{4}$	0	$\frac{3}{8}$	$\frac{1}{8}$
	TB	$\frac{1}{2}$	$\frac{5}{8}$	1	$\frac{1}{2}$
	BT	0	0	0	0
	BB	$\frac{1}{4}$	$\frac{5}{8}$	0	$\frac{3}{8}$

Figure 10.3.4

I if he plays *TB* and *II* plays *LR*, for example, is

$$\tfrac{1}{8}(\text{top left element of } \mathbf{A}_{11}) + \tfrac{1}{4}(\text{top right element of } \mathbf{A}_{12})$$
$$+ \tfrac{1}{4}(\text{bottom left element of } \mathbf{A}_{21}) + \tfrac{3}{8}(\text{bottom right element of } \mathbf{A}_{22})$$
$$= \tfrac{1}{8}\cdot 0 + \tfrac{1}{4}\cdot 0 + \tfrac{1}{4}\cdot 1 + \tfrac{3}{8}\cdot 1 = \tfrac{5}{8}.$$

So the expected pay-off matrix for the normal form of the Bayes equivalent game is as shown in Figure 10.3.4.

Nash equilibria of this game are easy to identify. For *TB* strictly dominates *BT* and *TT* for *I*, *LL* strictly dominates *LR* for *II* and then *TB* strictly dominates *BB* for *I*, so *LL* strictly dominates *RL* for *II*. Eliminating strictly dominated strategies in this way reduces the normal form to

	LL	*RR*
TB	$\frac{1}{2}$	$\frac{1}{2}$

and so *I* plays *TB* in all Nash equilibria, whereas *II* uses any mix of *LL* and *RR*. These are the Bayesian equilibria of the original incomplete information game.

10.3.2. Sequential Equilibria

This concept for equilibria in games of incomplete formation (but perfect recall) is the analogue of subgame perfection for games of complete information [see §10.2.3] and so sequential equilibria cannot be identified from the normal form. Introduced by Kreps and Wilson (1982), it requires that we broaden our definition of an equilibrium for games with incomplete information so that it is not just a strategy profile but also a set of probability distributions concerning players beliefs about the game tree. As with equilibrium strategies, these beliefs must be specified at all points of the game tree including information sets that will never be reached when equilibrium strategies are employed. They must also satisfy certain consistency requirements in that each player's beliefs must be updated from their original subjective priors according to Bayes' rule [see II, 16.4] whenever possible (otherwise they are assigned arbitrarily).

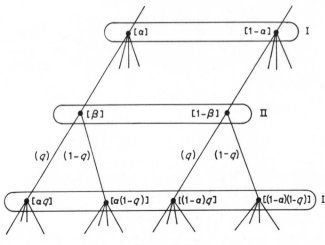

Figure 10.3.5

EXAMPLE 10.3.2.   In the game tree extract shown in Figure 10.3.5, player I has beliefs $[\alpha]$, $[1-\alpha]$ at his first information set h_1 and player II has beliefs $[\beta]$, $[1-\beta]$ at her information set h_2. Then if II employs the mixed strategy $(q, 1-q)^{\mathrm{T}}$ at h_2 in a sequential equilibrium, the equilibrium beliefs of player I at his second information set h_3 must, by Bayes' rule, be $[\alpha q]$, $[\alpha(1-q)]$, $[(1-\alpha)q]$, $[(1-\alpha)(1-q)]$ *unless* $p = 0$. But, if $p = 0$, player I never reaches h_3 in equilibrium, so the beliefs at h_3 in a sequential equilibrium with $p = 0$ are assigned arbitrarily.

The analogy with subgame-perfect equilibria comes from the requirement that sequential equilibrium strategies be *sequentially rational* in that, at every point of the game tree, decisions must be part of an optimal strategy for the remainder of the game. Kohlberg and Mertens (1986) describe this property by saying that players of subgame-perfect equilibria possess a *backwards induction rationality* [see §10.8.2]. But subgame perfection is an inadequate definition for a solution in games with incomplete information as the following example of Selten (1975) shows:

EXAMPLE 10.3.3.   The game in Figure 10.3.6 has two types of Nash equilibria, as the reader may verify by applying the methods of §10.2.2 to the normal form of the game (Figure 10.3.7). These are:

$$Type\ 1: \quad \hat{p}_1 = \hat{p}_2 = 0 \quad 0 \le \hat{p}_3 \le \tfrac{1}{4},$$
$$Type\ 2: \quad \hat{p}_1 = \hat{p}_3 = 1 \quad 0 \le \hat{p}_2 \le \tfrac{2}{3}.$$

They are also subgame perfect since the game has no proper subgames, but the type 2 equilibria are not sensible. For if $\hat{p}_3 = 1$, so III chooses l, II should play

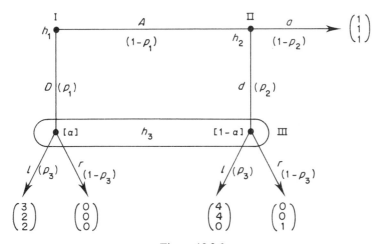

Figure 10.3.6

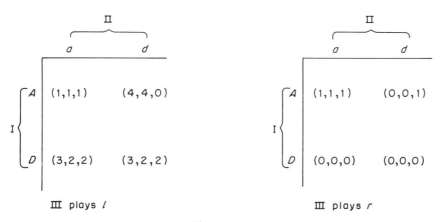

Figure 10.3.7

d, i.e. $\hat{p}_2 = 1$ *if she were called upon to move.* The reason equilibria exist with $\hat{p}_2 < 1$ when $\hat{p}_3 = 1$ is that *II* never gets to move in these equilibria since *I* always plays D ($\hat{p}_1 = 1$).

The fact that subgame-perfect equilibria are not always reasonable in games with incomplete information motivated Selten (1975) and then Kreps and Wilson (1982) to redefine the equilibrium concept by assuming that players choices are guided by conditional pay-offs at each information set, rather than expected pay-offs over the whole game. Kreps and Wilson's notion of sequential equilibria is to proceed by backwards induction, maximizing expected pay-offs conditional on having reached the information set, where the expectations are calculated over players consistent beliefs. We use the example above to illustrate the sequential equilibrium concept. For a formal definition see Kreps and Wilson (1982).

In Example 10.3.3 a sequential equilibrium is a pair $(\alpha; \pi)$ where α defines player III's beliefs at h_3 and $\pi = (\hat{p}_1, \hat{p}_2, \hat{p}_3)$ defines a strategy profile as shown in Figure 10.3.6. Proceeding by backwards induction we first choose p_3 to maximize III's expected pay-off given α:

$$E_3(\pi) = 2\alpha p_3 + (1 - \alpha)(1 - p_3)$$

The dependency of the optimal choice $\hat{p}_3$ on α is shown in Figure 10.3.8(a). Now at h_2 player II chooses p_2 to maximize

$$E_2(\pi) = 4p_2 \hat{p}_3 + (1 - p_2)$$

and the dependency of $\hat{p}_2$ on $\hat{p}_3$ is shown in Figure 10.3.8(b). The induction now proceeds to I who chooses p_1 to maximize

$$E_1(\pi) = 3p_1 \hat{p}_3 + (1 - p_1)[4\hat{p}_2 \hat{p}_3 + (1 - \hat{p}_2)]$$

given these constraints on $\hat{p}_2$ and $\hat{p}_3$. Now from Figure 10.3.8(b) we know that $\hat{p}_3 > \frac{1}{4}$ implies $\hat{p}_2 = 1$ and so arg max $E_1(\pi) = 0$, and that $\hat{p}_3 \leq \frac{1}{4}$ implies that $E_1(\pi) = 3p_1 \hat{p}_3 + (1 - p_1)$ so again $\hat{p}_1 = 0$. Hence every sequential equilibrium must be of the form $(\alpha; 0, \hat{p}_2(\alpha), \hat{p}_3(\alpha))$ and finally we must find α which are consistent with such strategies. Clearly the only α which is consistent with $\hat{p}_1 = 0$ is $\alpha = 0$, *unless $\hat{p}_2 = 0$* also, in which case α could be positive. But not all $\alpha > 0$ are consistent with $\hat{p}_2 = 0$, since by Figure 10.3.8(b) we must have $\hat{p}_3 \leq \frac{1}{4}$ which is only consistent with $\alpha \leq \frac{1}{3}$ by Figure 10.3.8(a). It is clear from these figures that the only sequential equilibria are

$$Type\ 1: \quad (\tfrac{1}{3}; 0, 0, \hat{p}_3) \quad \text{where } 0 \leq \hat{p}_3 \leq \tfrac{1}{4}$$

and

$$Type\ 2: \quad (\alpha; 0, 0, 0) \quad \text{where } 0 < \alpha < \tfrac{1}{3}.$$

The strategies in these sequential equilibria are the 'reasonable' subgame-perfect equilibrium strategies mentioned at the beginning of the example, and

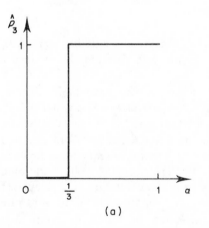

(a)

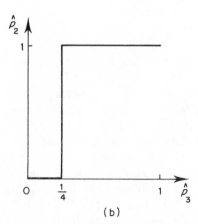

(b)

Figure 10.3.8

the sequential concept sucessfully excludes unreasonable behaviour. Kreps and Wilson (1982) showed that sequential equilibria always exist in finite games with incomplete information but perfect recall. Indeed more often than not there is an infinite number of them, their multiplicity being compounded by the continuum of possibilities for the arbitrary assignments of beliefs in information sets that are off the equilibrium path.

10.3.3.   Trembling Hand Perfect Equilibria

These were introduced by Selten (1975) as a superior solution concept in games with incomplete information since, like sequential equilibria, they distinguish between reasonable and unreasonable behaviour in a way that subgame-perfect equilibria do not. Kreps and Wilson (1982) showed that the set of all trembling hand perfect equilibrium strategies is contained in the set of all sequential equilibrium strategies, hence players of trembling hand perfect equilibria possess 'backwards induction rationality'. Like sequential equilibria they cannot be identified from the normal form—unless one considers the *agent normal form* as in Selten (1975). For almost all games the two concepts coincide but there may be some, usually isolated, points that are sequential but not trembling hand perfect equilibrium strategies. For example, the game in Figure 10.3.9, *Kohlberg's Dalek*, has the sequential equilibria shown. All of these except the point X are trembling hand perfect equilibria.

The difference between trembling hand perfect equilibria and sequential equilibria lies in the way players assign probabilites to reaching decision nodes which are off the equilibrium path. Instead of the consistent beliefs in sequential equilibria, in trembling hand perfect equilibria we assume that although players *want* to play their Nash equilibrium strategies, their hands 'tremble' as they do so and with a very small probability another strategy is employed. For example,

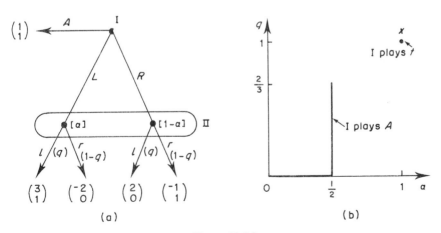

Figure 10.3.9

if a player's Nash equilibrium strategy involves choosing action A with probability p, but his 'rationality' breaks down with probability ε and he chooses A with probability q instead, then the total probability with which he chooses A is $p^* = (1 - \varepsilon)p + \varepsilon q$. We assume that p, q and ε are all independent [see II, 6.6] and that $\varepsilon < p^* < 1 - \varepsilon$, since ε is very small. A game where each player makes mistakes of this type is called a *perturbed game* of the original game and in this game all information sets are reached with positive probability in equilibrium. The equilibrium strategies of this game depend on the size of the random errors, or 'trembles', which need not be the same for all players and a *trembling hand perfect equilibrium* of the original game is the limit of equilibria in the perturbed games as all trembles tend to 0. Harsanyi and Selten (1988) have refined this concept to that of *uniformly perfect equilibria* which are trembling hand perfect equilibria in which all players make mistakes with a probability of at least ε, the *same* for all players. As usual we illustrate these concepts by example: for formal definitions see Selten (1975) and Harsanyi and Selten (1988).

EXAMPLE 10.3.4.   In *Selten's Horse*, the game of Example 10.3.3, all type 1 Nash equilibria are trembling hand perfect but none of the type 2 equilibria are. Hence the sequential and trembling hand perfect equilibrium concepts coincide in this game. To show that $(0, 0, \frac{1}{4})$ is trembling hand perfect, consider a perturbed game of Selten's horse where each player deviates from rationality with the same probability $\varepsilon < \frac{1}{4}$. From Figure 10.3.6 we see that player *II*'s expected pay-offs conditional on reaching his information set are $E_3(r) = p_2^*(1 - p_1^*)$ and $E_3(l) = 2p_1^*$, where p_1^* and p_2^* define the strategies employed by *I* and *II* in the perturbed game. If $E_3(r) > E_3(l)$ then *II* should choose r, but in this perturbed game he only plays r with probability $1 - \varepsilon$ and with probability ε he chooses l by mistake. Similarly we can show that

$$2p_1^* > p_2^*(1 - p_1^*) \quad \text{implies } p_3^* = 1 - \varepsilon,$$
$$2p_1^* < p_2^*(1 - p_1^*) \quad \text{implies } p_3^* = \varepsilon, \qquad\qquad (10.3.1)$$
$$2p_1^* = p_2^*(1 - p_1^*) \quad \text{implies } \varepsilon < p_3^* < 1 - \varepsilon.$$

Now consider player *II*. If $p_3^* > \frac{1}{4}$ he should choose $p_2 = 1$ in equilibrium but instead plays $p_2^* = 1 - \varepsilon$ in the perturbed game. Similarly

$$p_3^* > \tfrac{1}{4} \quad \text{implies } p_2^* = 1 - \varepsilon,$$
$$p_3^* < \tfrac{1}{4} \quad \text{implies } p_2^* = \varepsilon, \qquad\qquad (10.3.2)$$
$$p_3^* = \tfrac{1}{4} \quad \text{implies } \varepsilon < p_2^* < 1 - \varepsilon.$$

Finally for player *I*, equilibrium strategies in the perturbed game are

$$3p_3^* > 4p_2^*p_3^* + (1 - p_2^*) \quad \text{implies } p_1^* = 1 - \varepsilon,$$
$$3p_3^* < 4p_2^*p_3^* + (1 - p_2^*) \quad \text{implies } p_1^* = \varepsilon, \qquad\qquad (10.3.3)$$
$$3p_3^* = 4p_2^*p_3^* + (1 - p_2^*) \quad \text{implies } \varepsilon < p_1^* < 1 - \varepsilon.$$

Now suppose that $p_3^* < \frac{1}{4}$ so that $p_2^* = \varepsilon$ by (10.3.2) and since $\varepsilon > \frac{1}{4}$ we have $3p_3^* > 4p_2^* p_3^* + (1 - p_2^*)$, so by (10.3.3) $p_1^* = 1 - \varepsilon$. Thus $2p_1^* > p_2^*(1 - p_1^*)$ and by (10.3.1) $p_3^* = 1 - \varepsilon$, which is a contradiction because $\varepsilon < \frac{1}{4}$. Similar reasoning shows that p_3^* cannot be greater than $\frac{1}{4}$ and so it must equal $\frac{1}{4}$. Now (10.3.3) yields $p_1^* = \varepsilon$ and (10.3.1) implies that $2p_1^* = p_2^*(1 - p_1^*)$ so $p_2^* = 2\varepsilon/(1 - \varepsilon)$. Thus the (subgame-perfect) equilibrium in this perturbed game is $(\varepsilon, 2\varepsilon/(1 - \varepsilon), \frac{1}{4})$ which tends to $(0, 0, \frac{1}{4})$ as $\varepsilon \to 0$.

This proves that $(0, 0, \frac{1}{4})$ is a trembling hand perfect equilibrium, indeed a *uniformly* perfect equilibrium for Selten's horse. To show that other type 1 Nash equilibria are trembling hand perfect it is necessary to assume that the size of player *II*'s trembles is not ε but $2\varepsilon/(1 - \varepsilon)$, where $\varepsilon < p/2$ for some p in $(0, \frac{1}{4})$, so these are *not* uniformly perfect. For this and a proof that type 2 Nash equilibria are *not* trembling hand perfect see Selten (1975). This game has a unique uniformly perfect equilibrium, $(0, 0, \frac{1}{4})$ and Harsanyi and Selten (1988) would say that this is the solution of the game [see §10.8.2 for further details].

10.4. REPEATED GAMES WITH COMPLETE INFORMATION

A *supergame* is a countable sequence of finite non-cooperative games played by a fixed set of players, where the constituent game may vary with time. We consider the case where the *same game G*, called the *stage game*, is played many times by the same players. This is called a *stationary supergame* or *repeated game*. There may be a finite or infinite number of repetitions, and after each the players are informed of each others moves. Note that this is very different from knowing opponents *strategies*: in repeated games strategies are very complicated as we shall see below, and all we assume in this section is that opponents *actions* are observed after every stage, not that players know which mixed strategies of *G* (i.e. *stage strategies*) prescribed the action that was taken.

In this section we assume that there is complete information in *G* and leave the repetition of stage games with incomplete information to §10.5. We do not consider stochastic games or other non-stationary supergames and it is assumed throughout that opponents do not change between repetitions.

10.4.1. Finitely Repeated Games

We introduce the basic concepts using a 2×2 stage game *G* repeated *n* times, denoting the repeated game G_n. The generalization to larger stage games and more than two players is straightforward.

For each player, the *pay-off* in G_n is the average of his pay-offs at every stage. Strategies in G_n are more complicated: suppose that *I* has pure strategies *T* and *B* and that *II* has pure strategies *L* and *R* in *G*. Then an example of a pure strategy for *I* in G_n is

stage 1: T
stage 2: T if *II* moved L at stage 1; B if *II* moved R at stage 1
stage 3: T if *II* moved L at stage 1 and stage 2 or R at stage 1 and stage 2;
           otherwise B
  $\vdots$
stage n: B

That is, the pure strategies in G_n may be conditioned on any past moves in the game, so even for small n there are vast numbers of pure strategies available: already in G_3 there are 2^7 pure strategies for each player. Mixed strategies are therefore extremely complicated, but they are equivalent to behavioural strategies since there is perfect recall in G_n (Kuhn's theorem, Theorem 10.2.1) and this allows some simplification of the analysis.

As usual it is much easier when the stage game is two-person zero-sum. Recall from §10.2.2 and Owen (1982) that the pay-off to *I* when he plays any maximin strategy (equivalently any Nash equilibrium) is at least the value v of the game, and that *II* can hold him to exactly v (and her pay-off to exactly $-v$) if she employs one of her minimax strategies. Now if the stage game is repeated n times, there is an optimal strategy profile in the repeated game where player *I* plays a maximin strategy and player *II* plays a minimax strategy at every stage of the game:

THEOREM 10.4.1.   *Let G be a two-person zero-sum game with value v. Then G_n also has value v and strategies which consist of player I employing a maximin strategy and player II playing a minimax strategy of G at each stage are optimal strategies in G_n.*

EXAMPLE 10.4.1.   Let G be a 2×2 zero-sum game with pay-off matrix

$$\begin{pmatrix} 1 & 0 \\ 0 & 2 \end{pmatrix}$$

so *I* has a unique maximin strategy $(\frac{2}{3}, \frac{1}{3})$ and player *II* a unique minimax strategy, also $(\frac{2}{3}, \frac{1}{3})$, and $v = \frac{2}{3}$ [see I, §13.2.3]. It would be a lengthy task to solve G_2 directly since there are already eight pure strategies for each player in G_2. But by Theorem 10.4.1 we know that G_2 has value $\frac{2}{3}$, and an optimal strategy pair is for each player to play $(\frac{2}{3}, \frac{1}{3})$ at each stage. This solution in behavioural strategies is quite general (by Kuhn's theorem) and although it is not unique, it is equivalent and interchangeable with all other optimal strategy pairs since G_2 is a two-person zero-sum game [see §10.2.2].

Kuhn's theorem also allows us to apply the technique of backwards induction to find subgame-perfect Nash equilibria in finitely repeated *non*-zero-sum games. The following example illustrates the general idea. It is called the *prisoner's dilemma* and is the standard game for analysing cooperative behaviour in a competitive environment.

EXAMPLE 10.4.2. Two thieves are apprehended on suspicion of a crime but the police have no evidence. The suspects are questioned separately and simultaneously with no means of communication either before or during questioning. Each has two pure strategies: A (keep quiet) and B (confess). The normal form and pay-off region are shown in Figure 10.4.1 where $T > R > P > S$. The 'dilemma' arises because of the ordering of the pay-offs: if you think the other player is keeping quiet, then you are tempted with the pay-off T to confess, but if you both confess you get only the punishment pay-off P. It would be better for you to cooperate with each other and both keep quiet so you both get the reward R... but then there is always the temptation for your partner to 'grass' on you, in which case you'll get S for being a 'sucker'. You cannot trust your partner to keep quiet so you confess. So does he, since the game is symmetric, and there is a unique Nash equilibrium $\{B, B\}$.

In the one-shot game the cooperative play which results in the Pareto optimal outcome (R, R) is not a Nash equilibrium. However, if this game is repeated a large number of times, it is possible that players would evolve some sort of cooperation. For example, they could begin by moving $\{B, B\}$ in the initial stages but then one could 'hold out the olive branch' by moving A even though he expects his opponent to move B. If the other player then responds by moving A, a sequence of cooperative $\{A, A\}$ moves could ensue... until one decides to renege on this tacit agreement by moving B and getting the temptation pay-off T. Now if the stage-game is repeated an infinite or unknown number of times it is possible that this type of play could be sustained as a Nash equilibrium [see §10.4.2]. But when n is finite and known to both players they will both play B at the last stage since this is the unique optimal play, independent of previous moves. Having determined optimal play at the last stage, similar reasoning implies that $\{B, B\}$ is optimal at the penultimate stage, and proceeding

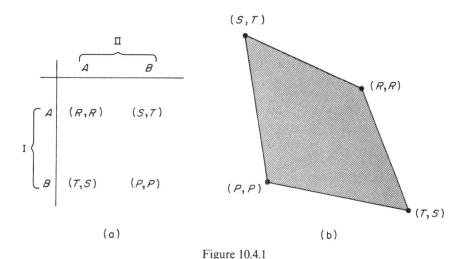

Figure 10.4.1

by backwards induction we find that there is a unique subgame-perfect equilibrium in G_n, with both players confessing at every stage.

When n is very large, empirical studies do not support this conclusion (see Luce and Raiffa (1957)) so in applications of the prisoner's dilemma, such as disarmament or oligopoly theory, it seems more reasonable to suppose that n is either infinite or unknown. In this case almost anything can be a Nash equilibrium, as we shall now see.

10.4.2.   Infinitely Repeated Games without Discounting

Let $\sigma = \{\sigma_i\}$ be a strategy profile in an infinitely repeated game G_∞ so σ_i denotes a strategy for player i. An intuitive definition for the pay-off in G_∞ under σ is $\lim (\Sigma a_r)/n$ as $n \to \infty$, where a_r denotes the pay-off at stage r under σ. When G is a two-person zero-sum game with value v and $\{\sigma_1, \sigma_2\}$ are maximin strategies, this *limit of average pay-offs* is v. There are two possible definitions of value in infinitely repeated zero-sum games: the *limit of value* is the value of the n-stage repeated game G_n as $n \to \infty$ and the *value of limit* is a limiting average pay-off in G_∞. In the two-person case with complete information the two approaches yield the same value, but this is not always so: the value of limit need not exist when either or both players have incomplete information (see Zamir (1973)).

When G is not two-person zero-sum the limit of average pay-offs does not always exist and so there may be problems defining pay-offs in G_∞. However, assuming that player's pay-offs can be suitably defined, it turns out that almost any pay-off can be supported by equilibrium strategies in G_∞. For a formal statement of these results we need the following:

DEFINITION 10.4.1.   A pay-off vector in the stage game G is *feasible* if it lies in the convex hull of pure strategy pay-off vectors in G, and *individually rational* if its ith component is at least as great as player i's minimax pay-off.

Feasible pay-offs may be achieved by the use of *correlated strategies* in G (see Aumann (1974)). But it is possible that some correlated strategies do not have pay-offs inside the feasible region as defined above [see §10.8.1]. An example of a game where the use of correlated strategies enlarges the pay-off region exactly to the set of feasible pay-offs is the Battle of the sexes [see Example 10.2.8]. The pay-off regions under the use of uncorrelated and correlated stategies are shown in Figure 10.4.2.

Notice that individually rational pay-offs are at least as great as a player's *minimax* pay-off, being the smallest pay-off that the other players can restrict this player to. If strictly dominated strategies are excluded from the game, the minimax pay-off is at least as great as the maximin pay-off (i.e. the largest pay-off that he can guarantee himself). For example consider a two player normal form game, where player I has pay-off matrix (a_{ij}). Denote by $\bar{a}_i$ and

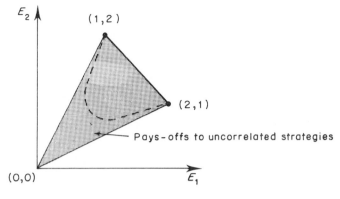

Figure 10.4.2

a_i the maximum and minimum values of row i. Then if $\min_i \bar{a}_i < \max_i \underline{a}_i$, say $\bar{a}_l < \underline{a}_k$, we have $a_{lj} \leq \bar{a}_l < \underline{a}_k \leq a_{kj}$ for all j, and so player I's kth pure strategy strictly dominates his lth pure strategy.

We can now state the fundamental result for Nash equilibria in G_∞, which Aumann (1981) termed the *folk theorem* for infinitely repeated games, since its origins are unknown.

THEOREM 10.4.2.   *All feasible individually rational pay-off vectors in G can be supported by Nash equilibria in G_∞.*

The proof of this result requires a player to employ *punishment strategies* if his opponent deviates from the equilibrium path. For example, in order to sustain (R, R) as a Nash equilibrium pay-off in the infinitely repeated prisoner's dilemma, the two players need to cooperate in playing A at every stage. Now if one player defects at some stage of G_∞ and plays B, his opponent must punish him by choosing B for the rest of the game. This threat prevents any defection since the one-period gain is outweighed by future losses. However, this is not a *credible threat* since the punishment hurts the punisher as much as the defector: both would prefer to return to cooperative play as soon as possible after one player defects and so this strategy profile is not a subgame-perfect Nash equilibrium in the infinitely repeated prisoner's dilemma.

However, it is not the case that subgame-perfection leads to a reduction in the equilibrium *pay-offs* of G_∞: Rubinstein (1977) proved the *perfect folk theorem*, that Theorem 10.4.2 also holds for subgame-perfect Nash equilibria in G_∞. Its proof also requires the use of punishment strategies, but now the defector has to punish the punisher if he fails to implement the punishment! Assuming mixed punishment strategies are used, how can the defector tell whether he is being punished? All he observes are the actions at each stage. But since the actions at any particular stage do not affect the pay-offs in G_∞, there is no cost incurred by making as many observations on his opponents behaviour as he likes, so deviation can be detected with arbitrarily high probability.

10.4.3. Infinitely Repeated Games with Discounting

We have seen that modelling infinitely repeated games without time dependence is not satisfactory: even if the pay-offs can be adequately defined, there are infinitely many subgame-perfect Nash equilibria in G_∞. So it is customary to introduce a *discount rate* δ after every stage in G_∞ and to denote the infinitely repeated game $G_\infty(\delta)$. Then the pay-off under a strategy profile σ is the *discounted present value* of all future pay-offs: $\sum_r \delta^r a_r$ where the first stage is labelled 0 for ease of notation. Equivalently we could introduce a *stopping probability* p after every stage in G_∞ and the analysis is identical to $G_\infty(\delta)$ with $\delta = 1 - p$, with the interpretation of pay-off being the *expected value* of all pay-offs under σ: $\sum(1 - p)^r a_r$ [see II, §8.1]. A discount rate $\delta < 1$ is used when players have *time-preferences* (see Fishburn and Rubinstein (1982)) because they are impatient, and a stopping probability $p > 0$ is used in an uncertain environment, so preferences are represented by the usual von-Neumann and Morgenstern expected utility value [see §10.1]. We phrase the following in terms of $G_\infty(\delta)$, but all results can be interpreted in the equivalent game with stopping probabilities.

When δ is small, punishment strategies are not effective, since the pay-offs at future stages of $G_\infty(\delta)$ are small. Hence cooperative play cannot necessarily be sustained as a Nash equilibrium. However, when δ is sufficiently large we have the *folk theorem with discounting*:

THEOREM 10.4.3.   *There exists a* $\delta^* \in (0, 1)$ *such that all feasible strictly individually rational pay-off vectors in G can be supported by Nash equilibria in* $G_\infty(\delta)$ *provided* $\delta \geq \delta^*$.

Notice that the theorem refers only to pay-offs that are *strictly* greater than players minimax pay-offs in G and a counterexample of Forges, Mertens and Neymann (1986) shows that this condition cannot be dispensed with. For the proof of Theorem 10.4.3 one can use the notion of *balanced temptation equilibria*, which are punishment strategies where the ratio of the one-period gain to the per-period later loss is the same for all players (Friedman (1977)). Abreu's (1983) idea of *simple strategies*—where deviation by a single player is punished uniformly by all players—can also be used to show that cooperative behaviour can be supported as Nash equilibria in $G_\infty(\delta)$.

Finally we mention that Theorem 10.4.3 remains valid if one substitutes 'subgame-perfect Nash' for 'Nash' equilibria, provided there are only two players or, if there are more than two players, that the feasible set of pay-offs has the same dimension as the number of players (Fudenberg and Maskin (1986)).

10.5. REPEATED GAMES WITH INCOMPLETE INFORMATION

In the previous section we saw how complicated the analysis of the repeated play of a finite non-cooperative game G can be, because of the enormous number of pure strategies for each player in the repeated game. The situation becomes even more complex when one or more players have incomplete information about the pay-offs. We need to consider the strategic use of information: How much information should the informed players reveal and when should they reveal it? What should the uninformed players believe about revealed information? We have already noted that it makes a difference whether players are assumed to observe the one-shot strategies employed by their opponents, called their *stage strategies* or, as in §10.4, just the actions prescribed by opponents stage strategies after every stage. Although the latter case is more realistic, the complexity of our analysis is greatly reduced if we assume that players know the actual stage strategies used after every stage. Thus we assume that after stage n and before stage $n+1$ the probability distribution used over the different pure strategy choices in G at stage n by each player is common knowledge, and the strategies for the repeated game are gradually revealed as the game progresses.

The theory of two-person zero-sum repeated games with incomplete information is far more complete than that of repeated non-zero-sum and more person games, and it is natural to make this division of material rather than to distinguish between finite and infinite repetitions as we have done in the previous section. Indeed many results hold for both finitely and infinitely repeated games when there is incomplete information about the stage game G.

10.5.1. The Two-person Zero-sum Case

First consider the case of the simplest form of incomplete information: Suppose **A** and **B** are $n \times m$ matrices defining two-person zero-sum games, and that either **A** or **B** is played repeatedly but only player I knows for certain which one it is. Player II has a probability distribution over these, believing that **A** is played with probability p, and this is common knowledge. Denote by $G_n(p)$ the Bayes equivalent of the n stage repeated game [see §10.3.1]. So in the game $G_n(p)$, chance moves first choosing **A** or **B** with probabilities p and $1 - p$ respectively. Only player I is informed of chance's choice and the chosen game is played n times. Let $v_n(p)$ denote the value of $G_n(p)$ and $u(p)$ denote the value of the game when I ignores his information. Thus $u(p)$ is the value of the game with pay-off matrix $p\mathbf{A} + (1 - p)\mathbf{B}$, or equivalently the value of the repeated game with this pay-off matrix since the two are equal [see §10.4.2].

THEOREM 10.5.1. $v_n(p)$ *is the least concave function that is* $\geq u(p)$.

Zamir (1971) has shown that the theorem also holds for $G_\infty(p)$, whether we use the limit of value or value of limit as the definition of $v_\infty(p)$ [see §10.4.2].

That $v_n(p) \geq u(p)$ is clear since information cannot hurt in a zero-sum game (but this is not always so in non-zero sum games). The proof that $v_n(p)$ is concave [see IV, §15.2] is quite complicated but that given by Aumann (1981) provides a useful construction for finding maximin strategies in $G_n(p)$. In general these are a particular type of *partially revealing strategies* which we describe in Example 10.5.2. But first we introduce the two extreme cases of information revelation: a *non-revealing strategy* for I in $G_n(p)$ is the repeated play of a maximin strategy for him in the game $p\mathbf{A} + (1-p)\mathbf{B}$—by ignoring his information he cannot reveal it—and a *completely revealing strategy* for I in $G_n(p)$ is one that enables player II to infer exactly which game is being played. It is not always the case that player I should employ a non-revealing strategy even in a zero-sum situation.

EXAMPLE 10.5.1.   (Aumann and Maschler (1966))

$$\mathbf{A} = \begin{pmatrix} 1 & 0 \\ 0 & 0 \end{pmatrix}, \qquad \mathbf{B} = \begin{pmatrix} 0 & 0 \\ 0 & 1 \end{pmatrix}$$

and $p = \frac{1}{2}$. Label player I's pure stage strategies H and T and those of player II, L and R respectively. Suppose I uses a non-revealing strategy in $G_n(\frac{1}{2})$, so he ignores his information and the stage game is equivalent to the game with pay-off matrix

$$\begin{pmatrix} \frac{1}{2} & 0 \\ 0 & \frac{1}{2} \end{pmatrix}.$$

This has value $\frac{1}{4}$ [see I, §13.2] so this is his pay-off in $G_n(\frac{1}{2})$ and in $G_\infty(\frac{1}{2})$. Can player I get more than this by revealing his information? In this case his completely revealing strategy is to play H at every stage if the game is $\mathbf{A}$, and T at every stage if the game is $\mathbf{B}$. After the first stage II will know that player I's first stage strategy was the pure strategy H (or T), and, since this is optimal only if $\mathbf{A}$ (or $\mathbf{B}$) is the game, she will infer from this exactly which game is being played. So thereafter she can always play R (or L) and so hold I to a pay-off of 0 at every stage. Hence the expected pay-off to I is $1/2n$ in $G_n(\frac{1}{2})$ and, since $n \geq 2$, he does better to use his non-revealing strategy.

But it is not always the case that player I should conceal his information. Suppose

$$\mathbf{A} = \begin{pmatrix} -1 & 0 \\ 0 & 0 \end{pmatrix} \quad \text{and} \quad \mathbf{B} = \begin{pmatrix} 0 & 0 \\ 0 & -1 \end{pmatrix},$$

so I's non-revealing strategy is a maximin strategy for

$$\begin{pmatrix} -\frac{1}{2} & 0 \\ 0 & -\frac{1}{2} \end{pmatrix}$$

at every stage. This yields the average pay-off of $-\frac{1}{4}$, whereas by using his completely revealing strategy he can guarantee himself an expected pay-off of 0 in $G_\infty(\frac{1}{2})$. In this case the completely revealing strategy is actually a maximin strategy for I and $v_\infty(\frac{1}{2}) = 0$.

EXAMPLE 10.5.2.

$$A = \begin{pmatrix} 3 & 0 & 2 \\ 3 & 0 & -2 \end{pmatrix}, \quad B = \begin{pmatrix} 0 & 3 & -2 \\ 0 & 3 & 2 \end{pmatrix}.$$

If I plays his completely revealing strategy he should play H, his first pure strategy, if the game is A and his second pure strategy T if the game is B. But after this II can hold him to a pay-off of 0 at every stage, since she will know which game is being played, and the completely revealing strategy only guarantees I a pay-off of 0 in $G_\infty(p)$. I does not necessarily do any better by concealing his information, for his non-revealing strategy is to use his maximin strategy for the game

$$\begin{pmatrix} 3p & 3(1-p) & 4p-2 \\ 3p & 3(1-p) & 2-4p \end{pmatrix}.$$

Now the value of this game is the function $u(p)$ of Figure 10.5.1, so if $p = \frac{1}{2}$ player I still gets only 0 by using his non-revealing strategy.

However, Theorem 10.5.1 tells us that there is a strategy for I that guarantees him at least that pay-off indicated by the dotted line in Figure 10.5.1 (this is the least concave function $\geq u(p)$). If $p \leq \frac{2}{7}$ or $\geq \frac{5}{7}$, then player I can do no better than to conceal his information, but if p lies between $\frac{2}{7}$ and $\frac{5}{7}$ then player I can guarantee himself at least $\frac{6}{7}$ if he partially reveals his information as

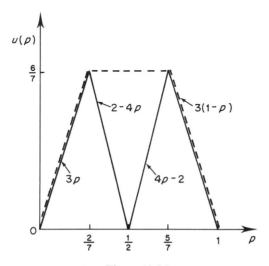

Figure 10.5.1

follows: Before the game is played, player I prepares two coins C_A and C_B. Then he tosses the relevant coin before each stage and takes the action prescribed (i.e. plays H if the coin comes up heads and T if it comes up tails). The preparation of these coins, the probability of a head for each and the outcome of the toss is common knowledge, but only player I knows which coin is being used (i.e. C_A if the game is A and C_B if the game is B). Denote by $p(H|A)$ and $p(H|B)$ the probability of a head for C_A and C_B respectively. Since I knows player II's subjective prior probability p that the game is A, he can calculate $\alpha \in [0, 1]$ such that $p = \alpha(\frac{2}{7}) + (1 - \alpha)(\frac{5}{7})$ and then use α and p to fix $p(H|A)$ and $p(H|B)$ as $5\alpha/7p$ and $2\alpha/7(1 - p)$ respectively. The reason for choosing these weights is that, knowing these, player II can update her subjective prior probability p (also denoted $p(A)$, below) that A is the game after knowing the result of the toss, using Bayes rule [see II, §16.4]. Now since the weights have been chosen so that $p(H) = \alpha$, these posteriors are

$$p(A|H) = p(H|A)p(A)/p(H) = \tfrac{5}{7},$$

and

$$p(A|T) = p(T|A)p(A)/p(T) = \tfrac{2}{7}.$$

Knowing these posteriors, player I has expected pay-off matrix $\frac{5}{7}A + \frac{2}{7}B$ if the coin comes up heads and, since he must play H the relevant row is $(\frac{15}{7}, \frac{6}{7}, \frac{6}{7})$. Similarly if the coin comes up tails, then player I has expected pay-off matrix $\frac{2}{7}A + \frac{5}{7}B$ under II's posteriors and the relevant row is $(\frac{6}{7}, \frac{15}{7}, \frac{6}{7})$ since he must play T. In either case he gets at least $\frac{6}{7}$, so I has chosen the weights of the coins and a method of partially revealing his information to guarantee himself the least concave function $\geq u(p)$.

This construction is not specific to the above example. In the more general case of Figure 10.5.2, I can guarantee a pay-off of the dotted line by using either his non-revealing strategy or, if $p_1 < p < p_2$, by finding $\alpha \in [0, 1]$ such that $p = \alpha p_1 + (1 - \alpha)p_2$ and setting $p(H|A) = \alpha p_2/p$ and $p(H|B) = \alpha p_1/(1 - p)$. This ensures that $p(H) = \alpha$ and so player II's posterior probabilities will be p_2 (that

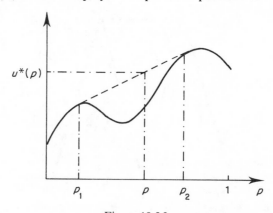

Figure 10.5.2

the game is **A** given that the coin showed heads) and p_1 (that the game is **A** given that the coin showed tails). Now, whatever the result of the toss, I has guaranteed himself an expected pay-off of $u^*(p) = \alpha u(p_1) + (1 - \alpha)u(p_2)$ under these posteriors.

In the above we have not considered optimal strategies for uninformed players since this is far more difficult to analyse and the reader is referred to Aumann (1981) for a review of literature in this area. Now we give only a brief introduction to the case of two-sided incomplete information, where players have no private initial information and are given a *signal* after every stage of the repeated game. That is, according to the actions of the players and the game being played, a referee announces a given letter of the alphabet. In Kohlberg and Zamir (1974) it is shown that the infinitely repeated game has a value if these signals are deterministic and symmetric (i.e. the same for both players) and their results were extended to the case of random signals (i.e. probability distributions on a given alphabet) by Forges (1982). The results depend on the reduction of the repeated game to a particular type of *stochastic game*—a multi-stage game where the game played at each stage changes according to given transition probabilities—viz. one with absorbing states [see II, §18.3.2]. This is beyond the scope of this chapter but the following example illustrates the general idea:

EXAMPLE 10.5.3.   Suppose that one of the first pair of zero-sum games in Example 10.5.1 is chosen to be played at every stage of an infinitely repeated game, each with probability $\frac{1}{2}$, and that neither player knows which game is played. Both players get the same signals according to the *signalling matrices*

$$\mathbf{S_A} = \begin{pmatrix} a & b \\ c & d \end{pmatrix}, \qquad \mathbf{S_B} = \begin{pmatrix} a & b \\ c & e \end{pmatrix}.$$

For example, if both players play their second pure strategy at a given stage and if the game being played is **A**, then both players receive the signal 'd' from the referee. Now if either 'd' or 'e' is announced, both players know for sure which game is being played and so they can ensure a pay-off of 0 at every subsequent stage (since this is the value of both games). If any other signal is announced, then no further information is revealed so the same game is repeated again. Therefore the game is equivalent to the infinitely repeated game

$$\begin{pmatrix} \frac{1}{2} & 0 \\ 0 & 0* \end{pmatrix}$$

(where * denotes an absorbing state) which has value 0.

10.5.2.   The Non-zero-sum Case

Results in the theory of repeated non-zero-sum games with incomplete information are rather sparse. However, there is some interesting work on specific games and some general results for the simplest general class of these games—non-zero-sum repeated games with two players, one-sided incomplete information and only two possible states of nature (i.e. one of two games is the stage game and only one of the players knows which it is). It is also assumed that players stage strategies are known after every stage and not just the actions prescribed by these strategies. Sorin (1983) showed that such games always have a Nash equilibrium—a problem outstanding for almost 20 years!—and Hart (1983) has characterized the Nash equilibrium pay-offs in the spirit of the folk theorems of the previous section.

Now to the more specific results: In §10.4 it was shown that the unique (subgame-perfect) Nash equilibrium in the finitely repeated prisoner's dilemma was the worst possible 'non-cooperative' equilibrium where both players confess at every stage. One way of modifying the game so that the Pareto optimal cooperative behaviour which is empirically observed *is* a Nash equilibrium, is to make the number of repetitions infinite. Another way is to introduce incomplete information into the finitely repeated version, and this was demonstrated by Kreps *et al.* (1982).

They considered two types of incomplete information: (1) uncertainty about the 'rationality' of one of the players, and (2) two-sided incomplete information about the pay-offs in the stage game. In model (1) player I knows his strategy in the repeated game but player II does not—all she observes is his action after every play of the game—but she *does* know that player I is either rational in Savage's sense [see §10.8.1] (in which case his optimal play is to confess at every stage) or, with very small probability $\delta > 0$ he plays 'TIT FOR TAT' [see §10.6.3]. Since only actions and not the entire stage strategies are observed, sequential equilibrium is the appropriate equilibrium concept [see §10.3.2]. Kreps *et al.* showed that there exist sequential equilibria with an upper bound to the number of stages that player II will confess. This upper bound depends only on δ and not on the number of repetitions. So cooperation *could* be optimal in the repeated prisoner's dilemma when players have even the smallest doubts about the rationality of their opponents!

They provide a similar result in model (2). Specifically, suppose that player I originally assesses a small probability $\delta_2 > 0$ that the pay-offs to player II in the stage game are such that $T < R$, so that cooperation *is* optimal for player II provided player I cooperates. Similar player II believes at the outset with probability $\delta_1 > 0$ that player I prefers to cooperate when it is met with cooperation. Suppose that each player knows their own pay-offs but that neither player knows for certain the pay-offs of their opponent. Then there exists a sequential equilibrium in which both players cooperate at every stage except the last few—as in model (1) the end play is very complex. The result in model (2) depends on the idea of *reputation:* if either player fails to cooperate at any stage then his opponent knows for sure that $T > R$ for him (and so it is optimal for him to confess) and the non-cooperative equilibrium ensues. So even though it is very likely that $T > R$ for both players, it is in their interest to keep their opponents guessing, by cooperating at every stage. For otherwise an endless stream of confessing is bound to follow, which yields the worst possible outcome $\{P, P\}$ at every stage whatever the parameters of the game. Note that this result requires two-sided incomplete information: if player I knew for sure that $T > R$ for player II (so that II will confess even if I keeps quiet) then the only sequential equilibrium is the non-cooperative one. By the same token, there will exist sequential equilibria without long run cooperation since reputational effects may be ignored by one or both of the players.

The first detailed study of such reputational effects was for the finitely repeated chain-store game with one-sided (and two-sided) incomplete information about pay-offs (Kreps and Wilson (1982a). As with the one-shot chain-store game [see Example 10.2.10] there exists a unique Nash equilibrium which is also subgame-perfect in the finitely repeated version, in which player I plays A (the monopolist aquiesces) and player II plays E (the potential entrant enters the market) at every stage. But now suppose that the monopolist is either weak or strong but it is only opimal for him to aquiesce if he is weak—the stronger monopolist prefers to fight entry—and only the monopolist knows for sure which type his is. Then in every sequential equilibrium of this game even the weak monopolist will fight at every stage, except the last few. This is because it is in the weak monopolist's interest to maintain the reputation of being strong since then the potential entrant will avoid a challenge for fear of the predatory response.

Milgrom and Roberts (1982) derive similar results for the repeated chain-store game with incomplete information about the rationality of the monopolist. The entrant entertains a small probability δ that the monopolist is irrational, playing some fixed strategy instead of what is optimal. In this case the rational monopolist may adhere to that strategy during the initial stages of the game since it is in his interest to generate this reputational effect.

These results are of interest to economists as the prisoner's dilemma demonstrates how cooperation can occur in antagonistic environments and the chain-store game explains entry deterrence in monopolies. But we are still a very long way from the classification of all possible solutions in general finitely repeated non-zero-sum games with incoomplete information.

10.6. EVOLUTIONARY GAMES

An *evolutionary game* models the dynamic behaviour of an infinite population of animals. We consider successive generations of the population when all animals die at the end of the period having reproduced asexually according to their *Darwinian fitness*. Maynard Smith (1966) defines (mean) Darwinian fitness as the average number of immediate descendants per animal in a large population, but there are many other possible definitions which take account of the reproductive sucess of an animal's offspring as well as its individual reproductive success (see Ewens (1979)). Animals are not rational in the sense that they choose an optimal strategy in any given situation. Instead each animal is endowed with a strategy for playing a fixed finite game G, which it passes on (with or without mutation) to its descendants. In each generation animals are paired randomly, each animal meeting exactly one opponent and the game G is played, after which the Darwinian fitnesses determine the number of descendants.

The population may be either *monomorphic*, where each animal is endowed with the same pure or mixed strategy $\mathbf{p}$, or *polymorphic*, where animals carry different pure or mixed strategies. Polymorphic populations may be characterized by a probability vector $\mathbf{p}_t$, each component $\mathbf{p}_{it}$ being the probability that the ith pure strategy is played by the population at time t. The relative success of the different pure strategies, measured in terms of Darwinian fitness, determines the number of descendants of the animals which played them, so the dynamic trajectory of $\mathbf{p}_t$ characterizes the evolution of the population. This will be described in more detail in the next section. In monomorphic populations, however, there is no obvious transmission mechanism that will change the strategy $\mathbf{p}$ over time, since all animals are bound to pass the same strategy onto their descendants—unless they mutate. So in monomorphic populations we only consider strategies which are stable against invasion by a smalll number of mutants and do not consider their evolution over time.

10.6.1. Symmetric Conflicts

Maynard Smith and Price (1973) defined an equilibrium of an evolutionary game to be an *evolutionary stable strategy* (ESS). This is a strategy which describes a monomorphic population which cannot be invaded by a small number of mutants.

DEFINITION 10.6.1.   Consider a monomorphic population all playing the strategy $\mathbf{p}$ except for a small proportion ε of mutants which play $\mathbf{q}$. Denote by $E(\mathbf{x}, \mathbf{y})$ the expected pay-off to an $\mathbf{x}$ strategist when it plays against a $\mathbf{y}$ strategist. Then $\mathbf{p}$ is an ESS if and only if

$$(1 - \varepsilon)E(\mathbf{p}, \mathbf{p}) + \varepsilon E(\mathbf{p}, \mathbf{q}) > (1 - \varepsilon)E(\mathbf{q}, \mathbf{p}) + \varepsilon E(\mathbf{q}, \mathbf{q}) \qquad (10.6.1)$$

or, since ε is very small, if and only if

$$E(\mathbf{p}, \mathbf{p}) \geqslant E(\mathbf{q}, \mathbf{p}) \qquad (10.6.2a)$$

and

$$E(\mathbf{p}, \mathbf{q}) > E(\mathbf{q}, \mathbf{q}) \quad \text{when} \quad E(\mathbf{p}, \mathbf{p}) = E(\mathbf{q}, \mathbf{p}). \qquad (10.6.2b)$$

In the notation of §10.2.2, the condition (10.6.2a) says that an ESS is a symmetric Nash equilibrium and (10.6.2b) is an additional stability requirement.

DEFINITION 10.6.2.   Let $\mathbf{p}_t$ be a probability vector which characterizes a polymorphic population at time t. If $\mathbf{p}_t$ converges to a probability vector $\mathbf{p}$

which is stable under small perturbations as $t \to \infty$, when $\mathbf{p}$ is called a (*genetically*) *stable polymorphism*.

By definition stable polymorphisms cannot be invaded by small numbers of mutants, so this is the corresponding equilibrium concept in polymorphic populations. The relationship between stable polymorphisms and the corresponding ESSs is in general quite complex, but in 2×2 symmetric evolutionary games the two are equivalent. We illustrate this result with the following *hawk–dove* game:

EXAMPLE 10.6.1. Consider the conflict between two animals which are competing for a resource with pay-off V, i.e. the animal who wins it adds V to its Darwinian fitness. Each animal could carry one of two pure strategies H (hawk) or D (dove). If two hawks meet, they fight over the resource and risk incurring an injury which will reduce their Darwinian fitness by C: so the victor gains $V - C$, but the injured party loses C. The probability of winning is $\frac{1}{2}$ for each hawk and so the expected pay-off to each hawk in this situation is $\frac{1}{2}V - C$. If two doves meet, they share the resource without fighting, so each adds $V/2$ to their Darwinian fitness. Finally if hawk meets dove, the dove retreats, leaving the entire resource to the hawk. The pay-off matrices for this game are shown in Figure 10.6.1, using the standard notation for 2×2 non-zero-sum games in normal form [see §10.2.2].

$$
\begin{array}{c|cc}
 & H & D \\
\hline
H & (\tfrac{1}{2}V-C,\tfrac{1}{2}V-C) & (V,0) \\
\\
D & (0,V) & (\tfrac{V}{2},\tfrac{V}{2})
\end{array}
$$

Figure 10.6.1

Note that the two pay-off matrices are related:

$$
\mathbf{A} = \begin{pmatrix} \tfrac{1}{2}V - C & V \\ O & \tfrac{1}{2}V \end{pmatrix}
$$

and $\mathbf{B} = \mathbf{A}^{\mathrm{T}}$. Because of this it makes no difference which 'player' is ascribed to which animal (notice that we have omitted the usual 'player' labels from the normal form) and the game is *symmetric*.

Consider an infinite polymorphic population of hawks and doves described by the probability vector $\mathbf{p}_t = (p_t, 1 - p_t)^{\mathrm{T}}$, so that p_t denotes the proportion of hawks in the population at time t and $1 - p_t$ the proportion of doves. Each animal engages in one contest and, since the probability that a given animal meets a hawk or a dove is p_t and $1 - p_t$ respectively, after the contests the change in Darwinian fitness will be $\frac{1}{2}p_t(V - 2C) + (1 - p_t)V$ for hawks and $(1 - p_t)V/2$ for doves. These are the two components $(\mathbf{Ap}_t)_1$ and $(\mathbf{Ap}_t)_2$ of the 2×1 vector $\mathbf{Ap}_t$. Each animal then reproduces asexually, and the growth in the proportion of hawks in the next generation is given by the proportional increase in their Darwinian fitness compared to the average increase in Darwinian fitness in the whole population. That is

$$(p_{t+1} - p_t)/p_t = ((\mathbf{Ap}_t)_1 - \mathbf{p}_t^{\mathrm{T}}\mathbf{Ap}_t)/\mathbf{p}_t^{\mathrm{T}}\mathbf{Ap}_t \qquad (10.6.3)$$

This describes the dynamics of the popuation of hawks and doves (the corresponding equation for the growth in the proportion of doves has $(\mathbf{Ap}_t)_2$ on the right-hand side). For which values of V and C does $\mathbf{p}_t$ converge to a stable polymorphism? This question is addressed in Maynard Smith (1982), where it is shown that for all initial states of the population, $p_t \to 1$ if $V > 2C$ and $p_t \to V/2C$ if $V \leqslant 2C$.

Now suppose that the population is monomorphic, so that each animal employs the strategy $\mathbf{p} = (p, 1 - p)^{\mathrm{T}}$. Is there an ESS for a population playing this game? Since $V/2 < V$, a glance at the normal form tells us that a population of doves can always be invaded by hawks, so the pure strategy D is not an ESS. On the other hand, if $\frac{1}{2}V - C > 0$, then no mutants which play dove with even the smallest probability can invade a population of hawks, i.e. if $V > 2C$ then H is an ESS. To find any strictly mixed ESS, $\mathbf{p} = (p, 1 - p)^{\mathrm{T}}$ with $0 < p < 1$, we use the fact that ESSs are symmetric Nash equilibria in combination with Theorem 10.2.3, which tells us that, in any ESS $\mathbf{p}$,

$$E(\sigma_1, \mathbf{p}) = E(\sigma_2, \mathbf{p}) = \ldots = E(\sigma_k, \mathbf{p}) = E(\mathbf{p}, \mathbf{p}),$$

where $\sigma_1, \sigma_2, \ldots, \sigma_k$ are the pure strategies to which $\mathbf{p}$ assigns positive probability. This result is known as the *Bishop–Cannings* theorem in the evolutionary games literature (see Bishop and Cannings (1978)). Thus for $\mathbf{p}$ to be a strictly mixed ESS in the hawk–dove game we must have $E(H, \mathbf{p}) = E(D, \mathbf{p})$, that is

$$\tfrac{1}{2}(V - 2C) + (1 - p)V = (1 - p)V/2$$

and so if $V \leqslant 2C$ we have an admissible mixed ESS, $p = V/2C$.

Hence the probability vectors which describe ESSs in monomorphic hawk–dove populations are identical to the probability vectors which characterize stable polymorphisms. But this is not true in general, as we see below.

Now consider a general symmetric evolutionary game in a polymorphic population. This is defined by (i) two $n \times n$ pay-off matrices $\mathbf{A}$ and $\mathbf{B}$, where $\mathbf{B} = \mathbf{A}^{\mathrm{T}}$, and (ii) a dynamic on the mixed strategy $\mathbf{p}_t$ which characterizes the population, usually assumed to be that of *genetic inheritance*, an example of which was given in (10.6.3). Under genetic inheritance the growth in the probability p_i that the ith pure strategy is used equals the difference between the pay-off to i strategists and the average pay-off to the population at time t:

$$(p_{i,t+1} - p_{i,t})/p_{i,t} = (\mathbf{A}\mathbf{p}_t)_i - \mathbf{p}_t^T \mathbf{A}\mathbf{p}_t. \tag{10.6.4}$$

Notice that (10.6.4) differs from (10.6.3) in the denominator $\mathbf{p}_t^T \mathbf{A}\mathbf{p}_t$ of the right-hand side. Taylor and Jonker (1978) show that this does not alter the qualitative behaviour of symmetric evolutionary games, but in asymmetric conflicts it does make a difference [see §10.6.2].

The dynamic (10.6.4) refers to systems in discrete time, but we may also consider the evolution of the population in continuous time, in which case it is usual to employ the function rather than the subscript notation. The continuous time analogue of (10.6.4) is

$$\dot{p}_i(t)/p_i(t) = (\mathbf{A}\mathbf{p}(t))_i - \mathbf{p}(t)^T \mathbf{A}\mathbf{p}(t). \tag{10.6.5}$$

In continuous time, stable polymorphisms correspond to asymptotically stable points (i.e. sinks) of the dynamical system, and ESSs to linear sinks [see §6.3 and IV, §7.11]. Thus the strategies which define ESSs are also probability vectors characterizing stable polymorphisms, but not vice versa in general. When $n = 2$ then ESSs and stable polymorphisms are equivalent (in both discrete and continuous time—and ESSs always exist—see Maynard Smith (1982)). However, when $n > 2$ and the dynamics are discrete, there may be no connection between ESSs and stable polymorphisms and no guarantee of existence. The properties of ESSs and their relationship with stable polymorphisms are quite complicated and the interested reader should consult Haigh (1975), Bishop and Cannings (1978) and Zeeman (1980) for further details.

10.6.2.   Asymmetric Conflicts

Symmetric evolutionary games assume that all animals have the same pure strategy sets and corresponding pay-offs, so they do not allow animals in the same population to adopt different roles (such as male and female) which may have different pure strategies available to them. A nice example of an evolutionary game with asymmetric strategies and pay-offs is given in Schuster and Sigmund (1981), where animals are able to adopt either the male or the female role, the male's pure strategies being 'faithful' or 'philandering' and the females also having two pure strategies but these are different—to be 'coy' or 'fast'. This is the *biological battle of the sexes* described in Example 10.6.3.

So now let us model the evolution of such a population, and suppose that each animal in the infinite population may assume the role of either player in an *asymmetric game*. That is, the normal form has $n \times m$ pay-off matrices $\mathbf{A}$ and $\mathbf{B}$ where $\mathbf{B} \neq \mathbf{A}^T$ (indeed they may not even be square matrices). In each generation contestants are paired in the usual random way, but now another (independent) random device assigns one animal of the pair to the role of player I and the other to player II, the probability of a given animal assuming either role being $\frac{1}{2}$. Animals carry a *pair* of strategies, one for each player and, when it comes to the contest, use only one of these strategies depending on which role it is assigned to.

The expected pay-off to an animal carrying the strategy pair $\{\mathbf{p}, \mathbf{q}\}$ (i.e. it plays $\mathbf{p}$ as role *I* and $\mathbf{q}$ as role *II*) when it plays against an animal carrying the strategy pair $\{\mathbf{x}, \mathbf{y}\}$ is therefore

$$E(\{\mathbf{p}, \mathbf{q}\}, \{\mathbf{x}, \mathbf{y}\}) = \tfrac{1}{2}\mathbf{p}^T \mathbf{A}\mathbf{y} + \tfrac{1}{2}\mathbf{x}^T \mathbf{B}\mathbf{q}. \qquad (10.6.6)$$

The definition of an ESS in asymmetric conflicts is a generalization of the definition in symmetric conflicts with the expected pay-off function (10.6.6), following Selten (1980). That is, an ESS is a strategy pair $\{\mathbf{p}, \mathbf{q}\}$ which characterizes a monomorphic population that cannot be invaded by a small number of mutants.

DEFINITION 10.6.3.   Consider a monomorphic population where all animals carry the strategy pair $\{\mathbf{p}, \mathbf{q}\}$ into which a small proportion ε of mutants, which carry $\{\mathbf{x}, \mathbf{y}\}$, attempt to invade. Then $\{\mathbf{p}, \mathbf{q}\}$ is an *evolutionary stable strategy* (ESS) if and only if

$$(1 - \varepsilon)E(\{\mathbf{p}, \mathbf{q}\}, \{\mathbf{p}, \mathbf{q}\}) + \varepsilon E(\{\mathbf{p}, \mathbf{q}\}, \{\mathbf{x}, \mathbf{y}\})$$
$$> (1 - \varepsilon)E(\{\mathbf{x}, \mathbf{y}\}, \{\mathbf{p}, \mathbf{q}\}) + \varepsilon E(\{\mathbf{x}, \mathbf{y}\}, \{\mathbf{x}, \mathbf{y}\}), \qquad (10.6.7)$$

and since ε is very small this is equivalent to

$$E(\{\mathbf{p}, \mathbf{q}\}, \{\mathbf{p}, \mathbf{q}\}) \geqslant E(\{\mathbf{x}, \mathbf{y}\}, \{\mathbf{p}, \mathbf{q}\}) \qquad (10.6.8a)$$

and

$$E(\{\mathbf{p}, \mathbf{q}\}, \{\mathbf{x}, \mathbf{y}\}) > E(\{\mathbf{x}, \mathbf{y}\}, \{\mathbf{x}, \mathbf{y}\}),$$

if

$$E(\{\mathbf{p}, \mathbf{q}\}, \{\mathbf{p}, \mathbf{q}\}) = E(\{\mathbf{x}, \mathbf{y}\}, \{\mathbf{p}, \mathbf{q}\}). \qquad (10.6.8b)$$

That is, an ESS is a stable, symmetric Nash equilibrium.

A result of Selten (1980) states that only pure strategies can be ESSs in asymmetric conflicts. The following example illustrates the reason.

EXAMPLE 10.6.2.   Consider the asymmetric game with normal form

$$\mathbf{A} = \begin{pmatrix} a & 0 \\ 0 & b \end{pmatrix}, \qquad \mathbf{B} = \begin{pmatrix} c & 0 \\ 0 & d \end{pmatrix},$$

where $\mathbf{A} \neq \mathbf{B}$. Although $\mathbf{A}$ and $\mathbf{B}$ are diagonal, this is the stage game of a general 2×2 asymmetric evolutionary game, as we shall see below. There are obvious pure Nash equilibria according to the signs of a, b, c and d which may be found by the usual arguments [see §10.2.2]. A mixed Nash equilibrium pair for the stage game is $\{\hat{\mathbf{x}}, \hat{\mathbf{y}}\}$, where $\hat{\mathbf{x}} = (d/(c + d), c/(c + d))^T$ and $\hat{\mathbf{y}} = (b/(a + b), a/(a + b))^T$, provided that a, b, c and d are such that these are probability vectors. Now suppose that $\hat{\mathbf{x}}$ is properly mixed and consider a population carrying $\{\hat{\mathbf{x}}, \hat{\mathbf{y}}\}$ except for a small proportion of mutants which carry $\{\mathbf{e}, \hat{\mathbf{y}}\}$, where $\mathbf{e}$ denotes the pure strategy $(1,0)^T$. It is easy to verify that the expected pay-off to every animal, whether it be a mutant or not, whether it plays a mutant or not, is $\frac{1}{2}(ab/(a+b) + cd/(c+d))$.

Thus the stability condition (10.6.8b) is violated and $\{\hat{\mathbf{x}}, \hat{\mathbf{y}}\}$ cannot be an ESS. Since either $\hat{\mathbf{x}}$ or $\hat{\mathbf{y}}$ must be properly mixed in a mixed ESS, and, since a similar argument holds for the invasion of $\{\hat{\mathbf{x}}, \hat{\mathbf{y}}\}$ by the mutants $\{\hat{\mathbf{x}}, \mathbf{e}\}$ there are no mixed ESSs in this game. The reason is that there is no difference between $\{\hat{\mathbf{x}}, \hat{\mathbf{y}}\}$ and $\{\mathbf{e}, \hat{\mathbf{y}}\}$ in an asymmetric evolutionary game because the strategies $\hat{\mathbf{x}}$ and $\mathbf{e}$ never meet each other in practice. This is a general result for $n \times m$ asymmetric conflicts which is not true for symmetric conflicts (as we have seen in Example 10.6.1).

Now consider the evolution of a polymorphic population in continuous time. So the population is characterized by a pair of probability vectors $\{\mathbf{p}(t), \mathbf{q}(t)\}$, where $p_i(t)$ denotes the probability that animals play the ith pure strategy in role I ($i = 1, \ldots, n$) and $q_j(t)$ is the probability that the jth pure strategy is used in role II ($j = 1, \ldots, m$), at time t. The generalization of the genetic inheritance dynamic (10.6.5) to the asymmetric case is

$$\dot{p}_i(t)/p_i(t) = (\mathbf{Aq}(t))_i - \mathbf{p}(t)^T \mathbf{Aq}(t) \tag{10.6.9a}$$

and

$$\dot{q}_j(t)/q_j(t) = (\mathbf{p}^T(t)\mathbf{B})_j - \mathbf{p}(t)^T \mathbf{Bq}(t). \tag{10.6.9b}$$

EXAMPLE 10.6.3. It is easy to verify that the genetic inheritance dynamics (10.6.9) are invariant under addition of a constant to any column of $\mathbf{A}$ and/or any row of $\mathbf{B}$, hence for 2×2 asymmetric conflicts we may assume without loss of generality that $\mathbf{A}$ and $\mathbf{B}$ are diagonal as in Example 10.6.2. This is then the general 2×2 asymmetric game, and its dynamic were investigated by Schuster and Sigmund (1981): let $\mathbf{p}(t) = (p(t), 1 - p(t))^T$ and $\mathbf{q}(t) = (q(t), 1 - (t))^T$ characterize the population at time t, so the dynamics (10.6.9) for the game of Example 10.6.2 may be written

$$\dot{p}(t) = p(t)(1 - p(t))((a+b)q(t) - b)$$

and

$$\dot{q}(t) = q(t)(1 - q(t))((c+d)p(t) - d).$$

Assuming there exists a mixed Nash equilibrium $(\hat{x}, \hat{y})$ and linearizing about this point [see §6.3] gives the Jacobian

$$\begin{pmatrix} 0 & (a+b)cd/(c+d)^2 \\ (c+d)ab/(a+b)^2 & 0 \end{pmatrix}$$

with eigenvalues $\pm \lambda$ where $\lambda = ((abcd)/(a+b)(c+d))^{\frac{1}{2}}$ [see I, 7.2].

Several cases may be considered according to the signs of a, b, c and d, but there are only two that yield interesting dynamics. The first is when $a, b \leqslant 0$ and $c, d \geqslant 0$ with not both of a and b nor c and d being zero. This game is called the *biological battle of the sexes*. The eigenvalues $\pm \lambda$ are purely imaginary and the invariant

$$p(t)^d (1 - p(t))^c \, q(t)^{-b} (1 - q(t))^{-a}$$

defines a family of closed orbits in $[0, 1] \times [0, 1]$ with time average $(\hat{x}, \hat{y})$. So the system is stable but not asymptotically stable and $(\hat{x}, \hat{y})$ is the centre of the flow [see §6.3.2] (Figure 10.6.2).

The other interesting case is when a, b, c and d are all $\geqslant 0$ with again not both of a and b nor c and d being zero. The eigenvalues are real and of opposite sign so (x, y) is saddle-point unstable [see §6.1.1 and §6.3.2] (Figure 10.6.3).

Recall that the inclusion of $\mathbf{p}(t)^{\mathrm{T}} \mathbf{A} \mathbf{p}(t)$ in the denominator of the right-hand side of (10.6.5) made no difference to the evolution of a symmetric conflict.

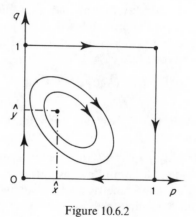

Figure 10.6.2

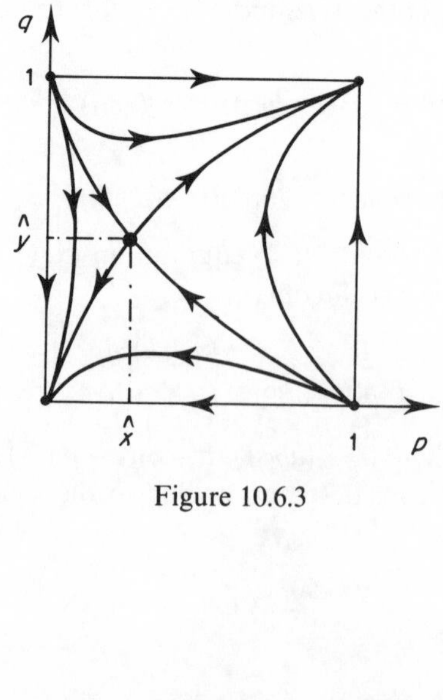

Figure 10.6.3

However, the time paths of asymmetric conflicts *do* change when the genetic inheritance dynamics (10.6.9) are normalized by the factors $p(t)^T Aq(t)$ and $p(t)^T Bq(t)$. An example of this is given in Maynard Smith and Hofbauer (1987), who show that the biological battle of the sexes evolves to a limit cycle [see §6.3.6] when the genetic inheritance dynamics are normalized in this way.

To summarize the general method of analysing asymmetric evolutionary games: first locate all Nash equilibria of the stage game as these will be fixed points of the flow. The only ESSs for the system are the pure Nash equilibria which are also linear sinks. Although mixed Nash equilibria will not be ESSs they may still be asymptotically stable, and to investigate these we linearize the dynamics about this point and use the methods of Chapter 6 of this volume.

10.6.3. The Evolution of Cooperation and Refinements of the ESS Concept

In Example 10.4.2 we saw that the unique subgame-perfect equilibrium in the finitely repeated prisoner's dilemma was for both players to confess at every stage. However, empirical studies do not support this conclusion: when the number of repetitions is sufficiently large there is some evidence to suggest that players will cooperate, simultaneously keeping quiet for several successive moves. Such cooperative play can be supported as equilibria in infinitely repeated games but not in finitely repeated games (with complete information). However, what if the finitely repeated game was played again and again by successive generations of a population? Is it possible for cooperative play to *evolve* in such a situation?

Axelrod (1984) simulated the evolution of the finitely repeated prisoner's dilemma under the usual genetic inheritance dynamic (10.6.4). He called for computer programs to be submitted which played the 200-move prisoner's dilemma with a pure strategy. Examples of these are: 'TIT FOR TAT' which plays A (keep quiet) on the first move and thereafter copies whatever the opponent played on the previous move; 'DOWNING' which estimates the probability of one's opponent cooperating (i.e. playing A given that it plays A) after every move, starting with the initial probability $\frac{1}{2}$ and thereafter updating using Bayes rule [see II, §16.4], then it selects the move which maximizes its expected pay-off; 'JOSS' which is like 'TIT FOR TAT' but which sneaks in the occasional defection (i.e. playing B); and 'ALLD' which plays B (confess) at every stage. Fourteen programs were submitted and each was represented in the initial population in equal numbers. The population then played the 200-move prisoner's dilemma and evolved according to the usual dynamic. After several generations 'TIT FOR TAT' (which is a nice, cooperative strategy) become the dominant program.

Axelrod then organized another tournament for programs to play the infinitely repeated prisoner's dilemma with discount rate δ less than but near to 1. This attracted more entries, 63 in all, and because the cooperative programs fared better in the first tournament, more 'nice' programs were submitted than 'nasty' ones. For all five variations of the game (with different values of the pay-off parameters P, R, S and T) 'TIT FOR TAT' was again the most successful program. However, this does not mean that 'TIT FOR TAT' is an ESS for the infinitely repeated prisoner's dilemma: only a few of the possible pure strategies were represented in the tournament, to say nothing of possible mixed strategies! Axelrod showed that 'TIT FOR TAT' is a symmetric Nash equilibrium provided that $\delta \geq \max\{(T-R)/(T-P), (T-R)/(R-S)\}$, but it does not satisfy the stability condition (10.6.2b). Indeed any 'nice' strategy, which prescribes playing A at every stage against itself and against 'TIT FOR TAT' can invade a population of 'TIT FOR TAT', given some initial random drift. It may be that 'TIT FOR TAT' is a *limit ESS*, a concept introduced by Selten (1983) which is similar to his notion of trembling hand perfection [see §10.4.3] and which allows different types of mutation to occur simultaneously.

However, 'ALLD', the nastiest strategy of all, *is* an ESS since any strategy which ever cooperates gets the 'suckers' pay-off S. So how could 'TIT FOR TAT' ever get established in a population of nasty strategies? Suppose that when mutation occurs the mutants are located in a cluster, so that they have a probability μ greater than the usual probability of playing each other. In this situation a stable symmetric Nash equilibrium is called a *Grafen ESS* (see Hines and Maynard Smith (1978)). 'TIT FOR TAT' again fails the stability requirement so it is not a Grafen ESS. However, the probability μ does not need to be very large for a μ-cluster of 'TIT FOR TAT' mutants to invade a population of 'ALLD'. Axelrod shows that this will occur provided that $\mu > ((P-S)(1-\delta))/(R-S(1-\delta)-P\delta)$.

Thus we envisage a cyclic evolution in this population when clustering is allowed: beginning with a nasty population playing 'ALLD', 'TIT FOR TAT' can invade in μ-clusters when μ is sufficiently large. However, once 'TIT FOR TAT' is established, any other 'nice' strategy can invade the population. But if this nice strategy lacks the robustness of 'TIT FOR TAT' it will be invaded by a population of nasty mutants such as 'ALLD', and the cycle is complete.

10.7. BARGAINING THEORY

Until the 1980s most of the literature on bargaining modelled the process as a cooperative game [see §10.1]. It began with Nash (1950) who defined the *bargaining problem* as a cooperative game in which the pre-play communication takes the form of bargaining over possible agreements in the set of possible outcomes X. As with all cooperative games it assumes the existence of an independent arbitrator who enforces the agreement made between the players. Nash was aware that this assumption is unrealistic, that the model *assumes* rather than *explains* cooperative behaviour, so he outlined the following problem, known as the *Nash Program*: is it possible to implement the solution to the bargaining problem as optimal pay-offs in a non-cooperative game? Since there are no strategies in cooperative games, there are many possible non-cooperative games one could associate with the bargaining prob em. Nash attempted to solve the problem with the *Nash demand game* [see §10.7.2] but it was not until the 1980s, when interest in the Nash program revived with Rubinstein's non-cooperative model of the bargaining process, that the program was completed satisfactorily. Moulin (1982) provided non-cooperative implementations of the Nash and the Kalai–Smorodinsky bargaining solutions, and following that the Nash program was completed by Binmore, Rubinstein and Wolinsky (1986). A detailed account of the whole Nash program is given in Binmore and Dasgupta (1987). Here we provide only a brief overview of these results, beginning with cooperative bargaining theory, since this is the object of the Nash program.

10.7.1. The Axiomatic Approach

Consider two players with utility functions $u_i: X \to \mathbb{R}$ ($i = 1,2$) such that the *bargaining set* $\beta = \{u_1(\mathbf{x}), u_2(\mathbf{x}) | \mathbf{x} \in X\}$ is compact and convex [see IV, §§2.8 and 15.2]. Take $\mathbf{s} \in \beta$ and call it the *status quo* point, which may be interpreted as the utilities of outcomes when the players fail to agree. An *arbitration scheme* is a function $f: \beta \times \{\mathbf{s}\} \to \beta$ which identifies points in the bargaining set as solutions to the bargaining problem. Nash (1950) showed that such a unique point exists provided the arbitration scheme satisfies the following axioms:

A1 $f(\beta, \mathbf{s})$ is Pareto optimal [see §10.1], feasible and individually rational [see Definition 10.4.1];

A2 $f(\Phi\beta, \Phi\mathbf{s}) = \Phi f(\beta, \mathbf{s})$ for affine $\Phi: \mathbb{R}^2 \to \mathbb{R}^2$ [see IV, §5.13];

A3 $f(\beta, \mathbf{s}) = f(\beta', \mathbf{s})$ for all $\beta' \subseteq \beta \ni f(\beta, \mathbf{s}) \in \beta'$.

In that case $f(\beta, \mathbf{s})$ is called the *Nash bargaining solution* and is given by

$$f(\beta, \mathbf{s}) = \operatorname{argmax}(u_1(\mathbf{x}) - s_1)^\tau (u_2(\mathbf{x}) - s_2)^{1-\tau} \exists (u_1, u_2) \geq (s_1, s_2) \quad (10.7.1)$$

The parameter τ is called the index of *bargaining power* since it gives a measure of players utility in the Nash bargaining solution (Figure 10.7.1).

EXAMPLE 10.7.1. Consider the 2×2 game shown in Figure 10.7.2. Without further information it is customary to take player's values as their status quo points in the Nash bargaining solution, since these are the most they can guarantee themselves when there is no cooperation. These are $v_1 = v_2 = \frac{1}{4}$, and give the individually rational, feasible bargaining set of Figure 10.7.2 (b). Since there is no reason to suppose that players have asymmetric bargaining power, we set $\tau = \frac{1}{2}$ and the (symmetric) Nash bargaining solution is the point $(3, 3)$.

Nash (1953) suggested that, instead of taking the status quo points to be players' values, it is more realistic to identify $\mathbf{s}$ with the pay-off to *optimal threat strategies* for the following reason: in Example 10.7.1, what can player *I* do if

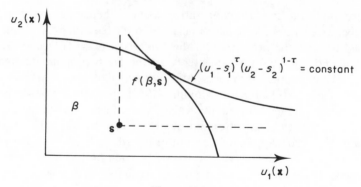

Figure 10.7.1

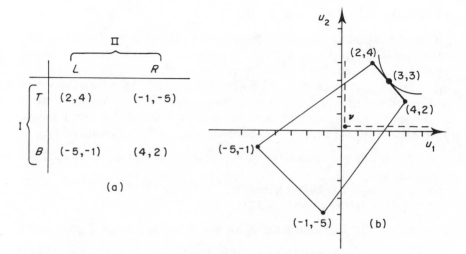

Figure 10.7.2

II threatens to play *L*? If he plays *T* then *II* gets her maximum payoff of 4, and if he plays *B* then he gets his minimum pay-off of -5. Player *II* is in a more powerful position than player *I*, who has no comparable threat—yet both players receive the same in the Nash bargaining solution. To obtain a bargaining solution which reflects the differences in player's threats, Nash had the idea of augmenting the bargaining game with a non-cooperative game in which players choose their threat strategies. The status quo point of the bargaining game is then taken as the threat outcome and the Nash bargaining solution calculated accordingly. This augmented game is called the *Nash threat game*.

THEOREM 10.7.1.   *There exists at least one pair of optimal threat strategies in the Nash threat game, and all such pairs lead to the same Nash bargaining solution.*

The determination of optimal threat strategies can be quite complicated, particularly if the Pareto optimal boundary of the bargaining set has several linear segments. However, when the Pareto optimal boundary is of the form $\alpha u_1 + u_2 = \beta$, then it can be shown that the optimal threats are solutions to the *zero-sum* game $\alpha(a_{ij}) - (b_{ij})$ (see Thomas (1984)).

EXAMPLE 10.7.2. In Example 10.7.1 the pareto optimal boundary is $u_1 + u_2 = 6$ and so optimal threat strategies are the solutions to the zero-sum game

$$\begin{pmatrix} -2 & 4 \\ -4 & 2 \end{pmatrix}.$$

These are $\mathbf{p}^* = (1,0)$ and $\mathbf{q}^* = (1,0)$ so the optimal threat strategies are T for I and L for II. Thus the status quo point is $(2,4)$, which is therefore also the solution of the Nash threat game.

Nash's third axiom, called the *independence of irrelevant alternatives*, is the most contentious. For it implies that the two bargaining sets in Figure 10.7.3 yield the same solution: although player II can get more in Figure 10.7.3(a) the Nash bargaining solution does not reflect this. Most research on the axiomatic approach to bargaining has used other assumptions in place of *A3*. For example the *Kalai–Smorodinsky* bargaining solution replaces *A3* by a monotonicity condition which yields the solution illustrated in Figure 10.7.4. There are many other axioms which can be used to replace the independence of irrelevant alternatives (see Roth (1979)) but the Nash bargaining solution remains the most widely used, probably because it is the most mathematically tractable. The main problem with applications of the Nash bargaining solution is the identification of the two parameters, the status quo point $\mathbf{s}$ and the index of bargaining power τ. It was not until a non-cooperative implementation of the

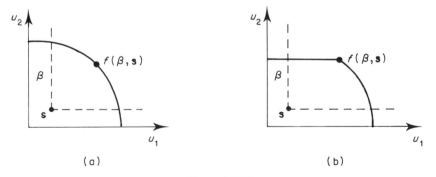

Figure 10.7.3

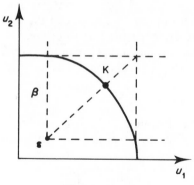

Figure 10.7.4

Nash bargaining solution was given by Binmore, Rubinstein and Wolinsky (1986) that we had sufficient information for model specific interpretations of **s** and τ.

10.7.2.   The Strategic Approach

The unrealistic assumption of the existence of an independent arbitrator casts doubt on the axiomatic approach as an explanation of cooperative behaviour. All that it can do is to tell us which outcomes are likely when we already assume cooperation will occur. The strategic approach to bargaining attempts to implement the axiomatic solutions as equilibrium pay-offs in a non-cooperative game, so that we can dispense with the assumption of cooperation.

In the *Nash demand game* the two players simultaneously announce demands d_1 and d_2 which are points on their utility scales. If there exists $\mathbf{x} \in X$ such that $u_1(\mathbf{x}) \geq d_1$ and $u_2(\mathbf{x}) \geq d_2$, then each player gets what he demanded, otherwise they receive their status quo (or threat) pay-offs. In general there are an infinite number of inequivalent Nash equilibria in this game: any Pareto optimal, individually rational and feasible point in β is a Nash equilibrium pay-off. To counteract this difficulty, Nash differentiated between equilibrium outcomes by 'smoothing' the game, and studying the behaviour of the equilibrium points in the smoothed game as the degree of smoothing tends to zero. The type of smoothing that he introduced—a small probability ε that player's receive only their status quo pay-offs even when their demands are feasible—is related to Selten's 'trembles' [see §10.3.3]. In his 1953 paper Nash showed that the Nash bargaining solution was the only 'necessary' limit of the equilibrium points of the smoothed game as $\varepsilon \rightarrow 0$, so in this he pre-empted Selten's concept of trembling hand perfection.

The Nash demand game is only one of many possible non-cooperative games that could be associated with a bargaining game and, moreover, its strategies

do not correspond to moves in a negotiation process. A non-cooperative model
of the bargaining process itself was introduced by Ståhl (1972). Two players
seek to make an agreement in the feasible set of outcomes $X = \{\mathbf{x} \in \mathbb{R}^2 | x_1, x_2 \geq 0$
and $x_1 + x_2 = 1\}$. Their strategies consist of alternating offers $\mathbf{x} \in X$ at discrete
points in time $t = 0,\ T,\ 2T,\dots$ which the other player either accepts or refuses.
At $t = 0$ player I makes an offer $\mathbf{x}$ to II (so I gets x_1 and II gets x_2). If II
accepts, the games ends with pay-offs $u_1(\mathbf{x})$ and $u_2(\mathbf{x})$, but if she refuses, then it
becomes her turn to make an offer to I at $t = T$, which he can either accept or
refuse. This process of alternating offers continues until agreement is reached.
However, players have no incentive to agree! In this model there is neither
uncertainty nor costs incurred by delay..., players lose nothing by refusing an
offer, the game could continue indefinitely, and any $\mathbf{x} \in X$ could be supported
by a Nash equilibrium strategy pair.

Rubinstein (1982) extended this *Ståhl bargaining model* by introducing *time
preferences* in such a way that there is a unique subgame-perfect equilibrium
outcome. Suppose players have preferences over $X \times \mathbb{R}$ such that $(\mathbf{x}, t) \gtrsim (\mathbf{x}, s)$
if $t \leq s$. Fishburn and Rubinstein (1982) showed that such preferences can be
represented by separable utility functions of the form $\delta^t u(\mathbf{x})$, where the discount
rate $\delta \in [0, 1]$, provided that in addition to the usual conditions of completeness,
transitivity, monotonicity and continuity for the representation of preferences
by utility functions [see §10.1], these time preferences are stationary and there
exists a 'time indifferent' agreement $\mathbf{g}$, i.e. $(\mathbf{g}, t) \sim (\mathbf{g}, s)$ for all t, s. The functions
$u: X \to \mathbb{R}$ are continuous, monotonic increasing and, for given δ, unique up to
affine transformations. Moreover $u(\mathbf{g}) = 0$ and there exists a δ^* such that for
all δ greater than δ^* the corresponding u are concave [see IV, §15.2].

We can now introduce *Rubinstein's game*: suppose δ_1 and δ_2 are sufficiently

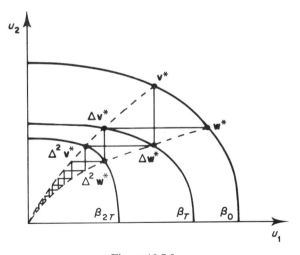

Figure 10.7.5

large that u_1 and u_2 are concave. Then the bargaining sets

$$\beta_t = \{\delta_1^t u_1(\mathbf{x}), \delta_2^t u_2(\mathbf{x}) | \mathbf{x} \in X\}$$

are compact and convex. Assuming that $\delta_1 \geq \delta_2$, the bargaining sets at $t = 0$, T and $2T$ are illustrated in Figure 10.7.5. There $\Delta(x, y) = (\Delta_1 x, \Delta_2 y)$, where $\Delta_i = \delta_i^T$ is player i's discount rate per period. Rubinstein's game is identical to the infinite Ståhl bargaining model described above, except that the bargaining set shrinks after every stage as in Figure 10.7.5. Because of the dynamic nature of the game, the appropriate solution concept is the subgame-perfect Nash equilibrium, and Rubinstein (1982) showed that there is a unique subgame-perfect equilibrium outcome, which is $\mathbf{w}^*$ if player I offers first and $\mathbf{v}^*$ if player II offers first, and both offers are accepted in the first stage of the game. The intuition behind this result lies in the stationary nature of the game and the fact that in any subgame-perfect equilibrium both players must be indifferent between acceptance and rejection. Hence if we put $\mathbf{v}^* = (v_1, v_2)$ and $\mathbf{w}^* = (w_1, w_2)$ then $v_1 = \Delta_1 w_1$ and $w_2 = \Delta_2 v_2$ as shown in Figure 10.7.5.

A convenient characterization of the subgame-perfect equilibrium outcome pair is given in Figure 10.7.6: a little algebra yields

$$\mathbf{v}^* = (\Delta_1 a\alpha, b\beta) \quad \text{and} \quad \mathbf{w}^* = (a\alpha, \Delta_2 b\beta),$$

where a and b are the intercepts shown and

$$\alpha = (1 - \Delta_2)/(1 - \Delta_1 \Delta_2) \quad \text{and} \quad \beta = (1 - \Delta_1)/(1 - \Delta_1 \Delta_2).$$

We now discuss the completion of the Nash program by Binmore, Rubinstein and Wolinsky (1986). They showed that the subgame-perfect equilibrium outcome pair in Rubinstein's game tends to the Nash bargaining solution as the time between offers approaches zero. To see this, note that as $T \to 0$, for

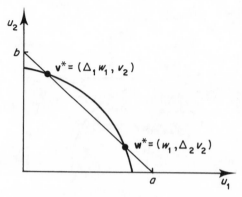

Figure 10.7.6

example

$$\alpha \to \ln \delta_2 / (\ln \delta_1 + \ln \delta_2) = \tau \quad \text{and} \quad \beta \to 1 - \tau$$

by de l'Hôpital's rule [see IV, Corollary 3.4.4], so the limiting value of both $\mathbf{v}^*$ and $\mathbf{w}^*$ is $(a\tau, b(1 - \tau))$ which is the generalized Nash bargaining solution with bargaining power τ and status quo point $(0,0) = (u_1(\mathbf{g}), u_2(\mathbf{g}))$. These characterizations of τ and $\mathbf{s}$ tell us how to make an appropriate choice of these parameters in applications of the Nash bargaining solution: $\mathbf{s}$ corresponds to players' utilities of a 'time indifferent' agreement and τ increases with δ_1, so the more 'impatient' is player I, the lower is his bargaining power.

However, these interpretations are only applicable to a model where friction arises because of costly delay in the bargaining process. If the motivation to form an agreement comes instead from a risky environment, then Binmore, Rubinstein and Wolinsky showed that the status quo should be taken as players 'fall-back' utility levels and $\tau = \lambda_2/(\lambda_1 + \lambda_2)$, where λ_i is related to player i's subjective assessment of the probability that the bargaining process breaks down: between times t and $t + h$ this is $\lambda_i t + o(h)$, i.e. player i thinks that the negotiation breakdowns follow a Poisson process of rate λ_i [see II, §20.1].

In the above model the breakdowns *must* occur exogenously, that is the players are not allowed to 'opt out' of the negotiations as one of their strategies. Shaked and Sutton (1984) extended Rubinstein's model to allow players the option of breaking negotiations, in which case they both get their *outside option* pay-off. They showed that these outside options only have an effect on the bargaining process if one or both of the players can get more from their outside option than they would when no such options are available. Thus, when formulating applied models of a bargaining process, outside options act as a constraint on the Nash bargaining solution and should not be taken as the status quo point.

Most of the solutions of the axiomatic approach to bargaining have straightforward generalizations to n-person bargaining problems, the difference being that players can now form coalitions so that they have options which are not always available to individuals. For the non-cooperative implementation of the three-person Nash bargaining solution, see Binmore's 'Bargaining and Coalitions' in Roth (1984), and for the implementation of the n-person Nash and n-person Kalai–Smorodinsky bargaining solutions see Moulin (1982).

10.7.3. Bargaining with Incomplete Information

Rubinstein (1985) extended his model of the bargaining process to the case where player II can be one of two types and player I does not know which type player II actually is. This is a standard form of incomplete information and an appropriate solution concept is the sequential equilibrium [see §10.3.2]. The usual problem of multiplicity occurs with the set of sequential equilibria in Rubinstein's game with incomplete information. However, Rubinstein argues that many of these are unreasonable, and by making additional requirements on players beliefs and on equilibrium behaviour he restricts the solution concept to that of *bargaining sequential equilibria*, which are (typically) unique.

The Bayesian equilibrium concept [see §10.3.1] was applied to a more general bargaining model with incomplete information by Harsanyi and Selten (1972). Consider the case where both players can be one of several types, the probability that player *I* is of type *i* being λ_i and that of player *II* being type *j*, μ_j. The Bayes equivalent of this game is a multi-person game of complete information where chance moves first, selecting the types of each player according to the appropriate probability distributions [see §10.3.1]. For the remainder of the game the players follow a bargaining procedure where information about their types may or may not be revealed, the Bayesian equilibria being called *completely revealing*, *partially revealing* or *non-revealing equilibria* according to the extent to which player's choice of strategies signals their type to the other player [see §10.5.1]. Typically there are a multiplicity of such equilibria in Harsanyi and Selten's model, but they identify the 'solution' outcome as that feasible pay-off $(\mathbf{x}, \mathbf{y})$ which maximizes

$$\Pi(x_i - s_i)^{\lambda_i} \Pi(y_j - r_j)^{\mu_j}, \tag{10.7.2}$$

where x_i is the pay-off and s_i the status-quo point for player *I* type *i*, and y_j is the pay-off and r_j the status-quo point for player *II* type *j*.

This is just the multi-person Nash bargaining solution with bargaining powers λ_i and μ_j. Harsanyi and Selten do not attempt to justify this choice strategically—instead they introduce an arbitration scheme to their bargaining game with incomplete information which yields (10.7.2) as the unique solution criterion. In Chapter 8 of Binmore and Dasgupta (1987), Binmore considers a non-cooperative implementation of Harsanyi and Selten's bargaining solution which is based on the Nash demand game. However, it does *not* select (10.7.2) as the unique solution except in certain special cases of Harsanyi and Selten's model, for example when all Bayesian equilibria are of the non-revealing type, or when there is only one-sided incomplete information.

10.8.   RATIONAL PLAY IN NON-COOPERATIVE GAMES

Most of the game theoretic literature requires that a 'solution' to the game be a Nash equilibrium, defined in §10.2 as a set of strategies, one for each player such that each player's strategy is optimal (in the sense that it maximizes his expected pay-off) *given* that the other players are employing their optimal strategies. Thus Nash equilibrium play requires that players maximize utility subject to *consistent* expectations about other player's behaviour. These expectations need not be *rational*—but what do we mean by this? Very many meanings have been ascribed to this term in the literature and the object of this section is to provide a brief account of the different notions of rationality in non-cooperative games.

In §10.8.2 we discuss various refinements of Nash equilibria according to different assumptions about the 'rationality' of player's expectations. In extensive form games we have already encountered the *backwards induction rationality* of subgame-perfect Nash equilibria [see §10.2.3] and refinements of this in extensive games with incomplete information [see §§10.3.2 and 10.3.3]. In normal form games we shall distinguish between the multiplicity of Nash equilibria by imposing certain robustness criteria which require an *iterated dominance rationality* in player's behaviour. Refinements of Nash equilibria were developed separately in extensive form and normal form games, but are unified by the rationality of *strategic stability*, introduced by Kohlberg and Mertens (1986).

The list of possible refinements of the Nash equilibrium concept in §10.8.2 seems endless—we could introduce many more games, with new idiosyncracies, requiring yet *another* tailor-made solution—but doesn't this cast doubt on the whole conventional approach to analysing games? Surely this plethora of additional requirements on player's behaviour has been introduced for the wrong reasons, i.e. for mathematical simplicity and/or to reduce the set of possible solutions. Not only this, the conventional approach ignores one of the most important features of a non-cooperative game, the 'intelligence' of one's opponent. The most recent advances in game theory are in the realms of *bounded rationality*, where opponents are modelled as *finite automata*—computer programs with bounded memories. The solution concept is again based on a Nash equilibrium (or one of its refinements) but, when players have bounded rationality, multiplicity may be less of a problem. A brief account of some major developments in this area is given in §10.8.3. But first we mention two alternative approaches to modelling rational play which are based on Savage's notion of rationality rather than the Nash equilibrium concept.

10.8.1. Rationalizable Strategic Behaviour and Bayesian Rationality

The solution concept of *rationalizability* was developed separately by Bernheim (1984) and Pearce (1984). It depends on the notion of *individual rationality* formalized by Savage (1954)—not to be confused with Definition 10.4.1! In Savage's sense an individual is rational if he optimizes subject to a subjective probability distribution over uncertain events which is consistent with all his information. Bernheim defines a *rationalizable strategy* to be one that is optimal for an individually rational player who regards the other player's actions as uncertain events in a game where the individual rationality of all players is common knowledge [see §10.3.1]. Nash equilibria are also rationalizable strategies—those in which players believe their opponents are playing Nash equilibria—but rationalizability allows players to believe *any* consistent play by their opponents, provided that they themselves then optimize according to this belief. So there may be rationalizable strategies which are not Nash equilibria and clearly multiplicity can be a problem. Because of this

Bernheim and Pearce go on to develop refinements of the rationalizability concept, including *subgame rationalizable strategies* in extensive form games. Bernheim shows that if no player is indifferent between the various outcomes in an extensive form game, then the unique subgame rationalizable strategy profile is the subgame-perfect Nash equilibrium, but this is the exception rather than the rule—in general rationalizable strategies are even more numerous than Nash equilibria.

We now turn to the results of a recent paper by Aumann (1987) which not only introduces a viable alternative to the use of Nash equilibrium as the solution concept for non-cooperative games, but also illuminates the notion of *correlated strategies*. These were defined by Aumann (1974) as 'mixed' strategies in which two (or more) players condition their choice of pure strategy on the same randomizing device. This *appears* to assume that the players must have some form of pre-play communication and, since this contradicts our usual assumptions for modelling non-cooperative games [see §10.1], we have not discussed correlated strategies up to now (except in relation to the feasible pay-off region in §10.4.2). As usual we introduce the concepts by example—for a more rigorous definition and further examples see Aumann (1974, 1987).

EXAMPLE 10.8.1.   In the economic battle of the sexes [see Example 10.2.8], the Pareto optimal boundary of the pay-off region marked as a heavy line on Figure 10.4.2 cannot be achieved by the use of ordinary strategies—except of course the end points. However, if the two players correlate their pure strategy choices by tossing a coin with Prob(heads) $= p$ and then both playing C if it comes up heads and F if it comes up tails, then the expected pay-off vector will be a point on this boundary. For example if $p = \frac{1}{2}$, then each player expects the pay-off $\frac{3}{2}$. The correlated strategy can be represented by the probability vector $(p, 1 - p, 0, 0)^T$ which denotes the distribution of player's choices over the joint pure strategy set $(\{C, C\}, \{F, F\}, \{F, C\}, \{C, F\})$.

In the above example there are other correlated strategies which yield expected pay-offs inside the Pareto optimal boundary (indeed all points in the feasible region are pay-offs to correlated strategies) but these need not be correlated *equilibria*. They are only correlated equilibria if no player has the incentive to deviate from the strategy (i.e. not to follow the action prescribed by the toss of the coin) assuming his opponent does not deviate.

Aumann (1987) links the use of correlated equilibria with the concept of *Bayesian rationality*, a concept which combines the notions of Bayesian equilibria and rationalizability. In Harsanyi's model of a game with incomplete information [see §10.3.1] it is assumed that the set Ω of 'states of nature' over which players have prior beliefs includes only the unknown parameters of the game, such as opponent's pay-off matrices. But, in Aumann's model, Ω includes *all* objects of uncertainty (i.e. not only the parameters of the game being played but also the other player's choice of actions in each game). As in Harsanyi's model, these prior beliefs are common to all players. This common prior assumption was referred to in §10.3.1 as the 'Harsanyi doctrine', but there Ω did not include other players actions—which is the traditional 'game-theoretic' view. On the other hand, the concept of rationalizability introduced above adhered to the 'Bayesian view' that players can form subjective priors over other player's actions, but it did not employ the common prior assumption. Aumann synthesizes the two concepts to define a *Bayes rational* player to be one who maximizes his expected pay-off under his posterior probability distribution over Ω (i.e. he is individually rational in Savage's sense) where all player's posterior distributions are based on the same prior. Of course, player's posteriors may differ if they have different information which they use to update their (common) prior.

Aumann proved that the strategies chosen in a finite game with incomplete information, where each player is Bayes rational and where this is common knowledge, will constitute a correlated equilibrium and that every correlated equilibrium can be obtained in this way (for an appropriate parametrization of the common prior). It tells us that the use of correlated strategies does *not* require explicit pre-play communication between players who discuss how to coordinate their choices to depend on the outcome of a randomizing device. Rather, the probability distribution which defines the correlated equilibrium reflects the player's uncertainty about Ω. Hence we can use the correlated equilibrium as a solution concept in *non*-cooperative finite games if we assume that it is common knowledge that all players are Bayes rational.

EXAMPLE 10.8.2.   (Aumann, 1987) In the following three-player game, player *I* chooses the row, *II* the column and *III* the matrix:

	L	R	L	R	L	R
T	$\begin{pmatrix}0,0,3$	$0,0,0\end{pmatrix}$	$\begin{pmatrix}2,2,2$	$0,0,0\end{pmatrix}$	$\begin{pmatrix}0,0,0$	$0,0,0\end{pmatrix}$
B	$\begin{pmatrix}1,0,0$	$0,0,0\end{pmatrix}$	$\begin{pmatrix}0,0,0$	$2,2,2\end{pmatrix}$	$\begin{pmatrix}0,1,0$	$0,0,3\end{pmatrix}$
		A		B		C

Suppose that it is common knowledge that player *III* believes that his opponents will pick either (T, L) with probability p, or (B, R) with probability $1 - p$, where $\frac{1}{3} < p < \frac{2}{3}$. Then *III* will pay B with certainty and this is a correlated equilibrium, yielding a pay-off of 2 to each player. Such a belief could be based on information like 'player *I* and player *II* are brother and sister, so are likely to act in a similar fashion', it does not require that players *I* and *II* *collude*, coordinating their strategy choices by the toss of a coin. All that is necessary to rationalize this equilibrium is that it is common knowledge that player *III* has this belief.

Unfortunately this does not help us to identify a unique solution, indeed correlated equilibria are even more numerous than Nash equilibria: any convex combination of Nash equilibria outcomes is a correlated equilibrium outcome and there may even exist correlated equilibria outcomes which lie outside this convex hull. For instance, in the example above no player can get more than 1 in any Nash equilibrium, yet they each get 2 in the correlated equilibrium described. So this example demonstrates how, when there are more than two players and subsets of these can employ correlated strategies, the inability of the excluded players to coordinate their strategy choices (either because they cannot condition them on the same randomizing device or because they lack some predetermined behavioural conditioning, depending on one's interpretation of correlated strategies) can improve the pay-off for everybody. Aumann (1974) gives an example of how a similar sort of 'partial' correlation can lead to pay-offs outside the feasible region in two-player games.

However, Aumann's work on correlated equilibria has reconciled certain practical and conceptual difficulties with the traditional view of strategy randomization: is it really feasible that players condition serious decisions on the toss of a coin, particularly when their only reason to do so is to prevent their opponents from deviating from equilibrium play? Harsanyi's concept of a *disturbed game* [see §10.8.2] shows how the probability distributions which define a mixed strategy Nash equilibrium can be thought of as reflecting incomplete information about opponents pay-offs. With this incomplete information, players can only estimate what their opponents will do, but given these estimates each player has a unique optimal *pure* strategy—so nobody really randomizes. Aumann extends this argument to explain randomization as an expression of ignorance about opponents actions, so that it can be rationalized even without introducing incomplete information about pay-offs. However, this view requires that in *mixed* strategy equilibria each player i must know exactly what every other player believes about his (i's) actions, but this need not be the case in correlated equilibria (see Aumann (1987) for an example which illustrates this).

We end this section by mentioning the relationship between the two approaches to modelling rational play described here: if we drop the common prior assumption from the definition of Bayes rational we would have the concept of rationalizability—except that it does not allow players to perceive the other's strategies as correlated—and so rationalizable outcomes are similar to those that occur in *subjective correlated equilibria*, i.e. correlated equilibria where players have different probability distributions over the randomizing event that determines their correlated actions.

10.8.2. Refinements of the Nash Equilibrium Concept

One of the many disadvantages with using the simple Nash equilibrium as a solution concept is that many Nash equilibria are highly sub-optimal as, for example, the unique Nash equilibrium in the prisoner's dilemma [see Example 10.4.2]. So Pareto optimality [see §10.1] is one possible criterion for distinguishing between a multiplicity of Nash equilibria. Aumann has suggested two refinements using this criterion: *twisted equilibria* are Nash equilibria in which no player can unilaterally decrease the other player's pay-offs (Aumann (1961)) and *strong equilibria* are Nash equilibria which are also in equilibrium against any subset of players. This is regardless of the redundant player's actions, and so strong equilibria must also be Pareto optimal (see Aumann (1959)). Shubik (1981) distinguishes between the consistent expectations of all Nash equilibria mentioned above and what he terms *rational expectations*, which are expectations held by player's about their opponents behaviour in Pareto optimal Nash equilibria. (The reader should not confuse these with the 'rational expectations' of economic theory introduced by Muth (1961).) However, there is no guarantee of existence of any of these 'solutions' based on the additional optimality requirement and many other criteria have been proposed. Shubik gives an overview of about 30 different refinements of Nash equilibria which are specific to a particular class of games. This is also the subject of the books by van Damme (1983), Harsanyi and Selten (1988) and Selten (1988a).

We have already discussed refinements of Nash equilibria in extensive form games in some detail earlier in this chapter. When games have some dynamic structure, the rationale for refining our solution concept is usually that certain Nash equilibria entail 'unreasonable' behaviour at some point in time, i.e. the Nash equilibria prescribes a suboptimal choice at some node of the game tree or some stage of the repeated game (but *if* the Nash equilibrium is played that point will never be reached). If a Nash equilibria requires such 'implausible' behaviour off the equilibrium path, then it is not a subgame-perfect Nash equilibrium. An example of this is the equilibrium $\{F, O\}$ in Example 10.2.10.

We now concentrate on Nash equilibrium refinements in normal form games. From §10.2.3 we know that the Nash equilibrium concept is not sufficiently restrictive because it admits the use of weakly dominated strategies and that the extra requirement of subgame perfection reduces the solution set precisely because it excludes these weakly dominated strategies. However, the static nature of normal form games precludes the use of backwards induction as an additional rationality requirement on equilibrium play, so how can we exclude Nash equilibria which use weakly dominated strategies in normal form games? We

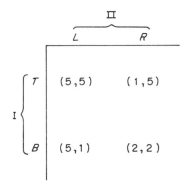

Figure 10.8.1

shall distinguish between Nash equilibria according to their degrees of *robustness* under different types of perturbations. Two aspects of normal form games can be perturbed separately: the pay-offs and the strategies. We have already encountered Nash equilibria which are stable under small perturbations in strategies: these are Selten's trembling hand perfect equilibria [see §10.3.3]. Although trembling hand perfect equilibria always exist there are some drawbacks with using trembling hand perfection as a solution concept in normal form games. Firstly they may be highly sub-optimal. For example, in the game of Figure 10.8.1 the trembling hand perfect equilibrium $\{B, R\}$ is pareto dominated by the Nash equilibrium $\{T, L\}$ which is not trembling hand perfect. But more important, the set of trembling hand perfect equilibria may not be invariant under the addition of strictly dominated strategies to the normal form. So although players have backwards induction rationality when they play trembling hand perfect equilibria (since trembling hand perfect equilibria are also sequential equilibria [see §10.3.3]), they need not possess *iterated dominance rationality* (a term introduced by Kohlberg and Mertens (1986) to describe players of equilibria which are invariant under successive deletion of dominated strategies). The following example illustrates this problem:

EXAMPLE 10.8.3   (van Damme (1983)) The game of Figure 10.8.2 has two pure Nash equilibria $\{T, L\}$ and $\{B, R\}$, but the latter is not 'sensible' because it assumes that both players completely ignore all parts of the pay-off matrices to which their opponents' strategies assign zero probability, that is $\{B, R\}$ is not trembling hand perfect (Figure 10.8.3).

However, if we add strictly dominated strategies to the normal form of Figure 10.8.2 it is possible that $\{B, R\}$ will be trembling hand perfect in the augmented game. For example, in Figure 10.8.3, D is strictly dominated by T and B for player I and A is strictly dominated by both L and R for player II. Now, the Nash equilibrium $\{B, R\}$ *is* trembling hand perfect in this game *because it is possible that both players believe that the mistakes A and D occur with larger probability than the mistakes T and L* and in this case it *is* optimal to play $\{B, R\}$.

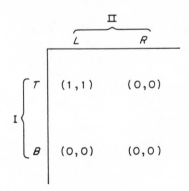

Figure 10.8.2

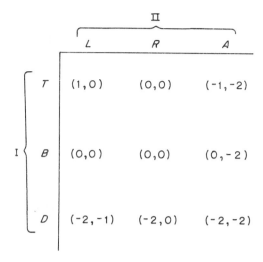

Figure 10.8.3

Making the mistake of playing a strictly dominated strategy is very 'costly' and Myerson (1978) used this idea to introduce a refinement of the trembling hand concept which excludes the undesirable property of possibly enlarging the solution set on adding strictly dominated strategies to the game. In Myerson's *proper equilibria* the players are assumed to make mistakes in a 'rational' way, that is the more costly a mistake (i.e. the more 'dominated' a strategy) the lower the probability of making it. Myerson shows that every normal form game has at least one proper equilibrium and that every proper equilibrium is also trembling hand perfect.

There are other refinements of Nash equilibria which require stability under small perturbations in strategies, for example the *persistent equilibria* of Kalai and Samet (1982) and *strictly perfect equilibria* of Okada (1981). But these do not always exist for normal form games and since space is limited we do not discuss these here.

A different type of robustness is required in *essential equilibria* (Wu Wen-Tsun and Jiang Jia-He (1962)). These are equilibria which are stable under small perturbations in the pay-off matrices for all players. Unfortunately these do not always exist, and even when they do they are not necessarily Pareto optimal. Harsanyi (1973a) also perturbs pay-off matrices in his definition of a *disturbed game*, his rationale for doing this being uncertainty about opponents pay-offs. His result that all Nash equilibria of such games are necessarily pure provides the interpretation of mixed strategy equilibria mentioned in §10.8.1, where randomization is rationalized by incomplete information about opponents pay-offs. He then defines the *regular equilibria* of the original game to be those Nash equilibria which are limits of the Nash equilibria in disturbed games as the uncertainty about opponents pay-offs decreases to zero. He shows that for 'almost all' normal games (i.e. all but a set of measure zero [see IV,

Definition 4.9.9]) all Nash equilibria are also regular.

The basis of all these stability requirements for Nash equilibria in normal form games is that no player should have the incentive to deviate from equilibrium play because they increase their pay-offs in so doing. For instance, in Example 10.8.3, $\{T, L\}$ is a trembling hand perfect stable equilibrium. But given that both players can increase their pay-offs by deviating from $\{B, R\}$ this cannot be either trembling hand perfect or stable. Not all Nash equilibria are 'stable': this term is usually applied to those equilibria in which players have more than just no incentive to deviate—they must have the incentive to actually *play* the Nash equilibrium also. Such Nash equilibria are called *self-enforcing* or *strategically stable*.

EXAMPLE 10.8.4.   (Aumann and Maschler (1972)) The game of Figure 10.8.4 has a unique Nash equilibria, the mixed strategies $\hat{\mathbf{p}} = (\frac{3}{4}, \frac{1}{4})^T$ and $\hat{\mathbf{q}} = (\frac{1}{2}, \frac{1}{2})^T$. No player has an incentive to deviate from this: if I plays $\hat{\mathbf{p}}$ then II gets $\frac{3}{4}$ whatever strategy she plays, so she may as well play $\hat{\mathbf{q}}$, and if II plays $\hat{\mathbf{q}}$ then I gets $\frac{1}{2}$ whatever strategy he plays so he may as well play $\hat{\mathbf{p}}$. But although no player has the incentive to deviate, neither do they have the incentive to *play* the Nash equilibrium, so it is not self-enforcing. Indeed they may prefer to play their maximin strategies which *guarantee* them $\frac{1}{2}$ and $\frac{3}{4}$ respectively against *any* strategy their opponent chooses and not just the equilibrium strategies $\hat{\mathbf{p}}$ and $\hat{\mathbf{q}}$.

This example illustrates a general problem in games where the only Nash equilibria are properly mixed: by Theorem 10.2.3, each player gets the same pay-off whatever he plays against the Nash equilibrium strategy of his opponent, so if his equilibrium pay-off is the same as his value he may as well play his maximin strategy, which guarantees him this pay-off even if his opponent deviates from the Nash equilibrium. However, the maximin strategy profiles are not, in general, Nash equilibrium. Hence properly mixed equilibria in which no player has the (pay-off) incentive to deviate may still be unstable. This problem was addressed by Harsanyi (1973a) and Kohlberg and Mertens (1986)

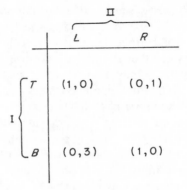

Figure 10.8.4

who introduced the concepts of regular and strategically stable equilibria respectively. Both these classes of equilibria possess all the robustness properties that one could ask for, being stable under perturbations in both strategies and pay-offs—so it is not surprizing that they do not always exist!

It is becoming increasingly clear that we should rethink this whole approach to analysing non-cooperative finite games—this is the topic of §10.8.3. But before embarking on this we mention the most successful of all Nash equilibrium refinements, the equilibrium selected by the *tracing procedure* of Harsanyi and Selten (1988), its 'success' being that it always identifies a unique solution in a finite game with perfect recall. Uniformly perfect equilibria were introduced in §10.3.3. In finite games with perfect recall they always exist, but are not necessarily unique. Now, for a given set of prior expectations about the player's choices, the (logarithmic) tracing procedure selects exactly one of these as the solution of the game (but when there is more than one uniformly perfect equilibrium, different priors give rise to different solutions). Let us introduce some formality to clarify the concept—for a rigorous definition and some informative examples the reader should consult Chapter 4 of Harsanyi and Selten (1988).

Suppose that each player has the *same* initial expectations about other player's choices in a finite n-person game G with perfect recall. Therefore the expectations about player i's choice of mixed strategy can be represented by a probability distribution over player i's pure strategies, i.e. a probability vector $\mathbf{p}_i$. Given the initial profile of expectations $\mathbf{p}_1, \mathbf{p}_2, \ldots, \mathbf{p}_n$, each player then determines his 'best response', characterized by the strategy profile $\sigma_1, \sigma_2, \ldots, \sigma_n$. Of course this strategy profile is not necessarily a Nash equilibrium of G, because the initial expectations are not necessarily an equilibrium—it is merely a tentative strategy plan based solvely on these expectations. The tracing procedure assumes that each player gradually reassesses their initial expectations in the light of what they know will be the other player's best replies, i.e. that player i gives increasing weight to the strategies $\sigma_1, \ldots, \sigma_{i-1}, \sigma_{i+1}, \ldots, \sigma_n$ and decreasing weight to $\mathbf{p}_1, \ldots, \mathbf{p}_{i-1}, \mathbf{p}_{i+1}, \ldots, \mathbf{p}_n$. At the same time he changes his own tentative strategy choice to using a new σ_i which is his best response to his current reassessed expectations. He also knows that the other players will change their tentative strategy plans in a similar way, which will require further changes in his expectations about their behaviour and so necessitate reassessment of his best response. This process continues until both the profile of expectations about other players' strategy choices and the profile of best response strategies converge to a unique uniformly perfect equilibrium of G.

A variety of solution concepts which may be applied to non-cooperative finite games has been described in this section, along with the different definitions of 'rationality' which underpin them. But there is a school of thought which finds the whole theory of equilibrium selection rather implausible. Can one really apply one such solution concept universally? Harsanyi and Selten advocate the uniformly trembling hand perfect equilibrium as the 'real solution', but in some games this would predict totally unreasonable behaviour.

EXAMPLE 10.8.5. (Binmore (1984)). If the uniformly perfect equilibrium concept is applied to Rosenthal's centipede, player *I* plays down at node 1 and all other nodes are never reached in equilibrium play (Rosenthal (1981)). However, player *II* still has to determine her 'rational' play at all subsequent decision nodes. At node 50, for example, she bases her strategy choice on the assumption that, instead of playing optimally at all previous nodes, player *I* has played across, thus making 25 uncorrelated but identical random errors. But surely she would get to a stage where she notices something *systematic* about these errors, possible due to a fundamental misunderstanding of the rules of the game? If you agree, then you will concede that the definition of rationality inherent in uniformly perfect equilibria (i.e. backwards induction rationality) is just not appropriate in this game.

Such considerations have fostered a new approach in the theory of non-cooperative games, in which the game model is augmented to include a model of players with a particular type of finite automaton. So now we give a brief account of some of these recent models of the procedural aspects of players' decision processes.

10.8.3. Bounded Rationality

The term 'bounded rationality' has encompassed a wide range of procedural aspects of decision making, in economic contexts notably by Simon (1978). Here we give a brief account of a specific type of bounded rationality which has recently found game-theoretic applications: instead of choosing strategies, players choose a *finite automaton*, i.e. a computer program which executes a fixed strategy to play the game for them. These automata have finite memory and the amount of memory they require determines an operational cost which is subtracted from their pay-off in the game. Players are rational in that they seek a finite automaton which maximizes their expected pay-off with the minimum possible operational cost.

An elegant introduction to the use of finite automata in two-person infinitely repeated games with complete information is given by Rubinstein (1986). He assumes that players choose a particular type of finite automaton called a *Moore machine*, with a particular type of operational cost, both defined below. The pay-off to the player is the limit of the average pay-offs in the infinitely repeated game (assuming this can be suitably defined [see §10.4.2]) minus the operational cost of the chosen machine. Machines observe the actions executed by the other player's machine, but since the stage strategy spaces S_1 and S_2 are assumed to be finite this is equivalent to assuming that stage strategies are observed. (In §10.5 we saw that this is not the general case.)

DEFINITION 10.8.1.   A *Moore machine* for player i, denoted M_i, consists of n states $\{q_1, q_2, \ldots, q_n\} = Q_i$, one of which is labelled $*$ to denote the initial state, an *output function* $\lambda_i: Q_i \to S_i$ and a *transfer function* $\mu_i: Q_i \times S_j \to Q_i$ such that $\mu_i(q^t, \sigma_j^t) = q^{t+1}$, where $q^t \in Q_i$ is the state of the machine and σ_j^t is the opponents stage strategy at stage t of the game (so $\sigma_j^t = \lambda_j(q^t)$ for some $q^t \in Q_j$).

In simple games, Moore machines can be diagramatically represented by drawing discs to represent the states in Q_i, lines between these states to represent the transfer function according to different opponents strategies in S_j, and labelling these states with a subscript denoting the value of the output function.

EXAMPLE 10.8.6.   Figure 10.8.5 illustrates Moore machines which play some well-known strategies for the infinitely repeated prisoner's dilemma [see §10.6.3].

(a) This machine plays 'ALLD'. The algebraic representation is

$$Q = \{q_B\}; \qquad q^* = q_B; \qquad \lambda_i(q_B) = B; \qquad \mu_i(q_B, A) = \mu_i(q_B, B) = q_B.$$

(b) This is 'TIT FOR TAT':

$$Q = \{q_A, q_B\}; \qquad q^* = q_A; \qquad \lambda_i(q_A) = A \quad \text{and} \quad \lambda_i(q_B) = B;$$
$$\mu_i(q_A, A) = \mu_i(q_B, A) = q_A \quad \text{and} \quad \mu_i(q_A, B) = \mu_i(q_B, B) = q_B.$$

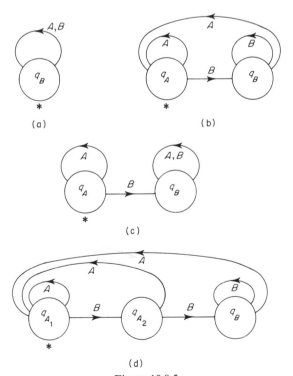

Figure 10.8.5

(c) This machine plays 'FREIDMAN', i.e. it plays A until its opponent plays B and B thereafter:

$$Q = \{q_A, q_B\}; \qquad q^* = q_A; \qquad \lambda_i(q_A) = A \quad \text{and} \quad \lambda_i(q_B) = B;$$
$$\mu_i(q_A, A) = q_A \quad \text{and} \quad \mu_i(q_A, B) = \mu_i(q_B, A) = \mu_i(q_B, B) = q_B.$$

(d) 'TIT FOR TWO TATS' (play B after opponent plays B twice consecutively, otherwise A):

$$Q = \{q_{A_1}, q_B, q_{A_2}\}; \qquad q^* = q_{A_1};$$
$$\lambda_i(q_{A_1}) = \lambda_i(q_{A_2}) = A \quad \text{and} \quad \lambda_i(q_B) = B;$$
$$\mu_i(q_{A_1}, A) = \mu_i(q_B, A) = \mu_i(q_{A_2}, A) = q_{A_1},$$
$$\mu_i(q_{A_1}, B) = q_{A_2} \quad \text{and} \quad \mu_i(q_{A_2}, B) = \mu_i(q_B, B) = q_B.$$

The operational cost, which reflects the computational complexity of a Moore machine can be defined in various ways. In some contexts the number of states $|Q_i|$ would be the most natural measure, but in a two- (or more) player game the number of states actually used by a Moore machine depends on the machine(s) it is playing. Moreover since the number of states of each machine is finite, after a certain time there must be a pair of states (one for each machine) that is returned to by the machines. So after an introductory sequence of pairs of states, a finite cycle of state pairs will be repeated infinitely often. The number of states that M_1 uses in this cycle when playing the machine M_2 is denoted $C_1(M_1, M_2)$ (and $C_2(M_1, M_2)$ denotes the number of states used by M_2). Rubinstein uses this *maintenance cost* $C_1(M_1, M_2)$ as the measure of computational complexity (i.e. he allows players to drop redundant states to avoid their maintenance costs). However, the solution concept that Rubinstein uses for this game requires more than just that players choose a machine to maximize their pay-off Π with the minimum of cost (given that the other player does this)—he also requires that all states of the chosen machines be used infinitely often. Hence in a solution $\{M_1^*, M_2^*\}$ to the game, $C_1(M_1^*, M_2^*)$ is equal to the number of states in M_1^* anyway. He also assumes that maintenance costs are of secondary importance in players utilities: they first seek machines which maximize pay-offs and from those choose the one(s) with least maintenance cost. Hence the solution is a Nash equilibrium in a game where the strategy sets are sets of possible machines, players preferences over the sets of all machine pairs have the lexicographic utility function $U_i(\Pi_i(M_1, M_2), C_i(M_1, M_2))$, and in any solution $\{M_1^*, M_2^*\}$ all states of both machines are used infinitely often.

Rubinstein shows that, with this definition of a solution, M_1 and M_2 must have full coordination in timing the switches of the stage strategies, a result which has far-reaching implications. For example, in 2×2 infinitely repeated games, all the stage pay-offs in a solution must lie in one of the two diagonals of the matrices.

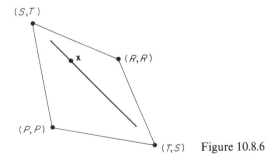

Figure 10.8.6

EXAMPLE 10.8.7. (Rubinstein (1986)). All feasible strictly individually rational outcomes which lie along the line segment joining (S, T) and (T, S) in Figure 10.8.6 can be supported by solutions when players choose Moore machines to play the infinitely repeated prisoner's dilemma [see Example 10.4.2]. A pair of Moore machines which support such an outcome **x** is shown in Figure 10.8.7.

Here $\mathbf{x} = \alpha(T, S) + (1 - \alpha)(S, T)$ and $n, m \in \mathbb{Z}_+$ are such that $\alpha = n/(n + m)$. M_1^* plays a strategy in which opponents must play A for n consecutive stages before the machine will also play A. Then it guarantees A for m stages before returning to the initial non-coperative part of the cycle. M_2^* starts with guaranteed cooperation (playing A) for n stages and then plays B until his opponent has played A for m consecutive moves before returning to the cooperative phase. In this solution $\{B, A\}$ will be played n times followed by $\{A, B\}$ m times in a cycle length of $n + m$. This yields the average pay-off vector **x**.

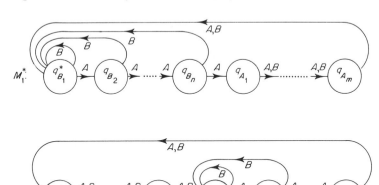

Figure 10.8.7

Binmore (1987) chooses a different type of finite automata to model the procedural rationality of players, using *universal Turing machines* of such high complexity that they reprogram themselves as they learn from experience (see also Canning (1988)). This *self-correction* takes the form of an algorithm applied to its basic *guessing rule*, a rule with which the machine is endowed and which it uses to guess its opponent's choice of strategy. Having guessed this, it can then predict its own behaviour. Now since the machines are universal, they assume that the opponent machine has access to all the same reasoning processes, so its opponent will predict its own behaviour in a similar fashion. The guessing algorithm requires that a machine updates its basic guessing rule to give some (small) weight to its opponent's prediction of its own behaviour, although most weight is still given to its original guess of the opponent's behaviour. However, this changes its prediction of its own behaviour, and the opponent will operate likewise. So both machines have to update their rules again, and Binmore's machines will iterate in much the same way as do rational players using Harsanyi and Selten's tracing procedure [see §10.8.2].

Although the guessing algorithm will eventually converge to a unique limit guessing rule (depending on the profile of basic guessing rules) the finite automaton cannot cope with the infinite number of necessary iterations, so Binmore endows them with a stopping rule for the guessing algorithm, which tells them when they are sufficiently close to the limit rule. This must be random, otherwise the self-correcting machines would get involved with another infinite iteration—of improving the stopping rule!

The limit rules predict a strategy profile σ which will be a Nash equilibrium if the game is static. But in a dynamic game, since the guessing algorithm is based on the machine's beliefs about the guessing algorithm of its opponent, these beliefs are 'vulnerable to Bayesian updating' during the course of the game. Now if it is common knowledge that all machines have the same objective input, then σ will again be a Nash equilibrium, otherwise Binmore asserts that it will be a correlated equilibrium.

C. O. A.

REFERENCES

Aumann, R. J. (1959). Acceptable Points in General Cooperative *n*-person Games, *Contributions to the theory of games IV* (*Annals. of Math. Studies* **40**), 287–324.

Aumann, R. J. (1961). Almost Strictly Competitive Games, *Jour. Soc. for Industrial App. Math.* **9**, 544–550.

Aumann, R. J. (1976). Agreeing to Disagree, *Annals of Stats.* **4**, 1236–1239.

Aumann, R. J. (1981). A survey of Repeated Games, in *Essays in Game Theory and Mathematical Economics in Honour of Oskar Morgenstern*. Wissenschaftsverlag, Bibliographisches Institute, 11–42.

Aumann, R. J. (1987). Correlated Equilibria as an Expression of Bayesian Rationality, *Econometrica* **55**, 1–8.

Aumann, R. J. and Maschler, M. (1966). Game Theoretic Aspects of Gradual Disarmament, *Mathematica* **ST-80**, 1–55.

Aumann, R. J. and Maschler, M. (1972). Some Thoughts on the Minimax Principle, *Man. Sci.* **18**, 54–63.

Axelrod, R. (1984). *The Evolution of Cooperation*, Basic Books.

Bernheim, B. D. (1984). Rationalizable Strategic Behaviour, *Econometrica* **52**, 1007–1028.

Binmore, K. G. (1987). Remodelled Rational Players, *London School of Economics* STICERD D. P. 87/149.

Binmore, K. G. and Dasgupta, P. (1987). *The Economics of Bargaining*, Basil Blackwell.

Binmore, K. G., Rubinstein, A. and Wolinsky, A. (1986). A Non-cooperative Implementation of the Nash Bargaining Solution, *Rand Journal of Economics* **17**, 176–188.

Bishop, D. T. and Cannings, C. (1978). A Generalized War of Attrition, *J. Theor. Biol.* **70**, 85–124.

Canning, D. (1988). Rationality and Game Theory when Players are Turing Machines, *London School of Economics* STICERD D.P. 88/183.

Davis, M. (1974). Some Further Thoughts on the Minimax Principle, *Man. Sci.* **20**, 1305–1310.

Dresher, M., Shapley, L. S. and Tucker, A. W. (eds) (1964). *Advances in Game Theory*, Annals of Mathematical Studies **52**, Princeton University Press.

Ewens, W. J. (1979). *Mathematical Population Genetics*, Springer-Verlag (Biomathematics, Vol. 9).

Fishburn, P. C. and Rubinstein, A. (1982). Time Preference, *Int. Econ. Rev.* **23**, 677–694.

Forges, F. (1982). Infinitely Repeated Games of Incomplete Information: Symmetric Case with Random Signals, *Int. J. Game Theory* **11**, 203–213.

Forges, F. (1985). Correlated Equilibria in a Class of Repeated Games with Incomplete Information, *Int. J. Game Theory* **14**, 129–150.

Forges, F., Mertens, J. F. and Neymann, A. (1986). A Counterexample to the Folk Theorem with Discounting, *Econ. Letters* **20**, 7.

Friedman, J. (1977). *Oligopoly and the Theory of Games*, North Holland.

Fudenberg, D. and E. Maskin (1986). The Folk Theorem in Repeated Games with Discounting and Incomplete Information, *Econometrica* **54**, 533–554.

Haigh, J. (1975). Game Theory and Evolution, *Adv. App. Prob.* **7**, 8–11.

Hammerstein, P. (1984). The Biological Counterpart in Non-cooperative Game Theory, *Nieuw Archief Voor Wiskunde* **4**, 137–149.

Harsanyi, J. C. (1967–68). Games of Incomplete Information played by Bayesian Players, Parts I, II and III, *Management Science* **14**, 159–182, 320–334, 486–502.

Harsanyi, J. C. (1973a). Games with Randomly Disturbed Pay-offs: A New Rationale for Mixed Strategy Equilibrium Points, *Int. J. Game Theory* **2**, 1–23.

Harsanyi, J. C. (1973b). Oddness of the Number of Equilibrium Points: A New Proof, *Int. J. Game Theory* **2**, 235–250.

Harsanyi, J. C. (1977). *Rational Behaviour and Bargaining Equilibrium in Games and Social Situations*, Cambridge University Press.

Harsanyi, J. C. and Selten, R. (1972). A Generalized Nash Solution for Two-person Bargaining Games with Incomplete Information *Management Science* **18**, 80–106.

Harsanyi, J. C. and Selten, R. (1988). *A General Theory of Equilibrium Selection in Games*, Massachusetts Institute of Technology Press.

Hart, S. (1982). Non-Zero-Sum Two-Person Repeated Games with Incomplete Information, *Math. Op. Res.*

Hines, W. G. S. and Maynard Smith, J. (1979). Games Between Relatives, *J. Theor. Biol.* **79**, 19–30.

Kalai, E. and Samet, D. (1982). Persistent Equilibria in Strategic Games, *Int. J. Game Theory*, **13**, 129–144.

Kohlberg, E. and Mertens, J. F. (1986). On the Strategic Stability of Equilibria, *Econometrica* **54**, 1003–1037.

Kohlberg, E. and Zamir, S. (1974). Repeated Games of Incomplete Information: The Symmetric Case, *Ann. Stats.* **2**, 1040–1041.

Kreps, D. and Wilson, R. (1982a). Reputation and Imperfect Information, *J. Econ. Theory* **27**, 253–279.

Kreps, D. and Wilson, R. (1982b). Sequential Equilibria, *Econometrica* **50**, 863–894.

Kreps, D., Milgrom, P., Roberts, J. and Wilson, R. (1982). Rational Cooperation in the Finitely Repeated Prisoner's Dilemma, *J. Econ. Theory* **27**, 245–252.

Kuhn, H. W. and Tucker, A. W. (eds) (1953). *Contributions to the Theory of Games*, Vols I–IV, Annals of Maths. Studies, Princeton University Press.

Luce, R. D. and Raiffa, H. (1957). *Games and Decisions*. Wiley.

Maynard Smith, J. (1966). *The Theory of Evolution*, Pelican.

Maynard Smith, J. (1982). *Evolution and the Theory of Games*, Cambridge University Press.

Maynard Smith, J. and Hofbauer, J. (1987). The Battle of the Sexes: A Genetic Model with Limit Cycle Behaviour, *Theor. Pop. Biol.* **32**, 1–4.

Maynard Smith, J. and Price, G. R. (1973). The Logic of Animal Conflict, *Nature* **246**, 15–18.

Mertens, J. F. (1982). Repeated Games: An Overview of the Zero-Sum Case, in *Advances in Economic Theory* Ed. W. Hildenbrand, Cambridge University Press.

Mertens, J. F. and Zamir, S. (1980). Minimax and Maximin of Repeated Games with Incomplete Information, *Int. J. Game Theory* **9**, 201–215.

Milgrom, P. and Roberts, J. (1982). Predation, Reputation and Entry Deterrence, *J. Econ. Theory* **27**, 280–312.

Moulin, H. (1982). Bargaining and Non-cooperative Implementation, *Ecole Polytechnique* D.P. A239-0282.

Muth, J. (1961). Rational Expectations and the Theory of Price Movements, *Econometrica* **29**, 315–335.

Myerson, R. (1978). Refinements of the Nash Equilibrium Concept, *Int. J. Game Theory* **7**, 73–80.

Nash, J. (1950). The Bargaining Problem, *Econometrica* **18**, 155–162.

Nash, J. (1951). Non-cooperative Games, *Ann. Math.* **54**, 286–295.

Nash, J. (1953). Two Person Non-cooperative Games, *Econometrica* **21**, 128–140.

Okada, A. (1981). On Stablility of Perfect Equilibrium Points, *Int. J. Game Theory* **10**, 67–73.

Owen, G. (1982). *Game Theory*, Academic Press.

Pearce, D. G. (1984). Rationalizable Strategic Behaviour and the Problem of Perfection, *Econometrica* **52**, 1029–1050.

Rosenthal, R. (1981). Games of Perfect Information, Predatory Pricing and the Chain-Store Paradox, *J. Econ. Th.* **25**, 92–100.

Roth, A. E. (1979). *Axiomatic Models of Bargaining*, Springer-Verlag Lecture Notes in Economics and Mathematical Systems.

Roth, A. E. (ed.) (1985). *Game Theoretic Models of Bargaining*, Cambridge University Press.

Rubinstein, A. (1977). Equilibrium in Supergames, CRIME, Hebrew University of Jerusalem, RM 25.

Rubinstein, A. (1982). A Perfect Equilibrium in a Bargaining Model, *Econometrica* **50**, 97–109.

Rubinstein, A. (1985). A Bargaining Model with Incomplete Information about Time Preferences, *Econometrica* **53**, 1151–1172.

Rubinstein, A. (1986). Finite Automata Play the Repeated Prisoner's Dilemma, *J. Econ. Theory* **39**, 83–96.

Savage, L. S. (1954). *The Foundations of Statistics*, Wiley.

Schuster, P. and Sigmund, K. (1981). Coyness, Philandering and Stable Strategies, *Animal Behaviour* **29**, 186–192.

Selten, R. (1973). A Simple Model of Imperfect Competition where 4 are Few and 6 are Many, *Int. J. Game Theory* **2**, 141–201.

Selten, R. (1975). Reexamination of the Perfectness Concept for Equilibrium Points in Extensive Games, *Int. J. Game Theory* **4**, 25–55.

Selten, R. (1978). The Chain-Store Paradox, *Theory and Decision* **9**, 127–159.

Selten, R. (1979). A Note on Evolutionary Stable Strategies in Asymmetric Animal Conflicts, *J. Theor. Biol.* **84**, 93–107.

Selten, R. (1983). Evolutionary Stability in Extensive Two-Person Games, *Math. Soc. Sci.* **5**, 269–363.

Selten, R. (1988a). *Models of Strategic Rationality*, Theory and Decision (library series C) Kluwer.

Selten, R. (1988b). Evolutionary Stability in Extensive Two-Person Games—Correction and Further Development, *Math. Soc. Sci.* **16**, 223–266.

Shaked, A. and Sutton, J. (1984). The Semi-Walrasian Economy, *London School of Economics* STICERD D.P. #98.

Shubik, M. (1981). Perfect or Robust Non-cooperative Equilibrium: A Search for the Philosophers Stone?, in *Essays in Game Theory and Mathematical Economics* in Honour of Oskar Morgenstern, Bibliographisches Institute.

Simon, H. A. (1978). On How to Decide What To Do, *Bell J. Econ.* **9**, 497–507.

Sorin, S. (1983). Some Results on the Existence of Nash Equilibria for Non-Zero-Sum Games with Incomplete Information, *Int. J. Game Theory* **12**, 193–205.

Team, F. (1988). Personal Communication.

Thomas, L. C. (1984). *Games, Theory and Applications*, Ellis Horwood.

Van Damme, E. (1983). *Refinements of the Nash Equilibrium Concept*, Springer-Verlag Lecture Notes in Economics and Mathematical Systems.

von Neumann, J. (1928). Zur Theorie der Gesellschaftsspiele, *Math. Annalen.* **100**, 295–320.

von Neumann, J. and Morgenstern, O. (1953). *The Theory of Games and Economic Behaviour* (2nd ed.), Princeton University Press.

Wu Wen-Tsun and Jiang Jia-He (1962). Essential Equilibrium Points of n-person Non-cooperative Games, *Sci. Sinica* **11**, 1307–1322.

Zamir, S. (1971). On the Relation between Finitely and Infinitely Repeated Games with Incomplete Information, *Int. J. Game Theory* **1**, 179–198.

Zamir, S. (1973). On the Notion of Value for Games with Infinitely Many Stages, *Ann. Stats.* **1**, 791–796.

Zeeman, E. C. (1979). Population Dynamics From Game Theory, *Proc. Int. Conf. Global Theory of Dynamical Systems*, Northwestern, Evanston.

CHAPTER 11

Population Structures

11.1. COHORTS

Consider a group of individuals. Some members of the group leave, others join it, and we observe the group at regular intervals, say yearly, and describe its transformations.

We shall use this model throughout this chapter, but it is clear that our demonstrations refer equally to any self-renewing aggregate, for instance to industrial replacement.

Let the initial group have $n(0)$ members of (nominal) age 0. In the simplest case, which we study first, we do not assume that any individuals join the group, but we assume that in the period from age $x - 1$ to age x a proportion q_x of individuals leaves it. We call q_x the loss rate, and $1 - q_x = p_x$ the survival rate. Let $p_{k+1} = 0$, hence $q_{k+1} = 1$. This means that the highest age an individual can reach, or the longest time he can remain in the group, is k. We call such a group a 'cohort'.

Denote the number of members in the cohort after x years by $n(x)$. Then, for a positive integer x, $n(x) = n(x - 1)p_x$. After x years the number of survivors in the cohort will be

$$n(x) = n(0)p_1 p_2 \ldots p_x \tag{11.1.1}$$

(to the nearest integer).

EXAMPLE 11.1.1. $n(0) = 1000$, and for

$x =$	1	2	3	4	5	6
$p_x =$	0·875	0·583	0·306	0·217	0·077	0·000
$n(x) =$	875	510	156	34	3	0

Now imagine that at the end of each year the population is joined by a new cohort of the same original age 0 and the same original size, subject also to the same loss rates as the initial cohort. This time the population will not disappear, and we want to establish the number of its members $1, 2, \ldots$ years after the entry of the first cohort.

Let $t = 0$ be the time of the entry of the first cohort, and $n_x(t)$ the number of those members at time t ($t = 0, 1, 2, \ldots$), who entered at time $t - x$. Their age is now x. Observe that $n_0(0) = n(0)$. We have, then,

$$n_0(t + 1) = n_0(t), \qquad n_1(t + 1) = n_0(t)p_1, \qquad n_2(t + 1) = n_0(t - 1)p_1 p_2$$

and generally

$$n_x(t + 1) = n(0)p_1 p_2 \ldots p_x. \tag{11.1.2}$$

After $t = k$, the population remains unchanged; the number of entrants will, in each year, equal the number of those who have just left, thus

$$n(0)(q_1 + p_1 q_2 + p_1 p_2 q_3 + \ldots + p_1 p_2 \ldots p_k) = n(0). \tag{11.1.3}$$

EXAMPLE 11.1.2.   Let $n(0)$ and the survival rates be those of Example 11.1.1. Then $n_x(t)$ is given in the following table:

$t =$	0	1	2	3	4	5	6	...
$x = 0$	1000	1000	1000	1000	1000	1000	1000	
1		875	875	875	875	875	875	
2			510	510	510	510	510	
3				156	156	156	156	
4					34	34	34	
5						3	3	
Totals	1000	1875	2385	2541	2575	2578	2578	

11.2.   CONSTANT POPULATIONS

The cohorts which we studied in §11.1 started with members who were all of the same age 0. In the present section we consider initial populations with members of various ages, and we ask how many members of age 0 must join the population, so as to keep the total membership constant. (In Example 11.1.2, the size of the population was changing from year to year.)

We denote again the number of members of age x at time t by $n_x(t)$. The number of new entrants at time $t + 1$, who replace the loss during the period $(t, t + 1)$, is given by

$$n_0(t + 1) = n_0(t)q_1 + n_1(t)q_2 + \ldots + n_k(t) \tag{11.2.1}$$

when $q_{k+1} = 1$. Also

$$n_x(t) = n_{x-1}(t - 1)p_x. \tag{11.2.2}$$

Given $n_x(0)$ for $x = 0, 1, \ldots, k$, we can compute $n_x(1)$ for $x = 1, \ldots, k$, from (11.2.2), then $n_0(1)$ from (11.2.1), then $n_x(2)$ for $x = 1, \ldots, k$ from (11.2.2) and $n_0(2)$ from (11.2.1), and so on, recursively.

EXAMPLE 11.2.1.   Let $\mathbf{n}(0)$ be the column vector $(500, 100, 100, 100, 100, 100)^T$ and the survival rates those of Example 11.1.1.

Table of $n_x(t)$

$t =$	0	1	2	3	4	5	6	7	8
$x = 0$	500	443	330	402	394	383	389	389	388
1	100	438	388	289	352	345	335	340	340
2	100	58	255	226	168	205	201	195	198
3	100	31	18	78	69	51	63	61	60
4	100	22	7	4	17	15	11	14	13
5	100	8	2	1	0	1	1	1	1
Totals	1000	1000	1000	1000	1000	1000	1000	1000	1000

We note that when $t \geq x$ (this is always the case when $t \geq k$), then

$$n_x(t) = n_{x-1}(t-1)p_x = n_{x-2}(t-2)p_{x-1}p_x = \ldots = n_0(t-x)p_1 p_2 \ldots p_x, \quad (11.2.3a)$$

and when $t \leq x$, then

$$n_x(t) = n_{x-1}(t-1)p_x = \ldots = n_{x-t}(0)p_{x-t+1}p_{x-t+2}\cdots p_x. \quad (11.2.3b)$$

It follows from (11.2.2a) that if $t \geq k$, then (11.2.1) can be written

$$n_0(t+1) = n_0(t)q_1 + n_0(t-1)p_1 q_2 + \ldots + n_0(t-k)p_1 p_2 \ldots p_k. \quad (11.2.4)$$

This is a difference equation for $n_0(t)$ (I, p. 446). Its characteristic equation is

$$y^{k+1} - y^k q_1 - y^{k-1}p_1 q_2 - \ldots - p_1 p_2 \ldots p_k = 0 \quad (11.2.5)$$

with roots $\lambda_0, \lambda_1, \ldots, \lambda_k$. One of the roots equals 1, since

$$1 - q_1 - p_1 q_2 - p_1 p_2 q_3 - \ldots - p_1 p_2 \ldots p_k = 0.$$

To solve the difference equation, we put

$$n_0(t) = d_0 \lambda_0^t + d_1 \lambda_1^t + \ldots + d_k \lambda_k^t,$$

where the coefficients $d_0, \ldots, d_k$ depend on $n_0(0), \ldots, n_0(k)$. For Example 11.2.1, (11.2.4) means

$$n_0(t+1) = 0.125n_0(t) + 0.365n_0(t-1) + 0.354n_0(t-2)$$
$$+ 0.122n_0(t-3) + 0.031n_0(t-4) + 0.003n_0(t-5).$$

For instance, when $t = 5$,

$$389 = 0.125 \times 383 + 0.365 \times 394 + 0.354 \times 402 + 0.122 \times 330$$
$$+ 0.031 \times 443 + 0.003 \times 500.$$

11.3.   THE STATIONARY POPULATION

In the last section we computed the vectors $\mathbf{n}(t) = (n_0(t), \ldots, n_k(t))^T$ recursively. Here we describe a method for finding these vectors without having to compute first all the vectors for smaller t.

We write (11.2.1) and (11.2.2) in matrix notation

$$\mathbf{n}(t + 1) = \mathbf{A}\mathbf{n}(t) \tag{11.3.1}$$

(and hence $\mathbf{n}(t + 1) = \mathbf{A}^2\mathbf{n}(t - 1) = \ldots = \mathbf{A}^{t+1}\mathbf{n}(0)$, a relation which we shall use later), where

$$\mathbf{A} = \begin{pmatrix} q_1 & q_2 & \cdots & q_k & 1 \\ p_1 & 0 & \cdots & 0 & 0 \\ 0 & p_2 & \cdots & 0 & 0 \\ \vdots & \vdots & & \vdots & \vdots \\ 0 & 0 & \cdots & p_k & 0 \end{pmatrix}.$$

This matrix has entries which add up to 1 in each column; it is 'stochastic' (I, Example 7.11.2) and so are all its powers.

The characteristic equation of $\mathbf{A}$ is

$$\begin{vmatrix} q_1 - \lambda & q_2 & q_3 & \cdots & q_k & 1 \\ p_1 & -\lambda & 0 & \cdots & 0 & 0 \\ 0 & p_2 & -\lambda & \cdots & 0 & 0 \\ \vdots & \vdots & \vdots & & \vdots & \vdots \\ 0 & 0 & 0 & \cdots & p_k & -\lambda \end{vmatrix} = 0.$$

By expanding the determinant, we have

$$(-1)^{k+1}(\lambda^{k+1} - q_1\lambda^k - p_1 q_2 \lambda^{k-1} - \ldots - p_1 p_2 \ldots p_k)$$
$$= (-1)^k(1 - \lambda)[\lambda^k + p_1\lambda^{k-1} + p_1 p_2 \lambda^{k-2} + \ldots + p_1 p_2 \ldots p_k] = 0. \tag{11.3.2}$$

Because all columns of the determinant add up to $1 - \lambda$, one root is $\lambda_0 = 1$. Since all coefficients within the square brackets are positive, any other root, if it is real, must be negative. A root can only be zero if one of the p_i is zero, which we assume not to be the case.

The $k + 1$ roots $\lambda_0, \lambda_1, \ldots, \lambda_k$ are the eigenvalues (I, 7.1) of the matrix $\mathbf{A}$. The set of eigenvalues is called the spectrum.

We know (Vajda (1978) p. 183) that no eigenvalue can have a modulus larger than 1. We assume now that λ_0 is the only eigenvalue with modulus 1, and that all eigenvalues are distinct.

Solving the system of linear equations in $x_{1i}, x_{2i}, \ldots, x_{(k+1)i}$,

$$\begin{aligned} (q_1 - \lambda_i)x_{1i} + q_2 x_{2i} + q_3 x_{3i} + \ldots + \quad q_k x_{ki} + x_{(k+1)i} &= 0, \\ p_1 x_{1i} - \lambda_i x_{2i} \qquad\qquad\qquad\qquad\qquad &= 0, \\ p_2 x_{2i} \quad - \lambda_i x_{3i} \qquad\qquad\qquad &= 0, \\ \ldots \qquad\qquad &= 0, \\ p_k x_{ki} - \lambda_i x_{(k+1)i} &= 0, \end{aligned} \tag{11.3.3}$$

we obtain the eigenvectors

$$\mathbf{v}_i = (x_{1i}, x_{2i}, \ldots, x_{(k+1)i})^{\mathrm{T}}.$$

These eigenvectors are not uniquely defined. When v_i is an eigenvector, then so is cv_i, provided $c \neq 0$.

We can make a precise statement about the eigenvector corresponding to the eigenvalue 1. It is easily seen that this eigenvector is (proportional to)

$$(1, p_1, p_1 p_2, \ldots, p_1 p_2 \cdots p_k)^{\mathrm{T}}. \tag{11.3.4}$$

We have $\mathbf{A}\mathbf{v}_i = \lambda_i \mathbf{v}_i$, and hence, by induction,

$$\mathbf{A}^t \mathbf{v}_i = \lambda_i^t \mathbf{v}_i \tag{11.3.5}$$

for any positive integer t.

If all eigenvalues are distinct, then it can be shown that the eigenvectors are linearly independent (I, Theorem 7.4.2). Hence any population structure can be expressed as a linear combination of the eigenvectors, thus

$$\mathbf{n}(0) = c_0 \mathbf{v}_0 + c_1 \mathbf{v}_1 + \ldots + c_k \mathbf{v}_k \tag{11.3.6}$$

and hence

$$\mathbf{n}(t) = \mathbf{A}^t \mathbf{n}(0) = c_0 \lambda_0^t \mathbf{v}_0 + c_1 \lambda_1^t \mathbf{v}_1 + \ldots + c_k \lambda_k^t \mathbf{v}_k. \tag{11.3.7}$$

As in §11.1, we are interested in finding

$$\lim_{t \to \infty} \mathbf{A}^t \mathbf{n}(0).$$

It follows from (11.3.7), and from our assumptions, that this limit equals $c_0 \mathbf{v}_0$. *Note*: It can be shown that a matrix of the type which we have here, that is a matrix (a_{ij}) such that

$$a_{ij} \geq 0, \quad \sum_i a_{ij} = 1, \quad a_{1j} > 0 \quad \text{and} \quad a_{i+1,i} > 0$$

cannot have any other eigenvalue of modulus 1 than λ_0, and that our second assumption, that of distinctness of the eigenvalues, is redundant.

If we start with $\mathbf{n}(0) = c_0 \mathbf{v}_0$, then the structure remains unchanged in time. We call such a structure 'stationary': $\mathbf{n}(0) = \mathbf{n}(1) = \mathbf{n}(2) = \ldots$.

EXAMPLE 11.3.1. Let the survival rates be as in Example 11.1.1. The characteristic equation of the matrix is then

$$(1 - \lambda)(\lambda^5 + 0.875\lambda^4 + 0.510\lambda^3 + 0.156\lambda^2 + 0.034\lambda + 0.003) = 0,$$

and its roots are given in the following table:

j	0	1	2	3	4	5
λ_j	1	-0.125	$-0.25 + iv$	$-0.25 - iv$	$-0.125 + iw$	$-0.125 - iw$

where

$$v = \left(\frac{5}{48}\right)^{\frac{1}{2}} \quad \text{and} \quad w = \left(\frac{7}{64}\right)^{\frac{1}{2}}.$$

The eigenvectors are (proportional to)

$\mathbf{v}_0$	$\mathbf{v}_1$	$\mathbf{v}_2$
1000	1000	1000
875	-7000	$-1313 - 1693i$
510	32664	$-763 + 2964i$
156	-79992	$2108 - 907i$
34	138656	$-1066 - 589i$
3	-85328	$34 + 227i$
2578	0	0

$\mathbf{v}_3$	$\mathbf{v}_4$	$\mathbf{v}_5$
1000	1000	1000
$-1313 + 1693i$	$-875 - 2315i$	$-875 + 2315i$
$-763 - 2964i$	$-3062 + 2701i$	$-3062 - 2701i$
$2108 + 907i$	$3125 + 1654i$	$3125 - 1654i$
$-1066 + 589i$	$271 - 2150i$	$271 + 2150i$
$34 - 227i$	$-459 + 110i$	$-459 - 110i$
0	0	0

If $\mathbf{n}(0) = (1000, 0, 0, 0, 0, 0)^{\mathrm{T}}$, then for

j	0	1	2	3	4	5
c_j	1	-0.001	$-0.07 + 0.06i$	$-0.07 - 0.06i$	0	0

If we divide the components of $\mathbf{v}_0$ given above by 2578, then we obtain (in rounded values)

$$388$$
$$339$$
$$198$$
$$61$$
$$13$$
$$1$$
$$1000$$

the vector, to which the structure tends in Example 11.2.1.

11.4.   EIGENVALUES AND EIGENVECTORS

It is convenient to insert here a few remarks about eigenvalues and eigenvectors.

The eigenvectors in §11.3 are so-called right-hand eigenvectors, while the left-hand eigenvectors are those which solve the system $(\mathbf{A}^{\mathrm{T}} - \lambda_i \mathbf{I})\mathbf{w}_i = 0$, where $\mathbf{I}$ is the identity matrix.

The spectrum of eigenvalues which solve $|\mathbf{A} - \lambda\mathbf{I}| = 0$ is the same as the spectrum of eigenvalues which solve $|\mathbf{A}^T - \lambda\mathbf{I}| = 0$ (I, Proposition 7.2.4), but the eigenvectors in the two sets are not necessarily the same.

Concerning the eigenvectors, we prove

THEOREM 11.4.1.   *If the eigenvalue λ_i differs from the eigenvalue λ_j, then the scalar product $\mathbf{w}_j^T \cdot \mathbf{v}_i$, that is*

$$w_{0j}^T v_{0i} + \ldots + w_{kj}^T v_{ki}$$

equals 0.

Proof.

$$\mathbf{w}_j^T \lambda_i \mathbf{v}_i = \mathbf{w}_j^T \mathbf{A}\mathbf{v}_i = \lambda_j \mathbf{w}_j^T \mathbf{v}_i, \quad \text{hence} \quad (\lambda_i - \lambda_j)\mathbf{w}_j^T \mathbf{v}_i = 0.$$

If $\lambda_i \neq \lambda_j$, then it follows that $\mathbf{w}_j^T \mathbf{v}_i = 0$.

When all columns of $\mathbf{A}$ add up to the same value, λ_0 say (as they do in our present case), then it is easily seen that the corresponding eigenvector of $\mathbf{A}^T$ consists of components which are all equal. It then follows from Theorem 11.4.1 that

THEOREM 11.4.2.   *The components of $\mathbf{v}_i$, $i \neq 0$, add up to zero.*

This is so in Example 11.3.1. Conversely, it follows from (11.3.7) that

THEOREM 11.4.3.   *If the components of the eigenvectors $\mathbf{v}_i$, $i \neq 0$, add up to zero, then the totals of any linear combination of such vectors and of $\mathbf{v}_0$ will develop in time as the totals of $c_0 \mathbf{v}_0$ develop.*

11.5.   GENERAL TRANSITION MATRICES

We construct now a more general case than the one we considered earlier, by not assuming any longer that the only distinguishing characteristic of the members of the population is their age. We shall now say that the members belong to 'states', and transitions between the states may be, for instance, changes of location, or perhaps promotions and demotions in a hierarchy. We define the states by subscripts $1, 2, \ldots$ and describe the transition rate from state i to state j by the transition matrix

$$\mathbf{P} = \begin{pmatrix} p_{11} & p_{12} & \cdots & p_{1k} \\ p_{21} & p_{22} & \cdots & p_{2k} \\ \vdots & \vdots & & \vdots \\ p_{k1} & p_{k2} & \cdots & p_{kk} \end{pmatrix}$$

We shall again keep the population constant, replacing losses by new entrants. If the present structure equals $(n_1, \ldots, n_k)^T$, then the loss from n_i will be, for example,

$$n_i(1 - p_{i1} - p_{i2} - \ldots - p_{ik}) = n_i w_i \tag{11.5.1}$$

and the total number of new entrants must equal $\sum_{i=1}^{k} n_i w_i$. If all new recruits join the same state, say state m, then we obtain the system

$$n_i(t + 1) = n_1(t)p_{1i} + n_2(t)p_{2i} + \ldots + n_k p_{ki} \quad \text{for } i \neq m,$$
$$n_m(t + 1) = n_1(t)(p_{1m} + w_1) + n_2(t)(p_{2m} + w_2) + \ldots + n_k(t)(p_{km} + w_k). \tag{11.5.2}$$

For instance, if all new entrants join state 1, then we find the eigenvalues of the matrix which transforms $n(t)$ into $n(t + 1)$ by solving

$$\begin{vmatrix} p_{11} + w_1 - \lambda & p_{21} + w_2 & p_{31} + w_3 & \cdots & p_{k1} + w_k \\ p_{12} & p_{22} - \lambda & p_{32} & \cdots & p_{k2} \\ \vdots & \vdots & \vdots & & \vdots \\ p_{1k} & p_{2k} & p_{3k} & \cdots & p_{kk} - \lambda \end{vmatrix} = 0.$$

EXAMPLE 11.5.1.   Let the transition matrix be

$$\mathbf{P} = \begin{pmatrix} 0.3 & 0.4 & 0.1 \\ 0.1 & 0.5 & 0.3 \\ 0.2 & 0.1 & 0.4 \end{pmatrix}$$

and hence the loss vector

$$\mathbf{w} = \begin{pmatrix} 0.2 \\ 0.1 \\ 0.3 \end{pmatrix}$$

then we have the system

$$n_1(t + 1) = (0.3 + 0.2)n_1 + (0.1 + 0.1)n_2 + (0.2 + 0.3)n_3,$$
$$n_2(t + 1) = \quad 0.4n_1 \quad + \quad 0.5n_2 \quad + \quad 0.1n_3,$$
$$n_3(t + 1) = \quad 0.1n_1 \quad + \quad 0.3n_2 \quad + \quad 0.4n_3,$$

$$|\mathbf{P}^{\mathrm{T}} - \lambda\mathbf{I}| = \begin{vmatrix} 0.5 - \lambda & 0.4 & 0.1 \\ 0.2 & 0.5 - \lambda & 0.3 \\ 0.5 & 0.1 & 0.4 - \lambda \end{vmatrix}$$

$$= \lambda^3 - 1.4\lambda^2 + 0.49\lambda - 0.09 = 0.$$

The eigenvalues are

$$\lambda_1^{(1)} = 1, \qquad \lambda_2^{(1)} = 0.2 + \sqrt{(0.05)}i, \qquad \lambda_3^{(1)} = 0.2 - \sqrt{(0.05)}i$$

and the eigenvectors are (proportional to)

$$\mathbf{v}_1^{(1)} = \begin{pmatrix} 392 \\ 362 \\ 246 \end{pmatrix}, \quad \mathbf{v}_2^{(1)} = \begin{pmatrix} -0.02 - 0.1118i \\ -0.07 + 0.0894i \\ 0.09 + 0.0224i \end{pmatrix}, \quad \mathbf{v}_3^{(1)} = \begin{pmatrix} -0.02 + 0.1118i \\ -0.07 - 0.0894i \\ 0.09 - 0.0224i \end{pmatrix}.$$

An initial structure of $(1000, 0, 0)^T$ would be represented by

$$\mathbf{n}(0) = \mathbf{v}_1^{(1)} + (-722 + 2602i)\mathbf{v}_2^{(1)} + (-722 - 2602i)\mathbf{v}_3^{(1)}.$$

Similarly, we obtain, when all new recruits join state 2,

$$\lambda^3 - 1\cdot3\lambda^2 + 0\cdot34\lambda - 0\cdot04 = 0$$

with eigenvalues

$$\lambda_1^{(2)} = 1, \qquad \lambda_2^{(2)} = 0\cdot15 + 0\cdot1323i, \qquad \lambda_3^{(2)} = 0\cdot15 - 0\cdot1323i,$$

and eigenvectors

$$
\begin{array}{ccc}
\mathbf{v}_1^{(2)} & \mathbf{v}_2^{(2)} & \mathbf{v}_3^{(2)} \\[4pt]
\begin{pmatrix} 162 \\ 541 \\ 297 \end{pmatrix} &
\begin{pmatrix} -0\cdot025 & -0\cdot0926i \\ -0\cdot110 & +0\cdot0794i \\ 0\cdot135 & +0\cdot0132i \end{pmatrix} &
\begin{pmatrix} -0\cdot025 & 0\cdot0926i \\ -0\cdot110 & -0\cdot0794i \\ 0\cdot135 & -0\cdot0132i \end{pmatrix}
\end{array}
$$

and, when they all join state 3,

$$\lambda^3 - 1\cdot5\lambda^2 + 0\cdot057\lambda - 0\cdot07 = 0$$

we find

$$\lambda_1^{(3)} = 1, \qquad \lambda_2^{(3)} = 0\cdot25 + 0\cdot0866i, \qquad \lambda_3^{(3)} = 0\cdot25 - 0\cdot0866i$$

with

$$
\begin{array}{ccc}
\mathbf{v}_1^{(3)} & \mathbf{v}_2^{(3)} & \mathbf{v}_3^{(3)} \\[4pt]
\begin{pmatrix} 193 \\ 263 \\ 544 \end{pmatrix} &
\begin{pmatrix} 0\cdot065 & -0\cdot0606i \\ -0\cdot150 & +0\cdot0346i \\ 0\cdot085 & +0\cdot0260i \end{pmatrix} &
\begin{pmatrix} 0\cdot065 & +0\cdot0606i \\ -0\cdot150 & -0\cdot0346i \\ 0\cdot085 & -0\cdot0260i \end{pmatrix}
\end{array}
$$

We notice again that the components of those eigenvectors which do not correspond to the eigenvalue 1 add up to zero.

We illustrate the development of an initial structure

$$(1000, 0, 0)^T$$

assuming that the new entrants join

(i) state 1, or (ii) state 2, or (iii) state 3

(i) $\begin{pmatrix} 1000 \\ 0 \\ 0 \end{pmatrix} \rightarrow \begin{pmatrix} 300 + 200 \\ 400 \\ 100 \end{pmatrix} \rightarrow \begin{pmatrix} 210 + 170 \\ 410 \\ 210 \end{pmatrix} \rightarrow \begin{pmatrix} 197 + 180 \\ 378 \\ 245 \end{pmatrix} \rightarrow \begin{pmatrix} 200 + 187 \\ 364 \\ 249 \end{pmatrix} \rightarrow \begin{pmatrix} 202 + 188 \\ 362 \\ 248 \end{pmatrix} \Rightarrow \begin{pmatrix} 392 \\ 362 \\ 246 \end{pmatrix}$

(ii) $\begin{pmatrix} 1000 \\ 0 \\ 0 \end{pmatrix} \rightarrow \begin{pmatrix} 300 \\ 400 + 200 \\ 100 \end{pmatrix} \rightarrow \begin{pmatrix} 170 \\ 430 + 150 \\ 250 \end{pmatrix} \rightarrow \begin{pmatrix} 159 \\ 383 + 167 \\ 291 \end{pmatrix} \rightarrow \begin{pmatrix} 161 \\ 368 + 174 \\ 297 \end{pmatrix} \rightarrow \begin{pmatrix} 162 \\ 365 + 176 \\ 297 \end{pmatrix} \Rightarrow \begin{pmatrix} 162 \\ 541 \\ 297 \end{pmatrix}$

(iii) $\begin{pmatrix} 1000 \\ 0 \\ 0 \end{pmatrix} \rightarrow \begin{pmatrix} 300 \\ 400 \\ 100 + 200 \end{pmatrix} \rightarrow \begin{pmatrix} 190 \\ 350 \\ 270 + 190 \end{pmatrix} \rightarrow \begin{pmatrix} 184 \\ 297 \\ 308 + 211 \end{pmatrix} \rightarrow \begin{pmatrix} 189 \\ 274 \\ 315 + 222 \end{pmatrix} \rightarrow \begin{pmatrix} 192 \\ 266 \\ 316 + 226 \end{pmatrix} \Rightarrow \begin{pmatrix} 193 \\ 263 \\ 544 \end{pmatrix}$

In each case, the structure tends towards the stationary one, that is the eigenvector corresponding to the eigenvalue 1.

Of course, the new entrants need not all join the same state, but may be distributed among the states in proportions $r_1, r_2, \ldots, r_k$. Then we have to solve the determinantal equation $|\mathbf{R}(\mathbf{r}) - \lambda\mathbf{I}| = 0$, where $\mathbf{R}(\mathbf{r})$ equals

$$\mathbf{R}(\mathbf{r}) = \mathbf{P}^{\mathrm{T}} + \mathbf{r}\mathbf{w}^{\mathrm{T}} = \begin{vmatrix} p_{11} + r_1 w_1 & p_{21} + r_1 w_2 & \cdots & p_{k1} + r_1 w_k \\ p_{12} + r_2 w_1 & p_{22} + r_2 w_2 & \cdots & p_{k2} + r_2 w_k \\ \vdots & \vdots & & \vdots \\ p_{1k} + r_k w_1 & p_{2k} + r_k w_2 & \cdots & p_{kk} + r_k p_k \end{vmatrix} = 0 \qquad (11.5.4)$$

EXAMPLE 11.5.2.   Let the p_{ij} be as in Example 11.5.1, and

$$r_1 = 0.3, \qquad r_2 = 0.5, \qquad r_3 = 0.2.$$

$$\mathbf{R} = \begin{pmatrix} 0.3 \\ 0.5 \\ 0.2 \end{pmatrix} = \begin{pmatrix} 0.3 + 0.2 \times 0.3 & 0.1 + 0.1 \times 0.3 & 0.2 + 0.3 \times 0.3 \\ 0.4 + 0.2 \times 0.5 & 0.5 + 0.1 \times 0.5 & 0.1 + 0.3 \times 0.5 \\ 0.1 + 0.2 \times 0.2 & 0.3 + 0.1 \times 0.2 & 0.4 + 0.3 \times 0.2 \end{pmatrix},$$

$$\begin{vmatrix} 0.36 - \lambda & 0.13 & 0.29 \\ 0.50 & 0.55 - \lambda & 0.25 \\ 0.14 & 0.32 & 0.46 - \lambda \end{vmatrix} = -\lambda^3 + 1.37\lambda^2 - 0.431\lambda + 0.061 = 0.$$

Again, 1 is an eigenvalue, and the corresponding (stationary) eigenvector is (proportional to)

$$236$$
$$441$$
$$323$$

If we start with $(1000, 0, 0)^{\mathrm{T}}$, then we observe the following development:

$$\begin{pmatrix} 1000 \\ 0 \\ 0 \end{pmatrix} \rightarrow \begin{pmatrix} 300 + 60 \\ 400 + 100 \\ 100 + 40 \end{pmatrix} = \begin{pmatrix} 360 \\ 500 \\ 140 \end{pmatrix} \rightarrow \begin{pmatrix} 186 + 49 \\ 408 + 82 \\ 242 + 33 \end{pmatrix} = \begin{pmatrix} 235 \\ 490 \\ 275 \end{pmatrix} \rightarrow \begin{pmatrix} 175 + 54 \\ 366 + 89 \\ 281 + 35 \end{pmatrix} = \begin{pmatrix} 229 \\ 455 \\ 316 \end{pmatrix}$$

$$\rightarrow \begin{pmatrix} 177 + 56 \\ 351 + 93 \\ 286 + 37 \end{pmatrix} = \begin{pmatrix} 233 \\ 444 \\ 323 \end{pmatrix} \rightarrow \begin{pmatrix} 179 + 56 \\ 348 + 94 \\ 285 + 38 \end{pmatrix} = \begin{pmatrix} 235 \\ 442 \\ 323 \end{pmatrix} \Rightarrow \begin{pmatrix} 236 \\ 441 \\ 323 \end{pmatrix}.$$

The Cayley–Hamilton theorem (I, Theorem 7.5.1) states that every matrix satisfies its own characteristic equation. We have $\mathbf{n}(t) = \mathbf{R}^t\mathbf{n}(0)$, and therefore

THEOREM 11.5.1.   *The values* $\mathbf{n}(t)$, $t = 0, 1, 2, \ldots$, *satisfy a difference equation, whose characteristic (reverse) polynomial* (I, 14.3.3) *is that characteristic equation.*

Also, the total number of new entrants is $\sum_{i=1}^{k} \mathbf{n}_i \mathbf{w}_i$, and hence:

THEOREM 11.5.2.   *The numbers of new entrants satisfy the same difference equation.*

Formula (11.2.4) exhibits a special case of this theorem.

EXAMPLE 11.5.2.   (Continued) Take $n(t)$ for $t = 1, 2, 3, 4$. We have indeed

$$\begin{pmatrix} 233 \\ 444 \\ 323 \end{pmatrix} - 1.37 \begin{pmatrix} 229 \\ 455 \\ 316 \end{pmatrix} + 0.431 \begin{pmatrix} 235 \\ 490 \\ 275 \end{pmatrix} - 0.061 \begin{pmatrix} 360 \\ 500 \\ 140 \end{pmatrix} = 0.$$

Concerning new entrants, take, for example, $n(t)$, $t = 2, 3, 4, 5$, and we find

$$188 - 1.370 \times 186 + 0.431 \times 178 - 0.061 \times 164 = 0.$$

11.6.   THE STABLE POPULATION

In many situations we cannot insist that the total population remain constant, because the number of new entrants depends on other considerations. For instance, in an animal species, it might depend on the fertility of the female members in the various states.

First, we consider a case where one eigenvalue, λ_1, is larger than 1, while the absolute values of all other eigenvalues are less than 1. Then (compare (11.3.7)), because

$$\mathbf{n}(t) = c_1 \lambda_1^t v_1 + c_2 \lambda_2^t v_2 + \ldots + c_k \lambda_k^t v_k, \tag{11.6.1}$$

we see that, with increasing t, the ratio $\mathbf{n}(t + 1)/\mathbf{n}(t)$ tends towards the constant λ_1. We say that the population tends towards a 'stable' one.

The total population at time t will be $\mathbf{n}(t)^{\mathrm{T}}\mathbf{e}$, where $\mathbf{e}$ is the vector whose every component equals 1. Moreover

$$\mathbf{n}(t + 1)^{\mathrm{T}}\mathbf{e} = \mathbf{A}\mathbf{n}(t)^{\mathrm{T}}\mathbf{e} = \begin{pmatrix} p_{11} n_1(t) + \ldots + p_{k1} n_k(t) \\ \vdots \qquad\qquad \vdots \\ p_{1k} n_1(t) + \ldots + p_{kk} n_k(t) \end{pmatrix} \cdot \mathbf{e}$$

$$= (p_{11} + \ldots + p_1 k) n_1(t) + \ldots + (p_{k1} + \ldots + p_{kk}) n_k(t).$$

If all columns of the transition matrix add up to the dominant eigenvalue, then the total population increases at this rate.

EXAMPLE 11.6.1.

$$\mathbf{P}^{\mathrm{T}} = \mathbf{A} = \begin{pmatrix} 0.4 & 0.7 & 1.2 \\ 0.8 & 0 & 0 \\ 0 & 0.5 & 0 \end{pmatrix}.$$

All columns add up to 1.2. The characteristic equations is

$$\lambda^3 - 0.4\lambda^2 - 0.56\lambda - 0.48 = 0$$

with eigenvalues

$$\lambda_1 = 1.2, \qquad \lambda_2 = -0.4 + \sqrt{0.24}\,i, \qquad \lambda_3 - 0.4 - \sqrt{0.24}\,i.$$

Observe that $|\lambda| = |\lambda_3| < 1$. The eigenvectors are (proportional to)

$$\begin{matrix} \mathbf{v}_1 \\ \begin{pmatrix} 1500 \\ 1000 \\ 417 \end{pmatrix} \end{matrix} \quad \begin{matrix} \mathbf{v}_2 \\ \begin{pmatrix} -0.5 + 1.25\sqrt{0.24}\,i \\ 1 \\ -0.5 - 1.25\sqrt{0.24}\,i \end{pmatrix} \end{matrix} \quad \begin{matrix} \mathbf{v}_3 \\ \begin{pmatrix} -0.5 - 1.25\sqrt{0.24}\,i \\ 1 \\ -0.5 + 1.25\sqrt{0.24}\,i \end{pmatrix} \end{matrix}$$

The components of $\mathbf{v}_2$ and of $\mathbf{v}_3$ add up to zero.

An initial population of

$$(1300, 1200, 1000)^T$$

would develop as follows:

$$\begin{pmatrix} 1300 \\ 1200 \\ 1000 \end{pmatrix} \rightarrow \begin{pmatrix} 2560 \\ 1040 \\ 600 \end{pmatrix} \rightarrow \begin{pmatrix} 2472 \\ 2048 \\ 520 \end{pmatrix} \rightarrow \begin{pmatrix} 3046 \\ 1978 \\ 1024 \end{pmatrix} \rightarrow \begin{pmatrix} 3832 \\ 2437 \\ 989 \end{pmatrix} \ldots$$

| Totals | 3500 | 4200 | 5040 | 1068 | 7258 |

The totals increase at a rate of 1.2 times the previous total, and in the limit each single component will increase at this rate, because no absolute value of any other eigenvalue reaches 1.

The facts are more complex if we have an eigenvalue with absolute value larger than 1, and another equal to -1, as in the following example.

EXAMPLE 11.6.2.   Let

$$\mathbf{A} = \begin{pmatrix} 1.5 & 1.5 \\ 2.5 & 0.5 \end{pmatrix}.$$

The eigenvalues are $\lambda_1 = 3$ and $\lambda_2 - 1$, while the eigenvectors are

$$\mathbf{v}_1 = \begin{pmatrix} 1 \\ 1 \end{pmatrix} \quad \text{and} \quad \mathbf{v}_2 = \begin{pmatrix} -3 \\ 5 \end{pmatrix}.$$

If we start with

$$\mathbf{n}(0) = 5\mathbf{v}_1 + \mathbf{v}_2 = \begin{pmatrix} 2 \\ 10 \end{pmatrix},$$

then the structures willl be

$$5\lambda_1^t \begin{pmatrix} 1 \\ 1 \end{pmatrix} + \lambda_2^t \begin{pmatrix} -3 \\ 5 \end{pmatrix},$$

that is

$$\begin{pmatrix} 2 \\ 10 \end{pmatrix} \rightarrow \begin{pmatrix} 18 \\ 10 \end{pmatrix} \rightarrow \begin{pmatrix} 42 \\ 50 \end{pmatrix} \rightarrow \begin{pmatrix} 138 \\ 130 \end{pmatrix} \rightarrow \begin{pmatrix} 402 \\ 410 \end{pmatrix} \cdots$$

| Totals | 12 | 28 | 92 | 268 | 812 |

Each total differs from three times the previous total alternately by -8 and by 8, thus exhibiting the periodic influence of $\lambda_2 = -1$, in addition to the influence of the dominant eigenvalue 3.

We add one more example, trivial in itself, but which shows clearly the periodicity introduced by eigenvalues of absolute value 1.

EXAMPLE 11.6.3.

$$\mathbf{A} = \begin{pmatrix} 0 & 1 & 0 \\ 0 & 0 & 1 \\ 1 & 0 & 0 \end{pmatrix}, \quad \mathbf{A}^2 = \begin{pmatrix} 0 & 0 & 1 \\ 1 & 0 & 0 \\ 0 & 1 & 0 \end{pmatrix}, \quad \mathbf{A}^3 = \begin{pmatrix} 1 & 0 & 0 \\ 0 & 1 & 0 \\ 0 & 0 & 1 \end{pmatrix},$$

and, in general,

$$\mathbf{A}^{3+t} = \mathbf{A}^t \quad \text{for } t = 0, 1, 2, \dots$$

An initial structure of $(10, 0, 0)^{\mathrm{T}}$ will develop thus:

$$\begin{pmatrix} 10 \\ 0 \\ 0 \end{pmatrix} \rightarrow \begin{pmatrix} 0 \\ 0 \\ 10 \end{pmatrix} \rightarrow \begin{pmatrix} 0 \\ 10 \\ 0 \end{pmatrix} \rightarrow \begin{pmatrix} 10 \\ 0 \\ 0 \end{pmatrix} \rightarrow \begin{pmatrix} 0 \\ 0 \\ 10 \end{pmatrix} \rightarrow \begin{pmatrix} 0 \\ 10 \\ 0 \end{pmatrix} \cdots$$

11.7.   THE SEMI-STATIONARY POPULATION

Let us now look at a particular interpretation of the meaning of a state. We define those members of the organization who have the same 'seniority' and who work in the same branch as being in the same state.

A member changes his or her state by increasing seniority from year to year, if their branch does not change, and possibly by transfer from one branch into another, where their seniority starts then with 0. Also, they may leave altogether.

We might now wish not only to keep the total population constant, but also the number of those who work in the same branch, irrespective of their seniorities.

If, then, a state is defined by (s, b), s being the seniority and b the branch, then we want to keep the total of those members constant whose parameter b is the same, adding over all values of the parameter s.

Formally, this means that we preserve not only the total of the entire population, but also, within it, the total of the combined populations of certain selected states. We call such a population 'semi-stationary'. (A stationary population is one in which all states are individually preserved.)

We shall study now the conditions for this to be possible, given transfer rates and, implicitly, loss rates.

We have seen that the number of new entrants after t years is $\mathbf{n}(t)^T \cdot \mathbf{w} = \mathbf{n}(0)^T (\mathbf{R}^T)^t \mathbf{w}$. If this number is to preserve the total of the first m states (any m states can be considered to be the first m), then it must replace the losses from these states. Therefore it must equal

$$\mathbf{n}(0)^T (\mathbf{R}^T)^t \cdot \mathbf{w}_{(m)}, \tag{11.7.1}$$

$$\mathbf{w}_{(m)} = (\mathbf{I} - \mathbf{P})\mathbf{e}_{(m)} \tag{11.7.2}$$

and $\mathbf{e}_{(m)}$ is the vector whose first m components equal 1, and the other components equal 0.

Hence a semi-stationary structure $\mathbf{n}(0)$ which preserves the total of the first m states combined satisfies

$$\mathbf{n}(0)^T (\mathbf{R}^T)^t (\mathbf{w} - \mathbf{w}_{(m)}) = 0 \quad (t = 0, 1, 2, \ldots). \tag{11.7.3}$$

It would clearly be impossible to establish the existence of such a structure if the infinite set (11.7.3) had to be tested. However, since every matrix satisfies its own characteristic equation [I, Theorem 7.5.1], every power higher than the kth depends linearly on the first k powers, and if $\mathbf{n}(0)$ satisfies (11.7.3) for $t = 1, 2, \ldots, k$, then it satisfies the whole set.

The equations of the set are homogeneous, and we know that a non-trivial solution exists, namely the stationary population. Therefore the rank of the set of the first k equations is at most $k - 1$. However, we are now interested in the case where the rank is less than $k - 1$, as in the following example.

EXAMPLE 11.7.1.   Let the transition matrix be

$$\mathbf{P} = \begin{pmatrix} 0\cdot2 & 0\cdot1 & 0\cdot6 & 0\cdot0 \\ 0\cdot2 & 0\cdot2 & 0\cdot0 & 0\cdot4 \\ 0\cdot3 & 0\cdot1 & 0\cdot2 & 0\cdot1 \\ 0\cdot3 & 0\cdot3 & 0\cdot2 & 0\cdot0 \end{pmatrix}$$

so that

$$\mathbf{w} = \begin{pmatrix} 0\cdot1 \\ 0\cdot2 \\ 0\cdot3 \\ 0\cdot2 \end{pmatrix}, \quad \mathbf{w}_{(2)} = \begin{pmatrix} 0\cdot7 \\ 0\cdot6 \\ -0\cdot4 \\ -0\cdot6 \end{pmatrix}, \quad \mathbf{w} - \mathbf{w}_{(2)} = \begin{pmatrix} -0\cdot6 \\ -0\cdot4 \\ 0\cdot7 \\ 0\cdot8 \end{pmatrix}.$$

When new entrants join state 1, then

$$\mathbf{R} = \begin{pmatrix} 0\cdot3 & 0\cdot4 & 0\cdot6 & 0\cdot5 \\ 0\cdot1 & 0\cdot2 & 0\cdot1 & 0\cdot3 \\ 0\cdot6 & 0\cdot0 & 0\cdot2 & 0\cdot2 \\ 0\cdot0 & 0\cdot4 & 0\cdot1 & 0\cdot0 \end{pmatrix}$$

and the set (11.7.3) is then (writing n_i for $n_i(0)$) (or rather the first four equations of the set)

$$
\begin{aligned}
6n_1 \quad + 4n_2 \quad - 7n_3 \quad - 8n_4 &= 0, \\
-20n_1 \quad\quad\quad + 18n_3 \quad + 28n_4 &= 0, \\
48n_1 \quad + 32n_2 \quad - 56n_3 \quad - 64n_4 &= 0, \\
-160n_1 \quad\quad\quad + 144n_3 \quad + 224n_4 &= 0.
\end{aligned}
$$

This set is seen to have rank 2, and two independent solutions are, for instance,

$$
\mathbf{v}_1 = \begin{pmatrix} 10 \\ 3 \\ 8 \\ 2 \end{pmatrix} \quad \text{and} \quad \mathbf{v}_2 = \begin{pmatrix} 1 \\ -1 \\ -2 \\ 2 \end{pmatrix}.
$$

Any linear combination fo $\mathbf{v}_1$ and $\mathbf{v}_2$ is also a solution and can serve as a semi-stationary structure, provided all its components are non-negative.

For instance, take $\mathbf{n}(0) = 43 \cdot 5(\mathbf{v}_1 + \mathbf{v}_2)$. This makes the total population equal to 1000, and we have the development (in rounded values)

$$
\begin{pmatrix} 478 \\ 87 \\ 261 \\ 174 \end{pmatrix} \rightarrow \begin{pmatrix} 243 + 179 \\ 143 \\ 374 \\ 61 \end{pmatrix} \rightarrow \begin{pmatrix} 243 + 196 \\ 126 \\ 340 \\ 95 \end{pmatrix} \rightarrow \begin{pmatrix} 243 + 191 \\ 131 \\ 350 \\ 85 \end{pmatrix} \rightarrow \begin{pmatrix} 243 + 192 \\ 130 \\ 348 \\ 87 \end{pmatrix}
$$

Totals	1000	1000	1000	1000	1000

and so on.

The total of the first two states combined are

$$565 \quad 565 \quad 565 \quad 565 \quad 565.$$

The structure tends to the stationary population, viz.

$$
43.5\mathbf{v}_1 = \begin{pmatrix} 435 \\ 130 \\ 348 \\ 87 \end{pmatrix}
$$

$$1000$$

It is desirable to have some indication of which set of states can be preserved, without having to solve k equations of the set (11.7.3) for all possible m. For this purpose we observe that, in the same way as we have proved Theorem 11.4.2, we can show that if the set of the first m components of the eigenvectors $v_1, \ldots, v_r$ add up to zero, and if $\lambda_1 = 1$, then any linear combination of $v_1, v_2, \ldots, v_r$ will preserve the total of its first m components. (Note that any m components can be made to be the first m.)

EXAMPLES 11.7.1.   (Continued) The eigenvalues are

$$\lambda_1 = 1, \qquad \lambda_2 = -0{\cdot}3, \qquad \lambda_3 = \sqrt{0{\cdot}08}, \qquad \lambda_4 = -\sqrt{0{\cdot}08},$$

with corresponding eigenvectors (proportional to)

$$
\begin{array}{cccc}
\mathbf{v}_1 & \mathbf{v}_2 & \mathbf{v}_3 & \mathbf{v}_4
\end{array}
$$

$$
\begin{pmatrix} 10 \\ 3 \\ 8 \\ 2 \end{pmatrix}
\begin{pmatrix} 1 \\ -1 \\ -2 \\ 2 \end{pmatrix}
\begin{pmatrix} 1 \\ -2 \\ 10\sqrt{0{\cdot}08} \\ 1 - 10\sqrt{0{\cdot}08} \end{pmatrix}
\begin{pmatrix} -1 \\ 2 \\ 10\sqrt{0{\cdot}08} \\ -1 - 10\sqrt{0{\cdot}08} \end{pmatrix}
$$

All columns of $\mathbf{R}$ add up to 1, and the components of all $\mathbf{v}_i$ $(i \neq 1)$ add up to 0.

The first two components of $\mathbf{v}_2$ also add up to 0, and therefore any linear combination of $\mathbf{v}_1$ and $\mathbf{v}_2$ will preserve the total of the first two states, a result which we have already reached above.

11.8.   AGEING AND PROMOTION

The rules for the development of population structures which we have so far considered were all very simple. In this section we exhibit an example of a somewhat more complex structure.

Consider a hierarchy divided into grades, and an initial population all of whose members are in grade 1 and are of 'age' 0. Of those individuals who reach age x, a ratio p_{x+1} will survive to reach age $x + 1$. Of those in grade g, who have been in this grade for s years, and who have thus 'seniority' s, a ratio $p_g(s)$ will be promoted to grade $g + 1$, independently of their age. No other transitions, and no entries take place at $t > 0$.

We shall write $P(x, x + m)$ for the product $p_{x+1}p_{x+2} \cdots p_{x+m}$, and $Q_g(s)$ for the product $q_g(0)q_g(1) \ldots q_g(s - 1)$, where $q_g(r) = 1 - p_g(r)$.

We want to determine the number $n_g(s, x)$ of those members in grade g who have reached age x and have been in their grade for s years.

Now $n_1(0, 0)$ is the number of members of the initial population, $n_1(s, x) = 0$ for $s \neq x$ and $n_1(s, s)$ is the number of those who entered s years ago at age 0 and have not yet been promoted.

Hence

$$n_g(s, x) = n_g(0, x - s)P(x - s, x)Q_g(s) \quad (s = 0, 1, \ldots, x). \tag{11.8.1}$$

The number of those just promoted from grade g into grade $g + 1$, at age x, equals $n_{g+1}(0, x)$. In the lower grade g they might have had seniority $0, 1, \ldots,$ or *x*. Therefore

$$n_{g+1}(0, x) = \sum_{s=0}^{x} n_g(s, x) p_g(s)$$

$$= \sum_{s=0}^{x} n_g(0, x - s) P(x - s, x) Q_g(s) p_g(s). \qquad (11.8.2)$$

Thus we have

$$n_2(0, x) = n_1(0, 0) P(0, x) Q_1(x) p_1(x),$$

$$n_3(0,x) = \sum_{s=0}^{x} n_2(0, x - s) P(x - s, x) Q_2(s) p_2(s)$$

$$= \sum_{s=0}^{x} n_1(0, 0) P(0, x - s) Q_1(x - s) p_1(x - s) P(x - s, x) Q_2(s) p_2(s)$$

$$= \sum_{s=0}^{x} n_1(0, 0) P(0, x) Q_1(x - s) Q_2(s) p_1(x - s) p_2(s)$$

and generally, by induction,

$$n_{t+1}(0, x) = \sum_{s_{t-1}} \cdots \sum_{s_2} \sum_{s_1} n_1(0, 0) P(0, x) Q_1(x - s_{t-1} - \ldots - s_1)$$

$$\times Q_2(s_2) \ldots Q_t(s_{t-1}) p_1(x - s_{t-1} - \ldots - s_1)$$

$$\times p_2(s_1) p_3(s_2) \ldots p_t(s_{t-1}) \qquad (11.8.3)$$

From (11.8.3) we obtain, using (11.8.1), $n_g(s, x)$ $(s = 0, 1, \ldots, x)$.

11.9. CONTINUOUS TIME

Consider a population with n_x members, all of age x. Let the proportional loss during the age interval $(x, x + \Delta x)$ be $\mu_x \Delta x$. Actuaries call μ_x the force of mortality at age x.

We have

$$n_{x+\Delta x} = n_x(1 - \mu_x \Delta x),$$

that is

$$\frac{n_{x+\Delta x} - n_x}{n_x \Delta x} = -\mu_x.$$

When Δx tends to 0, then

$$\frac{dn_x}{n_x dx} = \frac{d \ln n_x}{dx} = -\mu_x.$$

Hence

$$n_x = n_0 \exp\left(-\int_0^x \mu_t dt\right). \qquad (11.9.1)$$

Now let $n_x(t)$ be the total of a population with members aged x at time t. We wish to keep this population constant, with entries of age 0.

We see that

$$n_x(t) = n_0(t - x)\exp\left(-\int_0^x \mu_t dt\right) \quad \text{when } c \leqslant t \tag{11.9.2}$$

and

$$n_x(t) = n_{x-t}(0)\exp\left(-\int_{x-t}^x \mu_t dt\right) \quad \text{when } x \geqslant t. \tag{11.9.3}$$

Formulae (11.9.2) and (11.9.3) are clearly analogous to formulae (11.2.2a) and (11.2.2b) respectively.

We have also

$$n_0(t) = \int_0^0 n_x(0)\exp\left(-\int_x^{x+t} \mu_y dy\right)\mu_{x+t} dx$$

$$+ \int_0^t n_0(t - x)\exp\left(\int_0^x -\mu_y dy\right)\mu_x dx, \tag{11.9.4}$$

where the first integral counts the exits and replacements of members of the initial population at time t, and the second integral counts the exits and replacements of members who have entered, replacing earlier exits.

As regards the solution of the integral equation of the second kind in (11.9.4), we refer the reader to Feller (1941).

11.10. STATES IN CONTINUOUS TIME

We assume again, as we did in §11.2, that the membership of the population is divided into discrete states, but we consider now continuous flow between the states.

Let the rate of flow from state i into state j during the time interval $(t, t + \Delta t)$ be $p_{ij}\Delta t$, and write

$$1 - \sum_{j=1}^k p_{ij} = w_i.$$

Then the flow out of and into state j produces

$$n_j(t + \Delta t) = n_j(t)\left(1 - \sum_{\substack{i=1 \\ i \neq j}}^k p_{ji}\Delta t - w_j\Delta t\right) + \sum_{\substack{i=1 \\ i \neq j}}^k n_i(t)p_{ij}\Delta t.$$

If we add

$$n_j(t)p_{jj}\Delta t - n_j(t)p_{jj}\Delta t = 0$$

on the right-hand side, then we can replace

$$\sum_{\substack{i=1 \\ i \ne j}}^{k} \quad \text{by} \quad \sum_{i=1}^{k}$$

in the two places where this sum occurs. Then

$$n_j(t + \Delta t) = n_j(t)(1 - \Delta t) + \sum_{i=1}^{k} n_i(t)p_{ij}\Delta t,$$

that is

$$\frac{n_j(t + \Delta t) - n_j(t)}{\Delta t} = \sum_{i=1}^{k} n_i(t)p_{ij} - n_j(t).$$

When t tends to zero, we obtain the system

$$\frac{dn_1(t)}{dt} = n_1(t)(p_{11} - 1) + n_2(t)p_{21} + \ldots + n_k(t)p_{k1},$$

$$\frac{dn_2(t)}{dt} = n_1(t)p_{12} + n_2(t)(p_{22} - 1) + \ldots + n_k(t)p_{k2}, \qquad (11.10.1)$$

$$\ldots$$

$$\frac{dn_k(t)}{dt} = n_1(t)p_{1k} + n_2(t)p_{2k} + \ldots + n_k(t)(p_{kk} - 1).$$

If we want the total population to be constant throughout time, then we replace the waste by a flow of new entrants. Let these enter state i in proportion r_i. Then we replace in (11.10.1) p_{ij} by $p_{ij} + w_i r_j$. We obtain

$$\frac{dn_1(t)}{dt} = n_1(t)(p_{11} + w_1 r_1 - 1) + n_2(t)(p_{21} + w_2 r_1) + \ldots + n_k(t)(p_{k1} + w_k r_1),$$

$$\frac{dn_2(t)}{dt} = n_1(t)(p_{12} + w_1 r_2) + n^2(t)(p_{22} + w_2 r_2 - 1) + \ldots + n_k(t)(p_{k2} + w_k r_2),$$

$$\ldots$$

$$\frac{dn_k(t)}{dt} = n_1(t)(p_{1k} + w_1 r_k) + n_2(t)(p_{2k} + w_2 r_k) + \ldots + n_k(t)(p_{kk} + w_k r_k - 1).$$

$$(11.10.2)$$

Adding the equations of the system (11.10.2), we obtain

$$\frac{d\sum_{j=1}^{k} n_j(t)}{dt} = \sum_{i=1}^{k} n_i(t)\left[\sum_{j=1}^{k} (p_{ij} + w_i r_j) - 1 \right] = 0$$

because

$$\sum_{j=1}^{k} (p_{ij} + w_i r_j) = 1 \quad \text{for all } i.$$

This confirms that $\sum_{j=1}^{k} n_j(t)$ is independent of t. It also shows that the matrix on the right-hand side of (11.10.2) is singular, and has, therefore, an eigenvalue equal to 0. It can also be shown that all other eigenvalues must have negative real parts (Vajda (1978) App. I, 1.5).

In order to solve such a system of linear first-order differential equations, we proceed by a method analogous to that for the solution of a system of first-order linear difference equations, which we used in §11.3 (see IV, Theorem 7.9.4).

Find the eigenvalues of the matrix. They are, clearly, the same as those in §11.3, reduced by unity, while the eigenvectors are the same as those arrived at in that section.

If we denote the eigenvalues by $\lambda_1, \ldots, \lambda_k$, and the eigenvectors by $v_1, \ldots, v_k$, then

$$n_i(t) = c_1 \exp(-\lambda_1 t) v_{1i} + \ldots + c_k \exp(-\lambda_k t) v_{ki}, \qquad (11.10.3)$$

where the v_{ji} denote the jth component of v_j, and the coefficients c_i depend on the initial population structure.

EXAMPLE 11.10.1.   Let the transition matrix be

$$P = \begin{pmatrix} 0.2 & 0.2 & 0.3 \\ 0.2 & 0.6 & 0.1 \\ 0.1 & 0.2 & 0.6 \end{pmatrix}, \qquad w = \begin{pmatrix} 0.3 \\ 0.2 \\ 0.1 \end{pmatrix},$$

and assume that all new entrants join state 1. We have to solve the determinantal equation

$$\begin{pmatrix} -0.5 - \lambda & 0.3 & 0.2 \\ 0.2 & -0.4 - \lambda & 0.2 \\ 0.3 & 0.1 & -0.4 - \lambda \end{pmatrix} = 0.$$

The eigenvalues are

$$0, \qquad -0.7, \qquad -0.6,$$

and the corresponding eigenvectors are

$$\begin{pmatrix} 1 \\ 1 \\ 1 \end{pmatrix}, \qquad \begin{pmatrix} 1 \\ 0 \\ -1 \end{pmatrix}, \qquad \begin{pmatrix} 1 \\ 1 \\ -2 \end{pmatrix}.$$

For simplicity, and in order to concentrate on essentials, we have chosen a case where all eigenvalues are real. But even if the eigenvalues and the eigenvectors were complex, the numbers of members of the population would still turn out to be real, because conjugate complex values appear in pairs throughout.

Starting with $(1000, 0, 0)^T$, we find now

$$n_1(t) = 333 + 1000 \exp(-0.7t) - 333 \exp(-0.6t),$$
$$n_2(t) = 333 \qquad\qquad\qquad - 333 \exp(-0.6t),$$
$$n_3(t) = 222 - 1000 \exp(-0.7t) + 666 \exp(-0.6t).$$

We exhibit an extract from the computations of the developing structure.

$t =$	0	1	3	5	7	9	∞
$n_1 =$	1000	647	400	347	335	334	333
$n_2 =$	0	150	278	317	328	332	333
$n_3 =$	0	203	322	336	337	334	333

S. V.

REFERENCES

Feller, W. (1941). On the integral equation of renewal theory, *Annals of Mathematical Statistics*, **12**, 243–267.

Vajda, S. (1978). *Mathematics of Manpower Planning*. Wiley.

CHAPTER 12

Queueing and Related Theory

12.1. INTRODUCTION

It is the all-too-familiar experience of road users in congested urban areas to encounter tail-backs caused by unexpectedly heavy traffic, an accident, roadworks, a toll barrier, and so on. Such time wasters are matched in the world of air transport by the delays suffered by passengers due to late arrivals and departures of scheduled flights, and this experience is, unfortunately, often shared by train and bus passengers. Ports provide a range of similar situations, the victims being ships waiting to berth, or, if berthed, to unload or load cargo. These examples are drawn from the field of transport, but almost every human endeavour supplies instances where the same elements are present. The provision of health care in the public sector is a favourite whipping boy in the United Kingdom. We hear of old ladies waiting years for hip replacements and, at the other end of the scale, of the delays at out-patient clinics even when an appointment system is supposed to be in operation. Many banks and post offices have now gone in for the organization of a single serpentine queue to accommodate waiting customers in place of the old-fashioned régime where parallel queues lined up in front of the individual service points. The customers still complain, as it happens without justification: the advantage of the serpentine queue was proclaimed years ago by the theory described in this article; delay in application must be laid at the feet of management.

Queueing theory, also known as waiting line theory in the United States and the theory of mass service in the USSR, is concerned with the situations we have mentioned specifically and with the countless others that share the common property of customers, in a generalized sense, who arrive in a stream at a service centre, and demand services that the centre is supposed to provide. The central interest of the theory is in the delays suffered by customers, for whatever reasons, and in the accumulation of waiting customers said to form the queue. The customers themselves are primarily interested in systems which offer the minimum of delay: management has to concern itself with that too if it is at all interested in retaining customers, but also in the physical problems arising from the need to accommodate the large numbers that, under unfavourable circumstances, are obliged to wait. The theory strives to provide mathematical models and analysis which reveal the relation between the forces that drive the situation under study and the operational consequences, with a view to improving the functioning of the system. This is a primarily utilitarian objective in the true spirit of operational research.

Without doubt the honours go to A. K. Erlang, a Danish telephone engineer, for the first significant theoretical and practical developments which took place in the early years of this century. (It is interesting to observe that Erlang, revered by queueing theorists everywhere, is perhaps better known in his own country by the mathematical tables bearing his name used in Danish schools.) The Copenhagen Telephone Company required studies of the operations of telephone exchanges. At that time a caller (the customer) had to request connection to his destination through an operator and only if a line was free could the connection be made. Otherwise the call was said to be lost. A lost call could mean a lost or, at least dissatisfied, customer. To reduce the chance of loss, the obvious remedy is to instal more communication links, but this is a matter of investment which has to be balanced carefully against the capital outlay and the prospect of enhanced business. We shall refer more specifically to Erlang's work in later sections. His publications date from 1909 to 1923 and these are to be found in English in Brockmeyer, Halstrøm and Jensen (1948).

The communication industry has provided the motivation for much work in queueing theory and continues to do so in various applications, notably in connection with telecommunication and computer networks. Important developments associated with the names of Pollaczek and Khintchine took place in the 1930s, but it was after the Second World War that the subject's potential really captured the attention of mathematicians. An important review text by Saaty (1961) listed over 900 references and there is no doubt that by now the literature contains thousands. For this reason we shall limit ourselves to textbook references except in special cases that merit otherwise.

12.2. DEFINITIONS

Figure 12.2.1 represents a realization of a service system with a single service point. Customers wait in a body, or queue, and are served in order of arrival. This is the so-called first come, first served (FCFS) *queue discipline*. It is also

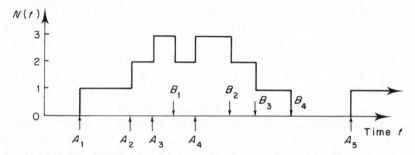

Figure 12.2.1. Realization of a first come, first served, single-server queueing model

called first in, first out (FIFO). The horizontal axis represents time t elapsed since the process began. This increases, as usual, to the right. The vertical axis represents *system state*, denoted by $N(t)$, defined to be the total number of cumstomers present in the system, waiting and being served. In Figure 12.2.1, $N(t)$ is a step function [see IV, Definition 2.3.4] with jumps at $A_1, A_2, \ldots$. At a jump point t, $N(t)$ is defined as $N(t+)$. Thus, for example, $N(B_1) = 2$, $N(A_4) = 3$. The points marked A_n $(n = 1, 2, 3, \ldots)$ indicate the instants in time (technically known as *epochs*) at which customers C_n arrive. As shown, the system is empty just before A_1. The corresponding customer C_1 thus begins service immediately and $N(t)$ jumps from zero to one. The sequence of points marked B_n $(n = 1, 2, 3, \ldots)$ indicate the epochs at which service of C_n terminates. At each epoch B_n, $N(t)$ decreases by one while at each epoch A_n it increases by one. Figure 12.2.1 shows the fluctuations in $N(t)$ over a short period. By the physical nature of the model, $N(t)$ is a non-negative integer.

The sequence of points (A_n) is the *arrival stream* and represents the external driving force of the system. The intervals between arrivals are positive, continuous random variables and we denote the time interval $A_{n+1} - A_n$ by T_{n+1}. This is called the *interarrival interval* between C_n and C_{n+1}. The T_n usually have a common distribution which we shall denote by $A(t)$ with probability density function $a(t)$ $(t \geq 0)$. The members of (T_n) are usually independent random variables [see II, Definitions 4.4.1 and 6.6.1] but they could be correlated [see II, §9.6] as when, for instance, a highly desirable item is on sale. In another special case, an appointment system, all T_n are equal and $A(t)$ has a special form.

The sequence (B_n) is triggered by the arrival process. When $N(B_n) > 0$ the intervals $S_{n+1} = B_{n+1} - B_n$ are the *service times* of C_{n+1}, positive continuous random variables with distribution function $B(t)$ and probability density $b(t)$. The S_n are usually modelled as independent random variables, but they need not be. In some cases they too can be constant. In the simplest situations, (T_n) and (S_n) are taken to be statistically independent, but they need not be so and, in practice, a cooperative server may tend to shorten S_n, that is work faster, when T_n decreases making the traffic heavier. The reverse may also happen. The arrival and service mechanisms can be non-stationary so that the distributions vary with time. This is realistic when a system is analysed over a long period under fluctuating traffic conditions. However, to avoid the complications caused by a non-stationary analysis, it is quite acceptable in most cases to divide time into periods during which conditions are stationary.

The period of time $B_4 - A_1$ is called a *busy period* for the obvious reason that the server was idle until A_1 and then worked continuously until B_4, again becoming idle. The period $A_5 - B_4$ is correspondingly known as an *idle period*. In connection with busy periods it is both the duration in time and the number of customers served that are of interest. *Output intervals* are also of concern. These are the time intervals separating successive departures from the system. During a busy period these are simply service times, but the interval between the end of a busy period and the first service completion after the beginning of the next is more complicated.

In this chapter we shall, above all, discuss $N(t)$ and *system time*. This is the time a customer spends in the system, that is it is the sum of queueing time and service time. For C_n the system time is

$$W_n = B_n - A_n. \tag{12.2.1}$$

Figure 12.2.2 helps to show that the sequence (W_n), random variables if S_n and T_n are, possesses a useful 'lack-of-memory' property: that is W_{n+1} depends only on W_n and none of the history of the system before A_n is needed to describe it.

The two cases can be summarized as

$$W_{n+1} = \max(W_n - T_{n+1} + S_{n+1}, S_{n+1}), \tag{12.2.4}$$

a formula noticed essentially by Lindley (1952) and put to effect by Lindley, Smith (1953), and later by Kingman (1962) in connection with 'heavy traffic', of which more later. The importance of (12.2.4) cannot be underestimated. It makes only general assumptions about $A(t)$ and $B(t)$ and does not exclude their possible correlation. It is, however, restricted to FCFS and a single-server system.

Let $f_n(w)$ be the probability density function of W_n, the system time of C_n. Now [see II, §16.2],

$$P(W_{n+1} < w) = P(W_{n+1} < w \mid W_n > T_{n+1})P(W_n > T_{n+1})$$
$$+ P(W_{n+1} < w \mid W_n < T_{n+1})P(W_n < T_{n+1})$$

(since $W_n - T_{n+1}$ is a continuous random variable, $P(W_n = T_{n+1}) = 0$ [see II, §10.1]). Thus

$$P(W_{n+1} < w) = P(W_n - T_{n+1} + S_{n+1} < w \mid W_n > T_{n+1})P(W_n > T_{n+1})$$
$$+ P(S_{n+1} < w)P(W_n < T_{n+1}).$$

Here

$$P(W_n < T_{n+1}) = \int_0^w P(T_{n+1} > w)f_n(w)\,dw.$$

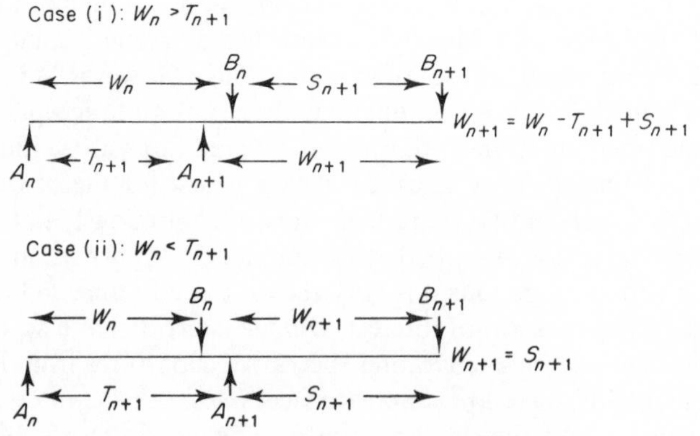

Figure 12.2.2. Structure of system time

It therefore follows from (12.2.4) that the sequence $(f_n(w))$ satisfies the integral difference equation

$$f_{n+1}(w) = b(w) \int_0^\infty A_c(x) f_n(x) dx + \int_0^w b(w-x) dx \int_0^\infty f_n(x+y) a(y) dy$$

$$\tag{12.2.5}$$

where

$$A_c(x) = \int_x^\infty a(y) dy \tag{12.2.6}$$

is the complementary distribution function of T. The expression $\int_0^\infty A_c(x) f_n(x) dx$ is $P(W_n < T_{n+1})$, the probability, namely, that C_{n+1} encounters an empty system. By defining $f_1(w)$, say as $b(w)$, implying that C_1 finds an empty system, $(f_n(w))$ can, in principle, be built up. Notice that the second term in (12.2.5) is the probability density function of the sum of two random variables, namely S_{n+1} and $W_n - T_{n+1}$, under the condition $W_n > T_{n+1}$.

A measure of *traffic intensity*, denoted in this article by r, is a fundamental requirement in the operational analysis of service systems. This parameter expresses the pressure of demand on the system, taking into account the system's capability of dealing with it. In a deterministic framework, if an appointment system were set up in which customers were told to arrive at ten-minute intervals when the service was known to need exactly fifteen minutes to satisfy each customer, then a queue build-up would be the consequence. When interarrival intervals and service times are random variables a build-up can be expected roughly when mean [see II, §8.1] service time $\hat{S}$ exceeds mean interarrival interval $\hat{T}$. This would point to the natural measure

$$r = \hat{S}/\hat{T}, \tag{12.2.7}$$

valid for both deterministic and stochastic systems.

With a stochastic system, an alternative approach is to examine the probability

$$p_0(t) = P[N(t) = 0 | N(0)] \tag{12.2.8}$$

that the system is empty at time t for a given finite initial value $N(0)$ of the system state. If as $t \to \infty$, $p_0(t) \to 0$ it can be concluded that the service in the long run is incapable of coping with demand and a queue build-up is practically inevitable. Viewed from this perspective the limit

$$p_0 = \lim_{t \to \infty} p_0(t) \tag{12.2.9}$$

provides another natural measure of traffic intensity. We shall see that (12.2.7) and (12.2.9) are sometimes equivalent.

The concept of *statistical equilibrium* enters naturally at this point. If p_0 exists and lies in the range $0 < p_0 < 1$, we say that the system is in statistical equilibrium, or that a *steady state* exists. This is equivalent to the statement that

$$p_n = \lim_{t \to \infty} p_n(t) \quad (n = 0, 1, \ldots), \tag{12.2.10}$$

where

$$p_n(t) = P[N(t) = n \mid N(0)] \qquad (12.2.11)$$

exists and is non-zero. The sequence (p_n) is called the set of equilibrium state probabilities.

One form of *queue discipline* has already been mentioned—FCFS. A variety of other possibilities exists, of which the most common are LCFS (or LIFO), and *random service*. The last speaks for itself; LCFS meaning last come, first served, and LIFO meaning last in, first out, are likewise unequivocal. All forms of queue discipline constitute a *priority discipline* of some kind according to which the next customer to be served is chosen according to some well-defined rule. FCFS and LCFS are clearly examples. *Head-of-the-line-priority* and *alternating priority* represent other kinds of discipline. In the former, it is to be imagined that each customer is allocated a ticket indicating his priority, a number, say, ranging from 0 to ∞. Whenever a customer completes service, that customer is selected to be served next who stands at the head of the queue with the highest priority rating number. Under alternating priority, customers of a certain priority class are served to exhaustion. Upon completion, service turns attention to that class with the highest priority out of those remaining, and so on.

Finally, a question of notation: $G/G/N_1/N_2$ designates a queueing system in which the interarrival intervals have a general (unspecified) distribution (the first G), the service times likewise (the second G), the number of service points is N_1, and the maximum number of customers allowed to be in the system is N_2. If N_2 is omitted, it can be assumed to be infinite—no limitation on system size. This is a flexible notation, due to Kendall (1951), and is often extended to describe variations on the basic models. the most common forms of G are M, or exponential (Markovian since such distributions are memoryless); D, or deterministic; and E_k, or 'Erlangian'—the distribution identified by Erlang as well-adapted to fit many observed distributions, and equivalent to the distribution of the sum of k independent exponential random variables [see II, §11.3.2].

12.3. EXPONENTIAL QUEUES

12.3.1. M/M/1/∞

We now outline the analysis of the simplest exponential system.

(a) *System State*

It is recalled that system state $N(t)$ is the non-negative integer-valued random variable illustrated in Figure 12.2.1. $N(t)$ is, in fact, a random walk, restricted in its extent to the non-negative integers, generated by the particular choices

$$a(t) = \lambda e^{-\lambda t}, \qquad b(t) = \mu e^{-\mu t}, \quad (t > 0), \qquad (12.3.1)$$

where λ and μ are positive constants. $a(t)$ generates a positive unit increment in $N(t)$; $b(t)$ generates a positive unit decrement. The physical interpretation of λ and μ is mean rate of arrival and service respectively. The exponential densities are the consequence of, and imply, that the random variables T and S are *Markovian*. This means that at whatever epoch an observer inspects the system when in operation, the time until the next arrival, and the residual service time, are independent of how long they have been in progress. Specifically, the probability that an arrival occurs in some small interval $(t, t + h)$ is given by $\lambda h + o(h)$, independently of the epoch t and of what happened before t. Likewise, the probability that a service in progress at t terminates in $(t, t + h)$ is $\mu h + o(h)$. Erlang identified the exponential distribution to be a good fit to observation of telephone traffic in a practically interesting range of circumstances, a most fortunate discovery because of the resulting theoretical simplifications. He also suggested a flexible and useful generalization, the so-called Erlang family of distributions, effectively the sum of exponentials.

Let

$$p_n(t) = P[N(t) = n \mid N(0)] \quad (n = 0, 1, 2, \ldots). \tag{12.3.2}$$

To find $p_n(t)$ the standard procedure is to concentrate attention on epoch $t + h$ and to establish $p_n(t + h)$ in terms of the probabilities of the possible states at epoch t and the transitions that have to take place to achieve $N(t + h) = n$. In this analysis h is small and positive and ultimately forced to tend to zero. Because it is small, all transitions other than $n + 1 \rightarrow n, n \rightarrow n$, and $n - 1 \rightarrow n(n > 1)$ have probabilities of order $o(h)$. For example, the probability of one arrival in $(t, t + h)$ is $\lambda h + o(h)$ and the probability of two is dominated by $\lambda^2 h^2$, which is $o(h)$ [see II, §20.4].

For $n \geq 1$ we get

$$p_n(t + h) = \lambda h p_{n-1}(t) + (1 - \lambda h - \mu h)p_n(t) + \mu h p_{n+1}(t) + o(h),$$

after collecting all the $o(h)$ terms together. Thus

$$\frac{p_n(t + h) - p_n(t)}{h} = -(\lambda + \mu)p_n(t) + \lambda p_{n-1}(t) + \mu p_{n+1} + \frac{o(h)}{h} \tag{12.3.3}$$

and, going to the limit as $h \rightarrow 0$, gives

$$\dot{p}_n(t) + (\lambda + \mu)p_n(t) = \lambda p_{n-1}(t) + \mu p_{n+1}(t) \quad (n \geq 1), \tag{12.3.4}$$

where the dot denotes differentiation with respect to t. The corresponding formula for $p_0(t)$ is

$$\dot{p}_0(t) + \lambda p_0(t) = \mu p_1(t). \tag{12.3.5}$$

The right-hand side of the equation preceding (12.3.3) is the sum of the three dominant probabilities that contribute to $p_n(t + h)$. All the comparatively negligible probabilities are incorporated in $o(h)$. Given $N(t) = n - 1$, we get $N(t + h) = n$ by means of an arrival in $(t, t + h)$ which has probability dominated by λh. Likewise $N(t) = n + 1 \to N(t + h) = n$ $(n \geq 1)$ requires a service completion (with probability dominated by μh); and the transition $N(t) = n \to N(t + h) = n$ is most likely to occur as a result of no service completion and no arrival, the probability of which is dominated by $1 - \lambda h - \mu h$. (12.3.5) is obtained in a similar manner. To have $N(t + h) = 0$ we need consider only the possibilities $N(t) = 0$ and $N(t) = 1$. In the former case the probability of no change is dominated by $1 - \lambda h$ since no service completion is possible, and, in the latter case, a service completion dominates the transition probability with contribution μh. Thus

$$p_0(t + h) = (1 - \lambda h)p_0(t) + \mu h p_1(t) + o(h),$$

and (12.3.5) follows. This analysis holds for any $t > 0$.

The set (12.3.4), supplemented by (12.3.5), can be solved in various ways. The quest for a solution occupied a good deal of attention in the 1950s. One method is to apply the Laplace transformation to (12.3.4) and (12.3.5) thereby reducing a set of differential-difference equations to pure difference equations of second order with 'constant coefficients'. These may be solved by a direct application of the theory of such equations, or by introducing the concept of generating functions. Having found a solution for the Laplace transforms, one has finally to return to the time domain, a task which is facilitated by one of the tables of Laplace transforms [see IV, §13.4] and their inverses. Another method which reveals more of the probabilistic structure of the problem is to deduce the time domain solution directly from the easily found solution of the random walk problem, which allows $N(t)$ to assume negative values. Details can be found in Conolly (1975). The solution involves modified Bessel functions of the first kind, $I_n(z)$, order n and argument z [see IV, §10.4.3]. With $N(0) = 0$, the full time-dependent solution is

$$p_n(t) = \sum_{m \geq 0} [\rho^{-m}q_{n+m}(t) - \rho^{-m-1}q_{n+m+2}(t)]$$

where $\rho = \lambda/\mu$ and

$$q_n(t) = e^{-(\lambda + \mu)t}\rho^{n/2}I_n[2t(\lambda\mu)^{1/2}] \quad (n \geq 0). \tag{12.3.6}$$

Exact time-dependent solutions are elusive animals in queueing theory, highly prized when captured.

As $t \to \infty$, physical considerations tell us that if traffic intensity exceeds a threshold level, the set $[p_n(t)]$ will all tend to zero, implying by total probability that in the long run the queue would be infinite (if there were space enough to accommodate it), an unattractive prospect for both customers and management. When, on the contrary, the service can contain demand, each $p_n(t)$ can in the long run be expected to tend to a positive value, independent of time, p_n say. The set (p_n) is easily found from (12.3.4) and (12.3.5). The members must satisfy the limiting form of those equations, namely with zero derivatives. Thus the p_n are given by

$$\lambda p_0 = \mu p_1, \qquad (\lambda + \mu)p_n = \lambda p_{n-1} + \mu p_{n+1} \quad (n \geq 1). \tag{12.3.7}$$

Successive substitution gives

$$p_n = \rho^n p_0 \tag{12.3.8}$$

with

$$\rho = \lambda/\mu = \hat{S}/\hat{T}, \tag{12.3.9}$$

and total probability requires that $\sum_{n \geqslant 0} p_n = 1$, giving

$$p_0^{-1} = \sum_{n \geqslant 0} \rho^n. \tag{12.3.10}$$

If $0 < \rho < 1$, the sum converges and

$$p_0 = 1 - \rho, \tag{12.3.11}$$

while if $\rho \geqslant 1$ the sum diverges and $p_0 = 0$. In this case it follows from (12.3.8) that $p_n = 0$ too for finite n. Thus, a necessary condition for a steady state is that $\rho < 1$. Otherwise the system is not in equilibrium.

By (12.3.9) $\rho = r$, the measure of traffic intensity proposed at (12.2.7). If $\rho < 1$, (12.3.11) is equivalent to $p_0 = 1 - r$ so that using either p_0 or r to measure traffic intensity is equivalent when $0 < \rho \leqslant 1$.

The set (12.3.7) is an expression of a conservation principle that, for exponential queues, can be a great help in the formulation of the steady-state equations. One can imagine a set of boxes labelled $0, 1, 2, \ldots$ containing heaps of sand, the quantity in box n being proportional to p_n. Due to random fluctuations, sand (probability) flows in and out of each box and if the system is in equilibrium the amount flowing in during a small interval of time must be balanced by the amount flowing out. We have noted that the only significant flow is between adjacent states (boxes). Figure 12.3.1 illustrates the flow in and out of the box labelled $n (\geqslant 1)$.

The significant flow of probability into box n is derived from boxes $n - 1$ (amount λp_{n-1} per unit time) and $n + 1$ (amount μp_{n+1} per unit time). This is balanced by flow out of box n: λp_n into box $n + 1$ and μp_n into box $n - 1$. The balance equation is thus

$$(\lambda + \mu)p_n = \lambda p_{n-1} + \mu p_{n+1}$$

and the argument applied to boxes labelled 0 and 1 gives the first member of (12.3.7). This is the *method of balance* introduced by Erlang. Such a conservation

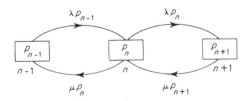

Figure 12.3.1. Illustrating the method of balance

principle holds for all such systems in equilibrium but can be usefully applied only for systems with memoryless driving mechanisms.

Finally it should be noted that (12.3.4) and (12.3.5) belong to the family of birth and death processes [see II, §20.4]. More exactly they represent an immigration/emigration model. λ and μ can be replaced by state dependent mean rates λ_n and μ_n without changing the formulation of the equations, but a time-dependent solution is not, in general, available. Such generalized exponential mechanisms are of practical interest in that they can model multiple service points as well as customer and service behaviour varying in response to the current system state.

(b) *System Time (First Come, First Served)*

The sequence of probability density functions $f_n(w)$ can be built up from (12.2.5) starting, say, with $f_1(w) = \mu e^{-\mu w}$. It is reasonable to suppose that, as $n \to \infty$, provided that the system settles down to an equilibrium, $f_n(w) \to$ **a** non-zero limiting density $f(w)$ corresponding to the system time W of a new customer who arrives when the effect of the initial conditions has 'worn off'. If $f_n(w)$ is replaced by its limit in (12.2.5) and the resulting integral equation solved with $a(t) = \lambda e^{-\lambda t}$, $b(t) = \mu e^{-\mu t}$, we get

$$f(w) = \mu(1 - \rho)e^{-\mu(1-\rho)w} \quad (0 < w < \infty). \tag{12.3.12}$$

The random variable W is also called the virtual system time, since it is the system time of a possibly imaginary customer who arrives at an arbitrary epoch when the system has settled to a steady state.

It is seen that the expected value ('mean value') $E(W) = \hat{W}$ is given by

$$\hat{W} = \frac{1}{\mu(1 - \rho)}, \tag{12.3.13}$$

while the mean value $\hat{N}$ of the system state N in equilibrium is

$$\hat{N} = \frac{\rho}{1 - \rho} \tag{12.3.14}$$

as can be derived easily from (12.3.8) and (12.3.11). It follows that

$$\hat{T}\hat{N} = \hat{W}, \tag{12.3.15}$$

a conservation result known as Little's formula (or result) (Little (1961)) which holds very widely.

Little's formula has been a great success in the literature, attracting much attention, as well as being useful. Its intuitive validity can be roughly seen as follows. Customers enter the system on average at rate $\lambda = \hat{T}^{-1}$. Thus, if the total number of customers present were thought of as passing through a large room (the system), they would be separated in time on average by amount $\hat{T}$, and the mean number present being $\hat{N}$ means that each customer spends on average $\hat{W} = \hat{N}\hat{T}$ in the system; this is (12.3.15).

Another equilibrium conservation result worth mentioning in passing is that the mean rate of exit from the system equals the mean rate of entry. Thus the mean value of the interval separating successive departures has to be $\hat{T}$. This does not mean that the output intervals have the same distribution as T, though it is the case for some systems, $M/M/1$ being one of them.

For $M/M/1$ the probability density function of virtual system time can be written down directly in terms of the distribution (p_n) and the density of sums of exponential service times. The result is

$$f(w) = \sum_{n \geq 0} p_n \mu e^{-\mu w} (\mu w)^n / n!, \tag{12.3.16}$$

which is easily seen to reduce to (12.3.12). The explanation for (12.3.16) is that if the new arrival finds the system empty, his system time is service time; if he finds that $N \geq 1$, his system time is the sum of residual service time, the combined service times of the $n-1$ customers ahead in the queue and his own service time. Since residual time has the same distribution in this case as complete service time, (12.3.16) follows.

12.3.2. $M/M/2/\infty$

Next we give a brief look at the effect on system state and time of increasing the number of servers from 1 to 2. The equilibrium state probabilities are

$$p_0 = \frac{1-\alpha}{1+\alpha}, \qquad p_1 = 2\alpha p_0, \qquad p_n = \alpha^{n-1} p_1 \quad (n \geq 1), \tag{12.3.17}$$

where

$$\alpha = \lambda/2\mu. \tag{12.3.18}$$

An exact time-dependent solution in terms of modified Bessel functions can be obtained.

The mean equilibrium system state is

$$\hat{N} = \frac{2\alpha}{1-\alpha^2}, \tag{12.3.19}$$

so that, by using Little's formula, we find mean system time to be

$$\hat{W} = \frac{1}{\mu(1-\alpha^2)}. \tag{12.3.20}$$

We can now see why it is more efficient to feed to servers by a single queue than to use two parallel queues. Suppose that two queues are formed. Each is fed by an arrival stream with mean intensity $\lambda/2$ and, since the mean service rate is μ, the mean system time is

$$\hat{W}' = \frac{1}{\mu(1-\alpha)}. \tag{12.3.21}$$

For fixed λ and μ, we see that

$$\frac{\hat{W}'}{\hat{W}} = 1 + \alpha \tag{12.3.22}$$

and mean wait is longer with two queues than one. The comparison is more striking still if made in terms of p_0. In terms of the p_0 appropriate to each system

$$\hat{W}' = \frac{1}{\mu p_0}, \qquad \hat{W} = \frac{(1+p_0)^2}{4\mu p_0}, \qquad \frac{\hat{W}'}{\hat{W}} = \frac{4}{(1+p_0)^2}. \tag{12.3.23}$$

This enables a comparison to be made for all admissible traffic levels measured by p_0 and thus restricted to equilibrium. Whether (12.3.22) or (12.3.23) is used depends on the nature of the study, but there is certainly an advantage.

12.3.3.   M/M/L/L

This is the telephone exchange model formulated and investigated by Erlang. L is the total number of lines available and no waiting is allowed. A caller who requests a connection when $N = L$ is rejected and said to be *lost*. One way of mitigating this situation is to increase L, but what improvement does that give? The equilibrium state probabilities are

$$p_n = \frac{\rho^n}{n!} p_0 \quad (\rho = \lambda/\mu, 0 \le n \le L)$$

with

$$p_0^{-1} = \sum_{n=1}^{L} \frac{\rho^n}{n!}. \tag{12.3.24}$$

The loss probability is

$$p_L = \frac{\rho^L}{L! \sum_{0}^{L} \frac{\rho^n}{n!}}. \tag{12.3.25}$$

Management questions are as follows!

 (i) What are the benefits of keeping p_L below a certain threshold?
 (ii) If traffic is measured by ρ, what value of L should be used so that p_L does not exceed an acceptable value?
 (iii) What compromises need be made in the light of capital and running costs for given L?

12.3.4.   M/M/∞

This is an exponential system with an infinity of service points. In practical terms this means that no customer has to wait. It is a remarkable system both practically and theoretically. Theoretically it is one of the elusive animals with an easy time-dependent form for $p_n(t)$, namely

$$p_n(t) = e^{-R(t)}(R(t))^n/n! \quad (n \geq 0) \qquad (12.3.26)$$

where

$$R(t) = \rho(1 - e^{-\mu t}). \qquad (12.3.27)$$

The system is always in equilibrium and as $t \to \infty$ the limiting values of $p_n(t)$ can also be deduced directly from §12.3.3 by letting $L \to \infty$.

From the practical viewpoint the model is useful in that it shows the upper limit of what can be done by adding an unlimited number of service points for given ρ. Thus, with a single server, $p_0 = 1 - \rho$, with two, $p_0(1 - \rho/2)/(1 + \rho/2)$, while if there were infinitely many $p_0 = e^{-\rho}$. For small enough ρ there is virtually no gain from adding extra service points, but as ρ increases the reduction in p_0 and other operational measures can be calculated precisely.

Another use of the model is to represent approximately the content of, say, a large supermarket, a car park, or the equivalent. Customers arrive in a Poisson stream, and, when their shopping time is known to be exponential, $(p_n(t))$ is the distribution of the number of customers present. Among other things this provides a criterion to address such questions as the size of the building.

12.4.   OTHER DEMAND AND SERVICE MECHANISMS

Although the assumption of Poisson demand and exponential service time is justified in a wide variety of applications, there are many exceptions and it is natural to enquire how the methods of analysis and results change in response to more general driving forces. We shall confine attention to single-server systems with no waiting limitations ($N_2 = 0$). The mechanisms briefly considered will be generalized exponential, M/G/1 and G/M/1.

12.4.1.   Generalized Exponential Systems

The systems we have in mind have are driven by state-dependent mechanisms. Thus, when $N(t) = n$, the mean arrival and service rates are λ_n, μ_n. μ_0 has to be defined as zero. Typical forms are $\lambda_n = \lambda/(n + 1)$ which embodies discouragement to the customer, and μn which encapsulates the concept of a cooperative server. A series of systematic investigations of the particular mechanisms cited was carried out in the 1960s and 1970s. See Conolly (1975) for a number of references and the *Journal of Applied Probability* during that period and later. Conolly and Chan (1977) give an intriguing account of the general properties of such systems; see also Chan and Conolly (1978) for an operational evaluation. The systems share features in common with their non-state-dependent relatives. For example, $p_0 = 1 - \lambda^*/\mu^*$, where λ^* and μ^* are appropriately defined mean *effective* arrival and service rates. Little's formula holds with λ^*. The key to the analysis, which enjoys Markov properties [see II, §19.2], is the definition of *equivalent* service time and *equivalent* interarrival interval.

12.4.2. M/G/1

The arrival pattern is exponential but service time s has an unspecified distribution with density $b(t)$. The Markov property that residual service time is independent of past history no longer holds. Consequently it is no longer possible to construct a description of the system state and time by analysis over an arbitrary small time interval $(t, t + h)$. Instead one recaptures the Markov property by concentrating on the epochs at which service begins and expressing the state probabilities at such epochs in terms of the states that must have existed at the epoch of the previous service beginning. The system state at such epochs is a Markov chain [see II, §19.3] embedded in the evolution of the process, even though it is not Markovian at an arbitrary epoch. This is the key to the analysis, and complicated expressions for $p_n(t)$ can in principle be constructed. For example, with the boundary condition $N(0-) = 0, N(0+) = 1$, so that at the initial epoch a service begins, it can be shown that

$$p_0(t) = e^{-\lambda t} \frac{d}{dt} \sum_{n \geq 1} \left[\frac{(\lambda t)^{n-1}}{n!} \int_0^t (t - s) b^{(n)}(s) ds \right] \qquad (12.4.1)$$

(Conolly (1975), Equation 5.2.3.11). $b^{(n)}(s)$ is the n-fold convolution of the density $b(t)$ with itself and so represents the probability density function of the sum of n service times. Another result concerning system state is the following. Let $\pi_n(z)$ be the Laplace transform of $p_n(t)$, that is

$$\pi_n(z) = \int_0^\infty e^{-zt} p_n(t) dt \quad (\text{Re } z > 0), \qquad (12.4.2)$$

and let

$$\Pi(x, z) = \sum_{n \geq 0} x^n \pi_n(z) \qquad (12.4.3)$$

be the generating function of $\pi_n(z)$ [see II, §12.1]. It is necessary to specify that x is such as to secure convergence. Then

$$[x - \beta(z + \lambda - \lambda x)] \Pi(x, z) = \frac{x^2 [1 - \beta(z + \lambda - \lambda x)]}{z + \lambda - \lambda x} - \frac{\beta(x + \lambda - \lambda x)(1 - x)\xi(z)}{z + \lambda - \lambda \xi(z)}.$$
$$(12.4.4)$$

Here $\beta(z)$ is the Laplace transform [see II, §11.3; IV, §13.4] of $b(t)$ and $\xi(z)$ is the unique zero in x with modulus less than 1 of

$$x - \beta(z + \lambda - \lambda x). \tag{12.4.5}$$

This formula is not particularly illuminating, but does in principle contain all information about $N(t)$ statistics for finite time and could lend itself to numerical treatment. The expression (12.4.5) crops up with as much frequency in the analysis of M/G/1 and G/M/1 models as modified Bessel functions do with pure exponential mechanisms.

The generating function $G(x)$ of equilibrium state probabilities p_n, namely

$$G(x) = \sum_{n \geq 0} x^n p_n \tag{12.4.6}$$

restricted to x for which the sum converges, can be obtained from (12.4.4) by use of so-called Tauberian Theorems, or directly. The following result is attributed to both Pollazcek and Khintchine (1932) independently.

$$G(x) = \frac{(1 - x)p_0}{1 - \dfrac{x}{\beta(\lambda - \lambda x)}}. \tag{12.4.7}$$

Here

$$p_0 = P(N = 0) = 1 - \rho, \tag{12.4.8}$$

where

$$\rho = \lambda \hat{S} \tag{12.4.9}$$

just as for M/M/1. (12.4.7) also gives the celebrated formula

$$E(N) = \rho + \frac{\lambda^2 b_2}{2(1 - \rho)}, \tag{12.4.10}$$

where $b_2 = E(S^2)$.

Little's formula can be evoked to get $E(W)$. This is popularly expressed in the form

$$\frac{E(W)}{\hat{S}} = 1 + \frac{\rho(1 + \text{Var}(S)/\hat{S}^2)}{2(1 - \rho)}. \tag{12.4.11}$$

This conveys useful information about how the distribution of S can effect mean system time. When S is exponential, $\text{Var}(S)/\hat{S}^2 = 1$ [see II, §11.2.1] and $E(W)/\hat{S} = 1 + \rho/(1 - \rho)$. If service time were constant, $\text{Var}(S) = 0$ and $E(W)/\hat{S} = 1 + (\rho/2)/(1 - \rho)$, which, not surprisingly is less than in the exponential case, but could one have guessed how much less? The queueing time component has been halved by this expedient. It is instructive to develop and expression for $\text{Var}(N)$ too. This can be done by differentiating (12.4.7) twice to get $E(N(N - 1))$ [see II, §12.1.6]. There is, however, no companion to Little's formula relating second moments of N and W.

For interest we quote the equilibrium system time formula for the probability density function $f(w)$ and its Laplace transform $\phi(z)$:

$$\phi(z) = (1 - \rho)\beta(z)/[1 - \lambda B_c^*(z)], \tag{12.4.12}$$

where $B_c^*(z)$ is the Laplace transform of $1 - B(t)$, and $B(t)$ is the distribution function of S:

$$f(w) = (1 - \rho)[b(w) + \sum_{n \geq 1} \lambda^n B_c^{(n)}(w) * b(w)]. \tag{12.4.13}$$

Here $B_c^{(n)}(w)$ is the n-fold [see II, §7.1] convolution of $1 - B(w)$ with itself, and the asterisk means the convolution of the functions it immediately separates.

12.4.3. G/M/1

This system is a kind of dual of M/G/1. Now it is the interarrival interval distribution $A(t)$ that is unspecified. The approach to the analysis follows the pattern of M/G/1: either a supplementary variable is included in the specification of system state, the expended part of the current interarrival interval in this case, to recapture the Markov property, or the embedded Markov chain approach is adopted, involving studying the system at those epochs at which an arrival occurs. This has to be done anyway to obtain a probabilistic description of the system at an arbitrary epoch. Using the supplementary variable method, one seeks the joint probability and probability density function $p_n(u, t)$, where, at time t, $N(t) = n$ ($n \geq 0$) and the elapsed time since the previous arrival is u. An initial condition has to be specified and it simplifies matters without losing touch with reality to suppose that $t = 0$ is an arrival epoch. It should be mentioned that the steady-state probabilities at an arbitrary epoch are different from those at an arrival epoch. For an arbitrary epoch we find

$$p_0 = 1 - \rho, \qquad p_n = \rho(1 - \xi_0)\xi_0^{n-1} \quad (n \geq 1), \tag{12.4.14}$$

where $\rho = (\mu \hat{T})^{-1}$, the equivalent of '$\lambda/\mu$', and ξ_0 is the smallest positive root of

$$x = \alpha[\mu(1 - x)], \tag{12.4.15}$$

which is less than 1 if $\rho < 1$ and equals 1 when $\rho \geq 1$. $\alpha(z)$ is the Laplace transform of the probability density function $a(t)$ of T. On the other hand, if

$$r_n = P(N = n \text{ at an arrival epoch}) \quad (n \geq 1). \tag{12.4.16}$$

it is found that

$$r_n = (1 - \xi_0)\xi_0^{n-1} \quad (n \geq 1). \tag{12.4.17}$$

Laplace transform results for time-dependent state probabilities are

$$\pi_0(z) = \frac{1}{z} - \frac{1 - \xi}{[1 - \alpha(z)](z + \mu - \mu\xi)},$$

$$\pi_n(z) = \frac{(1 - \xi)^2 \xi^{n-1}}{[1 - \alpha(z)](z + \mu - \mu\xi)} \quad (n \geq 1), \tag{12.4.18}$$

where $\xi(z)$ is the root in x with modulus less than unity of

$$x = \alpha(z + \mu - \mu x), \tag{12.4.19}$$

similar to (12.4.3). It is assumed that $\mathrm{Re}\, z > 0$.

There are simple results for system time in equilibrium. For example, relative to an arrival epoch,

$$f(w) = \mu(1 - \xi_0)e^{-\mu w(1 - \xi_0)}. \tag{12.4.20}$$

This is remarkable in retaining the exponential form and a hint at a property which obtains under heavy traffic conditions in more general circumstances [see §12.5.5].

12.5. SOME SPECIAL TOPICS

12.5.1. Machine Maintenance: A Closed Population

Under this heading is included the situation typified by a factory producing items using F machines. Each machine has probability $\lambda h + o(h)$ of breaking down in an arbitrary small time interval $(t, t + h)$. Repairs are carried out by a team of G mechanics each of whom requires exponentially distributed time to complete the job with, say, probability $\mu h + o(h)$ of doing so in $(t, t + h)$. The population of machines is *closed* in the sense that arrivals from the external world are excluded. A member machine of the population is, at any time, either requesting repair, or functioning normally.

The exponential mechanisms of arrival and service are quite realistic as well as convenient theoretically. One is interested in equilibrium solutions and among the practical objectives is the matching of machine reliability with enough mechanics so that production is not interrupted unacceptably. If N is the number of machines serviceable, $F - N$ are eligible for repair. If $F - N \leq G$, none has to wait. Balance can be used to formulate the equations for the equilibrium state probabilities $p_n = P(N = n)$. For illustration, let $G = 2$ and $F = 3$. The equations are

$$2\mu p_0 = \lambda p_1, \qquad (\lambda + 2\mu)p_1 = 2\mu p_0 + 2\lambda p_2,$$
$$(2\lambda + \mu)p_2 = 2\mu p_1 + 3\lambda p_3, \qquad 3\lambda p_3 = \mu p_2.$$

The solution is $p_1 = (2\mu/\lambda)p_0, p_2 = (2\mu^2/\lambda^2)p_0, p_3 = (2\mu^3/3\mu^3)p_0$ with

$$p_0^{-1} = 1 + \frac{2\mu}{\lambda} + \frac{2\mu^2}{\lambda^2} + \frac{2\mu^3}{3\lambda^3}.$$

12.5.2. Batch Queues

This heading describes situations where service can handle several customers simultaneously, where customers arrive in groups or batches, or both simultaneously. The sizes of the batches may be fixed or random. Batch service occurs when elevators (or lifts) arrive at a floor to accept passengers. The same situation occurs when trains and buses arrive at a predetermined station to pick up passengers within the limit of their capacity. Batch arrivals occur when a fixed, or random, number of clients arrive simultaneously asking for service. Clinics, where appointments at the same time are given for convenience, provide a classic example: letters arriving at a sorting centre for processing, or parties arriving at a restaurant provide another.

There is a considerable literature on this class of model, dating from the mid-1950s. A comprehensive survey is to be found in the book by Chaudhry and Templeton (1983).

12.5.3. Networks

The simplest example of a network is that of a customer whose global service consists of several consecutive stages. A medical instance is provided by the patient whose clinical investigation requires a visit to a consultant and then to the X-ray department followed by ECG, EEG, scanner, etc., and back to the consultant. Another example is a factory production line where a product requires a sequence of operations to be completed. These are not networks in the general sense where passage from the point of entry to exit may involve a number of alternative paths, but rather what is popularly known as a tandem queue (in Latin *tandem* means *at length*). A preferable description is *series*, or *sequential* queue. The theory rapidly becomes extremely complicated when there is a limit on waiting facilities between stages. When waiting is not permitted, or limited, the phenomenon of *blocking* is said to occur. This means that a customer who has completed the nth stage and is unable to pass to the next stage because the queue is full must remain at the nth service counter, blocking it from serving more customers, until there is space in the next queue.

A flavour of the heavy algebra entailed in the analysis of even exponential models can be found in papers by Langaris and Conolly (1984), and in the references. The matrix algebraic analytic apparatus proposed by Neuts is an effective representational and computational tool. There are many references in the current literature, but see Neuts (1981, 1989).

In the more general sense, networks of queues have been in the limelight since the mid-1950s and have more recently expanded through their application to computer and telecommunication networks. A network in this application may be regarded as a collection of nodes or service points. Customers enter the system at an entry node and then follow a deterministic or stochastic path through it, visiting some or all nodes, and then arrive finally at an output node from which to escape from the system. This is a so-called *open network*. A special case arises when the customers remain within the network, never leaving it. The network is then said to be *closed*. Taxicabs serving an urban area supply an example since they rarely leave their designated area of operation. The nodes are the cab ranks and the set-down points. Flexibility in terminology is required since the cabs, formally the 'customers', actually supply the service while it is the time on passage between nodes that constitutes service time. In this application the nodes may effectively be infinite in number if there is no restriction on set-down points. The machine maintenance model of §12.5.1 is another example of a closed network.

The first analysis of open networks is due to Jackson (1957). Suppose that the network has N nodes numbered $1, 2, \ldots, N$. Arrivals from outside to node i are Poisson with mean rate r_{0i}. Service at node i is exponential with parameter μ_i and, on completion, the customer proceeds to node j with probability r_{ij}, or leaves the system with probability r_{i0}. An unlimited queue is allowed at each node. The passage time from node to node is zero. Under conditions of a steady state with non-zero equilibrium state probabilities $p(n_1, n_2, \ldots, n_N)$ that n_i customers are present at node $i (i = 1, 2, \ldots, N)$, Jackson showed that $p(n_1, n_2, \ldots, n_N)$, has the *product form*

$$p(n_1, n_2, \ldots, n_N) = \prod_{i=1}^{N} (1 - \rho_i)\rho_i^{n_i},$$

where $\rho_i = \lambda_i/\mu_i$ and

$$\lambda_i = r_{0i} + r_{1i} + \cdots + r_{Ni} \qquad (12.5.1)$$

is the 'rate of flow' of customers into node i. This possibly surprising result shows that the network, so far as system state is concerned, behaves as though it were a set of independent M/M/1 queueing systems fed by an overall arrival rate

$$\lambda = \lambda_1 + \lambda_2 + \cdots + \lambda_N$$

of which component λ_i is fed to the ith system which has mean service rate μ_i. In fact the system in other respects is not like that and, for example, Disney (1981) showed that the internal flow between nodes is not Poisson. Mean waiting time at the nodes can be obtained by applying Little's formula.

The network becomes closed if $r_{0i} = r_{i0} = 0$ and again a product form solution is obtained under otherwise the same conditions.

For more recent work see Kelly (1979) and his more recent papers in the *Applied Probability* journals.

12.5.4. Heavy Traffic Analysis

This regime is of interest both from the practical and theoretical points of view. We are dealing by definition with a system in equilibrium where the demand can only just be contained by the service and management and clients want to know how long it takes to reach the steady state. Aside from this, $N(t)$ is generated by mechanisms in which the average number of demands and service completions in a given time interval are almost equal. Naively it might be supposed that the apparent balance favours system stability, but this is not the

case at all. Profound studies have been dedicated to the analysis of the features
of random walks of which Figure 12.2.1 is a specialized instance, revealing,
under a balanced regime, fascinating and unexpected behaviour beyond the
scope of this article.

In the context of queueing theory, parallel studies were initiated in the early
1960s, both in the USSR and in the West, into single and multiple server systems
under quite general assumptions and heavy traffic. A key paper is that of
Kingman (1962) which deals with the single-server system $G/G/1/\infty$, in
particular with the queueing time Q_n of customer C_n. Q_n is system time W_n less
service time S_n of C_n. Thus, by (12.2.4),

$$Q_{n+1} = \max(0, Q_n + U_{n+1}),$$

where

$$U_{n+1} = S_n - T_{n+1}. \tag{12.5.2}$$

It follows that

$$Q_n = \max(0, U_n, U_n + U_{n-1}, U_n + U_{n-1} + U_{n-2}, \ldots, U_n + U_{n-1} + \ldots + U_2)$$

and, since the U_n are independent,

$$Q_n = \max(0, U_2, U_2 + U_3, \ldots, U_2 + U_3 + \ldots + U_n). \tag{12.5.3}$$

Using an identity due to Spitzer, and the central limit theorem [see II, §17.3],
Kingman deduced that the random variable Q to which Q_n tends has
asymptotically an exponential distribution with mean

$$\tfrac{1}{2}[\mathrm{Var}(T) + \mathrm{Var}(S)]/(\hat{T} - \hat{S})$$

provided that

$$\alpha = \frac{\hat{T} - \hat{S}}{[\mathrm{Var}(S) + \mathrm{Var}(T)]^{\frac{1}{4}}}$$

is sufficiently small.

The rate of convergence of the distribution of Q_n to the exponential limit,
which is connected with α, is a problem of practical interest both in the
management of operational systems and when computer simulation is used to
study analytically intractable systems. Some discussion is given in Kingman's
paper in Smith and Wilkinson (1965).

12.5.5. Priority Disciplines

Allocation of priority to certain classes of customer is dictated by necessity
in many familiar situations. An example is provided by a hospital casualty
department where the arrival of a patient in critical condition pre-empts the
service. In the context of message transmission, certain types of 'life or death'
messages can, at a price, or otherwise, be given precedence, thus avoiding waiting

to be serviced in the normal way. There are many possible priority protocols: *pre-emptive* when the current customer's service is interrupted, *non-pre-emptive* when it is not, *alternating* when priority queues are dealt with in turn, and variations on these and others. The parameters of operational interest are the waiting times experienced by the various priority grades and the associated numbers of customers whose service is deferred by the priority system in operation.

It is obvious enough that to give priority to one class of customer can only be achieved at the expense of those with lower priority, unless service capacity can be increased. What is therefore of particular interest is the extent to which the service of non-priority customers is degraded or, equivalently, the extent to which their system time is exaggerated. This can be extremely serious, a point insufficiently dealt with in the literature which, on the whole, is dominated by algebraic details. It can happen under heavy enough traffic that the priority stream is in equilibrium and yet the lower priority streams are not and their prospect of ever receiving service is rather remote. Jaiswal (1968) expounds the theoretical aspects of some priority analysis. There is some discussion in Morse (1958), Gross and Harris (1985) and Kleinrock (1976). More recent work that attempts to come to grips with the realities of priorities is described in Dovletis (1988).

12.6.   CONCLUDING REMARKS

In this brief and selective review we have seen that queueing theory is a member of a family of stochastic models sharing general features and possessing a wide range of applications. It is for this reason popular and gives respectability to courses in operational research showing that the subject is concerned with models having a sound theoretical basis, but reasonably easy to grasp. In addition, the theory belongs to the field where fluctuations of sums of random variables are the concern, a field with a dignified history and literature, for which see Feller (1966), for example. It is accordingly highly suitable to form a component of a course in stochastic processes *per se*. Since the structure of a queueing situation is clearly defined in terms of the elementary events consisting of arrivals and departures, it forms a suitable basis on which to build courses in computer simulation, especially since there are many special cases with a complete analysis which can be used to verify simulation output and program fidelity.

Since the theory took shape at the hands of Erlang, its literature has expanded at least exponentially. The merit of many of the contributions is constantly called in question on the grounds of repetitiveness and lack of originality but, on the other hand, their richness testifies to the interest and amenability of the subject matter. It is indeed impossible to compile a short bibliography that does justice to the subject.

The queueing family embraces the related theories of reservoirs, inventory and insurance risk. An elegant, unified and concise account of these applications is given by Prabhu (1980).

Inventory theory is concerned with the quantity of material held in store by a warehouse or factory for sale or use in manufacture. The costs associated with the operation of an inventory policy are holding charges arising from the occupation of storage space (akin to rent), penalty costs arising from the inability to satisfy a demand, and reorder costs—for example a fixed charge may have to be paid for the delivery of an order, however big. A desirable inventory policy minimizes the associated costs. The inventory level of an item, N_{n+1}, at the $(n+1)$th reorder epoch is given by

$$N_{n+1} = N_n + A_n - B_{n+1}, \tag{12.6.1}$$

where A_n is the amount ordered by the warehouse at epoch n and B_{n+1} is the consumer demand between reorder epochs n and $n+1$. If N_n becomes negative, this means that demand has exceeded supply and various possibilities have then to be taken into account in determining the policy of reordering. In a queueing context, (12.6.1) might become

$$N_{n+1} = \max(0, N_n + A_n - B_{n+1}),$$

where N_n is the system size at the nth departure, A_n is the number of arrivals by the time of the next departure and $B_{n+1} = 1$. The parentage is plain.

Reservoirs are used to store water: they are lakes, artificial or otherwise, from which the outflow is controlled by a retaining wall, or dam, with sluices. The words dam and reservoir are often used interchangeably in the literature which is very large in its own right. The water stored from rainfall, melting snow, and so on, is released as required to meet demands by industry, agriculture and domestic needs. In the case of inventory models, it was the replenishment policy the could be regulated,, while demand could not be controlled directly. The objective of reservoir theory is concerned with the formulation of policies to regulate demand in the face of stochastic supply. Obviously an equation of type (12.6.1) appears.

The business of an insurance company presents similar features. A stock of money called the risk reserve is of central interest. This is replenished by premiums and depleted by claims of random amounts occurring at random time intervals. The concern here above all is to develop a policy wich avoids ruin, that is, satisfies the requirement that $N(t)$ remains positive over a prescribed time span. For further information see Seal (1969).

The family relationship between these members of the queueing family is clear enough. The descriptions given above are, of course, very sketchy.

A further area of topical interest in congestion theory is that of road traffic. At the beginning of this article we cited queueing examples of a conventional kind arising at airports, in traffic jams, and to these we could add toll booths, tunnels, roadworks and other restrictions. The problems of road traffic are very largely those of avoidance of congestion in the only too familiar sense. The theory is wider than that of queueing theory, and borrows from other disciplines such as fluid flow and diffusion. In the modelling of driver behaviour, psychology has a part to play. One basic problem concerns the delay to traffic or pedestrians at the intersection of a minor with a major road, or at a pedestrian crossing. This is a so-called static delay problem and has elements formally the same as encountered in queueing. At another level, moving delays are encountered and here a fundamental need is a description of the density of moving traffic as a function of speed, time and location that takes account of the interactions between the components of the traffic flow which are not points and are reluctant to collide with each other. Further problems arise from the formulation of physical traffic control policies using signals, islands and various ingenious physical devices. An interesting survey article for preliminary information is a paper by Weiss in Smith and Wilkinson (1965). There are specialized journals devoted to this field which the reader may wish to consult. A good introductry text is Ashton (1966).

B. W. C.

REFERENCES

Ashton, W. D. (1966). *The Theory of Road Traffic Flow*, Methuen and Wiley.

Brockmeyer, E., Halstrøm, H. and Jensen, A. (1948). The Life and Works of A. K. Erlang, *Trans. Dan. Acad. Tech. Sci., No. 2*, Copenhagen.

Chan, J. and Conolly, B. W. (1978). Comparative Effectiveness of Certain Queueing Systems with Adaptive Demand and Service Mechanisms, *Comp. and Op. Res.* **5**, 187–196.

Chaudhry, M. L. and Templeton, J. G. C. (1983). *A First Course in Bulk Queues*, Wiley.

Conolly, B. W. (1975). *Lecture Notes on Queueing Systems*, Ellis Horwood.

Conolly, B. W. and Chan, J. (1977). Generalised birth and death processes, *Adv. Appl. Prob.* **9**, 125–140.

Dovletis, G. (1988). Fundamental Results in Priority Queueing Theory, Ph.D. Thesis, University of London.

Feller, W. (1966). *An Introduction to Probability Theory and its Applications*, Wiley.

Jackson, J. R. (1957). Networks of Waiting Lines. *Ops. Res.* **5**, 518, 521.

Jaiswal, N. K. (1968). *Priority Queues*, Academic Press.

Kelly, F. P. (1979). *Reversibility and Stochastic Networks*, Wiley.

Kendall, D. G. (1951). Some Problems in the Theory of Queues, *J. Roy. Statist. Soc.* (B) **13**, 151–185.

Khintchine, A. Y. (1932). Mathematical Theory of a Stationary Queue, *Mat. Sb.* **39**, 73–84.

Kingman, J. F. C. (1962). On queues in heavy traffic, *J. Roy, Statist. Soc.* (B) **24**, 383–392.

Langaris, C. and Conolly, B. W. (1984). On the Waiting Time of a Two-stage Queueing System with Blocking, *J. Appl. Prob.* **21**, 628–638.

Lindley, D. V. (1952). Theory of queues with a single server, *Proc. Camb. Phil. Soc.* **48**, 277–289.

Little, J. D. C. (1961). A Proof for the Queueing Formula $L = \lambda W$. *Ops. Res.* **9**, 383–7.

Neuts, M. F. (1981). *Matrix-geometric Solutions to Stochastic Models*, The Johns Hopkins University Press.

Prabhu, N. U. (1980). *Stochastic Storage Processes*, Springer.

Saaty, T. L. (1961). *Elements of Queuing Theory*, McGraw-Hill.

Seal, H. L. (1969). *Stochastic Theory of a Risk Business*, Wiley.

Smith, W. L. (1953). Distribution of Queueing Times, *Proc. Camb. Phil. Soc.* **49**, 449–461.
Smith, W. L. and Wilkinson, W. E. (eds.) (1965). *Proceedings of Symposium on Congestion Theory*, University of N. Carolina Press.

BIBLIOGRAPHY

Cohen, J. W. (1982). *The Single Server Queue* (2nd edn), North-Holland. (An authoritative text.)
Gross, D. and Harris, C. M. (1985). *Fundamentals of Queueing Theory* (2nd edn), Wiley. (Reasonably thorough overview of most aspects of theory aimed at graduates and undergraduates.)
Khintchine, A. Y. (1960). *Mathematical Methods in the Theory of Queueing*, Griffin. (This is a translation from the Russian expounding this important author's contributions.)
Kleinrock, L. (1976). *Queueing Systems*, Vols. 1 and 2, Wiley. (The first volume deals with theory. The second is concerned with applications to computer networks and adopts an interesting engineering approach.)
Klimow, G. P. (1979). *Bedienungsprozesse*, Birkhäuser. (A German translation from a Russian original. Excellent, balanced theoretical coverage with a bibliography from East and West.)
Morse, P. M. (1958). *Queues, Inventories and Maintenance*, Wiley. (The first serious textbook by the well-known physicist and latter-day exponent of operational research. A physicist's approach with a clear view of application.)
Neuts, M. F. (1989). *Structured Stochastic Matrices of M/G/1 Type and their Applications*, Dekker.
Newell, G. F. (1982). *Applications of Queueing Theory* (2nd edn), Chapman and Hall. (An 'engineering' text by a transportation engineer: a quite individual approach using sensible approximations.)
Pollaczek, F. (1957). Problèmes Stochastiques posés par le Phénomène de Formation d'une Queue D'Attente à un Guichet et par des Phénomènes apparentés. *Mémorial de Sciences Mathématiques*, Fasc. 136. Gauthiers-Villars. (This is an exposition of most of the author's important and individualistic approach to queueing problems using contour integral representations.)
Takus, L. (1955). Investigation of waiting time problems by reduction to Markov processes. *Acta Math. Sci. Hung.*, **6**, 101–129. (A seminal paper followed by a long series of fundamental contributions to the mathematics of queueing theory.)

CHAPTER 13

Bootstrap Methods

13.1. INTRODUCTION

Bootstrap methods comprise a wide variety of techniques for assessing the statistical properties of estimators, tests, and so forth. A common feature is the use of simulation, either from the data themselves, or from a fitted model. These methods rely intrinsically on the computer. Data analyses are repeated many times, possibly thousands or millions, as a matter of routine—clearly impossible without modern high-speed computing equipment. The use of such methods is increasing and will become more common as the cost of computing continues to fall relative to the cost of a trained scientist. A second feature of bootstrap methods is that they often make it possible to relax distributional assumptions, whilst still enabling confidence intervals [see VIA, §4.2] with good theoretical properties, for example, to be calculated.

No attempt is made to give a comprehensive survey of the vast and rapidly-growing literature on bootstrap theory and applications, although references are given on specific points. General bibliographic comments and suggestions for further reading are relegated to §13.8.

Like all statistical procedures, bootstrap techniques must be used with careful attention to the assumptions which underlie them and the general principles underpinning good practice. Comments made throughout indicate some of the pitfalls to be avoided.

13.2. AN EXAMPLE

Consider the following $n = 20$ observations related to the determination of the speed of light:

$$29$$
$$29$$
$$28$$
$$27$$
$$26$$
$$25$$
$$24$$

$$
\begin{array}{ccccccc}
 & & & 24 & 34 & & \\
 & & & 23 & 33 & & \\
 & & 19 & 22 & 31 & & \\
-44 & & -2\ 16 & 21 & 30 & 40 &
\end{array}
$$

These are calculated from data collected by Newcomb and are quoted in Stigler (1977). Suppose that we wish to calculate a 95 per cent confidence interval for the true speed of light, based on these data. If we were willing to assume that the data are drawn independently from a normal population [see II, §11.4] with mean i.e. expectation θ and unknown variance, we would calculate the average and variance

$$
\bar{X} = \frac{1}{n} \sum_{i=1}^{n} X_i, \qquad s^2 = \frac{1}{n-1} \sum_{i=1}^{n} (X_i - \bar{X})^2
$$

of the sample, and use the t-distribution (see VIA, 4.5.1) to give the confidence interval

$$
\bar{X} \pm t_{19}(0.975) \frac{s}{\sqrt{n}}, \tag{13.2.1}
$$

or (13.50, 30.00), since $\bar{X} = 21.75$ and $s = 17.63$. However, two of the data are large and negative relative to the rest, which casts doubt on the assumption of normality. One possibility is to discard these observations as outliers, to calculate the interval without them. If we wish to not to do this, we can use an estimator designed to downweight their effect on the analysis. One such estimator is defined as the solution T to the equation

$$
\sum_{i=1}^{n} \psi(X_i, T) = 0, \tag{13.2.2}
$$

where

$$
\psi(x; \theta) = \begin{cases} -c & x - \theta < -c, \\ x - \theta & |x - \theta| < c, \\ c & x - \theta > c. \end{cases}
$$

The resulting estimate is the Huber M-estimate of location, which limits to $\pm c$ the influence any observation can contribute to $\sum \psi(X_i, \theta)$. The sample average, by contrast, solves equation (13.2.2), with $\psi(x; \theta) = x - \theta$; there is no limit in principle to the influence an extreme observation can have. We take $c = 6$ to avoid complications, so that in this particular case the observed value of T is $t_{\mathrm{obs}} = 25.62$.

The estimator T is more robust than the average $\bar{X}$ to the effects of non-normality, but its mean and variance are unknown except asymptotically in large samples. If T were approximately normally distributed with mean $\theta + b$, where b is the bias of T, and variance v, a calculation analogous to that leading to (13.2.1) would yield

$$
(T - b - z_{1-\alpha}\sqrt{v}, T - b + z_{1-\alpha}\sqrt{v}) \tag{13.2.3}
$$

as an approximate $(1 - 2\alpha) \times 100$ per cent confidence interval for the unknown θ. Here $\Phi(z_{1-\alpha}) = 1 - \alpha$, and $\Phi(\cdot)$ is the standard normal cumulative distribution function. The difficulty is calculation of the bias b [see VIA, Definition 3.3.2] and variance v without the use of asymptotic results of doubtful validity, or assumptions about the form of the distribution from which the data are drawn. The bootstrap finds estimates of b and v by the following algorithm.

(1) Draw a sample $X_1^*, \ldots, X_n^*$ randomly with replacement from the original data $X_1, \ldots, X_n$.

(2) Calculate T^* from $X_1^*, \ldots, X_n^*$ as the solution to the equation

$$\sum_{i=1}^{n} \psi(X_i^*; T^*) = 0.$$

(3) Repeat steps (1) and (2) S times independently to get estimates $T_1^*, \ldots, T_S^*$ of θ.

(4) Calculate the estimated bias

$$B^* = \frac{1}{S} \sum_{s=1}^{S} T_s^* - t_{obs} = \bar{T}^* - t_{obs}$$

and variance

$$V^* = \frac{1}{S-1} \sum_{s=1}^{S} (T_s^* - \bar{T}^*)^2$$

of T, and use these in place of b and v in (13.2.3).

Table 13.2.1 shows the resulting samples for an absurdly small bootstrap of size $S = 9$. The numbers in the central columns of the table give the frequency

data:	-44	-2	16	19	21	...	30	31	33	34	40	*M*-estimate 25.62
Sample			Frequency in bootstrap sample									T^*
1	4	1	0	3	2	...	0	0	1	0	1	21.78
2	1	2	3	0	1	...	1	1	0	1	0	22.60
3	3	0	3	0	0	...	0	3	0	0	0	24.00
4	1	2	1	2	1	...	1	0	0	1	0	22.62
5	1	0	0	0	1	...	4	4	0	1	0	27.41
6	1	2	1	1	0	...	0	2	1	1	1	25.50
7	1	1	0	3	0	...	1	1	1	1	2	26.64
8	0	2	2	3	1	...	1	0	0	0	1	23.21
9	1	1	2	2	1	...	1	2	0	0	1	23.42

Average of T^* for bootstrap samples	24.13
Variance of T^* for bootstrap samples	3.81

Table 13.2.1: A small bootstrap analysis of the Huber *M*-estimate of the location θ of the Newcomb data.

with which each of the original data occurs in each bootstrap sample. The first sample consists of -44 four times, -2 once, and so on. Each row of the table adds up to $n = 20$. The final column gives the value of T^* for each sample. The values of B^* and V^* are found from step (4) above to be $24.13 - 25.62 = -1.49$ and 3.81. The approximate 95 per cent confidence interval (13.2.3) for the true mean speed of light θ is (23.28, 30.94). In practical applications, more accurate estimates of b and v would be obtained by increasing the number S of bootstrap samples. In practice, values of S in the range 25–100 would give adequate estimates of bias and variance in a problem like this.

One assumption underlying the use of the bootstrap estimates $T_1^*, \ldots, T_S^*$ to find an approximate confidence interval based on (13.2.3) is that the distribution of T^* is approximately normal. This can be checked by plotting the ordered values of $T^* - t_{obs}$ against expected normal order statistics. The plot, given in Figure 13.1.1 for a bootstrap of size $S = 49$, shows that the distribution is perhaps slightly negatively skewed relative to the normal distribution, and in view of this a more reliable confidence interval for θ could be obtained with corrections which use the third and fourth moments of the $T_1^*, \ldots, T_S^*$.

Another possibility would be to use a distribution other than the normal to calculate the confidence intervals. One way to proceed is to see what distribution, if any will fit the bootstrapped values T_s^*. Use of the chi-squared distribution [see VIA, §2.5.4] rather than the normal would often be appropriate if the statistic being bootstrapped were a likelihood ratio statistic, for example, on the grounds that such a statistic has an approximate chi-squared distribution in large samples. The use of the normal distribution is natural in the Newcomb example because the distribution of T would be approximately normal for large values of n. This procedure is useful in many applications, but has theoretical

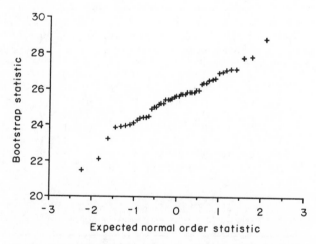

Figure 13.1.1 Plot of $S = 49$ ordered bootstrap statistics against expected normal order statistics; Newcomb data

disadvantages compared to direct use of the bootstrap distribution. The matter is discussed more fully in §13.7.

13.3   WHY THE BOOTSTRAP WORKS

13.3.1.   Generalities

At first sight the algorithm used in the previous section seems rather silly. How can generating samples from the observed data tell us more than the data do already?

Let

$$\hat{F}(x) = \frac{1}{n}\{\#X_i \le x\} \tag{13.3.1}$$

denote the empirical distribution function (see VIB, 14.2.2) of the data $X_1, \ldots, X_n$, whose true but unknown distribution function is F. The bootstrap, as described above, can be dissected into the following two ideas.

(a) The known empirical distribution function $\hat{F}$ is used to estimate the unknown true distribution function F; and

(b) theoretical calculations of the properties of statistical procedures are replaced by Monte Carlo simulations [see VIA, §2.2].

These two ideas are logically distinct. Suppose, for example, that Newcomb's data, plotted against expected normal order statistics, had looked like a sample from a normal distribution. Then we would have used the normal distribution with parameters $\bar{X}$ and s^2 to estimate F in (a), instead of $\hat{F}$. Our estimate of θ would have been $\hat{\theta} = \bar{X}$. If, however, we did not know how to calculate the mean and variance of $\hat{\theta}$, but could generate random samples from a normal distribution, we could find Monte Carlo estimates of the mean and variance of $\hat{\theta}$ by repeatedly generating independent random samples $X_1^*, \ldots, X_n^*$ from the normal distribution with mean $\bar{X}$ and variance s^2, and calculating the corresponding values of

$$\hat{\theta}^* = \frac{1}{n}\sum_{i=1}^{n} X_i^*.$$

We would then estimate the mean and variance of $\hat{\theta}$ using the values of the simulated $\hat{\theta}^*$, and use these estimates in calculations like that leading to (13.2.3). In fact, of course, Monte Carlo calculations such as these can be sidestepped completely by knowing that if F is normal, then $\hat{\theta}^*$ has a normal distribution with mean $\bar{X}$ and variance s^2/n. In bootstrap terms, the simulation just described in a parametric bootstrap, in which the parametric assumption is made that F is normal, but theoretical calculations are replaced by a simulation experiment.

The significance of idea (a) is that the empirical distribution function $\hat{F}$, which places probability $1/n$ at each of the observations X_i, is the non-parametric [see

VIB, §14] maximum likelihood estimate [see VIA, §6] of F. This means that the bootstrap algorithm works in a wide variety of problems under weak distributional assumptions. Together (a) and (b) show how to estimate characteristics of estimators, such as their bias and variance, under repeated sampling from F, with minimal distributional assumptions and without reliance on possibly unreliable asymptotic approximations.

13.3.2. Estimation of F

If we can assume that F has a particular parametric form, depending on parameter h is chosen so that f_n is not too bumpy (h too small) nor too smooth $F(\cdot; \hat{\beta})$, where $\hat{\beta}$ is the maximum likelihood estimate of β. If not, then the empirical distribution function $\hat{F}$ is the non-parametric maximum likelihood estimate of F, justifying its use when assumptions about the form of F are to be avoided.

Sometimes a smoothed estimate of F can give better results. Several possibilities have been investigated in the literature. In one, the probability density function f corresponding to F is estimated by a kernel density estimate (Silverman, 1986)

$$f_h(x) = \frac{1}{nh} \sum_{i=1}^{n} K\left(\frac{x - X_i}{h}\right), \tag{13.3.2}$$

where $K(\cdot)$ is a probability density function with mean zero and variance one. Instead of placing probability mass n^{-1} exactly at X_i this has the effect of diffusing the mass in the neighbourhood of X_i. Often $K(\cdot)$ is taken to be the standard normal density function $K(u) = (2\pi)^{-1/2} e^{-u^2/2}$. The smoothing parameter h is chosen so that f_h is not too bumpy (h too small) nor too smooth (h too large): $h = 1.06\sigma^{-1/5}$ is a standard recipe for the choice of h. Obviously σ, the standard deviation of the distribution F, must be replaced by an estimate based on the data at hand, such as σ_x.

As an estimate of f, f_h has the disadvantage that samples drawn from it have larger variance than those drawn from $\hat{F}$. Consequently the use has been suggested of the shrunk smoothed estimate

$$f_{h_s}(x) = \frac{h_s}{nh} \sum_{i=1}^{n} K\left\{\frac{h_s(x - \bar{X}) - (X_i - \bar{X})}{h}\right\}, \tag{13.3.3}$$

where $h_s = (1 + h^2/\sigma_X^2)^{-\frac{1}{2}}$, and σ_X^2 is the variance of the sample. The density (13.3.3) has mean $\bar{X}$ and variance σ_X^2. Once again h is a smoothing parameter. Sampling from (13.3.2) is achieved by selecting one of the X_i at random as before, and then adding to it a random variable with density $h^{-1}K(u/h)$. Sampling from (13.3.3) is similar, except that the value chosen is not X_i but the shrunk value $\bar{X} + (X_i - \bar{X})/h_s$, and that the random variable added has density $(h_s/h)K(h_s u/h)$.

Silverman and Young (1987) show that when smoothed estimates of F

improve on the use of $\hat{F}$, they do so in a rather subtle way; their paper should be consulted for details.

Smoothing can be advantageous when the number n of observations is rather small and the statistic being bootstrapped has singularities if too many of the X_i^* are equal, as in the next example.

EXAMPLE 13.3.1. Suppose that data arise in pairs $(U_1, V_1), \ldots, (U_n, V_n)$ and that we wish to bootstrap to find the standard error of the sample correlation coefficient

$$R = \frac{\sum_{i=1}^{n} (U_i - \bar{U})(V_i - \bar{V})}{\sqrt{\sum_{i=1}^{n} (U_i - \bar{U})^2 \sum_{i=1}^{n} (V_i - \bar{V})^2}},$$

where $\bar{U} = n^{-1} \sum U_i$ and $\bar{V} = n^{-1} \sum V_i$. The bootstrap samples consist of pairs $(U_1^*, V_1^*), \ldots, (U_n^*, V_n^*)$, sampled independently with replacement from the original pairs. The bootstrap value R^* of the correlation coefficient is undefined if all the pairs are equal, and $R^* = 1$ if exactly two of the original pairs occur in the bootstrap sample. If $n = 5$, one of these events occurs with probability 0.1. This type of singularity is avoided if the bootstrap pairs are sampled from a smoothed estimate of the joint distribution of (U, V).

13.4. MORE COMPLEX EXAMPLES

13.4.1. Two-sample Problem

Previously (see VIB, 14.6) various non-parameteric tests were described for use in the following situation:

$$U_1, \ldots, U_m \sim F, \qquad V_1, \ldots, V_n \sim G,$$

where F and G are two possibly different distributions on the real line [for '$\sim$' see VIA, §1.4.2(i)]. One estimate of any difference in location between the two distributions is the Hodges–Lehmann shift estimate

$$T = \text{median} \{V_j - U_i; i = 1, \ldots, m, j = 1, \ldots, n\}.$$

Having observed data and a corresponding value t_{obs} of T, we wish to calculate the variance of T—to see, for example, if the difference in location between F and G may be zero. In this situation we estimate F by the empirical distribution function $\hat{F}$ of the U_i, and G by that of the V_i, namely $\hat{G}$. A bootstrap replicate then consists of a sample $U_1^*, \ldots, U_m^*$ drawn independently with replacement from $\hat{F}$ and a similar sample $V_1^*, \ldots, V_n^*$ drawn from $\hat{G}$. We then calculate

$$T^* = \text{median} \{V_j^* - U_i^*; i = 1, \ldots, m, j = 1, \ldots, n\}$$

and repeat the process S times, eventually calculating bias and variance estimates as before.

This resampling scheme differs from the resampling of pairs (U_i, V_i) in Example 13.3.1. In that case resampling took place from the bivariate empirical distribution function of the original data, whereas here we resample separately from two univariate distributions. In each case we resample in such a way that we mimic as far as possible the way in which the original data were collected.

In the case of matched pairs (see VIB, 14.5), $n = m$, and each U_i is matched with a V_i. If U is the response of an experimental unit to a treatment, and V the response to a matched control unit, analysis is often based on the differences $D_i = U_i - V_i (i = 1, \ldots, n)$. Here the resampling scheme used for the correlation coefficient is suitable, again for the reason that the data were paired in the original sampling scheme: we resample pairs (U_i, V_i) or equivalently differences D_i if the analysis is to be based on them.

13.4.2.   Linear Regression

Suppose that independent data $Y_1, \ldots, Y_n$ follow a linear model (see VIA, 8) in which

$$Y_i = \mathbf{x}_i^T \boldsymbol{\beta} + \varepsilon_i,$$

where the $\mathbf{x}_i^T$ are known $p \times 1$ vectors of covariates, β is a $p \times 1$ vector of unknown parameters, and the errors ε_i are assumed to be independent with zero mean and unknown variance σ^2. In terms of vectors we write

$$\mathbf{Y} = \mathbf{X}\boldsymbol{\beta} + \boldsymbol{\varepsilon},$$

where $\mathbf{Y}$ and $\boldsymbol{\varepsilon}$ are $n \times 1$ vectors, and $\mathbf{X}$ an $n \times p$ matrix of covariates whose ith row is $\mathbf{x}_i^T$. The aim of an analysis may be to test hypotheses about the components of $\boldsymbol{\beta}$, to estimate $\boldsymbol{\beta}$ and σ^2 or some quantity derived from them, or to give confidence intervals for any of the quantities involved. Typically the assumption is made that the errors ε_i are normally distributed in order to base tests, confidence intervals and so forth on F-tests and the related paraphernalia of the normal-theory linear model. Here we suppose that we wish to perform such calculations without assuming that the errors are normal.

In designed experiments it is usual to fix the $\mathbf{x}_i$ before observing the responses Y_i, whereas in other cases the pairs $(Y_i, \mathbf{x}_i)$ may be collected together. In either case, the $\mathbf{x}_i$ are usually regarded as fixed for the purpose of the analysis, and this should be reflected in the bootstrap resampling scheme adopted. This is an example of the idea of conditioning at work: since knowledge of the design $\mathbf{D} = (\mathbf{x}_1, \ldots, \mathbf{x}_n)$ does not contribute to our knowledge of $\boldsymbol{\beta}$ or σ^2, any inferences drawn should be made conditional on the observed value of $\mathbf{D}$. Our bootstrap algorithm should not generate designs $\mathbf{D}^* = (\mathbf{x}_1^*, \ldots, \mathbf{x}_n^*)$ which are very different from $\mathbf{D}$. One way to achieve this is to find the usual least-squares estimates $\hat{\boldsymbol{\beta}} = (\mathbf{X}^T\mathbf{X})^{-1}\mathbf{X}^T\mathbf{Y}$ [see VIA, §8.2.2] and the corresponding residuals $e_i = y_i - \mathbf{x}_i^T\hat{\boldsymbol{\beta}}$. Bootstrap samples are generated by sampling e_i^* uniformly at random from

$e_1, \ldots, e_n$, and letting

$$Y_i^* = \mathbf{x}_i^T \hat{\boldsymbol{\beta}} + e_i^* \quad (i = 1, \ldots, n),$$

so that the resulting bootstrap estimate of $\boldsymbol{\beta}$ is $\hat{\boldsymbol{\beta}}^* = (\mathbf{X}^T\mathbf{X})^{-1}\mathbf{X}^T\mathbf{Y}^*$. The estimate of σ^2 is the corresponding residual sum of squares. There is some evidence that the standardized residuals $r_i = e_i/\sqrt{1 - h_{ii}}$, where h_{ii} is the ith diagonal element of the $n \times n$ matrix $\mathbf{X}(\mathbf{X}^T\mathbf{X})^{-1}\mathbf{X}^T$, should be resampled in place of the e_i. Division by $\sqrt{1 - h_{ii}}$ ensures that the quantities resampled have roughly equal variances. Hinkley (1988) discusses more complicated regression models and gives further references.

EXAMPLE 13.4.1.   As an example of a non-standard problem in regression, we consider the determination of the change-point in a two-phase regression model. That is, the assumed model is

$$Y_i = \begin{cases} \theta + \beta_0(\mathbf{x}_i - \gamma) + \varepsilon_i, & (i = 1, \ldots, \tau), \\ \theta + \beta_1(\mathbf{x}_i - \gamma) + \varepsilon_i & (i = \tau + 1, \ldots, n), \end{cases}$$

where $\mathbf{x}_1 < \mathbf{x}_2 < \ldots < \mathbf{x}_\tau \le \gamma < \mathbf{x}_{\tau+1} < \ldots < \mathbf{x}_n$ and the ε_i are assumed to be independent with constant variance σ^2. Figure 13.4.1 shows $n = 15$ data pairs $(y_i, \mathbf{x}_i)$ from an experiment on the relationship between blood factor VII and warfarin concentration. The solid line is the maximum likelihood estimate of the regression line when the errors ε_i are assumed normally distributed and $\beta_1 = 0$. The change of slope occurs at $(\hat{\gamma}, \hat{\theta})$. To fit the model involves least squares estimation of different lines in Figure 13.4.1 as $\tau = 2, \ldots, n - 2$, and some subsequent calculations to find the maximum likelihood estimates $\hat{\gamma}$ and $\hat{\theta}$. The estimator $\hat{\gamma}$ approaches its limiting normal distribution rather slowly. The data

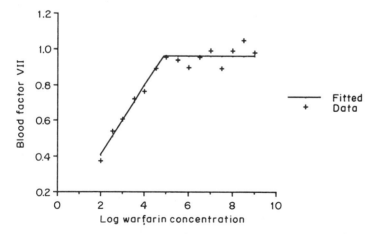

Figure 13.4.1. Data on blood factor VII and fitted change-point
regression model

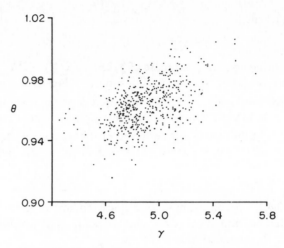

Figure 13.4.2. Distribution of $S = 500$ bootstrapped
change-points for data on blood factor VII

are taken from Hinkley (1971), who gives a detailed discussion of two-phase regression.

Residuals e_i were calculated for the model described above, and the variability of the estimated change-point was assessed by bootstrapping the procedure $S = 500$ times, assuming that $\beta_1 = 0$ throughout. Figure 13.4.2 shows the resulting values of $(\hat{\gamma}^*, \hat{\theta}^*)$, whose joint distribution is clearly not normal. The cluster of points towards the left of the figure corresponds to bootstrap samples where $\hat{t}^* = 5$, as opposed to $\hat{t} = 6$ in the original data. The asymptotic variance of $\hat{\gamma}$ calculated under the assumption of normal errors is 0.048, but the variance of the bootstrap values $\hat{\gamma}^*$ is 0.046. There is also a small positive bias of $\hat{\gamma}$.

13.4.3.   Dosage–Mortality Relationship

As a second regression example, we consider a biological assay, data from which are given in Table 13.4.1

At each dose level d_i of a poison, r_i out of m_i beetles died. The usual model for such experiments is to treat the number of deaths at the ith dose level as a binomial random variable with denominator m_i and probability p_i.

Log dose x_i	1.6907	1.7242	1.7552	1.7842	1.8113	1.8369	1.8610	1.8839
Number of beetles m_i	59	60	62	56	63	59	62	60
Number killed r_i	6	13	18	28	52	53	61	60

Table 13.4.1: Mortality of beetles on exposure to a poison.

The probability of death depends on the dose through the relationship $p_i = G(\alpha + \beta z_i)$, where $x_i = \log d_i$, and $G(\cdot)$ is a continuous cumulative probability distribution function. Often $G(\cdot) = \Phi(\cdot)$, the standard normal distribution function, but other possibilities are the logistic, for which $G(u) = 1/(1 + e^{-u})$ and the extreme-value, where $G(u) = 1 - \exp(-\exp u)$. The supposition is that different experimental units, in this case beetles, have tolerances distributed according to G, and that, if the dose given exceeds its tolerance $d = e^x$, a beetle will die. Different functions G correspond to different distributions of tolerances to the poison. The first of the models, called the probit, is discussed in VIA, 6.6. The model corresponding to the extreme-value distribution is known as the complementary log–log because of the form of $G^{-1}(\cdot)$.

One goal of analysis might be estimation of x_{50}, the value of x at which 50 per cent of the population would die. If $\hat\alpha$ and $\hat\beta$ denote the maximum likelihood estimates of α and β, the estimate of x_{50} is the solution of the equation

$$G(\hat\alpha + \hat\beta \hat x_{50}) = 0.5,$$

namely

$$\hat x_{50} = \frac{G^{-1}(0.5) - \hat\alpha}{\hat\beta} = -\frac{\hat\alpha}{\hat\beta}$$

in the case of the symmetric normal and logistic functions G. The standard theory of the maximum likelihood estimator yields an asymptotic variance for $\hat x_{50}$, as given in VIA, 6.6.6. However, the adequacy of this asymptotic variance is open to doubt if the m_i are small. When the probit model is fitted to the data given above, the variance given by asymptotic approximation is $(0.0038)^2$. The bootstrap can be used to find the adequacy of this, as follows.

(1) For each dose level, generate a binomial random variable R_i^* with denominator m_i and probability r_i/m_i.
(2) Calculate parameter estimates $\hat\alpha^*$ and $\hat\beta^*$, and hence $\hat x_{50}^*$, based on the simulated data.
(3) Repeat steps (1) and (2) S times, and calculate the estimated bias and variance of $\hat x_{50}$ as previously.

For the data given in the table above, and $S = 49$ bootstrap samples, the estimated bias and standard error of $\hat x_{50}$ were -0.007 and 0.0049. This leads to confidence intervals for the true x_{50} which are 25 per cent wider than those based on the asymptotic value. One reason for this diagreement is that the probit model does not fit the data particularly well. Figure 13.4.3 shows a plot of $w_i = \Phi^{-1}\{r_i/(m_i + 1)\}$ against log dose x_i. If the normal tolerance distribution were suitable, the graph would be approximately a straight line with slope β and intercept α, but Figure 13.4.3 seems to show definite curvature. If the complementary log–log model is used, $\hat x_{50} = 1.779$ with asymptotic standard error 0.0040. A bootstrap of size $S = 49$ yields $\sqrt{V^*} = 0.0043$, which is in closer agreement with the asymptotic value. The normal and extreme-value tolerance distributions are compared in Example 13.6.4.

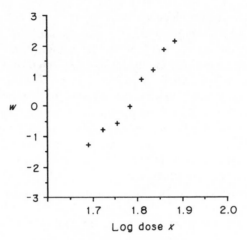

Figure 13.4.3. Plot of $\omega = \Phi^{-1}(r/m)$ against
log dose x for data on mortality of beetles

One lesson to be learnt from this is that although the bootstrap can give accurate estimates of variances and so forth in an automatic way, it is vital to be sure that the statistical model used is appropriate. The fit of a statistical model should be checked whatever methods are used to analyse it, as a rough calculation for a suitable model is more valuable than an exact calculation for an inadequate model.

A second point is more subtle. The simulated data here are generated from the binomial distribution with denominator m_i and probability r_i/m_i. But for the last column in Table 13.4.1 $r_i/m_i = 1$, and so $R_i^* = m_i$ in every bootstrap sample, although presumably $G(\alpha + \beta x_i) < 1$. Although the effect is small in this example, because the m_i are large and hence α and β are well-determined, it would be necessary to make some sort of adjustment in smaller samples. The problem is most acute in binary data, where $m_i = 1$ and $r_i = 0$ or 1. One way to tackle this would be to resample at random from all the data, but only to use samples whose design matrices were close to the original design in some suitable sense.

13.4.4. Autoregressive Time Series [see VIB, §18.8]

In each of the situations so far considered the observations have been regarded as independent. In bootstrapping time series and other stochastic processes, a key problem is to uncover underlying independent and exchangeable variables which can be resampled to form bootstrap realizations of the process. Apart from some rather simple examples, this is not yet well-understood, although Example 13.6.2 gives an example of bootstrapping point processes in two dimensions. One case where it is clear what to do is the the following.

EXAMPLE 13.4.2.   Suppose data are available from a first-order autoregressive process, in which $Y_0 = \varepsilon_0$ and $Y_t = \alpha Y_{t-1} + \varepsilon_t$, when $t = 1, \ldots, n$, where the ε_t are independent innovations with zero mean and variance σ^2. We find an estimate $\hat{\alpha}$ of α by maximum likelihood, and form residuals $e_t = Y_t - \hat{\alpha} Y_{t-1}$. The new bootstrap times series is $Y_0^* = \varepsilon_0^*$, $Y_t^* = \hat{\alpha} Y_{t-1}^* + \varepsilon_t^*$ for $t = 1, \ldots, n$, where the ε_t^* are sampled with replacement from the estimated innovations $e_0, \ldots, e_n$. This procedure extends readily to autoregressive processes of order p, whose structure is

$$Y_t = \sum_{j=1}^{p} \alpha_j Y_{t-j} + \varepsilon_t,$$

so that the errors ε_t can be reconstructed given estimates of the autoregressive coefficients α_j.

13.5.   EFFICIENT BOOTSTRAP METHODS

The accuracy of bootstrap estimates of bias, variance, and so forth depends on the size of the Monte Carlo simulation the scientist can afford. If the statistic of interest is expensive to calculate or computing resources are limited, more efficient methods for boostrap sampling may be useful. The idea is to reduce the number S of bootstrap samples required while retaining the same overall level of accuracy.

13.5.1.   Von Mises Expansions

The basis of understanding some of these ideas is the von Mises expansion. Let $\hat{F}$ denote the empirical distribution function of the data, as before, and let F^* denote the empirical distribution function based on a bootstrap sample $X_1^*, \ldots, X_n^*$. That is,

$$F^*(x) = \frac{1}{n} \#\{X_i^* : X_i^* \le x\}.$$

Let $T = t(F)$ denote the value of a statistic at distribution F; thus the sample value of T is $t(\hat{F}) = t_{\mathrm{obs}}$, and $T^* = t(F^*)$ is its value calculated from the bootstrap sample. The von Mises expansion of $t(F^*)$ about $\hat{F}$ has form

$$t(F^*) = t(\hat{F}) + \frac{1}{n} \sum_{j=1}^{n} L(X_j^*; \hat{F}) + \frac{1}{2n^2} \sum_{j=1}^{n} \sum_{k=1}^{n} Q(X_j^*, X_k^*; \hat{F}) + \ldots, \quad (13.5.1)$$

where the remaining terms are of order $n^{-3/2}$ in probability. The linear terms $L(X_j^*; \hat{F})$ in (13.5.1) have expected value zero under bootstrap sampling, i.e.

$$E^* L(X^*; \hat{F}) = \frac{1}{n} \sum_{j=1}^{n} L(X_j; \hat{F}) = 0,$$

since, under expectation E^* with respect to the resampling distribution, X^* takes each of the values $X_1, \ldots, X_n$ with probability n^{-1}. Similarly both marginal expected values of a quadratic term $Q(X_j^*, X_k^*; \hat{F})$ are zero, i.e.

$$E^*Q(X^*, y; \hat{F}) = E^*Q(y, X^*; \hat{F}) = 0,$$

for each fixed value of y.

Under bootstrap resampling the first term on the right of (13.5.1) is constant, while the second is an average on n independent terms each with mean zero, and so is of order $n^{-\frac{1}{2}}$ in probability. The third term can be regarded as an average of n terms, each of which is itself an average of n terms with mean zero, and so is of order n^{-1}. Thus (13.5.1) is analogous to a Taylor series expansion [see IV, §3.6] for the statistic $t(F^*)$, with inclusion of more terms bringing increasing orders of accuracy.

EXAMPLE 13.5.1.   Let $t(F) = \int x \, dF(x)$ be the mean of the distribution F. We interpret the integral in the Stieltjes sense [see IV, §4.8], and so

$$t(\hat{F}) = \int x \, d\hat{F}(x) = \frac{1}{n} \sum_{j=1}^{n} X_j = \bar{X},$$

the sample average, is the sample value of $t(F)$. Likewise $t(F^*) = \int x \, dF^*(x)$ is $\bar{X}^*$.

The linear term $L(y; F)$ is calculated as follows. Let

$$F_\varepsilon(x) = (1 - \varepsilon)F(x) + \varepsilon H_y(x) \tag{13.5.2}$$

be the distribution which chooses an observation from F with probability $1 - \varepsilon$ and takes value y with probability ε. We use $H_y(x)$ to denote the Heavyside function

$$H_y(x) = \begin{cases} 0 & x < y \\ 1 & x \geq y, \end{cases}$$

which corresponds to a random variable which always takes value y. Thus F_ε can be thought of as a perturbation of F which adds a small probability atom at y. We now define

$$L(y; F) = \left. \frac{dt(F_\varepsilon)}{d\varepsilon} \right|_{\varepsilon = 0}.$$

Thus in the case of the mean,

$$t(F_\varepsilon) = \int x \, d[(1 - \varepsilon)F(x) + \varepsilon H_y(x)] = (1 - \varepsilon)t(F) + \varepsilon y,$$

so $L(y; F) = y - t(F)$. Note that $EL(X, F) = 0$ when expectation is taken with respect to distribution F. It follows that $L(X_j^*; \hat{F}) = X_j^* - \bar{X}$, so (13.5.1) becomes

$$\bar{X}^* = \bar{X} + \frac{1}{n} \sum_{j=1}^{n} (X_j^* - \bar{X}),$$

which is exact.

Quadratic and higher-order terms are calculated analogously as

$$Q(y, z; F) = \left. \frac{\partial^2 t[(1 - \varepsilon_1 - \varepsilon_2)F + \varepsilon_1 H_y + \varepsilon_2 H_z]}{\partial \varepsilon_1 \partial \varepsilon_2} \right|_{\varepsilon_1 = \varepsilon_2 = 0},$$

and so forth. These are identically zero for the mean.

EXAMPLE 13.5.2.   Let

$$t(F) = \int x^2 \, dF(x) - \left[\int x \, DF(x) \right]^2,$$

the variance of the distribution F. Say $t(F) = \sigma^2$ and $\int x \, dF(x) = \mu$. Now

$$t(F_\varepsilon) = (1 - \varepsilon) \int x^2 \, dF(X) + \varepsilon y^2 - \left[(1 - \varepsilon) \int x \, dF(x) + \varepsilon y \right]^2,$$

so

$$L(y; F) = y^2 - \int x^2 \, dF(x) - 2 \left[y - \int x \, dF(x) \right] \int x dF(x) = (y - \mu)^2 - \sigma^2.$$

Similarly

$$Q(y, z; F) = -2(y - \mu)(z - \mu)$$

and cubic and higher-order terms in the expansion are zero. Again we see that $EL(X; F) = EQ(X, y; F) = EQ(y, X; F) = 0$ when X has distribution function F. For a given sample $X_1, \ldots, X_n$ we obtain

$$Q(X_j, X_k; \hat{F}) = -2(X_j - \bar{X})(X_k - \bar{X}).$$

EXAMPLE 13.5.3.   Suppose that an estimator is defined implicitly by the equation

$$\int \psi(x; T) d\hat{F}(x) = 0. \tag{13.5.3}$$

Examples of this are the M-estimate used in §13.2 or a maximum likelihood estimator, for which $\psi(x; \theta) = \partial l(\theta; x)/\partial\theta$, where l is the loglikelihood (see VI A, 6). The theoretical expression corresponding to (5.3) is

$$\int \psi\{x; t(F)\} dF(x) = 0, \tag{13.5.4}$$

which expresses more clearly the dependence of $t(F)$ on the distribution F. Since (13.5.4) is identically true for all distributions F, it follows that

$$0 = \int \psi\{x; t(F_\varepsilon)\} dF_\varepsilon(x)$$

$$= (1 - \varepsilon) \int \psi\{x; t(F_\varepsilon)\} dF(x) + \varepsilon\psi\{y; t(F_\varepsilon)\}, \tag{13.5.5}$$

where F_ε is defined at (13.5.2). We now differentiate (13.5.5) with respect to ε, using the chain rule [see IV, (3.28)], and set $\varepsilon = 0$ to obtain

$$0 = \frac{\partial t}{\partial \varepsilon} \int \frac{\partial \psi \{x; t(F)\}}{\partial t} dF(x) + \psi \{y; t(F)\}.$$

The linear approximation is

$$L(y; F) = \frac{\partial t}{\partial \varepsilon} = \frac{\psi \{y; t(F)\}}{-\int \dfrac{\partial \psi \{x; t(F)\}}{\partial t} dF(x)}.$$

For a maximum likelihood estimator [see VIA, §6.2] we find that

$$L(y; F) = I(\theta)^{-1} \frac{\partial l(\theta; y)}{\partial \theta},$$

where $I(\theta)$ is the information matrix [see VIA, §3.3] for a single observation from F. The empirical version of this is

$$L(X_i; \hat{F}) = \left\{ -\frac{1}{n} \sum_{j=1}^{n} \frac{\partial^2 l(\hat{\theta}; X_j)}{\partial \theta^2} \right\}^{-1} \frac{\partial l(\hat{\theta}; X_i)}{\partial \theta}.$$

The above calculations have taken for granted that a statistic T is sufficiently smooth as a function of F to possess the necessary derivatives. This is not the case with statistics such as the median [see II, Definition 10.3.3].

13.5.2. Bias Estimation

We can now calculate the leading term of the bias of $t(F^*)$. If we take expectations in (13.5.1) under bootstrap sampling in which the X_j^* are chosen uniformly with replacement from $X_1, \ldots, X_n$, we see that

$$E^* t(F^*) = t(\hat{F}) + \frac{1}{2n^2} \sum_{j=1}^{n} Q(X_j, X_j; \hat{F}). \tag{13.5.6}$$

The bias of $t(F^*)$ is the second term on the right-hand side, plus terms of smaller order in n. For the sample variance

$$t(F^*) = \frac{1}{n} \sum_{j=1}^{n} X_j^{*2} - \left(\frac{1}{n} \sum_{j=1}^{n} X_j^* \right)^2,$$

for example, the bias is

$$\frac{1}{2n^2} \times -2 \sum_{j=1}^{n} (X_j - \bar{X})^2,$$

so $t(F^*)$ has expected value $(n-1)^{-1} \sum (X_j - \bar{X})^2$ rather than $n^{-1} \sum (X_j - \bar{X})^2$. The estimated bias based on S bootstrap values $T_1^*, \ldots, T_S^*$ of T is approximately

$$\frac{1}{S}\sum_{s=1}^{S} T_s^* - t(\hat{F}) = \frac{1}{nS}\sum_{s=1}^{S}\sum_{j=1}^{n} L(X_{js}^*; \hat{F}) + \frac{1}{2n^2S}\sum_{s=1}^{S}\sum_{k=1}^{n} Q(X_{js}^*, X_{ks}^*; \hat{F}), \quad (13.5.7)$$

where X_{js}^* is the jth observation in the sth bootstrap sample. Expression (13.5.7) has linear terms L which we know from (13.5.6) should not be present. One way to make them disappear is as follows: instead of sampling randomly with replacement from the values $X_1, \ldots, X_n$, we choose bootstrap samples by taking observations randomly without replacement from S copies of the original data. In practice this can be done by putting S copies of $X_1, \ldots, X_n$ into a vector of length nS, randomly permuting the vector, and then reading off the S bootstrap samples from the permuted vector as successive blocks of size n. The effect of this resampling scheme is to balance the simulation so that each of the values X_i appears equally often and the sum of the linear terms in (13.5.7) is zero, as required. In terms of Table 13.2.1, the samples have been balanced so that the frequencies in each column sum to S as well as those in each row summing to n. In effect, the balancing scheme forces the first term on the right-hand side of (13.5.7) to take its expected value in an infinite bootstrap, $S = \infty$. The fact that this leading term is of higher order than the remainder reduces the variability of the bootstrap estimate of bias by an order of magnitude. An important point to note is that knowledge of the expansion (13.5.1) is not required to implement balanced resampling, only to understand its effects.

Other balanced resampling schemes are described by Graham *et al.* (1989) and Ogbonmwan and Wynn (1988).

13.5.3. Variance Estimation

The improved estimate of bias obtained through balanced sampling does not require explicit calculation of the terms in the von Mises expansion of T. However, if the terms can be calculated, they can be put to use to improve estimation of the variance of T. Let us term

$$T_L^* = t(\hat{F}) + \frac{1}{n}\sum_{j=1}^{n} L(X_j^*; \hat{F}) \qquad (13.5.8)$$

the linear approximation to T^*. This is often highly correlated with T^*, but it has known moments, and so can be used as a control variate, in the language of Monte Carlo simulation. Specifically, let $D^* = T^* - T_L^*$ be the difference between the statistic and its linear approximation. Then

$$E^*T^* = E^*T_L^* + E^*D^* = t(\hat{F}) + E^*D^*,$$

and

$$\text{Var}^*T^* = \text{Var}^*(T_L^*) + \text{cov}^*(T_L^*, D^*) + \text{Var}^*(D^*)$$

$$= \frac{1}{n^2}\sum_{j=1}^{n} L(X_j; \hat{F})^2 + \text{cov}^*(T_L^*, D^*) + \text{Var}^*(D^*).$$

In each of these expressions the leading term, which is of higher order than the remainder, has no simulation error. The other terms are estimated as follows. Suppose that the bootstrap statistics $T_1^*, \ldots, T_S^*$ have corresponding linear approximations $T_{L1}^*, \ldots, T_{LS}^*$, and that the differences are $D_s^* = T_s^* - T_{Ls}^*$. Then $E^* D^*$, $\text{cov}^*(T_L^*, D^*)$ and $\text{Var}^*(D^*)$ are respectively estimated by $\bar{D} = S^{-1} \sum D_s^*$, $S^{-1} \sum (D_s^* - \bar{D})(T_{Ls}^* - \bar{T}_L)$ and $(S-1)^{-1} \sum (D_s^* - \bar{D})^2$, where $T_L = S^{-1} \sum T_{Ls}^*$.

13.5.4. An Example

Consider again the Newcomb data of §13.2. Example 13.5.3 shows that the linear approximation for T is

$$L(X_i; \hat{F}) = \frac{\psi(X_i; t_{\text{obs}})}{-\dfrac{1}{n} \displaystyle\sum_{j=1}^{n} \dfrac{\partial \psi(X_i; t_{\text{obs}})}{\partial t}}$$

$$= \frac{n}{12} \psi(X_i; t_{\text{obs}}),$$

since

$$\frac{\partial \psi(x; t)}{\partial t} = \begin{cases} -1 & |x - t| < c \\ 0 & \text{otherwise,} \end{cases}$$

and we have set $c = 6$. Armed with this information, we can apply the methods outlined above to get improved estimates of the bias b and variance v of T. In $S = 199$ bootstrap samples the corelation between T^* and its linear approximation (13.5.8) was 0.98: the two are highly correlated.

				Percentile					
1	2	5	10	50	90	95	98	99
				Ordinary Bootstrap					
			Variances using raw percentiles						
6988	3130	1696	719	341	658	951	1789	2753
			With linear approximation						
1071	815	550	381	158	478	670	943	1170
				Permutation Bootstrap					
			Variances using raw percentiles						
8381	3265	1433	378	124	277	508	1289	2833
			With linear approximation						
564	419	266	165	5.5	128	215	343	450

Table 13.5.1: Variances $\times 10^4$ of percentile estimates for the Huber M-estimate of mean passage time of light for Newcomb data. Bootstrap size $S = 99$, replicated 50 times.

A small simulation study was performed to see the gains in estimating percentiles of the distribution of T. The variances of the estimated percentiles of the distribution of T were calculated for 50 replicates of each of four different bootstraps of size $S = 99$. The four methods used to calculate the percentile estimates were the raw percentiles of the ordinary bootstrap, a normal approximation using bias and variance estimates from the ordinary bootstrap together with the linear approximation method of §13.5.3, and similar estimates for the permutation bootstrap of §13.5.2, with and without the linear approximation. Table 13.5.1 shows the results. The effect of the linear approximation is fairly large, especially in the middle of the distribution where the permutation method also shows a good reduction in variance. Theoretical work by Hall (1988) shows that in general the reduction in variance between 2 and 4 of the table should be about a factor 1.75 in the middle of the distribution but only about 1.16 at the 2.5 per cent and 97.5 per cent points of the distribution.

13.6. BOOTSTRAP SIGNIFICANCE TESTS

One of the original role of bootstrap resampling was in testing statistical hypotheses [see VIA, §5] under weak assumptions about the mechanism generating the data. Frequently the alternative hypothesis is so ill-specified that the situation is what Cox and Hinkley (1974), Ch. 3, refer to as a 'pure significance test'. This is particularly the case in geometrical statistics, an important area of application. Suppose, then, that increasing values of a statistic T indicate increasing degrees of inconsistency between data and a null hypothesis H_0 concerning the data. If the observed value of T is t_{obs}, we aim to calculate the significance level

$$p_{obs} = \text{Prob}(T \geq t_{obs} | H_0).$$

The calculation is performed as if the null hypothesis were true. What this means depends on the context, but two broad approaches can be distinguished. In the first, the support of the resampling distribution is changed in some way to reflect H_0, whereas in the second the weights attached to different observations change.

EXAMPLE 13.6.1. Consider again the two-sample problem of §13.4.1. Suppose that we wish to test the hypothesis that $F = G$. If this is the case, then an estimate of the common distribution is found by combining the samples $U_1, \ldots, U_m$ and $V_1, \ldots, V_n$ into a single sample of size $n + m$ with empirical distribution function $\hat{F}_0$ say. Samples $U_1^*, \ldots, U_m^*$ and $V_1^*, \ldots, V_n^*$ are drawn independently from $\hat{F}_0$, and the bootstrap value

$$T^* = \text{median}\{V_j^* - U_i^*; i = 1, \ldots, m, j = 1, \ldots, n\}$$

of the statistic is calculated as before.

As a numerical illustration, consider the example discussed in VIB, 14.6, where the two independent random samples of the $m = 7$ observations

$$3.7 \quad -1.1 \quad 2.6 \quad 2.3 \quad 4.1 \quad 0.8 \quad 3.9$$

and the $n = 5$ observations

$$4.6 \quad 4.0 \quad 5.3 \quad 4.4 \quad 3.0$$

were used to illustrate two-sample non-parametric tests. The observed value of the Hodges–Lehmann statistic for these samples is $t_{obs} = -1.6$. In $S = 199$ bootstrap samples from the combined null hypothesis distribution $\hat{F}_0$, this value was exceeded 10 times, giving an estimated significance level of

$$\hat{p}_{obs} = \frac{10 + 1}{199 + 1} = 0.055.$$

This should be doubled for a two-sided test, and the result is to be compared with significance level 0.03 obtained for the Wilcoxon–Mann–Whitney test.

This example is a standard hypothesis-testing setting for classical non-parametric statistics. An example where modified support is used which falls outside such a setting is due to Kendall and Kendall (1980).

EXAMPLE 13.6.2.   We are presented with a set of points $X_1, \ldots, X_n$ in the plane. Three such points are said to be 'ε-collinear' if they form a triangle whose largest angle exceeds $\pi - \varepsilon$. There are $n(n-1)(n-2)/6$ such triplets, of which some number, T_ε say, are ε-collinear. The statistic T_ε serves as a test statistic

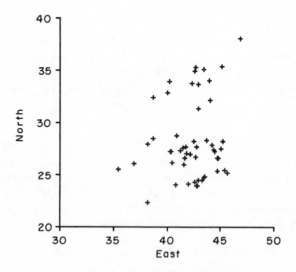

Figure 13.6.1. Sites of 52 standing stones

for the null hypothesis that the observed collinearities have arisen by chance alone, when the alternative is that there are too many to be due to chance.

In the setting originally considered, the points represent the positions of 52 standing stones in an area of Britain near Land's End. The observation had been made that there were 108 collinearities when ε was taken to be 40 seconds of arc, and it was desired to test if this number was too large to have arisen by chance, one possible alternative hypothesis being that the stones had been placed in lines in prehistoric times. The configuration of the stones is displayed in Figure 13.6.1.

In this and similar applications, the null hypothesis is very vague. A simple parametric null hypothesis would be that the data are generated by a spatial Poisson process. However, this hypothesis specifies complete absence of structure in the data under H_0, whereas the physical or topographical features of the area in which the stones are observed may limit where the stones may stand, thus restricting the possible configurations of stones with which the observed data should be compared. Moreover resampling spatial positions X_i^* with replacement from the observed positions is clearly inappropriate. Instead the following data perturbation scheme is used. Each of the original positions X_i is randomly moved by adding to it a bivariate normal random variable, with probability density function

$$f(x, y) = \frac{1}{2\pi\sigma_x\sigma_y} \exp[-\tfrac{1}{2}(x^2/\sigma_x^2 + y^2/\sigma_y^2)] \quad (-\infty < x, y < \infty).$$

The variances σ_x^2 and σ_y^2 are chosen large enough to destroy ε−collinearities in the original data, and so that $\sigma_x^2 = \zeta s_x^2$ and $\sigma_y^2 = \zeta s_y^2$, where s_x^2 and s_y^2 are the variances of the original data in the x- and y-directions. This has the effect of approximately preserving the ratio of principal standard deviations, and hence

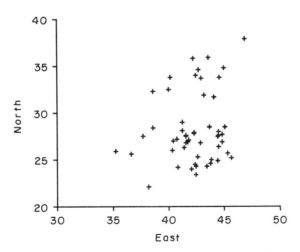

Figure 13.6.2. Sites of standing stones after perturbation of size $\zeta = 0.1$

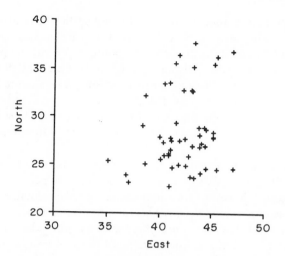

Figure 13.6.3. Sites of standing stones after perturbation of size $\zeta = 0.3$

the shape, of the data. Figures 13.6.2 and 13.6.3 show the effect of perturbations, where $\zeta = 0.1$ and 0.3. Clusters of the original points preserved when $\zeta = 0.1$ tend to be destroyed when $\zeta = 0.3$, which seems a little too large a value.

For the Land's End data it was found that, with a suitable choice of ζ, the average number of triplets under this perturbation scheme was 99, but with fluctuations commonly of the order of ± 25 or so. Thus the observed figure of 108 loses any claim to significance.

Among other instructive aspects of this example is the importance of ensuring that the data perturbation scheme generates datasets like the original. The observed data should be compared to other configurations with similar clusters of stones, areas where stones are sparse, and so forth, and the resampling scheme is designed to ensure that this is so. As in regression problems, this is an example of the conditionality principle at work.

A second possibility in resampling from a null hypothesis is to change the probabilities with which the original data are resampled.

EXAMPLE 13.6.3.   Suppose that we have univariate data $X_1, \ldots, X_n$, and that the null hypothesis is that these are drawn from a distribution F with mean μ_0. We choose to resample from a distribution $\hat{F}_0$ which is as 'close' as possible to the uniform distribution $\hat{F}$ on $X_1, \ldots, X_n$ but has mean μ_0. Under H_0 we suppose that

$$\mathrm{Prob}\,(X_j^* = X_i) = p_i,$$

say. One measures of the closeness of two distributions is the information or

Kullback–Leibler distance

$$d_F(F, G) = \int \ln \left\{ \frac{dG(x)}{dF(x)} \right\} dF(x).$$

If we take F in this expression to be the bootstrap null hypothesis distribution and G to be $\hat{F}$, we aim to minimize $\sum p_i(\ln p_i + \ln n)$ subject to the constraints $\sum p_i = 1$ and $\sum X_i p_i = \mu_0$. A standard application of Langrange multipliers [see IV, §5.15] yields that, provided $\min X_i < \mu_0 < \max X_i$,

$$p_i = \frac{\exp(\alpha X_i)}{\sum_{j=1}^{n} \exp(\alpha X_j)}$$

where α is chosen so that $\sum X_i p_i = \mu_0$. This is a so-called exponentially tilted distribution; $\hat{F}$ is obtained when $\alpha = 0$. The idea now is to test the hypothesis that the true mean of $X_1, \ldots, X_n$ is μ_0 by calculating $t_{obs} = \bar{X} - \mu_0$, and then generating S bootstrap samples from $\hat{F}_0$, calculating $T^* = \bar{X}^* - \mu_0$ for each. An estimate of the observed significance level

$$\hat{p}_{obs} = \frac{1}{S+1} [\#\{T_s^* : T_s^* \geq t_{obs}\} + 1]$$

is then found as in Example 13.6.1.

This idea is rather special and has been little exploited, although presumably it could be extended to statistics other than the mean. See Efron (1982), Ch. 10. A rather more complicated situation, in which the simulation is performed using probabilities determined by a null hypothesis, is in comparing non-nested models for a set of data (Wahrendorf, Becker and Brown (1987)).

EXAMPLE 13.6.4. One situation in which classical asymptotic theory is known to work only for very large samples is in the comparison of the fit to data of two models, neither of which may be obtained as a restriction of the other. Consider again the data in Table 13.4.1 on the deaths of beetles. Two possible models use the normal and extreme-value distributions of tolerances to the poison. Each of these models has two parameters, and neither is a restricted version of the other. One measure of the overall fit of a model is its deviance [see VIB, 11.2.3], which is 10.12 for the normal and 3.45 for the extreme-value tolerance distributions. Their difference, twice the difference in maximized loglikelihoods between the models with extreme-value and normal tolerance distributions, is $t_{obs} = 6.67$. This, which we take to be the test statistic, is known to be normally distributed in large samples (Cox (1961)), but its small-sample distribution is unknown. An unusual feature of this situation is the symmetry between the two competing hypotheses: it is desirable to know the level of significance of t_{obs} under both, as failure to achieve significance under either hypothesis would indicate that the data do not hold strong evidence for discrimination between the competing tolerance distributions. Accordingly the following bootstrap may be performed.

(1) Fit the probit model to the original data by maximum likelihood, giving fitted probabilities $\hat{p}_i = \Phi(\hat{\alpha}_p + \hat{\beta}_p x_i)$, where $\hat{\alpha}_p$ and $\hat{\beta}_p$ are the estimates of α and β under the model.

(2) Generate binomial data with denominator m_i and probabilities $\hat{p}_i$, and fit both probit and complementary log–log models to those data, giving a bootstrap statistic T^* as the difference in deviances between the probit and complementary log–log models.

(3) Repeat step (2) S times to give bootstrap statistics $T_1^*, \ldots, T_S^*$.

Here data are simulated on the basis of the null hypothesis, in this case the probit model. In the converse case in which the null hypothesis is that the tolerance distribution is the extreme-value, we change step (1) of the above algorithm to

(1') Fit the complementary log–log model to the original data by maximum likelihood, giving fitted probabilities $\hat{p}_i = 1 - \exp\{-\exp(\hat{\alpha}_c + \hat{\beta}_c x_i)\}$, where $\hat{\alpha}_c$ and $\hat{\beta}_c$ are the estimates of α and β under the model.

Figure 13.6.4 shows the results on performing each of these simulations with $S = 199$. Each set of simulated differences of deviances has been ordered and plotted against expected normal order statistics. The observed difference in deviances is shown at the left of the figure. Evidently t_{obs} lies in the centre of the distribution of simulations generated under the extreme-value tolerance distributions, but is in the tail of the those generated under the probit model. The respective observed significance levels are 48.5 per cent and 1.5 per cent.

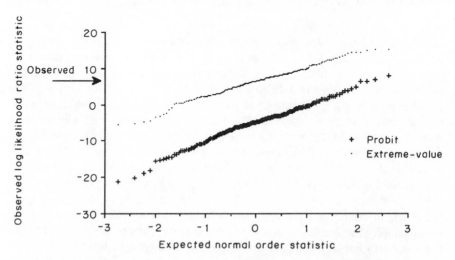

Figure 13.6.4. Normal order statistics plot for bootstrapped loglikelihood ratio statistic for competing models for data on mortality of beetles

There is strong evidence that the complementary log–log model gives the better fit to the data. Both distributions of ordered differences of deviances are roughly straight lines, which confirms that the differences of loglikelihoods have approximate normal distributions.

13.7.   CONFIDENCE INTERVALS

It is generally desirable to base confidence statements on pivotal statistics, whose distributions are known not to depend on the values of unknown parameters. Thus it may be advantageous to find a transformation of T, $h(T)$, say, which is more nearly pivotal than T itself. Sometimes a particular transformation will suggest itself by analogy with a parametric situation, as when T is a sample correlation coefficient, for example. Then $h(T) = \tanh^{-1} T$ [see IV, §§2.13 and 2.14] is approximately normal with mean $\tanh^{-1}\theta + \theta/[2(n-3)]$ and variance $1/(n-3)$ [see VIA, §2.7.3]. It would then be sensible to form confidence intervals for $\tanh^{-1}\theta$ based on the resampling distribution of $\tanh^{-1} T$. If $h(\cdot)$ is not suggested by analogy, the double bootstrap (Hinkley (1988)) can be used to find a pivotal transformation.

We assumed in previous sections that a normal approximation to the distribution of $T-\theta$ would be an adequate basis for confidence intervals, as may be the case if n is reasonably large. However, normal approximations do not exploit the full power of the bootstrap to account for effects such as skewness of the distribution of T, and numerous other approaches are available to find confidence intervals automatically from the distribution of bootstrap estimates. The simplest of these is the percentile method, in which the estimator T of θ is bootstrapped to give $T^*_{(1)} \le T^*_{(2)} \le \cdots \le T^*_{(S)}$, say. A $(1-2\alpha) \times 100$ per cent equi-tailed confidence interval for θ is then found simply by taking as its endpoints $2t_{\text{obs}} - T^*_{([(1-\alpha)(S+1)])}$ and $2t_{\text{obs}} - T^*_{([\alpha(S+1)])}$, where $[\cdot]$ denotes integer part. DiCiccio and Romano (1988) describe a number of improvements to this procedure which increase its accuracy at the expense of complicating it.

A second procedure is the bootstrap-t method. This relies on an estimate Q^* of the variance of T^*, calculable from $X^*_1, \ldots, X^*_n$. In particular, if the linear approximation (5.8) to T^* can be found, a suitable estimate is

$$Q^* = \frac{1}{n^2} \sum_{j=1}^{n} L(X^*_j; \hat{F})^2.$$

Rather than base confidence intervals on the ordered values of the T^*_S, we now order the values of the Student t-like [see VIA, §2.5.5] statistic $Z^* = (T^* - t_{\text{obs}})/\sqrt{Q^*}$ to get $Z^*_{(1)} \le Z^*_{(2)} \le \ldots \le Z^*_{(S)}$. This has the advantage of being closer to a pivotal quantity than is $T - t_{\text{obs}}$. The $(1-\alpha) \times 100$ per cent upper confidence limit for θ is then $t_{\text{obs}} - \sqrt{\hat{Q}} Z^*_{([\alpha(S+1)])}$, where $\hat{Q}$ is the estimated variance of $t_{\text{obs}} = t(\hat{F})$, based on the original sample $X_1, \ldots, X_n$. The $\alpha \times 100$ per cent lower confidence interval is $t_{\text{obs}} - \sqrt{\hat{Q}} Z^*_{([(1-\alpha)(S+1)])}$. The current evidence is that the bootstrap-t

method provides a reliable general method for bootstrap confidence intervals for scalar parameters, provided that an estimate of the variance of T can be calculated (Hall 1988b; see also the discussion). In order for confidence limits based on the percentile or bootstrap-t methods to be reliable, S needs to be much larger than the values used previously, for order 1000 or more.

EXAMPLE 13.7.1.   We turn yet again to the Newcomb data, and the Huber M-estimate of the mean speed of light, for which the $L(X_i; \hat{F})$ were calculated in §13.5.4. In this case the value of t_{obs} is 25.62, and $\hat{Q} = n^{-2}\sum L(x_i; \hat{F})^2 = (1.62)^2$. A bootstrap-$t$ 95 per cent confidence interval for the true mean speed of light was calculated from $S = 999$ bootstrap samples, for which it was found that $Z^*_{[0.025(S+1)]} = -1.98$ and $Z^*_{[0.975(S+1)]} = 1.71$. Note the asymmetry of these values due to the skewness of the original data. The 95 per cent confidence interval is $(25.62 - 1.62 \times 1.71, \ 25.62 - 1.62 \times 1.98)$, or $(22.85, 28.83)$. This contrasts with the interval $(22.70, 28.62)$ based on the assumption that T has a normal distribution and the use of (13.2.3) with estimates of bias b and variance v based on the same 999 bootstrap samples.

13.8.   BIBLIOGRAPHIC NOTES

Data perturbation was used by D. G. Kendall and others in the 1970s, but the bootstrap in the form described in §13.2 was introduced by Efron (1979), and his 1982 monograph is still the only book on the subject. A recent review of the subject, with many useful references, is Hinkley (1988). Efron and Tibshirani (1986) give an introduction which includes discussion of some fairly complicated applications, including non-parametric curve estimation and censored survival data. Wu (1986) and his discussants refer to much of what is known about resampling schemes in regression analysis. Young (1986) discusses the role of conditioning in geometrical applications, and gives other references to bootstrap significance tests. Banks (1988) refers to the literature on smoothed and Bayesian versions of the bootstrap. Work on efficient bootstrap methods is by Davison, Hinkley and Schechtman (1986), Graham *et al.* (1987), Johns (1988), Hall (1988a, 1989) and Efron (1989). Davison and Hinkley (1988) describe methods of completely removing Monte Carlo error from resampling plans for certain statistics. DiCiccio and Romano (1988) review work on bootstrap confidence regions; see also Tibshirani (1988). The construction in Example 13.6.3 suggests one of several possible ways to construct likelihood functions based on the bootstrap (see Owen (1988, 1989) and Hall (1987)).

ACKNOWLEDGEMENTS

I am grateful to David Hinkley, and to the Nuffield Foundation for financial support.

                                                                          A. C. D.

REFERENCES

Banks, D. L. (1988). Histospline smoothing the Bayesian bootstrap, *Biometrika*, **75**, 673–684.

Cox, D. R. (1961). Tests of Separate Families of Hypotheses, *Proc. 4th Berkeley Symp.* **1**, 105–123.

Cox, D. R. and Hinkley, D. V. (1974). *Theoretical Statistics*. Chapman and Hall.

Davison, A. C. and Hinkley, D. V. (1988). Saddlepoint Approximations in Resampling Plans, *Biometrika* **75**, 417–431.

Davison, A. C., Hinkley, D. V. and Schechtman, E. (1986). Efficient Bootstrap Simulation, *Biometrika* **73**, 555–666.

DiCiccio, T. J. and Romano, J. P. (1988). A Review of Bootstrap Confidence Intervals (with Discussion), *J. Roy. Statistic. Soc. B* **50**, 338–370.

Efron, B. (1979). Bootstrap Methods: Another Look at the Jackknife, *Ann. Statist.* **7**, 1–26.

Efron, B. (1982). *The Jackknife, the Bootstrap, and Other Resampling Plans*, Monograph 38, SIAM.

Efron, B. (1989). More Efficient Bootstrap Computations, *J. Amer. Statist. Assoc.* to appear.

Efron, B. and Tibshirani, R. (1986). Bootstrap Methods for Standard Errors, Confidence Intervals, and Other Measures of Statistical Accuracy, *Statistical Science* **1**, 54–77.

Graham, R. L., Hinkley, D. V., John, P. W. M. and Shi, S. (1987). Balanced Design of Bootstrap Simulations, Technical Report 48. Center for Statistical Sciences, The University of Texas at Austin.

Hall, P. G. (1987). On the Bootstrap and Likelihood-based Confidence Regions, *Biometrika* **74**, 481–493.

Hall, P. G. (1988a). Performance of Balanced Bootstrap Resampling in Distribution Function and Quantile Problems, Preprint, Australian National University.

Hall, P. G. (1988b). Theoretical comparison of bootstrap confidence intervals (with Discussion), *Annals of Statistics*, **16**, 927–985.

Hinkley, D. V. (1971). Inference in Two-phase Regression, *J. Amer. Statist. Assoc.* **66**, 736–743.

Hinkley, D. V. (1988). Bootstrap Methods (with Discussion), *J. Roy. Statistic. Soc. B* **50**, 321–337, 355–370.

Johns, M. V. (1988). Importance Sampling for Bootstrap Confidence Intervals, *J. Amer. Statist. Soc.* **83**, 709–714.

Kendall, D. G. and Kendall, W. S. (1980). Alignments in Two-dimensional Random Sets of Points, *Adv. Appl. Probab.* **12**, 380–424.

Ogbonmwan, S.-M. and Wynn, H. P. (1988). Resampling Generated Likelihoods, in *Statistical Decision Theory and Related Topics IV*, Eds. S. S. Gupta and J. O. Berger, vol. 1, Springer, 133–147.

Owen, A. B. (1988). Empirical Likelihood Ratio Confidence Intervals for a Single Functional, *Biometrika* **75**, 237–249.

Owen, A. B. (1989). Empirical Likelihood Ratio Confidence Intervals, *Ann. Statist.* to appear.

Silverman, B. W. (1986). *Density Estimation for Statistics and Data Analysis*. Chapman and Hall.

Silvermann, B. W. and Young, G. A. (1987). The Bootstrap: To Smooth or Not to Smooth? *Biometrika* **74**, 469–479.

Stigler, S. M. (1977). Do Robust Estimators Work with Real Data? (with Discussion), *Ann. Statist.* **5**, 1055–1098.

Tibshirani, R. (1988). Variance Stabilization and the Bootstrap, *Biometrika* **75**, 433–444.

Wahrendorf, J., Becker, H. and Brown, C. C. (1987). Bootstrap Comparison of Non-nested Generalized Linear Models: Applications in Survival Analysis and Epidemiology, *Appl. Statist.* **36**, 72–81.

Wu, C. J. F. (1986). Jackknife, Bootstrap and Other Resampling Methods in Regression Analysis (with Discussion), *Ann. Statist.* **14**, 1261–1350.

Young, G. A. (1986). Conditioned Data-based Simulations: Some Examples from Geometrical Statistics, *Int. Statist. Rev.* **54**, 1–13.

CHAPTER 14

Extreme Value Theory

14.1. INTRODUCTION

Extreme value theory is concerned with probability calculations and statistical inferences connected with extreme values in random samples and stochastic processes. There are many applications, and their number is growing all the time. Two of the main areas are environmental extremes such as river flow, wind speed, temperature and rainfall, and areas such as structural reliability and the strength of materials, where it is very often the case that the weakest component or part of a system is ultimately responsible for failure. Other areas where extreme value concepts are being increasingly applied include financial calculations such as probabilities of large insurance claims, monitoring of air pollution (ozone, acid rain, etc.) and some rather more novel ones such as horse races and athletics records. There are also connections with survival data analysis, since although the latter area is not strictly one to do with extreme values there is some overlap in the statistical concepts involved, such as the Weibull distribution. Finally, there are applications of extreme value theory in other areas of statistics, such as testing for outliers and change point problems.

The present review covers both the probabilistic and statistical sides of the subject, though with emphasis on the statistical. There are a number of recent books on extreme value theory, most of them concentrating on the probabilistic side of the subject. Galambos (1987) gives an all-round review at an intermediate mathematical level. Leadbetter, Lindgren and Rootzén (1983) give a rigorous development of the theory of extreme values in stochastic processes, in both discrete and continuous time. Such has been the pace of development in recent years that much of this is already out of date; for recent reviews see Lindgren and Rootzén (1987), Leadbetter and Rootzén (1988). Also at a rigorous mathematical level, but dealing with rather different aspects of the subject, is Resnick (1987). Whereas Leadbetter, Lindgren and Rootzén concentrate on stochastic processes, Resnick deals primarily with independent identically distributed random variables, developing the detailed theory of the joint distribution of extreme order statistics, its evolution in time, and from that the areas of records and extremal processes. He also gives the only comprehensive account of multivariate extremes to have appeared in book form so far. An entirely different perspective on extreme value theory is given by Aldous (1989), who disparages detailed technical proof and concentrates on giving intuitive arguments to support the approximations. The positive side of Aldous's book is that it takes a much broader view of the subject than any of the others. However, this writer takes the view that Aldous has underestimated much of the work actually being done in this field, and that a more patient approach might well pay off better in the long run. Other books include the conference proceedings edited by Tiago de Oliveira (1984a) and Hüsler and Reiss (1989).

Historically, the first developments of extreme value theory were in the papers of Dodd (1923), Fréchet (1927) and Fisher and Tippett (1928), the last named being the first to present the most celebrated result of the whole theory, the 'three types' theorem for the class of extreme value limit distributions. This was followed by works of von Mises (1936) and Gnedenko (1943), the latter providing the first rigorous proof of the three types theorem. The chain of results started by these authors was completed by de Haan (1970), who gave a complete solution to the domain of attraction problem. In applications, the biggest contributions were made by Weibull (1939, 1951) who was the first to emphasize the wide importance of extreme value concepts in the strength of materials, and Gumbel (1958), whose book is still regarded as something of a classic despite being long superseded in terms of the statistical techniques employed.

In recent years many developments have taken place in extreme value theory, but the gap between theory and applications is still rather wide. This chapter is concerned primarily with the extreme value distributions and the two major approaches to the statistics, namely the classical (Gumbel) approach of fitting the extreme value distributions directly to extreme data, and the more modern approaches based on the joint distribution of largest order statistics or on exceedances over thresholds. Also covered, if briefly, is the theory of multivariate extremes, an area under rapid development at the time of writing.

14.2.   THE EXTREME VALUE DISTRIBUTIONS

The 'three types' of extreme value distribution function for maxima are given in standardized form by the formulae

$$Type\ I: \quad H(x) = \exp\{-\exp(-x)\} \quad (-\infty < x < \infty),$$

$$Type\ II: \quad H(x) = \begin{cases} 0 & \text{if } x \le 0, \\ \exp(-x^{-\alpha}) & \text{if } x > 0, \end{cases}$$

$$Type\ III: \quad H(x) = \begin{cases} \exp[-(-x)^{\alpha}] & \text{if } x < 0, \\ 1 & \text{if } x \ge 0. \end{cases}$$

In II and III, α is any positive number. The three types are also often called the Gumbel, Fréchet and Weibull types respectively. They are the cornerstone of extreme value theory, and are very widely used as statistical models to be fitted to data on extreme values.

These distributions were originally derived as limiting forms for the maximum of a set of independent, identically distributed (henceforth i.i.d.) set of random variables. Suppose $X_1, X_2, \ldots, X_n$ are an i.i.d. sequence with distribution function

$$F(x) = \text{Prob}\{X \leq x\}.$$

Let $M_n = \max(X_1, \ldots, X_n)$. Then M_n has distribution function

$$\text{Prob}\{M_n \leq x\} = \text{Prob}\{X_1 \leq x, \ldots, X_n \leq x\} = F^n(x).$$

Limit theory is concerned with the behaviour of this in large samples, that is as $n \to \infty$. In direct form, however, the maximum of a sample simply tends to the right-hand endpoint of the distribution, whether that is finite or infinite. Therefore, the distribution is rescaled and we seek limits of the form

$$H(x) = \lim_{n \to \infty} F^n(a_n x + b_n) \tag{14.2.1}$$

for constants $a_n > 0$ and b_n. It is from this relation that the three types are derived.

More precisely, it can be shown that any limit H in (14.2.1) must satisfy

$$H^N(x) = H(A_N x + B_N), \tag{14.2.2}$$

where the constants $A_N > 0$ and B_N exist for each $N > 1$. The only non-degenerate solutions to this functional equation are Types I–III, where we define two distribution functions G and H to be of the same type if $H(x) = G(Ax + B)$ for some $A > 0$ and B. This means that the three types exhaust all the possible solutions of (14.2.2) except for location and scale transformations.

For statistical purposes, it is better to have a single family, and this may be achieved by defining

$$H(x; \mu, \sigma, \gamma) = \exp\left\{-\left[1 + \frac{\gamma(x - \mu)}{\sigma}\right]^{-1/\gamma}\right\} \tag{14.2.3}$$

defined when $1 + \gamma(x - \mu)/\sigma > 0$; $\sigma > 0$ and μ, γ arbitrary. The case $\gamma = 0$ is interpreted as the limit $\gamma \to 0$, that is

$$H(x; \mu, \sigma, 0) = \exp\left\{-\exp\left[-\frac{(x - \mu)}{\sigma}\right]\right\} \quad (-\infty < x < \infty). \tag{14.2.4}$$

Types II and III correspond respectively to $\gamma > 0$ ($\gamma = 1/\alpha$) and $\gamma < 0$ ($\gamma = -1/\alpha$). This defines the extreme value distributions as a single three-parameter form which can be estimated by any of the standard methods of estimation, including the method of maximum likelihood, to which most attention will be given in the discussion to follow.

14.3. DERIVATION OF THE EXTREME VALUE LIMIT

In this section we expand on the discussion in §14.2 by showing how the limit (14.2.1) may be derived in practice. The question being addressed is the following: given a sequence of independent random variables with a given distribution function F, how can we determine the constants a_n, b_n and the limit H such that (14.2.1) is satisfied? A thorough answer to this question requires the theory of regular variation (de Haan (1970); Resnick (1987); Bingham, Goldie and Teugels (1987)). We shall not go into details about that, preferring simple conditions which are sufficient for the vast majority of practical applications.

Loosely, when $1 - F(x)$ decreases polynomially slowly as $x \to \infty$, say $1 - F(x) \sim cx^{-\alpha}$ for some $c > 0$, $\alpha > 0$, then (14.2.1) holds with H of Type II form. Examples include the Pareto, Cauchy, t and F distributions. If there is a finite upper endpoint x^* and $1 - F(x^* - t) \sim ct^\alpha$ as $t \to 0$, then the Type III form arises. Most commonly, this is applied to minima in which everything is multiplied by -1 and the finite upper endpoint becomes a finite lower endpoint (often 0). Examples in this case include the beta and gamma distributions (with uniform and exponential as special cases) and of course the Weibull distribution itself, $F(x) = 1 - \exp[-(x/x_0)^\alpha]$, $0 < x < \infty$. In between, most distributions with exponentially decreasing tails, whether the endpoint is finite or infinite, are attracted to Type I. Examples include the normal, lognormal, reflected form of the lognormal (i.e. $\log(-X)$ is normal: this is a case with finite upper endpoint 0) and the upper tails of the gamma and Weibull distributions. The most important exceptions to the whole theory are discrete distributions such as Poisson and geometric. In these cases an exact limit theory is not available but it is possible to give lim sup and lim inf results (Anderson (1970)). Fortunately the need for such results in practice is comparatively rare, though the problem has arisen in connection with long runs in Bernoulli trials (Gordon, Schilling and Waterman (1986)).

More details will now be given, following an approach developed by Smith (1989a). The latter paper also gives second-order results and full proofs of the approximations. We start with the representation

$$-\log F(x) = \exp\left[-\int_{x_*}^x \frac{dt}{\phi(t)}\right] \qquad (14.3.1)$$

valid over $x_* < x < x^*$, where x_* and x^* are the lower and upper bounds (finite or infinite) on the range of X, and ϕ is a positive continuous function. Note that (14.3.1) requires only that F be continuously differentiable, since we may define $\phi(x) = -[F(x)\log F(x)]/F'(x)$. For $u < x^*$ fixed we write

$$\frac{-\log F(u + x\phi(u))}{-\log F(u)} = \exp\left[-\int_0^x \frac{\phi(u)}{\phi(u + s\phi(u))}ds\right]. \qquad (14.3.2)$$

If ϕ is differentiable, then

$$\frac{\phi(u + s\phi(u))}{\phi(u)} = 1 + \int_0^s \phi'(u + w\phi(u))dw.$$

An easy variant on the mean value theorem leads to

$$\frac{-\log F(u + x\phi(u))}{-\log F(u)} = [1 + x\phi'(y)]^{-1/\phi'(y)}, \qquad (14.3.3)$$

where y is between u and $u + x\phi(u)$.
   Suppose now

$$\lim_{x \to x^*} \phi'(x) = \gamma_0. \qquad (14.3.4)$$

Condition (14.3.4) is a unified form of the von Mises conditions which are well known to be widely applicable sufficient conditions for an extreme value limit (see, for instance, Section 1.4 of Resnick (1987), or Theorem 1.6.1 of Leadbetter, Lindgren and Rootzén (1983)). Then as $u \to x^*$ the limit in (14.3.3) is $(1 + \gamma_0 x)^{-1/\gamma_0}$. Defining b_n as the solution of $-n \log F(b_n) = 1$ and $a_n = \phi(b_n)$, we have

$$\lim_{n \to \infty} [-n \log F(a_n x + b_n)] = (1 + \gamma_0 x)^{-1/\gamma_0}$$

and hence

$$F^n(a_n x + b_n) \to \exp[-(1 + \gamma_0 x)^{-1/\gamma_0}]. \qquad (14.3.5)$$

This is in the form (14.2.1) with H given by (14.2.3) and $\mu = 0$, $\sigma = 1$.
   We might expect the foregoing approximation to be improved if we replace $\phi'(y)$ in (14.3.3), not by γ_0, but by $\phi'(u)$. Defining $\gamma_n = \phi'(b_n)$, this leads to

$$F^n(a_n x + b_n) \approx \exp[-(1 + \gamma_n x)^{-1/\gamma_n}]. \qquad (14.3.6)$$

Equation (14.3.6) is known as the 'penultimate approximation', in contrast to the 'ultimate approximation' (14.3.5). The idea of penultimate approximation is most clearly understood when $\gamma = 0$, since then the ultimate approximation is of Type I and the penultimate of Type II or III. In this form the idea was actually suggested by Fisher and Tippett (1928), and shown in many cases to improve upon the rate of convergence by Cohen (1982) and Gomes (1984). The idea of using a penultimate approximation when γ is not 0 is much more recent, but Gomes and Pestana (1986) and Smith (1989a) have given examples of its efficacy. The latter paper has also proposed higher-order approximations.
   The three main cases are as follows:

Case 1.   Polynomially decreasing tails. Suppose we have an expansion

$$-\log F(x) = Cx^{-\alpha}(1 + Dx^{-\beta} + o(x^{-\beta})), \qquad (14.3.7)$$

where $C > 0$, $\alpha > 0$, $\beta > 0$, and that we can differentiate (14.3.7) as often as required. Then

$$\phi(x) = \frac{x}{\alpha}\left[1 - \frac{D\beta x^{-\beta}}{\alpha} + o(x^{-\beta})\right],$$

$$\phi'(x) = \frac{1}{\alpha}\left[1 + \frac{D\beta(\beta - 1)x^{-\beta}}{\alpha} + o(x^{-\beta})\right].$$

Solving $-\log F(b_n) = 1/n$ gives

$$b_n = (Cn)^{1/\alpha}\left[1 + \frac{D(Cn)^{-\beta/\alpha}}{\alpha} + o(n^{-\beta/\alpha})\right],$$

$$a_n = \phi(b_n) = (Cn)^{1/\alpha}\left[1 - \frac{D(\beta-1)(Cn)^{-\beta/\alpha}}{\alpha} + o(n^{-\beta/\alpha})\right]\Big/\alpha,$$

$$\gamma_0 = 1/\alpha, \qquad \gamma_n = \frac{1}{\alpha} - \frac{D\beta(\beta-1)(Cn)^{-\beta/\alpha}}{\alpha^2} + o(n^{-\beta/\alpha}).$$

Case 2.   Finite upper endpoint. The corresponding treatment with $x^* < \infty$ and

$$-\log F(x^* - t) = Ct^\alpha(1 + Dt^\beta + o(t^\beta)) \tag{14.3.8}$$

is very similar, and left as an exercise for the reader.

Case 3.   For most distributions for which $\gamma = 0$, it is possible to simplify things slightly by substituting $1 - F$ for $-\log F$ in (14.3.1), so that the starting assumption is

$$1 - F(x) = \exp\left[-\int_{x_*}^x \frac{dt}{\phi(t)}\right] \tag{14.3.9}$$

valid over $x_* < x < x^*$. Since $1 - F$ is asymptotic to $-\log F$ as $F \to 1$, the two representations (14.3.1) and (14.3.9) are for most purposes interchangeable, the difference being of $O(1/n)$ in the probability calculations. In the two examples to be given, this is negligible compared with the main part of the approximation error.

If we do start from (14.3.9), so that

$$\phi(x) = \frac{1 - F(x)}{F'(x)},$$

we may define b_n by $n(1 - F(b_n)) = 1$ and then a_n, γ_n by

$$a_n = \phi(b_n) = \frac{1}{nF'(b_n)}, \qquad \gamma_n = \phi'(b_n) = -\frac{\phi(b_n)F''(b_n)}{F'(b_n)} - 1.$$

For example, for the standard normal distribution we have by Kendall and Stuart (1977), equation (5.68), $\phi(x) = x^{-1} - x^{-3} + 3x^{-5} - \ldots$, from which it follows that $\phi'(x) \to 0$, and we also have

$$a_n = (2\pi)^{1/2}n^{-1}\exp(b_n^2/2), \qquad \gamma_n = a_n b_n - 1.$$

The solution of b_n from $n\{1 - F(b_n)\} = 1$ is best performed numerically; the formula

$$b_n = (2\log n)^{1/2} - \frac{\log\log n + \log 4\pi}{(8\log n)^{1/2}}$$

is often quoted but is only an approximation which does not yield the best rate of convergence (Hall (1979)).

For example, with $n = 100$ the constants are $b_n = 2 \cdot 326$, $a_n = 0 \cdot 3752$, $\gamma_n = -0 \cdot 1271$. A commonly used measure of distance between two distributions with densities f and g is total variation distance,

$$v(f, g) = \sup \left[\left| \int_A f(x) dx - \int_A g(x) dx \right| \right],$$

where the supremum is taken over all measurable sets A. With this measure, the distance between the exact and approximate densities is $0 \cdot 039$ using the ultimate approximation, and $0 \cdot 0065$ using the penultimate approximation. The corresponding figures with $n = 1000$ are $b_n = 3 \cdot 090$, $a_n = 0 \cdot 2970$, $\gamma_n = -0 \cdot 0822$ with distances $0 \cdot 026$, $0 \cdot 0027$. These are direct calculations based on Smith (1989a). These figures confirm the considerable superiority of the penultimate approximation, and also that the relative gain in accuracy from using the penultimate approximation improves as sample size increases. The rates of convergence are $O((\log n)^{-1})$ and $O((\log n)^{-2})$ for the ultimate and penultimate approximations.

The lognormal distribution may be defined by the distribution function $F(x) = \Phi(\sigma^{-1} \log x)$ where $x > 0$, $\sigma > 0$ and Φ is the standard normal distribution function. In this case, if we start by defining B_n such that $n(1 - \Phi(B_n)) = 1$, then we may write

$$b_n = \exp(\sigma B_n), \qquad a_n = (2\pi)^{1/2} n^{-1} b_n \exp(B_n^2/2), \qquad \gamma_n = -1 + a_n b_n^{-1}(1 + B_n/\sigma).$$

For $\sigma = 1$ and $n = 100$, we find $b_n = 10 \cdot 240$, $a_n = 3 \cdot 842$, $\gamma_n = 0 \cdot 2481$ with total variation distances $0 \cdot 077$, $0 \cdot 0058$. For $n = 1000$ the corresponding figures are $21 \cdot 982$, $6 \cdot 529$, $0 \cdot 2148$, $0 \cdot 067$, $0 \cdot 0025$. The general comparisons are similar to the normal case, except that the rate of convergence is slower, and the distance between the ultimate and penultimate approximations, as measured by the value of γ_n, is very noticeable.

The logarithmic rate of decrease in such cases is often cited as argument that the extreme value approximations behave badly in moderate-sized samples. The preceding calculations are hardly detailed enough to refute such a notion, but they do suggest perhaps that the real issue is not the quality of the approximations themselves but the comparatively small improvement which results from increasing the sample size. The examples also confirm the practical superiority of the penultimate approximation.

14.4. EXCEEDANCES: THE GENERALIZED PARETO DISTRIBUTION

An alternative approach to the whole problem of extreme values is to formulate the basic results in terms, not of maxima over samples of fixed size, but of exceedances over high thresholds. Much current statistical methodology is based on this idea.

In this case we start directly from (14.3.9). The same reasoning as in §14.3 then leads to

$$\frac{1-F(u+x\phi(u))}{1-F(u)}=\exp\left[-\int_0^x \frac{\phi(u)}{\phi(u+s\phi(u))}\,ds\right]=[1+x\phi'(y)]^{-1/\phi'(y)} \quad (14.4.1)$$

for some y between u and $u+x\phi(u)$.

Now let us interpret (14.4.1), for $x \geq 0$, as the conditional probability that a random variable exceeds $u+x\phi(u)$ given that it exceeds u, for some fixed high threshold u. Under (14.3.4) we have the approximation

$$\frac{1-F(u+y)}{1-F(u)}\approx\left(1+\frac{\gamma y}{\sigma}\right)^{-1/\gamma} \quad (14.4.2)$$

where $\sigma=\phi(u)$ and γ is either γ_0 (the ultimate approximation) or $\gamma_u=\phi'(u)$ (the penultimate approximation).

The right-hand side of (14.4.2) is known as the generalized Pareto tail. The range of parameters is $\sigma>0$, $-\infty<\gamma<\infty$, and the range of y is $0\leq y<\infty$ if $\gamma\geq 0$, $0\leq y\leq -\sigma/\gamma$ if $\gamma<0$. The case $\gamma=0$ is interpreted as the limit $\gamma\to 0$, which is the exponential tail $\exp(-y/\sigma)$. The case $\gamma>0$ is just a reparametrization of the ordinary Pareto distribution, but the use of the generalized Pareto tail in the present context is due to Pickands (1975), who proved that it arises as a limit distribution for exceedances over thresholds under precisely the same conditions (on F) as the ordinary extreme value distributions arise as limit distributions for maxima in large samples.

The statistical importance of these results is that they justify the generalized Pareto distribution as a model for exceedances over thresholds, a theme developed in §14.9.

14.5.   STATISTICAL METHODS: INTRODUCTION

Most of the literature on the statistical aspects of extremes is concerned with the use of extreme value distributions as models for data. The classical reference is Gumbel (1958), but in recent years there have been many advances exploiting modern computational facilities. By far the most powerful and widely applicable technique is the method of maximum likelihood. There are no longer any serious obstacles to the computation of maximum likelihood estimators (Hosking (1985) has a Fortran algorithm) though there remain doubts over their small-sample properties (e.g. the simulations in Hosking, Wallis and Wood (1985)), and cases where even the large-sample properties are still not fully understood (Smith 1985a). In view of this a number of alternative techniques are still widely used.

The first technique to discuss is the Gumbel probability plot. Before computers, this was the main method of fitting the Type I distribution and an important starting point for the other types. Today, this role has been overtaken by automatic computational methods but the probability plot still has a role to play in judging the fit of the distribution and in detecting outliers.

Suppose we have a sample $X_1, \ldots, X_n$ from the distribution function (14.2.4). Let this be ordered $X_{1:n} \leq \ldots \leq X_{n:n}$. The idea is to plot $X_{i:n}$ against *reduced values* $y_{i:n}$ $(i = 1, \ldots, n)$ where

$$y_{i:n} = -\log[-\log(p_{i:n})]$$

and $\{p_{i:n}, i = 1, \ldots, n\}$ is a vector of *plotting positions*. There is a certain amount of controversy over how to choose the plotting positions (Cunnane (1978)); the ideal is that the $\{y_{i:n}\}$ should be the expected order statistics from a standardized ($\mu = 0$, $\sigma = 1$) Type I distribution but in practice approximations must be used. The choice $p_{i:n} = (i - 0 \cdot 5)/n$ may, however, be recommended as simple and of general applicability. The Gringorten formula $(i - 0 \cdot 44)/(n + 0 \cdot 12)$ is also often cited in the hydrology literature (see Cunnane) but produces almost identical results in practice.

If the Type I distribution is indeed a good fit, then the plot should be approximately a straight line with intercept μ and slope σ. This may be used to obtain rough estimates of the parameters. Outliers and lack of fit are usually evident from visual inspection of the plot. If the plot curves upwards, this usually indicates a Type II fit, while a downwards curvature is indicative of Type III. Therefore, the plot may also be a first step towards fitting these distributions. Probability plots to fit the generalized extreme value distribution have been considered (Arnell, Beran and Hosking 1986) but this gets very cumbersome; it is necessary to draw a separate plot for each γ and automatic methods of fitting are probably to be preferred.

Three examples are shown in Figure 14.5.1. Plot (a) is based on 35 annual maximum flows of the River Nidd in Yorkshire, England; the source of the data is the Flood Studies Report (NERC (1975)). This is a fairly typical example of a plot which roughly, though not at all precisely, follows a straight line, so we might expect the Gumbel model to be a reasonable fit. If the largest two observations are omitted, however, it is possible to detect a definite upward curvature. This turns out to be of some importance in the maximum likelihood analysis (§14.6). Figure 14.5.1(b) is based on annual maximum temperatures at Ivigtut, Greenland, 1875–1960; these data are taken from Tabony (1983). In this case there is a definite downward curvature, suggesting a Type III model, with the exception of a very prominent outlier at the upper end. Figure 14.5.1(c) is based on Sample 1 of Smith and Naylor (1987), which is a sample of strengths of glass fibres. For strength and failure time data, it is often thought that a two-parameter Weibull distribution, $\text{Prob}\{X \leq x\} = 1 - \exp[-(x/x_0)^\alpha], 0 \leq x < \infty$, is appropriate. In this case $Y = -\log(X)$ has a Gumbel distribution. Thus Figure 14.5.1(c) is in fact a Gumbel plot based on minus the logarithms of the fibre strengths. Again the plot has irregularities but with no obvious curvature or outliers, so we would expect the two-parameter Weibull distribution to be a reasonable fit.

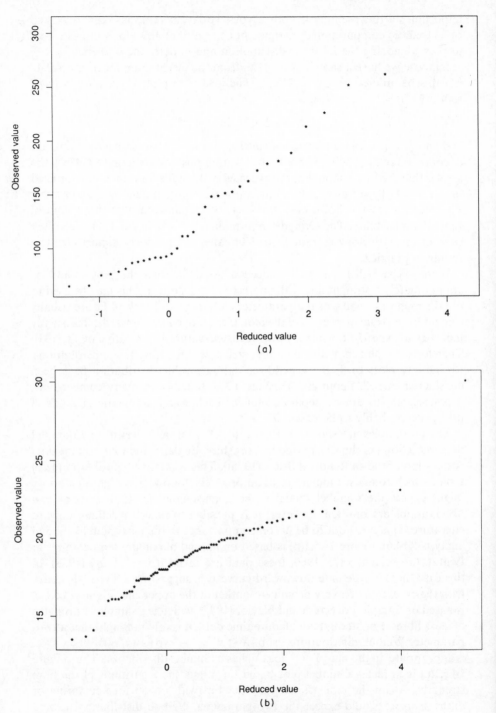

Figure 14.5.1. Gumbel probability plots (a) Nidd data (b) Ivigtut data (c) Negative log fibre strengths

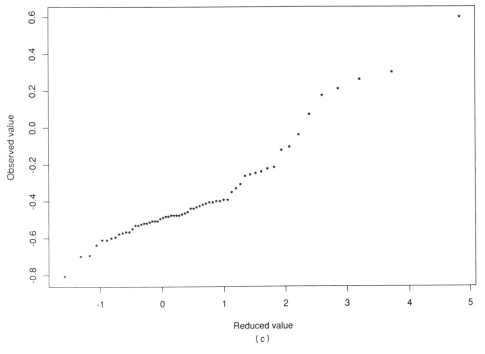

Figure 14.5.1. (*Contd.*)

Of other non-likelihood methods of fitting extreme value distributions, the method of moments is notoriously unreliable on account of the poor sampling properties of second- and higher-order sample moments. Jenkinson's sextiles method (see NERC (1975)) appears to have been superseded by subsequent suggestions. There is an extensive literature on linear combinations of order statistics (Mann, Schafer and Singpurwalla (1974) gave a comprehensive review of that), but these methods are confined largely to two-parameter families and their principal virtue, that they avoid iterative computation, is no longer relevant. Of all these more or less *ad hoc* methods, the one that has attracted most attention recently among hydrologists is the probability weighted moments method. This is described in detail by Hosking, Wallis and Wood (1985), though it had been suggested in the hydrology literature some years earlier. Suppose X is a random variable with distribution function F. In the form most relevant for extreme values applications, Hosking Wallis and Wood defined the rth probability weighted moment to be

$$\beta_r = E\{XF^r(X)\},$$

where $r = 0, 1, 2, \ldots$. For the generalized extreme value distribution, $F = H$, given by (14.2.3), it can be shown that

$$\beta_r = (r+1)^{-1}\{\mu + \sigma\gamma^{-1}[(r+1)^\gamma\Gamma(1-\gamma)-1]\},$$

where Γ is the gamma function. The idea, then, is to equate these to their sample values for $r = 0, 1, 2$. Hosking, Wallis and Wood discussed various ways of calculating the sample probability weighted moments and proposed a simple approximation to estimate the three parameters.

The method is simple to apply and performed well in simulation studies. However, until there is some convincing theoretical explanation of its properties, it is unlikely to be universally accepted. There is also the disadvantage that, at present at least, it does not extend to more complicated situations such as regression models based on the extreme value distributions.

14.6. ESTIMATION BY MAXIMUM LIKELIHOOD

Suppose $X_1,\ldots,X_n$ are a sample of independent observations from a distribution with density $f(x;\theta)$, where θ is a vector of unknown parameters. The log likelihood function is given by

$$l(\theta) = \sum_{i=1}^{n} \log f(X_i;\theta).$$

The maximum likelihood estimator of θ is that value $\hat{\theta}$ which maximizes $l(\theta)$. The matrix of second-order derivatives of $-l$, evaluated at $\hat{\theta}$, is called the observed information matrix. Its importance derives from the property that the inverse of this matrix is the approximate variance–covariance matrix of $\hat{\theta}$. Moreover the estimators are asymptotically efficient and normally distributed as the sample size tends to infinity. All these statements are standard statistical theory and covered in such books as, for example, Cox and Hinkley (1974). The use of observed information in this context has gradually supplanted the older technique based on Fisher or expected information; the theoretical justification for this is based on a paper of Efron and Hinkley (1978), and Prescott and Walden (1983) demonstrated by simulation its specific relevance for the extreme value distributions.

In the case of the generalized extreme value distribution, the regularity conditions for these results to be valid are satisfied provided $\gamma > -0.5$; when that condition is violated a variety of other results is known, too complicated to describe here (Smith (1985a)). Moreover, studies such as Johnson and Haskell (1974), Hosking, Wallis and Wood (1985) and Smith and Naylor (1987) have pointed to a number of difficulties that arise in small or moderate samples even when the regularity condition is satisfied. Nevertheless it is the only method apart from numerical Bayesian methods which may really be called a general method, in the sense that it applies to a whole range of other problems such as those involving regression or censored data. It is for this main reason that we concentrate on it.

Computational aspects have been considered by NERC (1975) and Prescott and Walden (1980, 1983); Hosking (1985) published a Fortran algorithm. These are all based on iterative procedures. The log likelihood is often far from the quadratic shape which makes iterative procedures work well, and so it is important to have reasonable starting values. Sometimes these are calculated by one of the non-iterative methods already mentioned. As far as the algorithm itself is concerned, Hosking used a Newton–Raphson procedure with some modifications: a maximum step-length is imposed, the Newton–Raphson step is replaced by a steepest-ascent step in regions where the log likelihood function is not convex, and a check is made to ensure that the Newton–Raphson step does not result in an infeasible parameter value or a decrease of the log likelihood function. Quasi-Newton procedures work just as well if not even better; the author has made extensive use of Algorithm 21 of Nash (1979), modified to retain an approximation to the observed information matrix for the calculation of approximate variances and covariances.

After the parameters are estimated, it is possible to check the fit of the model by doing a probability plot of residuals, similar in concept to the Gumbel probability plot already mentioned. The idea of this is to plot the order statistics $\{X_{i:n}, i = 1, \ldots, n\}$ against $\{F^{-1}(p_{i:n}), i = 1, \ldots, n\}$ where F^{-1} is the inverse of the fitted distribution function and $p_{i:n}$ may be defined to be $(i - 0.5)/n$ as before. In the case of the generalized extreme value distribution, $F^{-1}(p)$ is easily calculated to be

$$F^{-1}(p) = \begin{cases} \mu + \sigma[(-\log p)^{-\gamma} - 1]/\gamma & (\gamma \neq 0), \\ \mu - \sigma \log(-\log p) & (\gamma = 0). \end{cases}$$

The same formula is useful to calculate quantiles of the fitted distribution; standard errors of these may be obtained, approximately, by the delta method (Rao (1973), p. 388).

As examples, we consider the Ivigtut and Nidd data from the previous section. The Gumbel model fitted to the Ivigtut data yields estimates $\hat{\mu} = 17.77$ (standard error 0.26), $\hat{\sigma} = 2.19$ (standard error 0.17), with a negative log likelihood (NLLH) of 188.2. Fitting the generalized extreme value distribution yields $\hat{\mu} = 17.90$ (0.26), $\hat{\sigma} = 2.19$ (0.17), $\hat{\gamma} = -0.11$ (0.04), NLLH = 186.1. A standard way to test whether the drop in NLLH is significant is to multiply by 2 and then apply a χ^2 test; this relies on the fact that the asymptotic distribution of this statistic is χ^2 with degrees of freedom equal to the number of additional parameters, here 1. In this case we compute twice the difference in NLLH to be 4.2, which is significant at the 5 per cent level as a χ_1^2 variable. These calculations, however completely ignore the outlier. A residual plot (Figure 14.6.1(a)) shows not only that the outlier is well above the expected straight line, but that it has had a distorting effect on the fit, since the nearby observations lie below the fitted line. In Figure 14.6.1(b) the model was refitted without the outlier, though the outlier has been retained for the plot. In this case we have $\hat{\mu} = 18.14$ (0.26), $\hat{\sigma} = 2.15$ (0.18), $\hat{\gamma} = -0.36$ (0.07), clearly significant against the Gumbel model, but the plot shows the outlier to be a long way from its expected value under

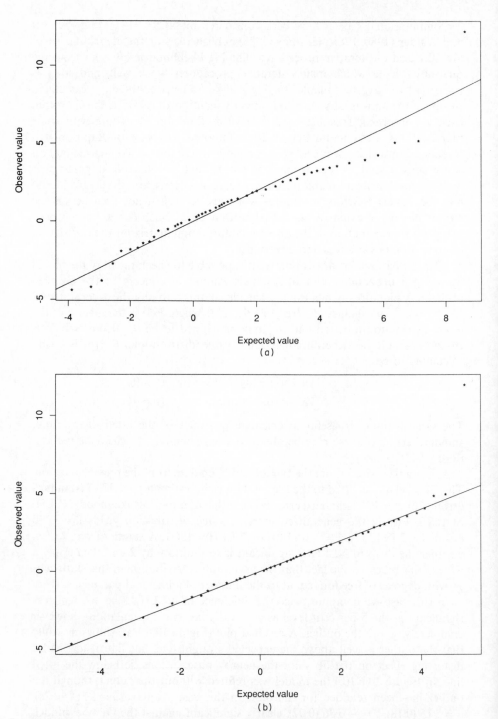

Figure 14.6.1. Residual plots (a) Ivigtut raw data (b) Ivigtut data with largest value dropped for model fit (c) Nidd data

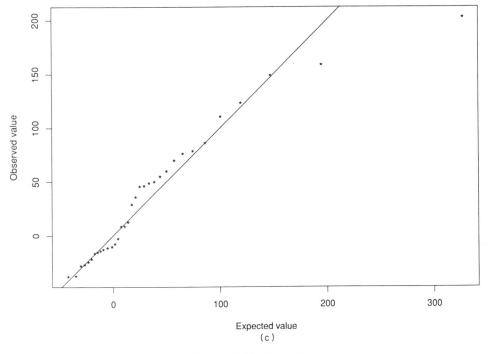

Figure 14.6.1. (*Contd.*)

the fitted model. Indeed the fitted model estimates the upper endpoint of the distribution to be $\hat{\mu} - \hat{\sigma}/\hat{\gamma} = 24 \cdot 3$, whereas the outlier is $30 \cdot 1$! It would be unrealistic to expect an easy solution to this difficulty; in other areas of statistics one might dismiss the outlier as a freak observation, but this hardly seems realistic here where the interest is specifically in extremes.

Outliers also have a part of play in the analysis of the Nidd data, though this time less obviously. A Gumbel fit yields $\hat{\mu} = 109 \cdot 9$ (standard error $7 \cdot 6$), $\hat{\sigma} = 42 \cdot 9$ ($6 \cdot 1$), NLLH $= 188 \cdot 4$, and a generalized extreme value fit gives estimates $\hat{\mu} = 103 \cdot 1$ ($7 \cdot 6$), $\hat{\mu} = 36 \cdot 1$ ($6 \cdot 6$), $\hat{\gamma} = 0 \cdot 32$ ($0 \cdot 22$), NLLH $= 187 \cdot 1$. In this case the generalized extreme value fit is not a significant improvement over the Gumbel, as judged by a χ^2 test for example, but the large estimated value of γ shows that the two fitted models are actually very different, so that if the Gumbel model were wrong it could lead to very seriously misleading estimates. For this reason I have preferred to retain the generalized extreme value model in this example, but then a residual plot (Figure 14.6.1(c)) shows that the largest two observations lie well below the plot; in other words they are not as extreme as the fitted model would lead us to expect. Although the effect is not as clear-cut as that with the Ivigtut data, it poses a similar dilemma—should we change our whole analysis of the problem on the basis of one or two extreme observations which do not fit the rest of the sample?

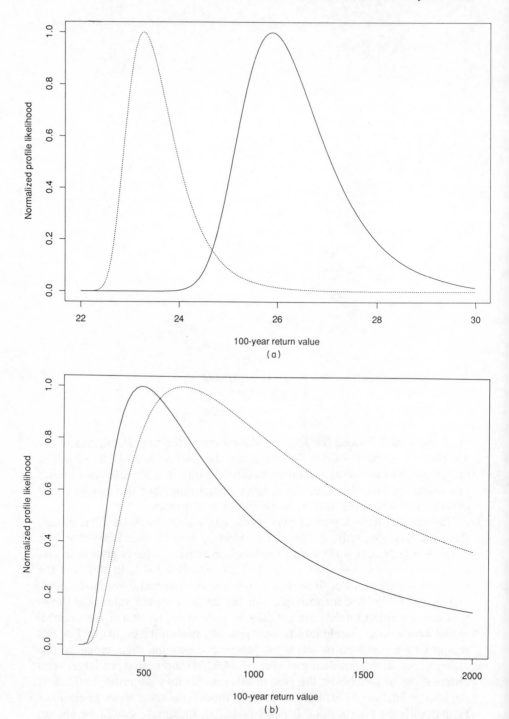

Figure 14.6.2. Profile likelihoods for return values (a) Ivigtut data with outlier (solid curve) and without (broken) (b) Nidd data with and without largest two observations

One way to try to resolve the difficulty is to consider the effect of the outlier on the estimation of return values. For present purposes we may define the N-year return value to be $F^{-1}(1 - 1/N)$. Take $N = 100$. For the Ivigtut data we estimate the 100-year return value to be 25·9 (standard error 0·8) in the analysis including the outlier, 23·1 (0·4) without it. In estimating return values, the typically asymmetric shape of the likelihood function makes confidence intervals based on standard errors unreliable. A better idea is to draw the profile likelihood: given N, for each x, fit the generalized extreme value distribution under the constraint that the N-year return value is x. The maximized likelihood value is the profile likelihood for x, and this can then be plotted against x. With the plot normalized to have maximum value 1, an appoximate 95 per cent confidence interval consists of all values having a normalized profile likelihood greater than 0·147 ($= e^{-1.92}$, 1·92 being half the 95 per cent point of the χ_1^2 distribution).

This idea does not help resolve the dilemma over the outlier in the Ivigtut data; as can be seen in Figure 14.6.2(a), the two profile likelihoods, with and without the outlier, are almost separate from each other. In the case of the Nidd data (Figure 14.6.2(b)), the effect of the suspect outliers is less critical, but in this case both profile likelihoods are very spread out and in particular give rise to excessively large upper confidence limits. Contrast the scales of Figures 14.6.2(a) and 14.6.2(b). In this case an element of engineering judgement would seem to be required.

These examples have been chosen to illustrate some of the difficulties that can arise from a routine application of extreme value distributions. The Ivigtut data represent just about the worst case of difficulty with an outlier: a sample that would indicate a very short-tailed distribution were it not for the enormous outlier. In the Nidd example the influence of the outliers is less critical, but they do suggest that there is some change in the distribution at the right-hand end of the sample, and a more detailed analysis based on exceedances over thresholds has confirmed that this is a significant effect (Davison and Smith (1990)). Hosking, Wallis and Wood (1985) and Hosking and Wallis (1987) have also published analyses of this data set.

14.7. MAXIMUM LIKELIHOOD IN MORE COMPLICATED SITUATIONS

The big advantage of maximum likelihood procedures is that they can be generalized, with very little change in the basic methodology, to much more complicated models in which trends or other effects may be present. Two examples will be given here; there are many others.

Let us first consider the simple case of an additive linear trend. A plausible model is that the annual maximum in year n, X_n say, has a generalized extreme value distribution with parameters μ_n, σ, γ, where $\mu_n = \alpha + \beta n$. this may be estimated by an easy extension of the method in §14.6: the log likelihood is a four-parameter function involving α, β, σ and γ, and is maximized by the same numerical techniques as described there.

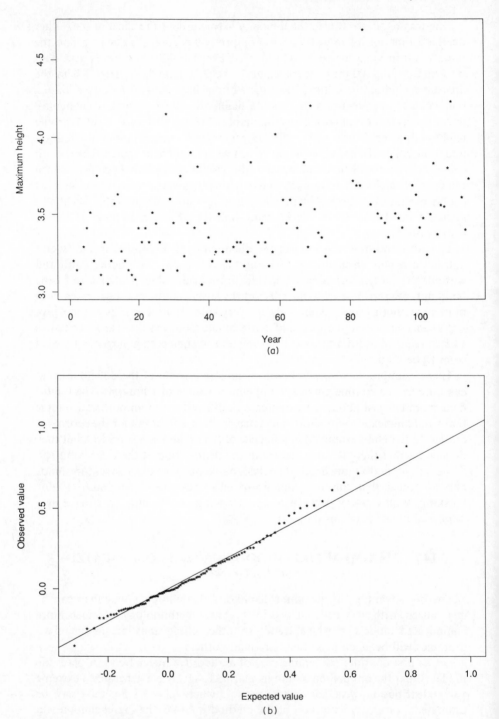

Figure 14.7.1. Sheerness annual maxima (a) Plot of raw data (b) Residuals after fitting
linear trend model

As an example of this analysis, Figure 14.7.1(a) shows a plot of annual maximum sea level at Sheerness in the Thames estuary. It appears from the plot that there is an increasing trend, and this is confirmed by the maximum likelihood analysis: the negative log likelihoods are respectively 4·16 for the Gumbel model without trend, 3.88 for the generalized extreme value model without trend, $-7\cdot81$ for the Gumbel model with a linear trend, $-8\cdot40$ for the generalized extreme value model with trend. In this case the preferred model would appear to be the Gumbel model with trend, leading to estimates $\hat{\alpha} = 3\cdot18$ (standard error 0·04), $\hat{\beta} = 0\cdot0031$ (0·0006), $\hat{\sigma} = 0\cdot19$ (0·01). Again there is some difficulty with an outlier (Figure 14.7.1(b) shows a residual plot), though its influence is not nearly as great as in the Ivigtut example §14.6.

There are many other examples of this sort of analysis. Smith (1986) studied a number of different trend models for sea levels in Venice; it is possible to put additional terms into the log likelihood corresponding to quadratic or cyclic trend terms, and hence to test for the presence of such trends. In the Venice data there is some suggestion of an 11-year cycle, which may be connected with sunspots. Athletics records have been considered by Smith (1988a). In this case the hypotheses of a linear trend in the records over time was tested against a number of alternatives. Contrary to what one might expect, a linear trend is very often the best model. This example required an additional extension of the methodology, in which ideas from the analysis of censored data were used to take account of the difference between a list of records and a set of annual maxima or minima.

A different kind of example, this time in a reliability context, is provided by Watson and Smith (1985). The data, part of an experiment performed by Dr Mark Priest in the Department of Materials Science and Engineering, University of Surrey, consist of strength tests of bundles of carbon fibres at four different lengths. For the present analysis we concentrate on two of the data sets, one concerned with tests on single fibres and the other with impregnated bundles consisting of 1000 fibres impregnated in epoxy resin.

Figure 14.7.2 shows Weibull plots of the data with fitted Weibull lines, computed by fitting a two-parameter Weibull distribution separately to each of the four samples within each data set. The Weibull plots here are a mirror image of the Gumbel plots described in §14.5; in this case $\log X_{i:n}$, the logarithm of the ith smallest order statistic in a sample of size n, is plotted against $\log[-\log(1 - p_{i:n})]$, where $p_{i:n} = (i - 0\cdot5)/n$ as before. In this case the interest lies in whether the four Weibull lines are parallel, implying that they have the same shape parameter w, and more especially in the *weakest link hypothesis*, which would imply a relationship of the form

$$\text{Prob}\{X > x|l\} = [1 - G(x)]^l \qquad (14.7.1)$$

where l is length and G is the distribution function for length 1. If we assume that all the distributions are two-parameter Weibull

$$\text{Prob}\{X \leq x|l\} = 1 - \exp[-(x/x_l)^{w_l}],$$

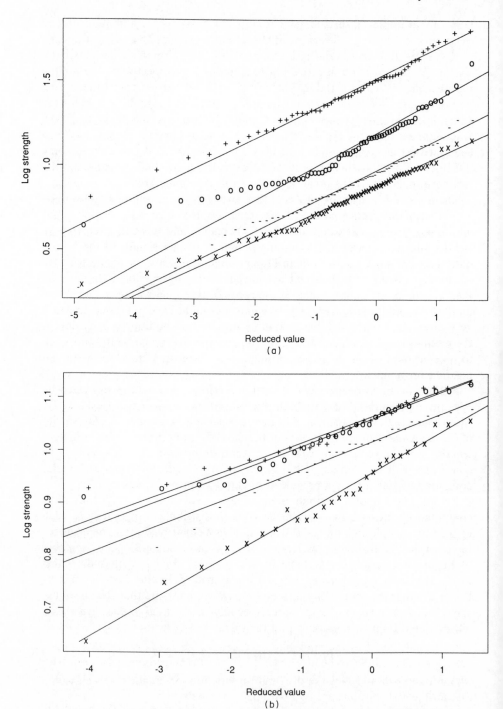

Figure 14.7.2. Weibull plots for bundles data (a) Single fibres, lengths 1, 10, 20, 50 mm (in order +, ○, −, ×) (b) Impregnated bundles, lengths 20, 50, 150, 300 mm (+, ○, −, ×)

then we may consider a sequence of nested hypotheses, of the form:

H_0: $\log(x_l) = -(1/w)\log l + \log(x_1)$, $w_l = w$,
H_1: $\log(x_l) = -\beta_1 \log l + \log(x_1)$, $w_l = w$,
H_2: $\log(x_l) = -\beta_1 \log l + \log(x_1)$, $\log(w_l) = -\beta_2 \log l + \log(w_1)$,
H_3: x_l, w_l unrestricted.

Here H_0 is weakest link and the others represent successive weakenings of that hypothesis. If in H_1 we write $\beta_1 = \alpha/w$ then what we have in effect is a weakest link hypothesis in l^α instead of l.

For the single fibres data of Figure 14.7.2, the negative loglikelihood values for hypotheses $H_0 - H_3$ are respectively 229·1, 227·7 ($\hat{\alpha} = 0.90$), 227·6 and 220·1. Thus there is reasonable support for the weakest link hypothesis against the l^α alternative, but a direct test of H_0 against H_3 yields a χ_6^2 statistic of 18·0, whereas the upper 99 per cent value of χ_6^2 is 16·81, thus casting doubt on whether any of the models improves on a simple Weibull fit to each length.

Similar analysis for the bundles data yields negative loglikelihoods $-21·8$, $-29·6$ ($\hat{\alpha} = 0.58$), $-31·5$ and $-35·8$, which in this case points towards H_1 as the most suitable model and decisively rejects H_0.

The l^α model was suggested by Watson and Smith and independently by V. Tamusz; it has now been observed in a number of studies but no clear physical explanation has emerged. A different approach was taken by Lindgren and Rootzén (1987); they developed a non-parametric test of (14.7.1) based on a proportional hazards analysis. This is potentially important because it separates out the weakest link effect, which is the important thing, from the somewhat arbitrary assumption of a Weibull distribution. However, their results were, if anything, even more negative about the weakest link effect than the results that have been given here.

14.8.   THE r-LARGEST ORDER STATISTICS METHOD

The methods of §§14.5–14.7 are generally applied when actual extreme values are observed, such as the annual maxima of a river flow or sea height. Such values are necessarily extracted from a much more detailed series, such as daily or hourly values. In such cases, it is natural to ask whether use can be made of the data other than annual maxima.

The methods to be discussed now are based on a fixed number of r largest values per year. The basis of the method is that the limiting joint distribution of the r largest order statistics from a sample of size N, for fixed r as $N \to \infty$, may be calculated by methods similar to those of §14.3. This may then be used as a statistical model to be fitted, again, by maximum likelihood. In cases where n years' data are available, the likelihood is constructed from the r largest values in each year.

The first paper to discuss this approach was Weissman (1978), based on a single sample of values ($n = 1$). For certain two-parameter models, Weissman was able to obtain closed-form estimators. The more general approach followed here was introduced by Gomes (1981), and developed with applications to oceanographic data by Smith (1986) and Tawn (1988a).

It is convenient to work in terms of densities rather than distribution functions. The limiting joint density of the r largest-order statistics $x_1 \geq \ldots \geq x_r$ corresponding to (14.2.4) for $r = 1$ is

$$f(x_1, \ldots, x_r; \mu, \sigma) = \sigma^{-r} \exp\left\{ -\exp\left[-(x_r - \mu)/\sigma \right] - \sum_{j=1}^{r} (x_j - \mu)/\sigma \right\}. \quad (14.8.1)$$

The same formula corresponding to (14.2.3) is

$$\sigma^{-r} \exp\left\{ -[1 + \gamma(x_r - \mu)/\sigma]^{-1/\gamma} - (1/\gamma + 1) \sum_{j=1}^{r} \log[1 + \gamma(x_j - \mu)/\sigma] \right\}. \quad (14.8.2)$$

The idea then is to apply (14.8.1) or (14.8.2) to the largest r values in each year, forming a likelihood function as usual by multiplying the densities for the different years together, and then maximizing numerically using the same general theory as §§14.6 and 14.7. The method has been presented with one year as the natural unit of time, but of course the idea is more generally applicable in situations where the data are divided into subsamples and the r largest values taken from each subsample.

Smith (1986) based all the analysis (14.8.1), applying the method to sea levels in Venice. Assuming σ to be constant over the sampling period, the analysis considered various time-dependent functions for μ as a function of n over the years 1931–1981. As expected, a linear increasing trend was found to be dominant. A method of residual analysis was also proposed. Essentially this consists of extending the probability plots idea, using the marginal distribution of the jth order statistic (for fixed j between 1 and r) and drawing a probability plot for each j.

Two principal methodological points arise from this method. One is the choice of r. In the Venice example it was found, using the probability plotting technique just mentioned, that the model fits well for $r = 1$, less so for $r = 5$, and quite badly for $r = 10$. Comparisons of standard errors of the parameter estimates, under the three models, show a sharp drop from $r = 1$ to $r = 5$ but much less of a drop going to $r = 10$. Based on a crude trade-off of fit and standard error, it was suggested that $r = 5$ is about right in practice, for this data set.

A second methodological issue concerns the effect of seasonality and serial dependence, which the foregoing analysis ignores, on these conclusions. If the seasonal effect is too weak to matter or so strong as to be concentrated all in one season (so that there is no substantial mixing of extreme values from different seasons) then the previous analysis should still be satisfactory; however, more study is needed of this point. In cases where the seasonality is such as to affect the results, it may be better to abandon this method in favour of one that takes direct account of seasonal variation [see §14.9]. Serial dependence tends to manifest itself in the clusters of large values corresponding, for example, to meteorological storms. In such cases it is still often reasonable to treat the *clusters* as independent and base the analysis on the *r* largest cluster maxima rather than the *r* largest values altogether.

Tawn (1988a) developed the theory of Smith (1986) further to encompass the model (14.8.2), and also discussed other aspects. In particular, in the cluster context just mentioned, he developed a method of identifying clusters from the concept of a standard storm length; the key point which appears to emerge here is that it does not matter too much exactly how the clusters are identified, so long as the standard storm length corresponds at least roughly with physical intuition, and in particular is not ignored altogether! Tawn also discussed an idea of using data from neighbouring sites to improve the estimation, though a full treatment of this latter concept requires the theory of bivariate extremes (see §14.10).

14.9. THRESHOLD METHODS

Threshold methods are methods based on exceedances over high thresholds. They have been well developed in the hydrology literature, often under the acronym of POT (peaks over thresholds), a terminology that encompasses the identification of clusters as mentioned at the end of §14.8. For example the Flood Studies Report (NERC (1975)) reviewed the literature up to that time and subsequent developments have included North's (1980) general model for seasonal variation. The discussion here is based on Davison (1984) and Smith (1984), plus the papers of Davison and Smith (1990) and Smith (1989b).

The simplest case is when the method is applied to i.i.d. observations. In this case, first a high threshold is selected and the observations over the threshold recorded. The number of exceedances provides an estimate of the probability that an observation is over the threshold; then the excess (or differences between observations over the threshold and the threshold itself) may be fitted by the generalized Pareto distribution introduced in §14.4. Estimates of extreme quantiles of the original distribution may be based on the fitted distribution for the excesses.

It would be desirable to have a graphical technique similar to the Gumbel probability plot of §14.5, both to assess the overall fit of the model and more specifically as an aid to choosing the threshold. A useful device in this connection is a 'mean residual life plot' in which the mean excess over a threshold u is plotted against u, for a wide range of values of u. The name comes from the use of a similar idea in survival analysis. This suggestion is based on the fact that, if Y is generalized Pareto with parameters σ and γ, then provided $\gamma < 1$, $u \geq 0$ and $\sigma + 0$ we have

$$E(Y - u \mid Y \geq u) = (\sigma + \gamma u)/(1 - \gamma).$$

Consequently we expect the plot to be roughly linear with intercept $\sigma/(1-\gamma)$ and slope $\gamma/(1-\gamma)$. In practice the plot tends to be very irregular at its right-hand end, when u is large and the number of exceedances of u small, but away from there it can be a very useful way of checking up on the appropriateness of the generalized Pareto distribution.

As an example of this kind of analysis, we consider some data arising from the US/Canada acid rain study. The particular data considered here are measurements of sulphate and nitrate levels in 504 rain storms over a 6-year period at Penn State, Pennsylvania. To what extent extreme values are relevant in this study is not entirely clear, but it is known that sulphate and nitrate levels are a good indicator of the damage caused by acid rain and it is of interest to know how large measurements can get. Mean residual life plots for the sulphate and nitrate data are shown in Figure 14.9.1. In both cases the plot behaves very erratically near its right-hand end, an inevitable feature of the way it is constructed, but of greater interest is the change in the nature of both plots near the value 100. This indicates that the behaviour below and above this level are somewhat different and suggests taking a threshold somewhere around that point.

For the sulphate data I took thresholds at 99·5, 109·5 and 119·5 (not 100, 110, 120 because of rounding in the data creating a large number of ties) with respectively 38, 33 and 22 exceedances. The fitted generalized Pareto models were $\hat{\sigma} = 30.3$ (standard error 8.1), $\hat{\gamma} = 0.31$ (0·22) for threshold 99·5, $\hat{\sigma} = 11.3$ (9·3), $\hat{\gamma} = 1.09$ (0·83) for threshold 109·5, $\hat{\sigma} = 32.0$ (11·2), $\hat{\gamma} = 0.36$ (0·29) for threshold 119·5. The unusual fit at threshold 109·5 is almost certainly a consequence of the fact that several data values were (presumably after rounding) exactly 110. This must have biased the fit in the lower tail. As further evidence of this, I also computed the probability weighted moments estimators suggested by Hosking and Wallis (1987). They agreed very well with the above values for thresholds 99·5 and 119·5 but not for 109·5.

Similar analysis was performed for the nitrate data, with thresholds 99·5, 139·5, 159·5 creating respectively 52, 35 and 21 exceedances. Parameter estimates are $\hat{\sigma} = 69.2$ (standard error 10·9), $\hat{\gamma} = -0.33$ (0·09) for threshold 99·5, $\hat{\sigma} = 30.8$ (7·7), $\hat{\gamma} = -0.01$ (0·18) for threshold 139·5, $\hat{\sigma} = 24.4$ (8·3), $\hat{\gamma} = 0.09$ (0·06) for threshold 159·5. All three cases agreed well with the corresponding probability weighted moments estimates but there is a clear trend for $\hat{\gamma}$ to increase with the threshold, suggesting that one must be cautious about the generalized Pareto fit in this case.

The general form of these mean residual life plots is not atypical of real data, especially the change of slope near 100 in both plots. Smith (1985b) observed similar behaviour in data on extreme insurance claims, and Davison and Smith (1990) used a similar plot to identify a change in the distribution of the threshold form of the River Nidd data. Such plots therefore appear to be an extremely useful diagnostic in this form of analysis.

In this example it was reasonable to treat the values in successive storms as i.i.d., but in most applications of threshold methods this will not be a valid

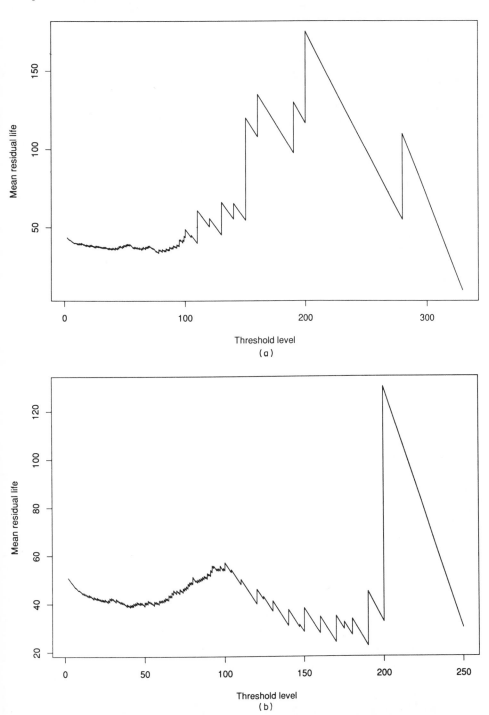

Figure 14.9.1. MRL plots for acid rain data (a) Sulphate level (b) Nitrate level

assumption. Indeed, one of the main advantages of threshold methods is that they deal much more flexibly with these issues than the traditional statistical methods based on annual maxima.

One way of dealing with this problem is to introduce covariates into the analysis. The generalized Pareto parameters σ and γ may be allowed to depend on covariates and regression models fitted to the exceedance data. As an example Davison (1984) and Davison and Smith (1990) have studied computer-simulated data on depositions from a nuclear power plant, in which relevant covariates include the nuclide decay rate, the deposition velocity, the washout coefficient, the distance from the source of the radiation to the point being studied and the time that has elapsed since the leakage occurred. The aim here is not so much to predict critical statistics such as return values, but more to find a concise and meaningful model to describe the pattern of high-level depositions. Fitting may again be by maximum likelihood; Davison and Smith describe a GLIM procedure.

Other important features, particularly when dealing with long time series, are seasonality and serial dependence in the data. The treatment of serial dependence has been much influenced by modern developments in the theory of extremes of stochastic processes; see for instance Leadbetter, Lindgren and Rootzén (1983), Leadbetter and Rootzén (1988) and Aldous (1989). In stationary discrete-time processes, these authors have developed extreme value theory assuming mixing conditions. The mixing conditions ensure that high-level exceedances, from values far apart in the series, are approximately independent. Under conditions of this nature, it has been shown that extremes from stationary stochastic processes are attracted to the same three limit types as in i.i.d. sequences. This is not to say, however, that the actual limiting distribution (rather than type of distribution) is the same. The most common relation is one of the form

$$\text{Prob}\,(M_n \leq x) \approx F^{n\theta}(x),$$

where M_n is the maximum of n values of the series and F the marginal distribution function. The parameter θ may take any value between 0 and 1 (including 0 and 1 themselves) and has the rough interpretation that $1/\theta$ is the mean cluster size; thus $\theta = 1$ is the case of no clustering (as with independent data), and $\theta = 0$ the opposite extreme where the clusters are of infinite expected size. Extensions to non-stationary series exist; Smith (1988b) showed how a mathematical technique due to Stein and Chen may be used to give short proofs of results based on Hüsler (1986) amongst other papers.

The statistical implication of these results is that the notion of clusters of high values is a very general one, and that the use of extreme value distributions for cluster maxima, in conjunction with some technique to identify clusters and to estimate θ, is of wide applicability in applying extreme value theory to dependent series.

The treatment of seasonality is rather more *ad hoc*, but Davison and Smith (1989) and Smith (1989b) have recommended a technique in which the year is broken up into independent seasons, with separate generalized Pareto distributions being fitted to each season. This is a different concept from that of North (1980), who essentially allowed for sinusoidal variation in building regression models for both the rate of exceedances and the mean excess. It is a critical decision how many seasons to use; too few and the seasonal effect will not be properly allowed for, but if too many are used there will not be enough exceedances in each season to allow good estimation of the parameters. Smith (1989b) applied this technique to analyse daily ozone data in Houston, Texas, while Davison and Smith (1989) gave a seasonal threshold analysis of the Nidd data which we have already seen in §14.6.

At the time of writing there are still many open questions about the applications of threshold techniques, but overall they offer a much more flexible approach to statistical modelling than the classical extreme value techniques and therefore seem certain to be the focus of much attention as these techniques continue to develop in the future.

14.10. MULTIVARIATE EXTREMES

Our discussion so far has been entirely concerned with extremes in a single series, whether or not the observations are independent. Multivariate extreme value theory is concerned with the joint distribution of extreme values in several series. It might be used, for example, to examine the dependence between flooding at two nearby sites, or between the breakdown of two components in a reliability network. If applied to the joint distribution of successive observations in a time series, it also has implications for the extreme value analysis of dependent time series, alternative to the cluster-identification techniques mentioned in §§14.8 and 14.9.

The first issue is how to define multivariate extremes. By analogy with the limit theory of §14.2, the traditional approach has been to base the definition on componentwise maxima. Thus if $X_1, \ldots, X_n$ represent i.i.d. p-vectors and $M_n = (M_{n,1}, \ldots, M_{n,p})$ is the vector of maxima of each component, then we seek normalizing constants $a_{n,p} > 0, b_{n,p}$ such that

$$\text{Prob}\left\{ \frac{(M_{n,1} - b_{n,1})}{a_{n,1}} \le x_1, \ldots, \frac{(M_{n,p} - b_{n,p})}{a_{n,p}} \le x_p \right\} \to H(x_1, \ldots, x_p) \quad (14.10.1)$$

for some non-degenerate p-variate distribution function H. This approach has been criticized as not always being an appropriate definition (Barnett (1976)), but the detailed theory developed by Resnick (1987), amongst others, has shown that conditions relevant for (14.10.1) are also relevant for a variety of other extreme value behaviour. Thus it seems entirely reasonable to focus attention on (14.10.1) as a starting point, and in the present review this will be the only approch considered.

An immediate consequence of (14.10.1) is that the univariate margins of H

are the classical extreme value distributions, and it follows that there is no loss of generality in restricting to a particular univariate extreme value distribution. In this regard different authors have followed different conventions; thus Tiago de Oliveira (1984b and many earlier papers) has assumed Gumbel margins, de Haan and Resnick (1977) formulated their theory in terms of Fréchet variables with shape parameter 1, and Pickands (1981) transformed everything to minima and assumed unit exponential margins. We follow Pickands here. Thus, the problem is equivalent to studying a particular class of multivariate exponential distributions for minima.

According to this theory a survivor function

$$S(y_1, \ldots, y_p) = \text{Prob}\{Y_1 > y_1, \ldots, Y_p > y_p\}$$

is multivariate extreme if it satisfies the three properties:

(a) Marginal exponentiality: $S(0, \ldots, 0, y, 0, \ldots, 0) = e^{-y}$ in whichever place the y lies.
(b) Homogeneity: $S(ay_1, \ldots, ay_p) = S(y_1, \ldots, y_p)^a$ for any $a > 0$ (this is equivalent to saying that $\log S$ is homogeneous of order 1).
(c) S is a valid survivor function, in the sense that it gives non-negative probability mass to any hyper rectangle in the p-dimensional positive orthant.

Condition (c) is usually the hardest to verify in practice!

We now focus on the bivariate case ($p = 2$), for which an alternative representation is

$$S(x, y) = \exp[-(x + y)A(y/(x + y))] \tag{14.10.2}$$

(Pickands (1981)) where the function A satisfies the following properties: $A(w)$ is a convex function on $w \in [0, 1]$ such that its graph lies entirely within the triangle in the $(w, A(w))$ plane bounded by the points $(0, 1)$, $(0.5, 0.5)$ and $(1, 1)$ (see Figure 14.10.1). Pickands also gave an integral representation for A and extended the definition to the multivariate case; this is equivalent, but not obviously so, to the integral representation given by de Haan and Resnick (1977). Following Pickands we call A the *dependence function*, though it should be noted that other authors on multivariates extremes, such as Deheuvels (1978) and Galambos (1987), have used the same expression to mean something quite different.

Pickands also proposed the following estimation procedure for A. Suppose $\{(X_i, Y_i), \ 1 \leq i \leq n\}$ denotes a bivariate sample, assumed to have been transformed to unit exponentiality. For each $w \in (0, 1)$ let

$$A_n(w) = n\left\{\sum_{i=1}^{n} \min[X_i/(1 - w), Y_i/w]\right\}^{-1}. \tag{14.10.3}$$

Then, for each w, $A_n(w)$ is a consistent and asymptotically efficient estimator of

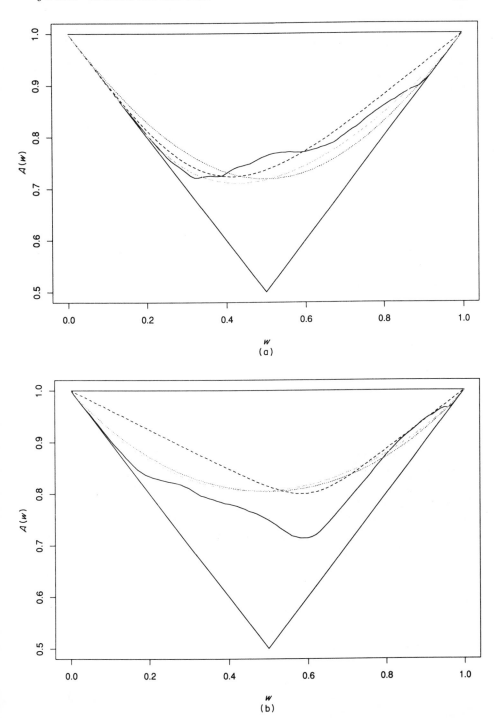

Figure 14.10.1. Bivariate dependence functions (a) Maximum temperatures at Worthing and Portland Bill (b) Mile best performances in successive years

$A(w)$. However, as a function of w, A_n does not have all the properties required of A: it is neither convex nor differentiable, whereas A is certainly required to be convex and will be differentiable as well whenever distribution is jointly continuous. Improved non-parametric estimators have been proposed by Pickands (1981), Smith (1985b) and in the unpublished Surrey Ph.D. thesis of H. K. Yuen; the latter is summarized in Smith, Tawn and Yuen (1990).

In addition to the non-parametric proposals, a number of parametric models have been considered. The best-known parametric model in this field is that of Marshall and Olkin (1967), but this has the disadvantage for statistical purposes that the distribution is not jointly continuous. Tiago de Oliveira (1984b) reviewed a number of other models, and recently Tawn (1988b) has made a number of new proposals.

Amongst the most promising parametric models appear to be:

(i) logistic model: $A(w) = [(1 - w)^{1/\alpha} + w^{1/\alpha}]^{\alpha}$, $(0 \leq \alpha \leq 1)$,
(ii) asymmetric logistic model: $A(w) = (1 - \theta)(1 - w) + (1 - \phi)w + \{[\theta(1 - w)]^{1/\alpha} + (\phi w)^{1/\alpha}\}^{\alpha}$, $(0 \leq \alpha \leq 1, 0 \leq \theta \leq 1, 0 \leq \phi \leq 1)$,
(iii) bilogistic model:

$$A(w) = \int_0^1 \max\left[(1 - \alpha)(1 - w)s^{-\alpha}, (1 - \beta)w(1 - s)^{-\beta}\right]ds \quad (0 \leq \alpha \leq 1, 0 \leq \beta \leq 1).$$

In all three cases the restrictions on the parameters arise from the need to satisfy condition (c) above. The logistic model is due to Gumbel and was found by Tawn to be the most successful of the various models in the existing literature. However, it has one feature which in some contexts may be a disadvantage, namely that it is symmetric in the two components. The asymmetric logistic model was suggested by Tawn as an alternative model which alllows for such asymmetry. The bilogistic model is another way to solve this problem. It was suggested by the alternative integral representation of de Haan (1984); note that when $\alpha = \beta$ it reduces to the logistic model. In other cases evaluation of the inegral depends on numerical solution of the equation $(1 - \alpha)(1 - w)s^{-\alpha} = (1 - \beta)w(1 - s)^{-\beta}$, but the author has found that this is a straightforward task easily incorporated within a numerical maximum likelihood procedure.

We give two examples of this procedure. The first uses annual maximum temperatures at Worthing and Portland Bill, tabulated in Tabony (1983). In both series the generalized extreme value distribution was fitted as a preliminary to transforming to unit exponentiality. Only those years for which values for both sites are available have been included in the following analysis.

Based on 77 transformed data points, the initial negative loglikelihood (NLLH) assuming independence is 154. The logistic model gave $\hat{\alpha} = 0.524$, NLLH $= 126.7$; strong evidence of dependence. The asymmetric model was found to be estimated at $\hat{\alpha} = 0.467$, $\hat{\theta} = 0.844$, $\hat{\phi} = 1.0$, NLLH $= 125.7$. This is not a significant improvement on the symmetric logistic model. However, the bilogistic model did better: $\hat{\alpha} = 0.682$, $\hat{\beta} = 0.183$, NLLH $= 122.2$. All three estimated dependence functions, together with the raw non-parametric estimate A_n, are plotted in Figure 14.10.1(a). The bilogistic model is the thick dashed curve and this does appear to follow the raw estimate most closely of the three parametric curves.

Our second example illustrates the possible application of these ideas to dependent processes. The data consist of a series of annual best performances in the mile race. Smith (1988a) fitted a generalized extreme value distribution with linear trend to the data from 1931–1985, treating successive years as independent. Earlier, however, Ballerini and Resnick (1985) showed that there is significant autocorrelation in this series. It is therefore logical to look for dependence in the context of an extreme value analysis. As in the previous example, the first step was to transform the series to unit exponentiality, using the fitted marginal extreme value distribution. Given that we have such marginal distribution, the likelihood function for successive pairs from the series behaves the same as if they were independent pairs from a bivariate exponential distribution. (This was pointed out by H. K. Yuen in his thesis.) Therefore, we can proceed in the same way as in the previous example, fitting the various bivariate extreme models to the pairs of successive transformed observations. The negative loglikelihoods for the independent, logistic, bilogistic and asymmetric logistic models are respectively 108, 100·6 ($\hat{\alpha} = 0.686$), 100·5 and 99·1. Thus in this case the logistic and bilogistic models are almost indistinguishable, while the asymmetric logistic model is better but still not significantly so. This is reflected in Figure 14.10.1(b), where it is seen that none of the fitted dependence functions is very close to the non-parametric estimate. The thin dashed curve (asymmetric logistic) best reflects the asymmetry of the non-parametric curve, but it is still not very close to it. Thus we conclude that there is significant evidence of dependence as measured by the logistic model, but the evidence for asymmetry (which corresponds to irreversibility in the original time series) is not significant.

The theoretical justification for maximum likelihood in these models has been considered in detail by Tawn (1988b). The models are regular in the interior of their parameter spaces, but non-regular on the boundaries. Thus, testing independence (which corresponds to $\alpha = 1$ in the logistic model, and any of $\alpha = 1$, $\theta = 0$, $\phi = 0$ in the asymmetric logistic) is a non-regular problem and special procedures must be used. The examples given here do not exhibit boundary behaviour except for the asymmetric model applied to the temperature data, where we found $\phi = 1$. Thus, with this exception, the use of standard likelihood interpretations seems justified.

There are still many open questions. For example, in the two examples just studied, the marginal distributions were fitted first and then a bivariate exponential model fitted to the transformed data. In theory it should be more efficient to fit all the parameters in a single large maximization of the likelihood, but the computational practicality and theoretical advantages of this have not so far been examined. It should also be reiterated that these results apply only to bivariate extremes, and it is a challenging problem to come up with a sufficiently broad class of models and estimation procedures for multivariate $(p > 2)$ data. Tawn (1989) has made a start on this problem.

14.11.   MISCELLANEOUS STATISTICAL APPLICATIONS

The emphasis in this chapter has been largely on the use of extreme value methods to fit tails of distributions and to predict extreme quantities such as return values. There are other areas of statistics, however, where extreme value theory has played a role. Extreme value arguments may be used to construct formal tests of whether suspected outliers in a sample are genuine outliers or just the extreme values that might be expected in a sample of the given size; Kimber (1982, 1983) has exploited this aspect. The sports-related applications also go beyond the records example of §14.10. For instance, Henery (1981), developing an earlier theory of Plackett (1975), gave formulae for the distribution of finishing order in a horse race. His model essentially assumed that running times are independent, but not identically distributed, random variables, an aspect that has not been covered here. Recently de Haan and Weissman (1988) have made a more rigorous study of the problem of determining which member of an independent set of random variables gives rise to the largest or smallest value.

Perhaps the most substantial current application of extreme value theory to another area of statistics is the use of the theory for extremes in continuous-parameter stochastic processes to study a class of problems related to non-regular behaviour of likelihood ratio tests in problems involving parameters non-identifiable under the null hypothesis. In such cases certain test statistics (including the likelihood ratio statistic) may be represented as an extreme of a likelihood function which, under certain conditions, may be treated as a continuous-parameter stochastic process. Davies (1977, 1987), used extensions of 'Rice's formula' to derive approximations to the distributions of these test statistics. The stationary case of this formula is dealt with in Section 7.2 of Leadbetter, Lindgren and Rootzén (1983), but Davies also needed non-stationary extensions. Another problem in similar spirit is testing for the existence of a change point. For example, let $X_1, \ldots, X_n$ be independent normal with $E(X_j) = \mu_j$ and known standard deviation, and consider the problem of testing $\mu_1 = \mu_2 = \ldots = \mu_n$ against the alternative that $\mu_1 = \ldots = \mu_r \neq \mu_{r+1} = \ldots = \mu_n$ for some 'change point' r. This is a non-regular problem because the two means are not separately identifiable when the null hypothesis is true. In this case Rice's formula is not applicable because the stochastic process to which it has to be applied turns out to be non-differentiable, but a number of other approaches have been developed; for a review see Siegmund (1988). These problems have been mentioned because they draw attention to applications of extreme value theory outside the usual and obvious applications to things like hydrological extremes. At the moment there are quite a number of problems of this type which are not covered by the Davies approach, so this is another active area of research.

ACKNOWLEDGEMENTS

Part of this chapter was written during a visit to the Department of Statistics, University of Chicago. I am grateful to the Department for the invitation and for the use of their computing facilities. Individuals who have helped me in one way or another with the examples are too numerous to name in full, but I would like to thank Mike Bader, David Cox, Anthony Davison, Harry Joe, David Pugh, Jonathan Tawn, Martin Van Montfort, Ishay Weissman and Sammy Yuen for contributions of one sort or another.

R. L. S.

REFERENCES

Aldous, D. J. (1989). *Probability Approximations via the Poisson Clumping Heuristic*, Applied Mathematical Sciences series no. 77, Springer Verlag.

Anderson, C. W. (1970). Extreme Value Theory for a Class of Discrete Distributions with Applications to some Stochastic Processes, *J. Appl. Prob.* **7**, 99–113.

Arnell, N., Beran, M. A. and Hosking, J. R. M. (1986). Unbiased Plotting Positions for the General Extreme Value Distribution, *J. Hydrology* **86**, 59–69.

Ballerini, R. and Resnick, S. (1985). Records from Improving Populations, *J. Appl. Prob.* **22**, 487–502.

Barnett, V. (1976). The Ordering of Multivariate Data (with Discussion), *J. R. Statist. Soc. A* **139**, 318–354.

Bingham, N. H., Goldie, C. M. and Teugels, J. L. (1987). *Regular Variation*, Cambridge University Press.

Cohen, J. P. (1982). Convergence Rates for the Ultimate and Penultimate Approximations in Extreme Value Theory, *Adv. Appl. Prob.* **14**, 838–854.

Cox, D. R. and Hinkley, D. V. (1974). *Theoretical Statistics*, Chapman and Hall.

Cunnane, C. (1978). Unbiased Plotting Positions—a Review, *J. Hydrology* **37**, 205–222.

Davies, R. G. (1977). Hypothesis Testing when a Nuisance Parameter is Present only under the Alternative, *Biometrika* **64**, 247–254.

Davies, R. G. (1987). Hypothesis Testing when a Nuisance Parameter is Present only under the Alternative, *Biometrika* **74**, 33–43.

Davison, A. C. (1984). Modelling Excesses over High Thresholds, with an Application, in *Statistical Extremes and Applications*, Ed. Tiago de Oliveira, Reidel, 461–482.

Davison, A. C. and Smith, R. L. (1990). Models for Exceedances over High Thresholds (with Discussion), *J. R. Statist. Soc. B* **52** (to appear).

Deheuvels, P. (1978). Caractérisation Complète de Lois Extrêmes Multivariées et de la Convergence des Types Extrêmes, *Publ. Inst. Statist. Univ. Paris.* **XXIII**, 1–36.

Dodd, E. L. (1923). The Greatest and Least Variate under General Laws of Error, *Trans. Amer. Math. Soc.* **25**, 525–539.

Efron, B. and Hinkley, D. V. (1978). Assessing the Accuracy of the Maximum Likelihood Estimator: Observed versus Expected Fisher Information (with Discussion), *Biometrika* **65**, 457–487.

Fisher, R. A. and Tippett, L. H. C. (1928). Limiting Forms of the Frequency Distributions of the Largest or Smallest Member of a Sample, *Proc. Camb. Phil. Soc.* **24**, 180–190.

Fréchet, M. (1927). Sur la Loi de Probabilité de l'Écart Maximum, *Ann. Soc. Polonaise Math. (Cracow)* **6**, 93.

Galambos, J. (1987). *The Asymptotic Theory of Extreme Order Statistics* (2nd edn), Krieger. (First edn published 1978 by John Wiley, New York.)

Gnedenko, B. V. (1943). Sur la Distribution Limite du Terme Maximum d'une Série Aléatoire, *Ann. Math.* **44**, 423–453.

Gomes, M. I. (1981). An *i*-dimensional Limiting Distribution Function of Largest Values and its Relevance to the Statistical Theory of Extremes, in *Statistical Distributions in Scientific Work*, Eds. C. Taillie *et al.* **6**, Riedel, 389–410.

Gomes, M. I. (1984). Penultimate Limiting Forms in Extreme Value Theory, *Ann. Inst. Statist. Math.* **36**, 71–85.

Gomes, M. I. and Pestana, D. D. (1987). Nonstandard Domains of Attraction and Rates of Convergence, in, *New Perspectives in Theoretical and Applied Statistics*, Eds. M. L. Puri, J. P. Vilaplana and W. Wertz, John Wiley, 467–477.

Gordon, L., Schilling, M. F. and Waterman, M. S. (1986). An Extreme Value Theory for Long Head Runs, *Prob. Th. Rel. Fields* **72**, 279–287.

Gumbel, E. J. (1958). *Statistics of Extremes*. Columbia University Press.

Haan, L. de (1970). *On Regular Variation and its Application to the Weak Convergence of Sample Extremes*, Mathematical Centre Tracts 32.

Haan, L. de (1984). A Spectral Representation for Max-stable Processes, *Ann. Probab.* **12**, 1194–1204.

Haan, L. de and Resnick, S. I. (1977). Limit Theory for Multivariate Sample Extremes, *Z. Wahrscheinlichkeitstheorie v. Geb.* **40**, 317–337.

Haan, L. de and Weissman, I. (1988). The Index of the Oustanding Observation Among *n* Independent Ones, *Stoch. Proc. Appl.* **27**, 317–329.

Hall, P. (1979). On the Rate of Convergence of Normal Extremes, *J. Appl. Prob.* **16**, 433–439.

Henery, R. J. (1981). Permutation Probabilities for Horse Races, *J. R. Statist. Sco. B* **43**, 86–91.

Hosking, J. R. M. (1985). Algorithm AS215: Maximum Likelihood Estimation of the Parameters of the Generalized Extreme-value Distribution, *Appl. Statist.* **34**, 301–310.

Hosking, J. R. M. and Wallis, J. R. (1987). Parameter and Quantile Estimation for the Generalized Pareto Distribution, *Technometrics* **29**, 339–349.

Hosking, J. R. M., Wallis, J. R. and Wood, E. F. (1985). Estimation of the Generalised Extreme-value Distribution by the Method of Probability-weighted Moments, *Technometrics* **27**, 251–261.

Hüsler, J. (1986). Extreme Values of Non-stationary Random Sequences, *J. Appl. Prob.* **23**, 937–950.

Hüsler, J. and Reiss, R.-D. (eds.) (1989). *Extreme Value Theory Proceedings, Oberwolfach 1987*. Lecture Notes in Statistics No. 51, Springer-Verlag.

Johnson, R. A. and Haskell, J. H. (1983). Sampling Properties of Estimates of a Weibull Distribution of use in the Lumber Industry, *Can. J. Statist.* **11**, 155–169.

Kendall, M. and Stuart, A. (1977). *The Advanced Theory of Statistics, Vol I: Distribution Theory* (4th edn), Charles Griffin.

Kimber, A. C. (1982). Tests for many Outliers in an Exponential Sample, *Appl. Statist.* **31**, 263–271.

Kimber, A. C. (1983). Discordancy Testing in Gamma Samples with both Parameters Unknown, *Appl. Statist.* **32**, 304–310.

Leadbetter, M. R. and Rootzén, H. (1988). Extermal Theory for Stochastic Processes, *Ann. Probab.* **16**, 431–478.

Leadbetter, M. R., Lindgren, G. and Rootzén, H. (1983). *Extremes and Related Properties of Random Sequences and Series*, Springer Verlag.

Lindgren, G. and Rootzén, H. (1987). Extreme Values: Theory and Technical Applications, *Scand. J. Statist.* **14**, 241–279.

Mann, N. R., Schafer, R. E. and Singpurwalla, N. D. (1974). *Methods for Statistical Analysis of Reliability and Life Data*, Wiley.

Marshall, A. W. and Olkin, I. (1967). A Multivariate Extremal Distribution, *J. Amer. Statist. Assoc.* **62**, 30–44.

Nash, J. C. (1979). *Compact Numerical Algorithms for Computers*, Adam Hilger.

NERC (1975). *The Flood Studies Report*. The Natural Environment Research Council.

North, M. (1980). Time-dependent Stochastic Models of Floods, *J. Hyd. Div. ASCE* **106**, 649–655.

Pickands, J. (1975). Statistical Inference using Extreme Order Statistics, *Ann. Statist.* **3**, 119–131.

Pickands, J. (1981). Multivariate Extreme Value Distributions, *Bull. I.S.I.* **XLIX** (Book 2), 859–878.

Plackett, R. L. (1975). The Analysis of Permutations, *Appl. Statist.* **24**, 193–202.

Prescott, P. and Walden, A. T. (1980). Maximum Likelihood Estimation of the Parameters of the Generalized Extreme Value Distribution, *Biometrika* **67**, 723–724.

Prescott, P. and Walden, A. T. (1983). Maximum Likelihood Estimation of the Parameters of the Three-parameter Generalized Extreme-Value Distribution from Censored Samples, *J. Statist. Comput. Simul.* **16**, 241–250.

Rao, C. R. (1973). *Linear Statistical Inference and Its Applications* (2nd edn), Wiley.

Resnick, S. (1987). *Extreme Values, Point Processes and Regular Variation*, Springer Verlag.

Siegmund, D. (1988). Confidence Sets in Change-point Problems, *Int. Statist. Rev.* **56**, 31–48.

Smith, R. L. (1984). Threshold Methods for Sample Extremes, in *Statistical Extremes and Applications*, Ed. Tiago de Oliveira, Reidel, 621–638.

Smith, R. L. (1985a). Maximum Likelihood Estimation in a Class of Nonregular Cases, *Biometrika* **72**, 67–92.

Smith, R. L. (1985b). Statistics of Extreme Values, *Bull. I.S.I.* **LI** (Book 4), Paper 26.1.

Smith, R. L. (1986). Extreme Value Theory Based on the *r* Largest Annual Events, *J. Hydrology* **86**, 27–43.

Smith, R. L. (1988a). Forecasting Records by Maximum Likelihood, *J. Amer. Statist. Assoc.* **83**, 331–338.

Smith, R. L. (1988b). Extreme Values for Dependent Sequences via the Stein–Chen Method of Poisson Approximation, *Stoch. Proc. Appl.* **30**, 317–327.

Smith, R. L. (1989a). Approximations in Extreme Value Theory. Submitted.

Smith, R. L. (1989b). Extreme Value Theory for Environmental Time Series: An Example Based on Ozone Data, To appear, *Statistical Science*.

Smith, R. L. and Naylor, J. C. (1987). A Comparison of Maximum Likelihood and Bayesian Estimators for the Three-parameter Weibull Distribution, *Appl. Statist.* **36**, 358–369.

Smith, R. L., Tawn, J. A. and Yuen, H.-K. (1990). Statistics of Multivariate Extremes. To appear, *Internat. Statist. Review*.

Tabony, R. C. (1983). Extreme Value Analysis in Meteorology, *The Meteorological Magazine* **112**, 77–98.

Tawn, J. A. (1988a). An Extreme Value Theory Model for Dependent Observations, *J. Hydrology* **101** 227–250.

Tawn, J. A. (1988b). Bivariate Extreme Value Theory—Models and Estimation, *Biometrika* **75**, 397–415.

Tawn, J. A. (1989). Modelling Multivariate Extreme Value Distributions, Submitted.

Tiago de Oliveira, J. (ed.)(1984a). *Statistical Extremes and Applications*, Reidel.

Tiago de Oliveira, J. (1984b). Bivariate Models for Extremes, in *Statistical Extremes and Applications*, Ed. Tiago de Oliveira, Reidel, 131–153.

Von Mises, R. (1936). La Distribution de la Plus Grande de *n* Valeurs, reprinted in *Selected Papers* II, Amer. Math. Soc., Providence, R.I. (1954), 271–294.

Watson, A. S. and Smith, R. L. (1985). An Examination of Statistical Theories for Fibrous Materials in the Light of Experimental Data, *J. Mater. Sci.* **20**, 3260–3270.

Weibull, W. (1939). A Statistical Theory of Strength of Materials, *Proc. Ing. Vetenskapsakad* **151**.

Weibull, W. (1951). A Statistical Distribution of Wide Applicability, *J. Appl. Mechanics* **18**, 293–297.

Weissman, I. (1978). Estimation of Parameters and Large Quantiles Based on the k Largest Observations, *J. Amer. Statist. Assoc.* **73**, 812–815.

Index